All-Russian Research-to-Practice Conference for Young Researchers, Postgraduate Students and Undergraduates 2016

Ecology and Safety in the Technosphere: Current Problems and Solutions

IOP Conference Series: Earth and Environmental Science
Volume 50

Yurga, Russia
17 - 19 November 2016

ISBN: 978-1-5108-3523-8
ISSN: 1755-1307

Printed from e-media with permission by:

Curran Associates, Inc.
57 Morehouse Lane
Red Hook, NY 12571

Some format issues inherent in the e-media version may also appear in this print version.

Printed by Curran Associates, Inc. (2017)

For permission requests, please contact the Institute of Physics
at the address below.

Institute of Physics
Dirac House, Temple Back
Bristol BS1 6BE UK

Phone: 44 1 17 929 7481
Fax: 44 1 17 920 0979

techtracking@iop.org

Additional copies of this publication are available from:

Curran Associates, Inc.
57 Morehouse Lane
Red Hook, NY 12571 USA
Phone: 845-758-0400
Fax: 845-758-2633
Email: curran@proceedings.com
Web: www.proceedings.com

Table of contents

Volume 50

Ecology and safety in the technosphere: current problems and solutions
17–19 November 2016, Yurga, Russian Federation

Accepted papers received: 12 December 2016
Published online: 10 January 2017

Preface

Papers

IOP Conf. Series: Earth and Environmental Science **50** (2017) 011001 doi:10.1088/1755-1315/50/1/011001

PREFACE

All-Russian research-to-practice conference for young researchers, postgraduate students and undergraduates "Ecology and safety in the technosphere: current problems and solutions" will make a contribution to the rethinking of the relationship between man and nature in today's world, to finding the best ways to overcome the negative effects of the environmental crisis and to solve security issues.

The conference aims at discussing research development and introducing innovations in the field of environmental and technosphere safety; searching for collaborative research areas and strengthening cooperation between young researchers and professionals from educational and scientific institutions in Russia; creating a platform for creative communication and exchange of experiences between young researchers and professionals in the issues of environmental and technosphere safety. It will also focus on intellectual and creative development of young researchers and specialists; demonstrating the best achievements of young researchers and professionals in scientific and technological activities; acquisition scientific debate and public speaking skills by young researchers; increasing the prestige of education, research and innovative activity in the field of environmental and technosphere safety; systematization of scientific data on the topical issues of ecological and technosphere safety, developing evidence-based approaches to their solution.

The results of the conference for young researchers, postgraduate students and undergraduates can be applied in making improvements in the curricula of students in the field of technosphere security; developing measures for the organization of schools for young researchers; presenting the research results for further discussion with colleagues and leading scientists in the priority areas; issuing new monographs, textbooks, laboratory exercises, lecture notes.

The Organizing Committee of the Conference thank all the participants of the Conference for their fruitful work and personal contribution to the development of these Conference Proceedings.

Dmitry A. Chinakhov, PhD,
Conference Convenor,
Vice-Chancellor, Yurga Institute of Technology
of National Research Tomsk Polytechnic University

Peer review statement

All papers published in this volume of *IOP Conference Series: Earth and Environmental Science* have been peer reviewed through processes administered by the proceedings Editors. Reviews were conducted by expert referees to the professional and scientific standards expected of a proceedings journal published by IOP Publishing.

IOP Conf. Series: Earth and Environmental Science **50** (2017) 012001 doi:10.1088/1755-1315/50/1/012001

Increasing Of Efficiency Of Environmentally Friendly Technology Of AWJ Of A Glass Fiber Plastic

G Barsukov, T Zhuravleva, O Kozhus

Orel State University named after I.S. Turgenev, Orel, Komsomolskaya street 95, Russia, 302026

E-mail: okozhus@mail.ru

Abstract. Special technological methods allowing to reduce the probability of stratifications spreading for the cut-off blank when applying environmentally friendly AWJ technology are developed by the authors of the article. As a result the quality of the item improves, its cost value decreases.

Introduction

One of the most widespread materials of polymeric group are glass fiber plastics. The glass fiber plastic is used for manufacturing of electrical distribution devices, panels which designs may contain a large number of through-elements (fastener holes, windows, slots of different configuration).

The condition of material, presence of micropores and microcracks, inherent stresses, discontinuity flaw of binding connection and other defects, connected with the process of manufacturing of sheet polymeric materials as well as their storage period, have an influence on the quality of machined surface.

The traditional methods of cutting which are not excluding manufacturing defects also have other shortcomings, such as low productivity; complexity of making details of a figured profile; significant waste of material because of irrational cutting, big allowances and overlapping. Usually, waste caused by technological features of machine processing, physical and chemical characteristics of material and their interference are considered in the manufacturing process by increasing the standard of a material consumption depending on the complexity of a details configuration, requirements for processing accuracy, batch size, etc.

To that end, environmentally friendly technology of waterjet cutting with liquid or hydro abrasive jet becomes widespread [1-4]. Absence of thermal influence (as the temperature in the cutting zone is 90 degrees) and power mechanical impact on the processing unit, as well as a small amount of material waste, distinguish this technology [5-10].

Main type of reject in the processing of glass fiber plastic by AWJ-method is stratification in the places where abrasive liquid jet cut into the material. The factors which significantly influence on the emergence of stratifications, are similar at the hydro abrasive cutting and processing by the edge tool, and can be divided into two groups:

1) the technological factors which depend on the modes of processing and technological parameters of the equipment;

2) a condition of material at the time of processing.

The influence of the first group of factors at the hydroabrasive processing of details made of glass fiber plastic was considered in [11 - 15] in detail. It is found out that the size of stratifications is highly affected by hydroabrasive jet outflow pressure during plunge cutting, abrasive grit, diameter of a nozzle outlet and its interference [11, 12].

A condition of material at the time of processing is understood as the presence or the absence of micropores or microcracks, inherent stresses, discontinuity flaw of binding connection and other defects, connected with the process of manufacturing of multilayer sheet materials as well as their storage period.

Though the influence of the first group of factors is subject of researches [16 - 19], it is difficult to predict the influence of the second group of factors in advance.

The complex of technological methods is developed to completely exclude or to reduce the influence of stratifications of material using AWJ, when the demanded parameters of quality of details made of glass fiber plastic with minimal labour inputs, costs of material and energy resources are ensured.

1. Development of complex of technological methods

1 Technological methods used for cutting of external contours [20]:

a) Increasing the length of the entry of a jet to the cutting-out blank contour (figure 1). Value of entry to a contour should be slightly larger than a size of stratification.

The consumption of the cut-out material increases, cutting becomes less rational. For more rational cutting it is possible to put the blanks as shown in figure 2. Plunge cutting for four blanks in this case is realized at one point, which saves the quantity of an abrasive necessary for plunge cutting for blanks 2, 3, 4 since the opening for jet entry is already made during plunge cutting for blank 1 at a point (1) and excludes emergence of stratifications. Cutting at a such positioning of blanks on a plate will look as shown in figure 3.

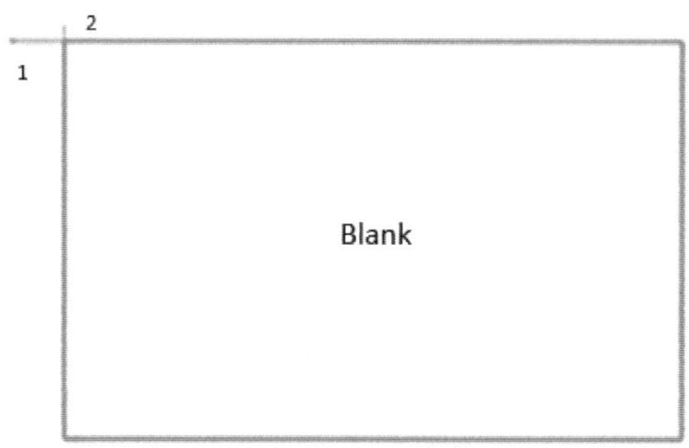

Figure 1. Technological method: increasing the length of entry to the blank outline for exclusion of reject. 1 – entry to the outline, 2 – leaving the outline.

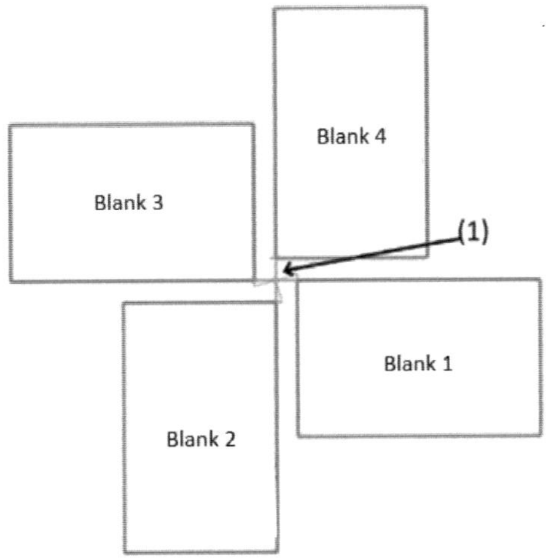

Figure 2. An example of disposition of a blanks with the increased length of entry to the contour.

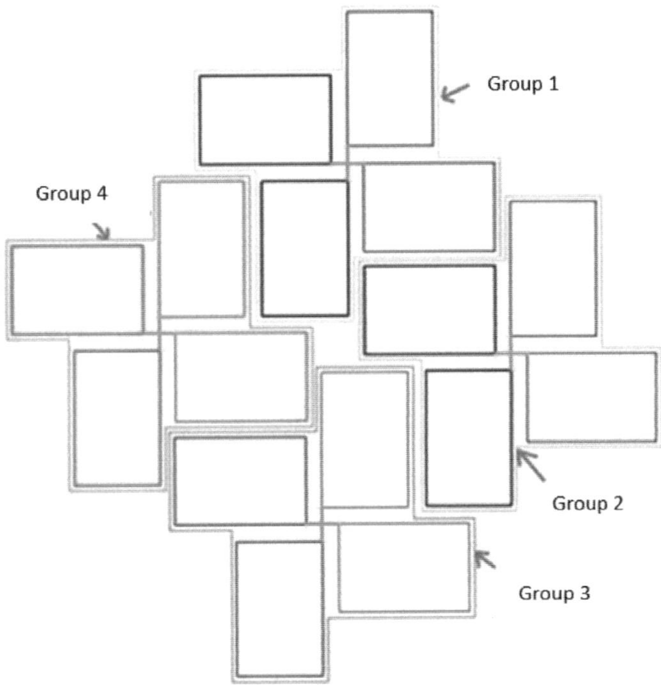

Figure 3. An example of disposition of a blanks on a sheet material.

b) To make plunge cutting outside the material. Cutting is made outside out of material at distance 0,5-1mm from the edge of the sheet.

Processing of the first group of blanks is shown in a figure 4.

As plunge cutting is made outside material, stratifications do not appear. Thus, it is possible to reduce the length of entry to a blank contour.

Processing of the second and the subsequent groups of blanks needs to be performed as follows: after cutting of the first group of blanks, the operator stops the process and extracts the cut out blanks. Plunge cutting for the second group of blanks has to be performed in the openings which remained from the taken blanks of the first row (figure 5). After cutting of the second group of blanks, the operator extracts them from a sheet; plunge cutting for cutting of the third group of blanks is made in the openings which remained from the second row. Thus, cutting of a sheet is made consistently by rows.

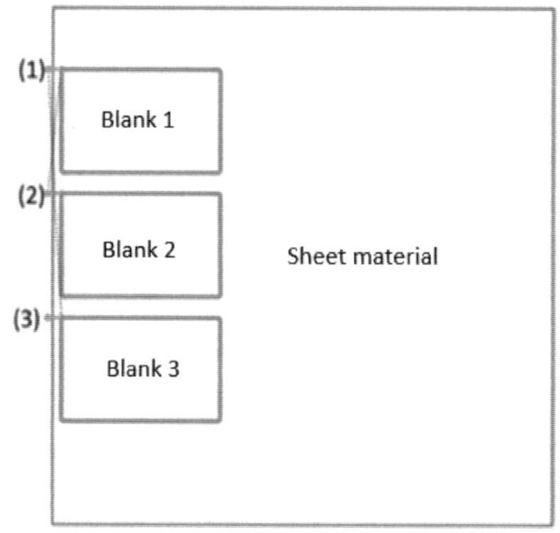

Figure 4. Cutting of the first group of blanks with plunge cutting at the points (1), (2), (3) outside the material.

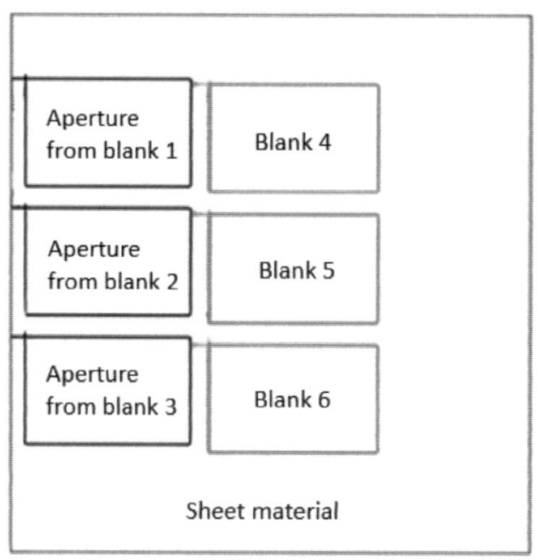

Figure 5. Processing of the second group of blanks with plunge cutting outside the material.

Such method of cutting has a shortcoming: the equipment downtime connected with a need of extraction of the cut-out details. It is expedient to use this method for processing of sheet material from 10 mm thick where time of a blank cutting is rather great, and also for cutting large-size blanks.

c) Use of the clipping cuts. Cutting of sheet material of relatively small thickness occurs with high enough speed (for example, from 1500 mm/min for STEF-1 with thickness of 6 mm, and to 10000 mm/min for STEF-1 with thickness of 1 mm). Therefore the technological method with plunge cutting into the openings from the taken blanks will lead to considerable downtime of the expensive equipment. In this case it is expedient to use the clipping cuts.

For this purpose the technological clipping cut (figure 6) is made before cutting of the first group of details.

Cutting of the first row of blanks is made after the clipping cut. Plunge cutting into material is made on the left of the clipping cut, and the contour of the blank is located to the right (figure 7). Thus, the defect does not reach further than the clipping cut at the emergence of stratifications during entry into the material.

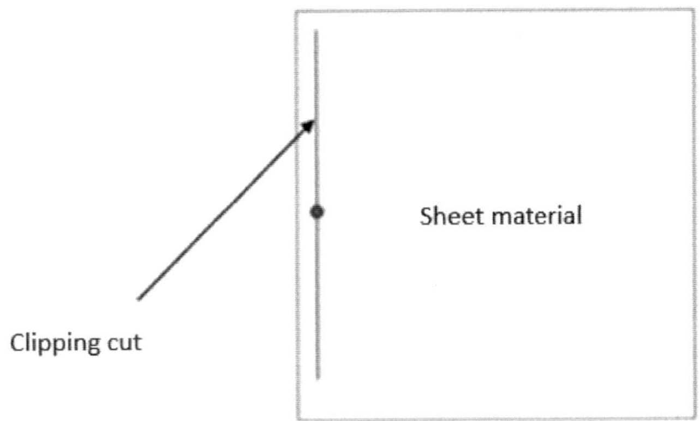

Figure 6. Execution of the technological clipping cut

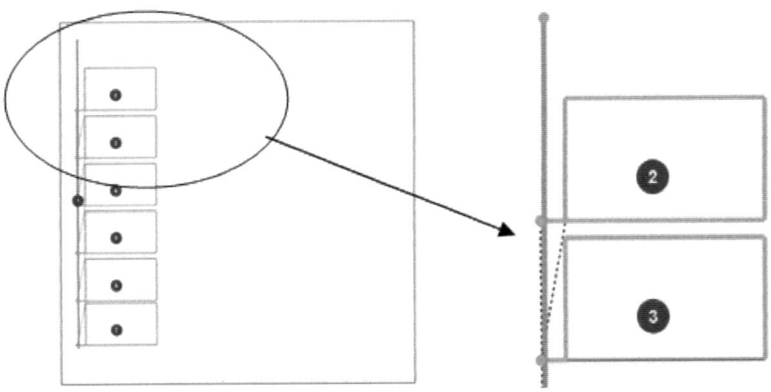

a) b)

Figure 7. Layout of the blanks using clipping cut: a - a general view; b - closer view.

Further, the second clipping cut and cutting of the subsequent row (fig. 8) are performed. It is possible to completely cut the material with the cut-off blanks and to remove it from an equipment desktop when performing of the second clipping cut; it is also possible to make the clipping cut as it is shown in fig. 8.

Before processing of the third row one more clipping cut is made, etc.

Figure 8. Cutting of the second group of blanks.

When using this technological method, it is important to make sure that the plunged cutting is made at a distance no more than 0,5-1 mm to the left of the clipping cut to avoid of hooking of an edge of material with a nozzle, that can lead to shift of sheet material or damage of a nozzle. It is connected with the fact that cutting thin-sheet materials up to 5 mm thick, very often results with the uplift of the cut-off material edge that can lead to collision of a nozzle and the edge and, correspondingly, to the damage of a nozzle and shift of sheet material during processing. Distance from a place of entry into the clipping cut has to be less than the outer diameter of a nozzle. It allows not to hook the rising material edge with the nozzle edge.

2. Summary

Authors of the article developed the following special technological methods for cuttings of external contours, cuttings of internal contours, a choice of a type of a basic surface for placing of material, a choice of the direction of execution of the contour and the sequence of cutting of blanks. Their use allows to decrease or to exclude the probability of distribution of stratifications on the cut-out blank using AWJ; that provides improvement of quality of items made of glass fiber plastic and decrease its prime cost.

References

[1] Tikhomirov R A, Petukhov E.N., Babanin V.F., Starikov I.D., Kovalev V.A. High-pressure jetcutting // Mechanical Engineering (1992)

[2] Ansari A.I., Hashish M. Effect of abrasive waterjet parameters on volume removal trends in turning // Journal of engineering for industry. (1995)

[3] Tikhomirov R.A. Waterjet cutting: process and equipment // Russian Engineering Research. (1997)

[4] Stepanov Yu.S., Barsukov G.V., Bishutin S.G. Technological Fundamentals for Efficiency Control of Hydroabrasive Cutting // 2nd International Conference on Industrial Engineering (ICIE-2016). Procedia Engineering. Volume 150, 2016, Pages 717–725.

[5] Bach F.W., Louis H., Versemann R., Schenk A., Characterization of a pure water-jet cleaning process - Process simulation, Strojniski Vestnik/Journal of Mechanical Engineering. (2006).

[6] Chillman A., Hashish M., Ramulu M., A novel approach to energy based evaluations of ultra highpressure waterjets, American Society of Mechanical Engineers, Pressure Vessels and Piping Division (Publication) PVP. (2010).

[7] Salenko A.F., Shchetinin V.T., Fedotyev A.N., Improving accuracy of profile hydro-abrasive cutting of plates of hardmetals and superhard materials Improving accuracy of profile hydro-abrasive cutting of plates of hardmetals and superhard materials, Journal of Superhard Materials.

(2014).
[8] Chillman A., Hashish M., Ramulu M., Potential of waterjet peening for mainstream industrial applications, in: Proceeding of BHR Group - 21st International Conference on Water Jetting: Looking to the Future, Learning from the Past. (2012).
[9] Hashish M., Erosion modes during AWJ lathe slotting, American Society of Mechanical Engineers, Manufacturing Engineering Division, MED. (1995).
[10] I. Ajmal, Hashish M., Volume removal trends in abrasive waterjet turning effect of abrasive waterjet parameters, American Society of Mechanical Engineers, Production Engineering Division (Publication) PED. (1993).
[11] Koenig, W. Machining of Fibre Reinfonced Plastics / W. Koenig, Ch. Wulf, P. Grass, H. Willersheild // Manufact. Tech., CIRP Annals, 1985.-Vol. 34.-P. 537-548.
[12] Hashish, M. Machining of advanced composites with abrasive-waterjets / M. Hashish // ASME Winter Annual Meeting, Chicago, Illinois, Nov.27-Dec.2 ,1988.
[13] Hashish, M. Characteristics of surfaces machined with abrasive waterjet/ M. Hashish // MD. - 1999 - Vol. 16. - P. 23 -32.
[14] Ho-Cheng, H. A failure analysis of water jet drilling in composite laminates / H. Ho-Cheng // Int. J. Mach. Tools Manufact., 1990.- Vol. 30.-P.423-429.
[15] Louis H., Pude F., Von Rad Ch., Versemann R. Abrasive water suspension jet technology fundamentals, application and developments // Welding in the World. (2007).
[16] M. Ramulu, T. Briggs, M. Hashish Quality and surface integrity of waterjet machined automotive composites // in: Proceeding of BHR Group - 22nd International Conference on Water Jetting (2014)
[17] Ansari A.I., Ohadi M.M., Hashish M., Effect of waterjet pressure on thermal energy distribution in the workpiece during cutting with an abrasive waterjet, American Society of Mechanical Engineers, Production Engineering Division (Publication) PED. (1988).
[18] Ohadi M.M., Ansari A.I., Hashish M., Experimental study of the role of thermal energy transfer in material removal processes using abrasive waterjets, American Society of Mechanical Engineers, Fluids Engineering Division (Publication) FED. (1988).
[19] Louis H., Pude F., Versemann R. Abrasive water suspension jet technology - Fundamen-tals, application and developments //Rivista Italiana della Saldatura (2007)
[20] Barsukov GV, Zhuravleva TA, Semes VV Patent RU №2577667 Method Waterjet cut-ting sheet layered polymeric materials.

Aerodynamical Probation Of Semi-Industrial Production Plant For Centrifugal Dust Collectors' Efficiency Research

Y I Buligin[a], M G Zharkova[b], L N Alexeenko[c]

Don State Technical University
Gagarin square 1, Rostov-on-Don, Russian Federation, 344000

E-mail: [a]bulyur_rostov@mail.ru, [b]mariazharkova@mail.ru, [c]puchinka78@mail.ru

Abstract. In previous studies, experiments were carried out on the small-size models of cyclonic units, but now there completed the semi-industrial pilot plant «Cyclone», which would allow comparative testing of real samples of different shaped centrifugal dust-collectors and compare their efficiency. This original research plant is patented by authors. The aim of the study is to improve efficiency of exhaust gases collecting process, by creating improved designs of centrifugal dust collectors, providing for the possibility of regulation constructive parameters depending on the properties and characteristics of air-fuel field. The objectives of the study include identifying and studying the cyclonic apparatus association constructive parameters with their aerodynamic characteristics and dust-collecting efficiency. The article is very relevant, especially for future practical application of its results in dust removal technology.

1. Introduction

In recent years the «Life Safety and Environment» department of Don State Technical University conducts researching works of aerodynamic characteristics and dust-collecting efficiency of different shaped cyclonic units.

If in previous studies experiments were carried out on the small-size models of cyclonic units [1-6], then now there completed the semi-industrial pilot plant «Cyclone», which would allow comparative testing of real samples of different shaped centrifugal dust-collectors and compare their efficiency. This original research plant is patented by authors in 2014-2015. [7, 8].

The aim of the study is to improve efficiency of exhaust gases collecting process, by creating improved designs of centrifugal dust collectors, providing for the possibility of regulation constructive parameters depending on the properties and characteristics of air-fuel field.

The objectives of the study include identifying and studying the cyclonic apparatus association constructive parameters with their aerodynamic characteristics and dust-collecting efficiency.

There why the article is very relevant, especially for future practical application of its results in dust removal technology.

Statement of research objectives and requirements for aerodynamic experiment

To achieve the goals and objectives there were formulated requirements for the experiment:

- to achieve tightness of completed semi-industrial experimental plant (first experimental studies of the plant must be carried out without hoppers);

- tests are carried out in accordance with the requirements of GOST 12.3.018-79 Safety security standards system, ventilation systems, aerodynamic testing methods;
- to identify characteristics of pressure changing, air velocity and its flow rate at different points in the cross sections of inlet and outlet pipes of cyclonic plant for different defined performance of exhaust blower;
- to obtain experimentally full, static and dynamic pressures at different points in the cross sections of inlet and outlet pipes of cyclonic plant for different defined performance of exhaust blower;
- to compare the calculated and experimentally determined coefficients of hydraulic resistance at the inlet and outlet of investigated cylindrical cyclone CN-15-300.
- to determine the flow regime at the plant (to calculate the Reynolds number).

Experimental plant (Fig. 1) is constructed in such way, that it could be possible to provide research of removable cyclone devices of various shapes. This is achieved by quick release at the area of inlet and outlet pipes and transition of lower cyclone cone into dust-collecting hopper. There is also the possibility of regulation of inlet pipe connecting highs with ductwork, feeding dusty gas into the cyclone plant.

Figure 1 – Experimental plant «Cyclone»: 1 – inlet pipe, 2 – outlet pipe, 3 – Cyclone CN-15-300 (through quick release clamps can be easily replaced to COK), 4 – hopper, 5 – round blower

2. Aerodynamical tests of the semi-industrial plant

With the aim to study aerodynamical characteristics of cyclonic apparatus precision differential micromanometers Testo 521-1 with a set of pressure Pitot tubes and impeller pocket anemometer Testo 410-2 were used.

For founding relations between aerodynamical characteristics and shape of testing cyclonic apparatus at the plant the next research was carried out.

To measure pressure and velocity of air flow in ductwork (channels) there must be chosen sectors with location of dimensional sections at distances for at least six hydraulic diameters D_h, m behind the place for the flow disturbance (taps, dampers, apertures etc.) and at least two hydraulic diameters before it. Coordinates of pressure dimension points and velocity (Fig. 2), as well as number of points are determined by the shape and size of dimensional circular section in line with GOST 12.3.018-79.

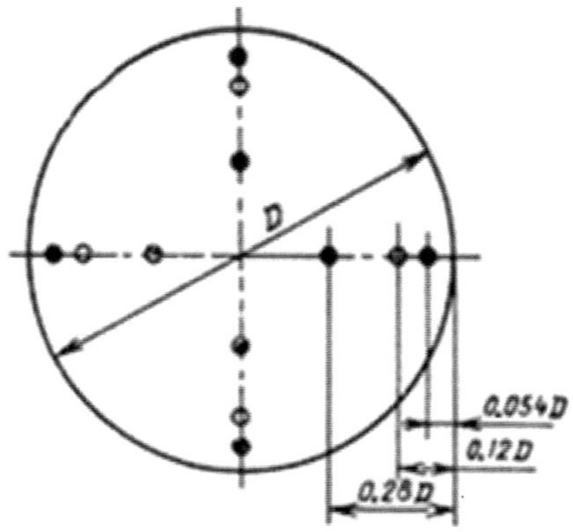

Figure 2 – Scheme of placing of pressure dimension points and velocity in cylindrical section ductwork

In our case the pipe diameter is 200 mm, consequently, dimension points will be placed at a distance of 0.12D from wall of pipe.

There were made some perforation at the inlet and outlet pipes of the experimental plant with D_{OTB} = 20 mm. In this perforation Pitot pipes were fixed with silicone plugs, connected to two differential micromanometers Testo 521-1. Next wasrunning up radial fan, mounted on the outlet connection. After 15 minutes working in the inlet and outlet of the plant were carried out measurements of full, static and dynamic pressure, rate and velocity of air flow. For objectivity of the study measurements were conducted 60 times.

3. Results and Discussion

Experimental results were analyzed and are presented in graphs (Fig. 3-8).

As seen from the graphs (Fig. 3), velocity of air flow in the outlet pipe for different performance of blower in any dimensional point greatly exceeds velocity of air flow in the inlet pipe (in 1.73...2 times). This is due to the blower VC 1446 (working on the hood) is placed structurally closer to the outlet pipe of cyclonic apparatus.

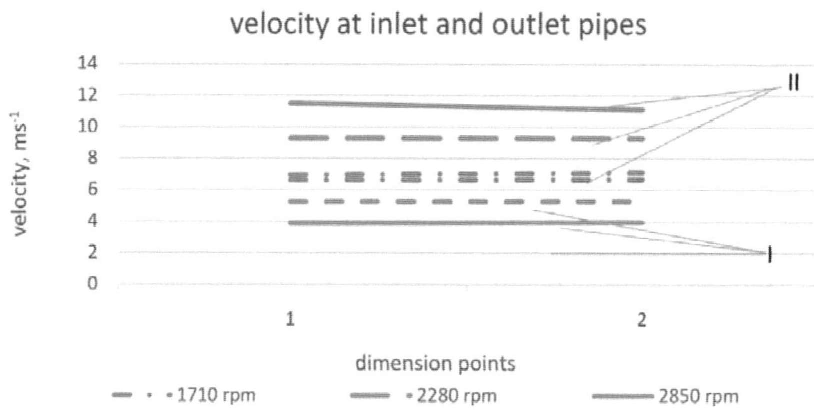

Figure 3 – Air flow velocity at inlet and outlet pipes dimension points of the experimental plant for different performance of blower

At the Fig. 4 presented relations of obtained experimentally airflow costs at the inlet and outlet respectively. These experiments made it possible to establish, that air cups in the direction of movement are absent and the plant has integrity. Costs value at the inlet and outlet pipes by dimension points and for different performance of blower practically have no differ. Wherein experiments were specially conducted without hoppers with cap on the cyclone CN-15 – 300.

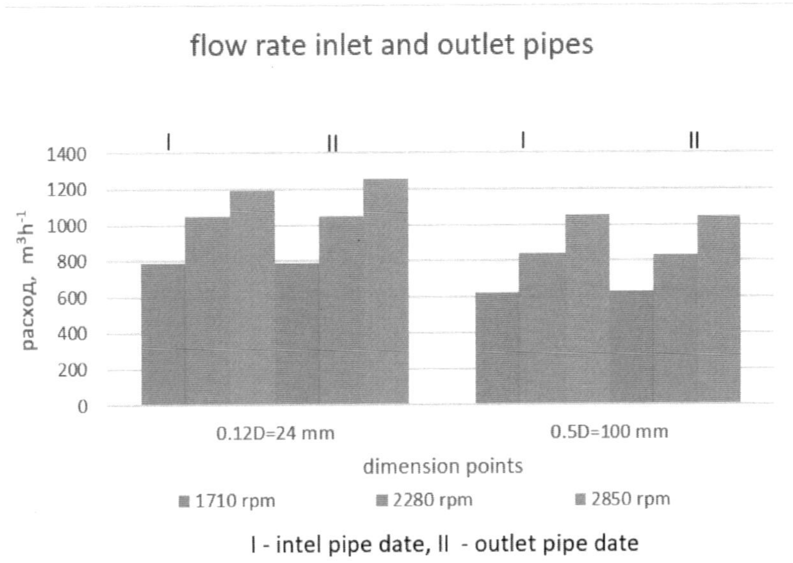

Figure 4 – Inlet and outlet air flow rate for the plant for different performance of blower.

At the Fig. 5 presented the results of measurement of full air flow pressures at the inlet and outlet pipes of the plant for different performance of blower.

The changing characteristics of full pressures shows, that in the outlet pipe of the cyclone vacuum forming goes on. This phenomenon is explained through vortex flow not only in the cyclone body, but also at the plant outlet, wherein pressure is distributed on the periphery of the vortex.

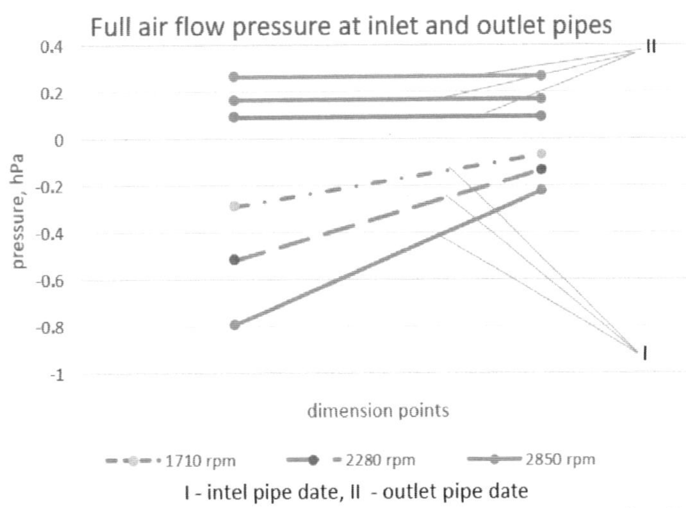

Figure 5 – Full air flow pressure at inlet and outlet pipes of cyclone CN-15- for different performance of blower

Basing on the experimental values of full pressures at the inlet and outlet of cyclone, and also measured velocities the hydraulic resistance coefficients of cyclone apparatus CN-15-300 were calculated. They are presented at Fig. 6. As cyclone CN-15-300 was produced under production conditions, it was important to determine KGF at the inlet of present apparatus and compare it with KGF of standard cyclone CN-15-300. Average value KGF at the inlet pipe is 4.12. For standard cyclone CN-15 KGF=4.30.

The Reynolds numbers have also been calculated according to the available experimental data. Reynolds number at the outlet pipe nearly 2 times more, than at the inlet pipe (7000) and it is up to 15000. The values of Reynolds numbers allow us to conclude, that in the experimental plant there occurs the turbulent movement of air.

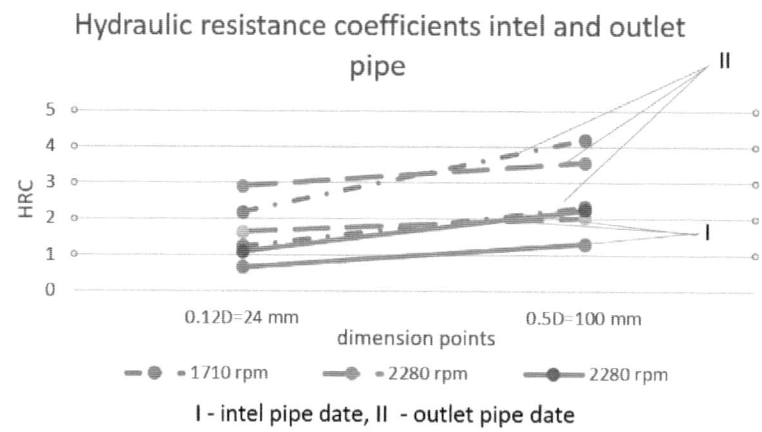

Figure 6 – Hydraulic resistance coefficients at the outlet of cyclone CN-15-300 for different performance of blower

4. Conclusion

The impermeability of completed semi-industrial experimental plant is reached, which allows to use it further in compare tests with different dust collectors.

The character of pressure change, air flow velocity and its rate at different points in the cross sections of inlet and outlet pipes of cyclonic plant for different defined performance of exhaust blower is identified.

Experimentally full, static and dynamic pressures at different points in the cross sections of inlet and outlet pipes of cyclonic plant for different defined performance of exhaust blower are obtained.

Coefficients of hydraulic resistance at the inlet and outlet of investigated cylindrical cyclone CN-15-300, which are consistent to data, given in the literary sources [9], [10] are found.

Turbulent air flows in the plant is obtained.

In future we suppose to conduct more detailed experimental studies, including cyclone apparatus dust collecting efficiency.

References

[1] Potyomkina A V. Theoretical and experimental researching of influencing the immersion depth of exhaust pipe on aerodynamic characteristics and effectivity of cyclonic equipment / Potyomkina A V, Lucash E S. // DSTU Reporter. Rostov-on-Don, 2010. T.10. 4(47). P.526-533

[2] Bulygin Y I et al., Features of Velocity and Pressure Fields Formation in the Centrifugal Dust-Collector 2014, Applied Mechanics and Materials, Volume 698, 542-545

[3] Meskhi B Ch et al., Cyclone with Controlled Parameters and Self-Emptying Bin for Air Dedusting in Machine Building Plants 2014, Applied Mechanics and Materials, 682, 46-52

[4] Bulygin Y I, Koronchik D A, Abuzyarov A A, Engineering of centrifugal dust-collectors based on parallel comparing tests applying computer modelling: 6th International Scientific Practical Conference on Innovative Technologies and Economics in Engineering IOP Conference Series: Materials Science and Engineering, 2015

[5] Panchenko O S, Abuzyarov A A, Synthesis and study of cyclonic apparatus communication design parameters with its aerodynamic characteristics and dust collecting efficiency / Materials of III Rostov Youth Scientific and Practical Conference « Youth initiative - 2012». – Rostov-on-Don: Publisher of the Southern Federal University, 2012. 164

[6] Panchenko O S, Sameshchenko A A, Sameshchenko E A, Legkonogikh A N. Improving the effectiveness of air dedusting of work areas of metalworking and woodworking industries / Perspektiva–2013: Materials of International scientific conference of students, graduate students and young scientists. – T. IV. – Nalchik : Kab.- Balk. un, 2013. 472

[7] Patent number 2506880 of 20.02.2014 . Dust collector / Meskhi B Ch, Mikhailov A N, Buligin Y I, Alexeenko L N, Denisov O V, Panchenko OS.

[8] Installation for dust collection: Patent 2557741 A47L 9/16/ Meskhi B.Ch., Buligin Y.I., Legkonogikh A N, Alexeenko L N, Koronchik D A, Denisov O V, Abuzyarov A A – № 2014136881/12; of. 11.09.14; pub. 27.07.15, St. № 21

[9] Lazarev V A, Cyclones and vortex scrubbers: Directory. – Nizhny Novgorod: «Company OZONE-NN», 2006. 320

[10] Lazarev V A. Using of cyclones in the composition of the recirculation woodworking aspiration systems // Engineering systems. AVOK Northwest, 2005. 2. P. 34 – 39

Factors Of Environmental Safety And Environmentally Efficient Technologies Transportation Facilities Gas Transportation Industry

Bogdan U Vasiliev

Saint-Petersburg mining university
Vasilevsky Island, 21 line d.2, Saint Petersburg, Russian Federation, 199106

E-mail: vasilev.bu@spmi.ru

Abstract. The stable development of the European countries depends on a reliable and efficient operation of the gas transportation system (GTS). With high reliability of GTS it is necessary to ensure its industrial and environmental safety. In this article the major factors influencing on an industrial and ecological safety of GTS are analyzed, sources of GTS safety decreasing is revealed, measures for providing safety are proposed. The article shows that use of gas-turbine engines of gas-compressor units (GCU) results in the following phenomena: emissions of harmful substances in the atmosphere; pollution by toxic waste; harmful noise and vibration; thermal impact on environment; decrease in energy efficiency. It is shown that for the radical problem resolution of an industrial and ecological safety of gas-transmission system it is reasonable to use gas-compressor units driven by electric motors. Their advantages are shown. Perspective technologies of these units and experience of their use in Europe and the USA are given in this article.

1. Introduction

The gas industry plays a strategic role in economy of Europe. About a half of total production and internal consumption of energy resources depends on natural gas. With refusal of coal the role of natural gas will only continue to grow in an energy balance.

The most powerful gas transportation system (GTS) functions for transportation of natural gas from places of its production to consumers in Europe. Main suppliers of natural gas are Norway, Great Britain, France, the United Arab Emirates and Russia. The main consumers of the Russian natural gas in Europe are Germany and Italy. The Figure 1 shows the scheme of the gas transportation system of Europe.

The stable development of the European countries depends on a reliable and efficient function of the gas transportation system [1, 2]. With high reliability of GTS it is necessary to en-sure its industrial and environmental safety. In this article the major factors influencing on an industrial and ecological safety of GTS are analyzed, sources of GTS safety decreasing is revealed, measures for providing safety are proposed.

Figure 1 – Scheme of the gas transportation system in Europe

2. Materials and methods

2.1. Structure of park of gas-compressor units The main elements of gas-transportation systems (GTS) are main gas pipe-lines (MGP) and compressor stations (CS), which consist of compressor shops. The compressor shop consists of gas-compressor unit (GCU), gas coolers, power supply system, water supply system, alarm system, fire extinguishing and room artificial climate system, connection points, shutdown safety valves, stop valves, dust collectors etc. The gas-compressor units are the largest consumers of energy resources and, in fact, are a power center of the gas transportation system.

(a) (b)

(c)

Figure 2 – Different types of gas-compressing units

The working body of the gas-compressing units is centrifugal superchargers (CS) which compress and transport the natural gas along a main gas pipeline. Power of centrifugal superchargers, now in use on the main gas pipelines, varies from 2.5 to 32 MW [3]. The following engine types are used for driving the centrifugal superchargers: gas turbine engines; electric motors; gas engine compressors.

Figure 2 (a) shows the external view of the Rolls-Royce (Great Britain) gas-compressor unit driven by a gas turbine, Figure 2 (b) is the Siemens (Germany) electric gas-compressor unit, Figure 2 (c) is the Caterpillar (USA) gas-compressor unit driven by a gas engine.

More than half of park of gas-compressor units in Europe have the gas turbine drive. For gas turbine gas-compressor units a natural gas is used as a main fuel, which is transported in main gas pipelines. Combustion of natural gas in a huge number at compressor stations leads to decrease in level of an industrial and ecological safety.

2.2. Industrial and environmental safety of gas-transportation system Level of an industrial and ecological safety of gas transportation system objects is in a direct connection with reliability of the system. Many of the measures directed to increase of gas transportation system reliability should be considered as a way of decrease in risks of defeat and death of people and environment violation, i.e. safety increase.

Efficiency and safety of the gas transportation system functioning is substantially determined by ecological compatibility of work of the system main objects, one of which is a compressor station. Exactly here the greatest number of the power-intensive equipment intended for ensuring technological process of natural gas transportation is concentrated, branched systems of technological communications function and a large number of service personnel are involved [4, 5].

The main sources of the polluting substances are the gas-compressor units driven by gas turbines, centrifugal superchargers, oil degasser and technological processes connected with their operation.

2.3. Main factors of gas-compressor unit impact on environment Let's observe the main factors of the gas-compressor unit impact on environment: emissions of harmful substances in the atmosphere; toxic waste; noise and vibration; thermal influence.

2.3.1. Emissions of harmful substances in the atmosphere. These are emissions of natural gas and emissions of combustion products (exhaust gases of the gas-turbine engine). Table 1 shows parameters of technical standards of emissions of gas-compressor unit polluting substances, received experimentally.

Emission of natural gas on compressor station happens at start-up and stop of the gas-compressor unit, and also during operation with tightness violation of its main parts.

The exhaust gases are created as a result of the natural gas combustion in the gas turbine of the gas-compressor unit. They consist of: combustion products – nitrogen, water vapor, carbon dioxide; nitrogen oxides; carbon dioxide; sulphur oxides; hydrocarbons (including not completely burned methane); soot.

At combustion of the gases containing hydrogen sulfide, sulfuric and sulphurous anhydrides, not burned hydrogen sulfide are also released into the atmosphere.

The Directive of the European Parliament and Council 2001/80/EU specifies maximum permissible values of emissions in relation to the gas-turbine engines of gas-compressor units certified before 11/27/2002 and accepted for operation before 11/27/2003: on nitrogen oxides of 50 mg/m3 when using as fuel natural gas with the content of inert gases no more than 20%; on nitrogen oxides of 75 mg/m3 for gas-turbine engines in cogeneration installations with the general efficiency more than 75%, for gas turbines in a combined heat and power plant (CHP plant) with the general average annual efficiency more than 55%.

2.3.2. Toxic waste. Gas-compressor unit has lubricating systems of the gas turbine engine and the centrifugal supercharger. hydraulic systems of the centrifugal supercharger shaft sealing which use different lubricating oil types with storage oil tanks for the engine and the centrifugal supercharger accordingly [6, 7].

During work of the gas-compressor unit toxic substances are emitted from oil tanks of the gas-turbine engine and the centrifugal supercharger, from the oil degasser of the supercharger sealing system.

The toxic substances arriving from a shaft oil sealing system of the centrifugal supercharger because of pressure difference in the supercharger and the degasser, which is under atmospheric pressure are emitted in the degasser. Toxic substances are emitted in atmospheric air from oil tanks of the gas-turbine engine and a centrifugal supercharger in which the oil heated in the engine and the supercharger circulates.

For gas-compressor units with power up to 10 MW inclusive, irrevocable losses of oil should not exceed: with stationary gas-turbine installations – 1.0 kg/h; with converted engines – 2.0 kg/h; for gas-compressor units with power more than 10 MW: with stationary gas-turbine installations – 1.5 kg/h; with converted engines – 2.5 kg/h.

2.3.3. Noise and other types of influence. Noise pollution of the atmosphere is an operation consequence of gas-compressor units and traffic (road transport. helicopters. etc.). If noise levels exceed health protection standards. Serious problems for service personnel arise.

Table 1 - Emissions of polluting substances of gas-compressor units

Parameter	Parameter values				
Power of gas-compressor unit [MW]	2 ÷ 4	6 ÷ 8	10 ÷ 13	16 ÷ 18	22 ÷ 31
Fuel gas consumption [m3/h]	1106 ÷ 1795	1824 ÷ 2872	3077 ÷ 5051	4720 ÷ 6593	6889 ÷ 10539
Temperature of combustion products [°K]	1113 ÷ 1270	947 ÷ 1295	1053 ÷ 1456	1130 ÷ 1456	1188 ÷ 1518
Concentration in dry combustion products:					
– nitrogen oxide [mg/m3]	35 ÷ 136	69 ÷ 202	69 ÷ 199	48 ÷ 179	50 ÷ 353
– carbonic oxide [mg/m3]	25 ÷ 82	33 ÷ 239	29 ÷ 275	48 ÷ 229	41 ÷ 441
Modified concentration:					
– nitrogen oxide [mg/m3]	50 ÷ 195	135 ÷ 490	100 ÷ 230	80 ÷ 250	50 ÷ 400
– carbonic oxide [mg/m3]	30 ÷ 130	90 ÷ 300	60 ÷ 300	80 ÷ 300	50 ÷ 500
Emission efficiency:					
– nitrogen oxide [g/s]	0.52 ÷ 1.67	2.58 ÷ 6.93	2.38 ÷ 7.81	3.29 ÷ 11.8	2.91 ÷ 26.65
– carbonic oxide [g/s]	0.41 ÷ 1.35	1.5 ÷ 6.66	1.81 ÷ 8	3.5 ÷ 14.5	3.59 ÷ 39.08
Specific emission on fuel gas unit:					
– nitrogen oxide [g/s]	1.39 ÷ 5.42	3.75 ÷ 13.62	2.78 ÷ 6.39	2.22 ÷ 6.95	1.39 ÷ 11.12
– carbonic oxide [g/s]	0.83 ÷ 3.61	2.5 ÷ 8.75	1.67 ÷ 8.34	2.22 ÷ 8.34	1.39 ÷ 13.9
Specific emission on thermal power unit:					
– nitrogen oxide [g/GJ]	41 ÷ 162	112 ÷ 407	83 ÷ 249	66 ÷ 207	41 ÷ 332
– carbonic oxide [g/GJ]	24 ÷ 108	75 ÷ 249	49 ÷ 174	66 ÷ 249	41 ÷ 415
Specific emission on unit of work:					
– nitrogen oxide [g/kWh]	0.47 ÷ 2.31	1.48 ÷ 5.88	0.86 ÷ 2.16	0.73 ÷ 2.59	0.43 ÷ 3.92
– carbonic oxide [g/kWh]	0.29 ÷ 1.22	0.9 ÷ 3.73	0.65 ÷ 2.73	0.8 ÷ 3.27	0.54 ÷ 5.12
Specific emission on gas supply unit:					
– nitrogen oxide [g/thousands of m^3]	6 ÷ 114	11 ÷ 24	10 ÷ 37	8 ÷ 18	5.28 ÷ 48.99
– carbonic oxide [g/thousands of m^3]	4.17 ÷ 49.22	5.40 ÷ 52.89	4.34 ÷ 32.50	9.06 ÷ 37.73	6.53 ÷ 61.23

Gas-compressor units have four main (primary) sources of noise: gas-turbine engine; compressor; centrifugal supercharger; fans. The main sources of noise are paths of suction and exhaust of gas turbine installations.

Let's examine basic reasons of a low effectiveness of the existing silencers of noise suppression systems: pattern incompatibility of absorption and emission spectrum; existence of acoustic bridges (bypassing the silencer) for a sound distribution; insufficiency of a sound-absorbing surface area;

defects of a design (assembly gaps. unreasonably large cross sections); short life of the used sound-absorbing material (foam rubber).

The noise suppression of the gas-compressor unit is carried out by four main methods. Let's examine these methods:

- dissipative noise suppression in paths of suction and exhaust. For noise reduction by this methods absorbers of resonant type are used of which elementary method is the limited air cavity connected by an opening (throat) to environment.
- use of jet silencers (resonators). The principle of operation of these silencers is «locking» of extending modes of sound vibrations and their reflection on the channel back to a noise source.
- active noise suppression in channels. The physical mechanism of noise reduction when using active methods consists, as well as in case of use of the usual soundproofing systems, in addition of fluctuations with various phases, at the same time however the secondary (compensating) field is created not in the passive way (for example by reflection), but is radiated by special electro-acoustic converters.
- combination. Entering of absorption elements of sound energy in jet silencers improves their parameters since sound reflection effect from jet type de-vices is weakened, and also levels of sound pressure in forming zones of not ex-tending modes of fluctuations are decreased. Sound-absorbing elements provide dissipative withdrawal (drain) of acoustic energy, transferring it to thermal energy. When a sound dissipation in natural absorbers is small, the absorbing elements are introduced into the jet silencers.

2.3.4. Thermal influence. Use gas-turbine engines in the gas-compressor unit drive leads also to thermal impact on environment and emergence of geo-ecological factors. Thermal influence of compressor station on environment is especially actually during the operation in the conditions of permafrost.

2.3.5. Energy saving problem. The gas consumption on own needs which first of all fall on fuel gas for gas-turbine engines of gas-compressor units can constitute up to 5% of transportation amount for 1000 km of the route. Therefore in absolute value in case of transportation distance more than 5000 km the gas consumption on own needs of compressor station will be very considerable. The solution of this problem has several main aspects: decrease in energy consumption for a gas compression due to use of gas-compressor unit with the efficiency high values of the gas-turbine engine and the centrifugal supercharger; decrease in pressure lost during gas transportation in a gas pipeline due to: use of pipes with an internal smooth covering; periodic cleaning of the pipeline internal space; use of high-performance equipment and gas piping of compressor station; decrease in the general energy consumption due to application of energy saving technology.

During development of gas transportation technology the gas energy saving is the same factor, as other important resource-saving factors connected with investments, metal investments and operating costs.

The choice of gas transmission technology first of all depends on a ratio between the cost of pipes and the equipment of compressor station on the one hand and the gas consumed for own needs, on the other hand. The final decision shall be made by means of optimization of the economic indicators characterizing efficiency of the project and the ecological safety of the project.

3. Results and discussion

3.1. The Electric gas-compressor units (EGCU) Use of the electric gas-compressor units radically ensures an industrial and ecological safety. Their use provides: exclusion of natural gas use as a fuel for the drive of the gas-compressor unit; exclusion of emissions of natural gas combustion products in the atmosphere; exclusion of necessity have a toxic oil treatment device; absence of compressor station thermal impact on environment; absence of noise and vibration pollution; high level of resource-saving for compressor station; increased level of compressor station reliability; possibility to create unmanned technologies of natural gas transportation.

3.2. Experience of use of electric gas-compressor units The world electric companies, such as ABB. MAN. Siemens, etc. are leaders in research, development and production of electric gas-compressor stations [2]. Electric gas-compressor stations were widely adopted in Europe. America and the countries of the Asia-Pacific Region [8]. Let's review some examples of use of electric gas-compressor stations in the different countries.

The German gas transmission company Thyssengas, is one of regional operators of main pipelines in Germany [9]. At compressor station Hünxe which is located in the federal land Northern Rhine-Westphalia gas-compressor units of 6.3 MW driven by a gas-turbine were used since 1991. In 2000 these gas-compressor units have been replaced with electric gas-compressor units for ensuring environmental friendliness and energy efficiency. The ALSTOM company installed on Hünxe compressor station the 6 MW adjustable electric motor drive of an alternating current and rotating speed 12700 RPM, made according to the HOFIM technology – high speed-oil free compression system (Figure 3 (a). Also, due to use of electric gas-compressor unit the compressor station Hünxe has been completely automated, including dimensions of compressor station were reduced and emissions of greenhouse gases and leak of oil from lubricating oil systems were excluded.

(a) (b)

Figure 3 – Electric gas-compressor units made according to the technologies HOFIM and MOPICO

Experience of the American gas transmission company Columbia Gas Transmission Corp can serve as other example of use of electric gas-compressor unit of the ALSTOM company at compressor station Rutledge. Originally, the Rutledge station has been developed for use of gas piston engines. However, the proximity of residential communities, noise and a large number of emissions of greenhouse gases from the working gas-compressor unit was not compatible. These circumstances have forced the company to modernize Rutledge compressor station and to replace gas-piston engines with electric gas-compressor unit. In this case the ALSTOM company used electrical gas-compressor unit made according to the MOPICO technology – motor pipeline compressor (Figure 3 (b)). This unit represents the hermetic compressor driven by an asynchronous engine. The electric motor which is installed in the unit rotates two centrifugal superchargers. Power of these unites is 3 MW, rotating speed is 10 000 RPM.

3.3. Technologies of electric gas-compressor units The HOFIM and MOPICO technologies are the main in the world according to which the leading electric companies produce electric gas-compressor units [10]. The main features of these units are the arrangement of the high-speed electric motor and compressors in the tight single case, and they are installed on common shaft which is supported by magnetic bearings. Thus, the drive and operating device are connected without gears and couplings.

4. Conclusion

Complexes driven by motors for natural gas transportation. undoubtedly, in the future will be used most widely in compressor stations. It is defined by the fact that electric gas-compressor units have for this purpose the most powerful objective arguments and extensive possibilities. Thus, in order that an electric motor was used as the main type of mover for the centrifugal supercharger it is necessary to realize the whole complex of actions which will provide uninterrupted power supply for compressor stations.

Underwater electric gas-compressor units of special designs are non-competitive technical means for a natural gas compression on a seabed at development of offshore fields. In this case the highest requirements should be imposed to the electric drive and whole unit concerning reliability of its functioning and providing a continuity of all technological processes of natural gas transportation.

The type of the gas-compressor unit drive shall be chosen based on the sys-tem and detailed analysis of technical and economical and other factors in case of construction and reconstruction new and old compressor stations respectively. The reliable and regular natural gas supply of consumers and ensuring development of economy of Europe shall be a main goal of construction and reconstruction of compressor stations.

References

[1] Cortinovisa A, Ferreaua H J, Lewandowskib D, Mercangöza M. Experimental evaluation of MPC-based anti-surge and process control for electric driven centrifugal gas compressors. Journal of process control, 2015, 34, pp 13–25.
[2] Kozyaruk A E, Vasilev B Y. Structure composition and control algorithms of high-efficiency electric drives of gas-compressor units. Russian Electrical Engineering, 2013, 84 (2), pp 94–102.
[3] Mardashov D V, Vasiliev B U. Development of well killing technology during well service on oil gas and condensate fields with carbonate reservoirs. International journal of applied engineering research, 2015, 10 (22), pp 43103–43105.
[4] Verma M, Parker D, Grinbaum I I, Nanney J. Converting to electric motors and adjustable speed drives: A case study of a 20.000 HP gas turbine driven compressor. IEEE Petroleum and Chemical Industry Committee Conference, 2015, pp 1–9.
[5] Wyszkowski M, Sivitskaya V. Effect of different substances on some properties of soil contaminated with heating oil. Journal of Ecological Engineering, 2015, 16 (1), pp 62–66.
[6] Carlo N Hamelinck, Roald A A, Suurs, Andre P C. Faaij International bioenergy transport costs and energy balance. Biomass and Bioenergy, 29 (2), 2005, pp 114–134.
[7] Sydney Thomas, Richard A Dawe. Review of ways to transport natural gas energy from countries which do not need the gas for domestic use. Energy, 2003, 28 (14), pp 1461–1477.
[8] Pascal Poudenx. The effect of transportation policies on energy consumption and greenhouse gas emission from urban passenger transportation. Transportation Research Part A: Policy and Practice, 2008, 42 (6), pp 901–909.
[9] Lin Gan. Globalization of the automobile industry in China: dynamics and barriers in greening of the road transportation. Energy Policy, 2003, 31 (6), pp 537–551.
[10] Anand B Rao, Edward S Rubin A. Technical, economic, and environmental assessment of amine-based CO_2 capture technology for power plant greenhouse gas control. Environmental sciences technology, 2002, 36 (20), pp 4467–4475.

IOP Conf. Series: Earth and Environmental Science **50** (2017) 012004 doi:10.1088/1755-1315/50/1/012004

Increasing The Use Of Natural Gas On Motor Transport By An Efficient Location Of The NGV RS (On The Example Of Tyumen)

E Chikishev, D Chainikov, I Anisimov

Motor transport operation, Industrial University of Tyumen
Volodarskogo Street 38, Tyumen, Russian Federation, 625000

E-mail: chikishev_e@mail.ru

Abstract. This article is devoted to increasing the use of natural gas on the basis of an efficient territorial and quantitative location of natural gas refuelling stations (NGV RS). The key factors which can affect it are offered. On the basis of the chosen factors on the example of Tyumen the option of a territorial location of 4 NGV RS capable to satisfy the demand for natural gas by motor transport in the short term is considered. Increased use of natural gas (NG) on motor transport allows to reduce the negative impact on the environment from harmful substances emitted with exhaust gases.

1. Introduction

In April 2016 many countries signed the Paris agreement to which the 21st conference (COP 21) preceded at the end of 2015 to the United Nations Framework Convention on Climate Change. This paper discusses the problem and importance of decreasing carbon dioxide in the atmosphere to slow the growth of global warming, since 2020. The purpose of this conference was to sign the international agreement on maintaining the average temperature increases of the planet to less than 2 °C, applicable to all countries. The agreement will replace one of the first global documents to regulate the negative impact of humans on the environment, the Kyoto Protocol (COP 3), which was accepted in Japan in 1997. The Paris Protocol was already appended by 175 countries, including Great Britain, China, Russia, the USA, France, etc., thus confirming the importance of environmental issues and climate change around the world.

At the same time, it is known that one of the main anthropogenic sources of environmental pollutants is motor vehicles, with over 1 billion vehicles worldwide and steady growth in this number each year.

One way to reduce the harmful effects of vehicles is to use more environmentally-friendly alternative energy sources, such as natural or oil gas, biofuels, electricity, hydrogen, etc. Application of these types of fuel is already widely used in various countries.

In Russia, the most urgent is the use of natural gas (NG). The country has the world's largest reserves of this fossil (24.6%) and is the second-largest producer worldwide (18%) [1]. Also on state and regional levels a number of legislative initiatives which in the near future have to increase consumption of NG on the motor transport in compressed (CNG), and liquefied (LNG) states is accepted. Countrywide construction of NGV RS has to play a leading role in increase of this hydrocarbon consumption.

For example, according to the program for the construction of compressed natural gas refueling stations (CNG RS)s and liquefied natural gas refueling stations (LNG RS)s developed by PJSC "Gazprom" and LLC "Gazprom gas fuel" it is planned to construct in Russia 2158 CNG RSs and 462 LNG RSs until 2030 [2]. This program also includes transferring 524 thousand vehicles to NG. The program contributes to the environmental and economic efficiency of vehicles use, increases energy security of the country and corresponds to the Russian state program "Energy efficiency and the development of renewable energies" from 15.04.2014, № 321 (edition. Resolutions of the Government of the Russian Federation from 09.10.2015 № 1079; from 07.12.2015 № 1339; from 25.05.2016, № 464) and the order of the Government RF № 767-r "On regulation of the relations in the sphere of the use of gas motor fuel, including NG as motor fuel" from 13.05.2013.

On the base of analysis of the submitted documents it is possible to note the next highlights. The purpose of the "Energy efficiency and the development of renewable energies" Program (an edition from 07.12.2015 No 1339) is – reliable providing the country with fuel and energy resources, increase in efficiency of their use and decrease in anthropogenic impact of fuel and energy complex on the environment. The aim of the subprogram "Development of the gas industry" is – the development of the gas industry, effectively providing the gas demand of the domestic market and the accomplishment of contract commitments on supply of NG for export, increasing the production and exports of LNG, the increase in the volume and depth of processing of raw material gas in the framework of diversification and leaving from export and raw model of an industry. A target indicators are including: a consumption of the compressed gas on to the motor transport work and the number of NGV RSs.

In the order of the Government RF No 767-r "On regulation of the relations in the sphere of the use of gas motor fuel, including natural gas as motor fuel" the harmonization of legal acts of the Russian Federation in the field of standardization with the relevant international documents in the sphere of the use of NG as motor fuel is offered; expediency of development of the technical regulation of the Customs Union in the sphere of use of NG is proved.

It is entrusted to develop a set of measures by 2020 in territorial subjects of the Russian Federation to bring the level of NG use on the public motor transport and transport of road municipal services in the following proportions: 1. In the cities with population more than 1 million people – to 50% of the total number of vehicles. 2. In the cities with population more than 300 thousand people – to 30% of the total number of vehicles. 3. In the cities and settlements with population more than 100 thousand people – to 10% of the total number of vehicles.

It is offered to develop the decrease mechanism (zeroing) of an import customs duty rate on the components required for the production of vehicles, using NG as motor fuel, as well as equipment and machinery, used for refueling vehicles with NG.

It is noted about the development of offers on subsidizing of the transfer of vehicles to NG in order to update the bus fleet, vehicles of road municipal services, etc.

On the basis of the submitted documents it is possible to conclude that growth of number of NGVs and gas NGV RSs will promptly grow in the nearest future. However today, for example, NGV RSs are concentrated generally in the European part of Russia and their quantity obviously insufficiently for growth of gas fuel consumption. Thus, expansion of NG use on the motor transport in eastern regions of the country is very urgent.

In this work we consider only one kind of NGV RSs – CNG RSs.

2. Increase in level of NG consumption on the motor transport

2.1 The factors influencing a numerical and territorial location of CNG RSs

Issues prospects for the NG use on road transport by the construction and infrastructure development of CNG RSs in the world and Russia are in many papers [3-15, etc.]. Also in sources [2, 4-8, 10] it is noted that in the combustion process of NG in the engines the smaller number of harmful substances emitted in comparison with traditional liquid motor fuels is allocated.

From the technical point of view weak branching of the CNG RSs in east Federal Districts of the country is possible to explain with several reasons among which the main ones:

1. Weak branching (or total absence) the gas transportation system in Eastern Siberia and in Far East. As is well-known the main part of the gas transportation system of the country is located in the western, central and southern parts, and also in the Urals. And supply of RSs with NG requires existence of the gas pipeline.

2. The significant extent of the specified FDs territory and the resulting problematic nature of the creation of so-called "blue corridors" (i.e., vehicles moving in different settlements and using NG mainly). As noted by professor Ya. S. Mkrtychan [12], the distance between CNG RSs should be in the range of 70–250 km.

However in view of the continuing arrangement of a number of new gas fields (Chayandinskoye, Kovyktinskoye, Kshukskoye, Nizhne-Kvakchikskoye, Sakhalinskoye, etc.), construction of the "Force of Siberia" gas transportation system and by commissionings of gas pipelines "Sobolevo — Petropavlovsk-Kamchatsky", "Sakhalin — Khabarovsk — Vladivostok", etc. it is possible to conclude that problems with gasification of the East are intensively solved. This may help, in the short term, increase the level of NG consumption by motor transport in these regions.

Thus, to determine the required numberof the CNG RSs in the cities and their territorial location it is offered to consider a number of factors:

1. Branching of the gas transportation system in the settlement (the number and a territorial location of gas distribution stations (GDS) and gas control points (GCP), length of gas pipelines and their capacity, distance from the main highways).
2. The fire safety requirements regarding remoteness of the CNG RSs from various objects of city infrastructure [16].
3. The number of the large highways passing through the settlement.
4. The number of potential motor transport which can be converted to run on CNG.
5. The availability and geographical location of existing CNG RSs.

2.2. Calculation of the CNG RSs optimum number and their territorial location on the example of Tyumen

In this work expansion of NG use by rational territorial and quantitative construction of the CNG RSs is considered on the example of the city of Tyumen.

The choice of the city of Tyumen is caused by the fact that the city is one of the largest transport hubs of the Ural Federal District. Through the city there passes one of two existing highways connecting the east and west of the country, and also its northern oil and gas areas – Khanty-Mansi Autonomous District and Yamalo-Nenets Autonomous District. Since 2016 the city of Tyumen is also included in the "Reduction of greenhouse gas emissions from motor transport in the cities of Russia" project. The project is directed to decrease in emissions in the atmosphere of greenhouse gases by means of improvement of planning and management of an urban transportation by creation of effective systems of monitoring and promotion of environmentally friendly transport modes. Project implementation will be enabled at the expense of the Ministry of transport of the Russian Federation and the United Nations Development Programme with the assistance of the Global Environment Facility.

As of August, 2016 in Tyumen operates one CNG RS, located in 12 km from the downtown. Let's choose the optimum number of the CNG RSs for Tyumen taking into account the above factors.

NG is supplied to the south of the Tyumen region from its northern regions on the main gas pipeline "Urengoy - Chelyabinsk I", "Urengoy - Chelyabinsk II". In the city oparates 2 GDS, about 30 GCP, the total length of gas pipelines more than 1000 km, and volume more than 1 million m^3 of gas a day (daily from compressor station-11 for the needs of Tyumen served 1 million m^3 of gas) (figure 1). In view of extensive branching of gas pipelines it is possible to locate the CNG RS in many of its parts. However according to [16, 17] the location of the CNG RS in the central parts of the city is almost impossible.

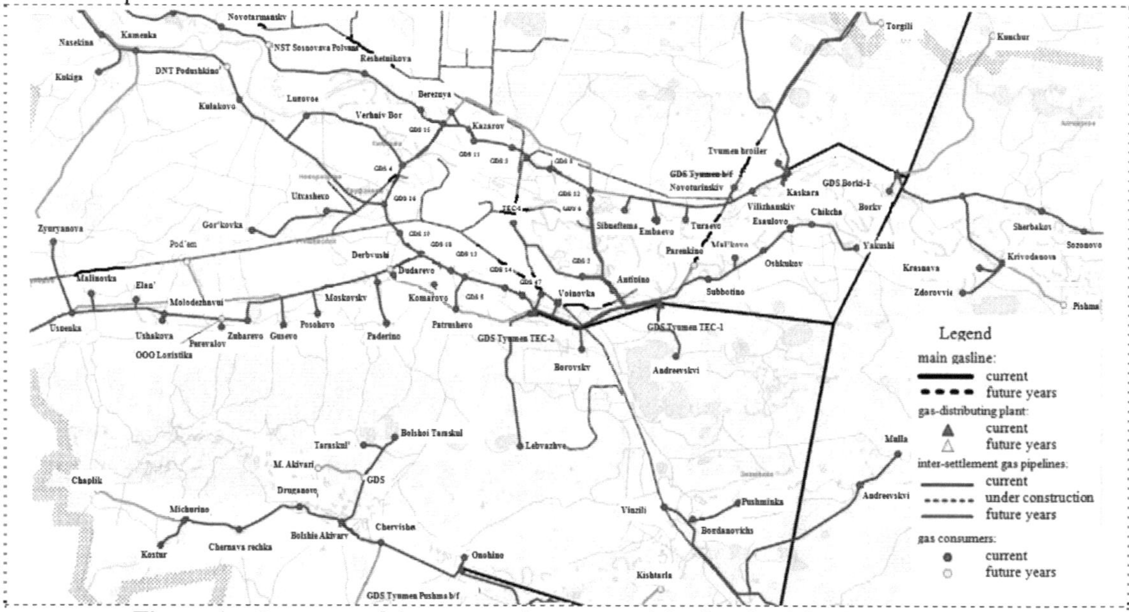

Figure 1 – Fragment of the gas supply system of the South of the Tyumen region

As a rule, diameter of gas pipelines from GDS to GCP is 120 mm, and from GCP to consumers of 35 mm. Pressure in gas pipelines inside the line from GDS to GCP makes 0,6 MPas, and after GCP to consumers, as a rule, moves under pressure of 0,05 MPas.

As presented in RD 3112199-1095-03 [18] and the order of the Government of the Russian Federation № 767-r in case of intercity transportations it is the most effective to use CNG first of all by buses of various capacity and transport of road municipal services.

When choosing a CNG station locations in the city of Tyumen it is offered to consider the territory adjacent to the following city highways:

P402 (Tyumen – Yalutorovsk – Ishim – Omsk) –adjacent to the south-east, on the way to the city and is called the Yalutorovsky tract.

P351 (Tyumen – Ekaterinburg) – adjacent to the west, on the way to the city and called Moscovsky tract.

P404 (Tyumen – Tobolsk – Surgut – Nefteyugansk –Khanty-Mansiysk) – adjacent to the east, on the way to the city and called Tobolsky tract.

Tyumen – Kurgan – adjacent to the south, on the way to the city and called Chervishevsky tract.

On these highways there passes a large number of transit vehicles, which potentially can also refuel at CNG RSs of the city of Tyumen.

Regarding determination of quantity of the potential vehicles which can be transferred to NG it is possible to note the following:

1. According to [19] transport service of the population in the city of Tyumen are carried out by 1162 units. The number of high-capacity buses M3 is 593, M2 – 569 units.

2. According to [20] for the main contract organizations of Tyumen 624 units of road municipal services vehicles are available. And in case of force majeure circumstances, the city administration signed contracts with a number of the organizations having the road and municipal services vehicles in number of 503 units.

That is 2289 units of vehicles in total. At the same time to select the optimal performance of CNG RS it is necessary to know their total daily CNG consumption. Fuel consumption is calculated on the example of the most characteristic brands of the cars which are operated in the city of Tyumen. Buses of the class M3 (MAZ-206) 20 m^3/100 km in city conditions. M2 Bus (Peugeot Boxer) consume about 11 m^3/100 km. Fuel consumption of the combined road machines (ED-405AG) makes about 50 m^3/100 km.

At the same time it is known that in the city of Tyumen each bus provides about 150 kilometers per day, and the road municipal services vehicles about 100 km.

Thus, knowing an approximate day run and an expense of CNG we will determine daily fuel consumption of the vehicle fleet (by modes of transport) as follows:

$$Q_{day} = 0.01 \text{ x } Hs \text{ x } A \text{ x } L_m, \quad (1)$$

where *Hs* – the base rate of NG consumption on a vehicle mileage, m^3/100 km; *A* – number of vehicles, units; L_m – day mileage of 1 vehicle, km.

Define daily fuel requirement of M3 buses:

$Q_{day\text{-}M3} = 0.01 \text{ x } 20 \text{ x } 593 \text{ x } 150 = 17790 \text{ m}^3$.

Define daily fuel requirement of M2 buses:

$Q_{day\text{-}M2} = 0.01 \text{ x } 11 \text{ x } 569 \text{ x } 150 = 9388.5 \text{ m}^3$.

Define daily fuel requirement of the road municipal services vehicles:

$Q_{day\text{-}RMS} = 0.01 \text{ x } 11 \text{ x } (624+503) \text{ x } 100 = 12397 \text{ m}^3$.

Next, we'll define the total daily fuel demand of the fleet:

$$Q_O = Q_{day\text{-}M3} + Q_{day\text{-}M2} + Q_{day\text{-}RMS}. \quad (2)$$

Substituting the values we get:

$Q_O = 17790 + 9388.5 + 12397 = 39575.5 \text{ m}^3$.

The received values represent only 4% of total daily quantity of the NG consumed by the city of Tyumen.

Except the considered transport potentially it is also possible to transfer to CNG the vehicles serving trade enterprises, private and office cars.

Further we will choose the CNG RS of necessary performance taking into account 4 large highways passing through Tyumen and the general potential day need for CNG. According to the classification performance of CNG stations, they are usually divided into 500, 250, 125 and so on of conventional refueling. Conventional refueling is the number of cargo vehicles refuels per day at the rate of 55 m^3 on 1 vehicle. For example, daily performance of CNG RS-250 and CNG RS -125:

$P_{CNG\ RS\text{-}250} = 250 \text{ x } 55 = 13750 \text{ m}^3$.

$P_{CNG\ RS\text{-}125} = 125 \text{ x } 55 = 6875 \text{ m}^3$.

Thus, at the maximum daily requirement of CNG 39575.5 m^3, calculated using the formula 2, it is enough to construct 2 CNG RSs – 250, and 2 CNG RSs – 125:

$P_{CNGRS\ TOTAL} = (250 \times 2 + 125 \times 2)55 = 41250 \quad m^3$.

The following places of CNG RSs location are offered:

1. On departure from Tyumen on the highway P402 near the TEC-2 (CNG RS-125).
2. On departure from Tyumen on the highway P351 next to the s. Derbyshy (CNG RS-250).
3. On departure from Tyumen on the highway near the P404 s. Yar (CNG RS-250).
4. On departure from Tyumen on the highway "Tyumen – Kurgan" near the village Patrushevo (CNG RS-125).

Schematically the location of the offered CNG RSs in Tyumen is shown in figure 2:

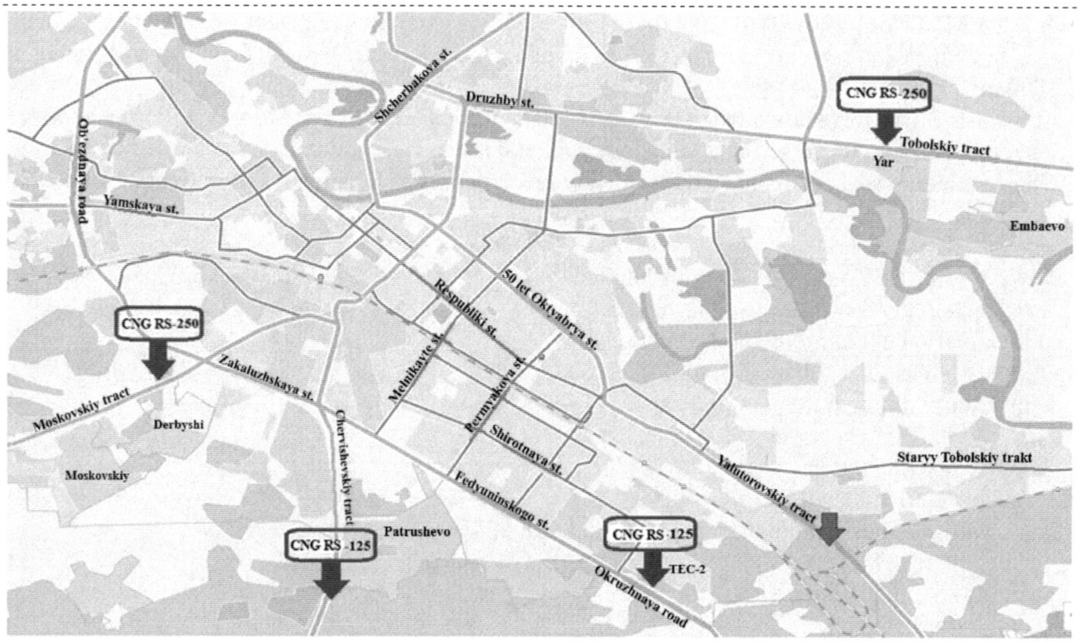

Figure 2 – Scheme of CNG RSs location in the city of the Tyumen (a red arrow shows already existing CNG RS, blue arrows show the offered locations of the CNG RSs)

3. Conclusions

The proposed location of the CNG RSs can satisfy the demand for NG as a city car owners and transit transport. The important point is that refueling stations it is offered to locate close to the city limits, in order to reduce the vacant run, and that their location corresponds to the fire safety requirements in terms of distance from the buildings and infrastructure. During the maximum operation of all 4 CNG RSs a load on gas transportation network will increase slightly – by only 4%. Respectively, there is no need to reconstruct it. However for supply the gas to the CNG RS it is necessary to equip additionally the gas transportation system with separate GCP. The rational territorial and quantitative location of the CNG RS in the city of Tyumen can lead to growth not only NG buses and road municipal services vehicles, but also NG cargo transport, and cars. In the long term the given CNG RSs can become a part of the All-Russian (connecting the western and eastern regions of Russia, and also Northern territories of Western Siberia) and international (connecting territories of Russia, Kazakhstan and China) "blue transport corridor". Increased use of NG will reduce environmental pollution by harmful substances emitted with exhaust gases.

References

[1] The total energy consumption. [Electronic resource]: https://yearbook.enerdata.ru/ (accessed July 20, 2016).

[2] Dementiev V V, Koklin I M. About the construction of CNG RSs and LNG RSs to expand the use of NGV market (2014) *Autogas filling complex + alternative fuel.* – Moscow: JSC Mechanical Engineering Publishing, 4 (85), pp. 34-50.

[3] Bondarenko E V, Filippov A A, Fazullin M R, Shailin R T. To a question of the metane gas stations network development needs in Orenburg (2012) *World of transport and technological machines* – 4(39). – pp. 15-23.

[4] Chikishev E M. Identification the reasons constraining gasification of road transport with natural gas (2016) *Autogas filling complex + alternative fuel.* – Moscow: JSC Mechanical Engineering Publishing, 1(106), pp. 3-8.

[5] Chikishev E, Ivanov A, Anisimov I, Chainikov D. Prospects of and problems in using natural gas for motor transport in Russia (2016) IOP Conf. Series: Materials Science and Engineering – Volume 142, 012110, doi: 10.1088/1757-899X/142/1/012110.

[6] Chinese D, Patrizio P, Bonotto M. A service station location model to explore prospects and policies for alternative transport fuels: A case of CNG distribution in Italy (2014) *Advances in Intelligent Systems and Computing* – Volume 262, pp. 71-84, doi: 10.1007/978-3-319-04630-3_6.

[7] Evstifeev A A. Methodology of rational construction and the continuous improvement of the regional CNG RSs network (2014) *Alternative Fuel Transport* – 3(39). – pp. 53-60.

[8] Farzaneh-Gord M, Deymi-Dashtebayaz M, Rahbari H R. Optimising Compressed Natural Gas filling stations reservoir pressure based on thermodynamic analysis (2012) *International Journal of Energy* - Volume 10, Issue 3, pp. 299-320, doi: 10.1504/IJEX.2012.046836.

[9] Frick M, Axhausen K W, Carle G, Wokaun A. Optimization of the distribution of compressed natural gas (CNG) refueling stations: Swiss case studies (2007) *Transportation Research Part D: Transport and Environment*, 12 (1), pp. 10-22, doi: 10.1016/j.trd.2006.10.002.

[10] Han J, Li Y. Beijing necessity of development of natural gas buses (2011) Advanced Materials Research / *2011 International Conference on Materials and Products Manufacturing Technology, ICMPMT* - Volume 339, Issue 1, pp. 509-516, doi: 10.4028/www.scientific.net/AMR.339.509.

[11] Koklin E M, Potapenko M C, Malenkina E F. Development of regional gas supply systems for providing consumers with gaseous motor fuel (2013) *Autogas filling complex + alternative fuel*. – Moscow: JSC Mechanical Engineering Publishing, 5(74), pp. 7-18.

[12] Mkrtychan Ya. S. The basic principles of creation of city network of motor transport gas supply (2013) *Alternative Fuel Transport* – 3(33). – C. 10-14.

[13] Ogunlowo O O, Bristow A L, Sohail M. Developing compressed natural gas as an automotive fuel in Nigeria: Lessons from international markets (2015) *Energy Policy* - Volume **76**, pp. 7-17, doi: 10.1016/j.enpol.2014.10.025.

[14] Pevnev N G, Banket M V, Bakunov A C. The prospects of infrastructure development of CNG use in Omsk (2014) *Alternative Fuel Transport* – 5(41). – pp. 7-11.

[15] Pronin E N, Samsonov R O, Malenkina E F. The international transport corridors with use of natural gas as motor fuel (2009) *Gas industry* – 11. – pp. 8-9.

[16] The order of Ministry of the Russian Federation for Civil Defense, Emergency Management and Natural Disasters Response of May 5, 2014 № 221 "About the approval of the rules set "Automobile refueling stations. Requirements of fire safety, 2014. – pp. 52.

[17] Gordienko D M, Malkin V L, Il'ichev V V, Smirnov E V. Features of fire danger of refueling stations with the presence of compressed natural gas (2015) *Fire safety* – **1**. – pp. 60-62.

[18] Guidance document 3112199-1095-03. Guidelines on the organization and operation of gas-cylinder vehicles that run on compressed natural gas. – Ministry of Transport, Road Department, JSC SaRIoMT - 2002. 59 p.

[19] Bus passions: tariffs raise – whether quality grows? [Electronic resource]: http://newsprom.ru/news/Obschestvo/ 208061.html (accessed August 18, 2016).

[20] Materials of the press service of the Administration head of the city of Tyumen [Electronic resource]: http://moi-portal.ru/columns/rabota/ 19674.html (accessed August 20, 2016).

IOP Conf. Series: Earth and Environmental Science **50** (2017) 012005 doi:10.1088/1755-1315/50/1/012005

Level Recession Of Emissions Release By Motor-And-Tractor Diesel Engines Through The Application Of Water-Fuel Emulsions

A Ivanov[1,a], E Chikishev[2]

[1] Technical systems in the agricultural sector, Northern Trans-Ural State Agricultural University ,Respubliki street 7, Tyumen, Russian Federation, 625003
[2] Motor transport operation, Industrial University of Tyumen, Volodarskogo street 38, Tyumen, Russian Federation, 625039

E-mail: [a]chikishev_e@mail.ru

Abstract. The paper is dedicated to a problem of environmental pollution by emissions of hazardous substances with the exhaust gases of internal combustion engines. It is found that application of water-fuel emulsions yields the best results in diesels where production of a qualitative carburetion is the main problem for the organization of working process. During pilot studies the composition of a water-fuel emulsion with the patent held is developed. The developed composition of a water-fuel emulsion provides its stability within 14-18 months depending on mass content of components in it while stability of emulsions' analogues makes 8-12 months. The mode of operation of pilot unit is described. Methodology and results of pilot study of operation of diesel engine on a water-fuel emulsion are presented. Cutting time of droplet combustion of a water-fuel emulsion improves combustion efficiency and reduces carbon deposition (varnish) on working surfaces. Partial dismantling of the engine after its operating time during 60 engine hours has shown that there is a removal of a carbon deposition in cylinder-piston group which can be observed visually. It is found that for steady operation of the diesel and ensuring decrease in level of emission of hazardous substances the water-fuel emulsion with water concentration of 18-20% is optimal.

1. Introduction

The most powerful sources in atmospheric air of hazardous substances are vehicles, including the cars and tractors using traditional hydrocarbon fuel. The emission of hazardous substances with the exhaust gases in the atmosphere is a result and a necessary condition of normal functioning of internal combustion engines. In Russia diesel is recognized as the most dangerous engine from the ecological point of view and in traditional completeness doesn't meet the modern requirements for an emission limitation of hazardous substances with the exhaust gases.

Considerable part of trucking facilities and practically all agricultural vehicles are completed with diesel engines. The world park of mobile farm vehicles contains more than 100 million units which are releasing annually more than 300 million tons of total number of hazardous substances with the exhaust gases. In all cases these emissions exert negative impact on working conditions, farmlands and products received from them.

In use technical condition of tractors, as a rule, worsens: energy consumption increases, indicators of working capacity, technical and ecological safety decrease. The fact that in Russia there are more than 50% of tractors which are over 10 years old and most of them operate strained to the limit of technical capabilities aggravates considerably an ecological situation.

In spite of the fact that the modern diesel is the efficient and long-life engine now, there is a possibility of further improvement of its characteristics by updating of operating process which quality depends on many factors: a gasdynamic condition of an air charge, parameters of the fuel equipment, quality of carburation, conditions of pressurization, physical and chemical indicators of fuels and so on, which opportunity is used not completely. Besides the formation of carbon dust in diesel causes introduction of the intensive radiation heat exchange increasing heat release rate of the engine's details.

Being adsorbents and further carriers of many substances, harmful to live organisms, the carbon dust particles which are present at the exhaust gases exert negative impact on health of the person and ecology of the environment.

Ecological and economic indicators of diesels in many respects depend on quality of carburation behavior and combustion. One of the directions of further development of diesels is increase in their specific capacity and decrease in harmful emissions in case of buildup in mean effective pressure and crankshaft speed that is connected with updating of engine cycle.

Thus, the solution of the main problems of diesel engine manufacturing (improvement of economic and ecological indicators, increase in specific capacity) requires improvement of quality of fuel combustion process. Therefore researches in this field are of current interest.

Application of alternative types of motor fuel for the purpose of increase in ecological safety of diesels' operation is the most effective now. The water-fuel emulsion belongs to one of them [1-7]. The best results are yielded by application of water-fuel emulsions in diesels where receiving a qualitative carburetion is the main problem for the organization of working process [8].

The water method of application in the form of an emulsion with the diesel fuel and various additives improving its properties is the simplest, cheap and available. This method allows saving diesel fuel to some extent doesn't require considerable costs for introduction of constructive changes and additions in diesel and can be realized for the engines which are already in operation. Most intensively, the researches connected with using water-fuel emulsions are conducted in two directions: 1) injection of water in the combustion chamber just before fuel inflaming; 2) preparation of a water-fuel emulsion with in advance selected optimum composition of water and fuel [9].

But, despite satisfactory indicators of operation of the engine on a water-fuel emulsion, mass application of emulsions is interfered by their insufficient stability that is they are inclined to breakdown to diesel fuel and water eventually. The main reasons are complexity of selection or high cost of development of emulsifiers and difficulties at the choice of a mode of emulsions' processing.

2. Methodology

Increase in stability of a water-fuel emulsion that is prevention to breakdown to diesel fuel and water was one of research problems for a long time. To estimate efficiency of the developed emulsion it is necessary to make comparative tests of operation of the diesel engine.

Therefore the general technique of pilot studies included:

1. Development of a water-fuel emulsion composition (selection of surfactants, water addition; test procedure of influence of surfactants for stability of a water-fuel emulsion);
2. Pilot studies of operation of the D-243 diesel engine using diesel fuel and a water-fuel emulsion.

Installation of equipment and devices, tests of the engine and the analysis of the exhaust gases were carried out taking into account requirements of state standards. At test procedure diesel fuel of the L-0,5 brand name, the M-10-G2 engine oil, surfactants and industrial water were used.

Frequency of carrying out experiments equaled to five.

For the purpose of increase in stability of a water-fuel emulsion the composition of the water-fuel emulsion containing diesel fuel, water and emulsifiers which remains stable within several months [10] was developed. Results of tests of stability of a water-fuel emulsion are presented in tab. 1.

Table 1 – Composition of a water-fuel emulsion and results of tests of its stability

Component of emulsion and its indicator	Content of a component, masses. % in emulsion			
	1	2	3	4
Diethanolamide of oleic acid	1.640	1.640	3.120	4.230
Diethanolamine soap of oleic acid	0.500	0.770	0.960	1.420
Diethanolamine	0.850	1.440	1.500	2.030
Monoester of oleic acid and diethanolamine	0.800	1.200	1.250	1.570
Oleate of sodium	0.190	0.150	0.100	0.080
Water	10	15	20	25
Diesel	The rest			
Stability, month	18	18	15	14

At the first stage there was an emulsifier preparation. Further process of its mixture with diesel fuel and water, and then treatment in a homogenizer. After that monitoring of stability of a water-fuel emulsion as by a visual method, and by means of devices was performed. The analysis of dispersion of a water-fuel emulsion under a microscope was made once a week. The water fuel emulsion was considered stable if its dispersion made 3-5 microns.

Dedication of components of an emulsion: diethanolamide of oleic acid is surfactant and provides droplet breakup; diethanolamine soap of oleic acid regulates optimum hydrophilic and lipophilic balance of emulsifier; diethanolamine facilitates receipt of a microemulsion; monoester of oleic acid and diethanolamine improves stability of an emulsion, reduces an emulsifier consumption; oleate of sodium is surfactant and increases stability of an emulsion.

The developed composition of a water-fuel emulsion provides its stability within 14-18 months depending on the mass content in it of components while stability of emulsions' analogues makes 8-12 months. It is noted that application of homogenizers for treatment of emulsions gives considerable effect concerning receiving high-disperse phases in comparison with mechanical treatment.

Pilot studies were conducted for the D-243 diesel engine of the Minsk motor plant (Belarus) which is widely used in cars and tractors (tab. 2).

Table 2 – Specification of the engine D-243

Indicator	Value
Type	Four-stroke, no turbo-supercharging
Number of cylinders	4
Location of cylinders	In-line, vertical
Displacement volume, l	4.75
Capacity, kW (hp)	60 (81)
Rotary speed, min^{-1}	2200
Torque capacity, Nwm	258
Mass, kg	448

On the diesel engine installation for delivery of a water-fuel emulsion was mounted. The schematic diagram of experimental installation is submitted in fig. 1. From capacity 3 with diesel fuel the fuel moves the fuel lift pump 8 via the filter of primary refining of fuel 5, the valve 7, the distributing valve 9 and the filter of fine purification of fuel 10 to the fuel pump of high pressure 11 (fig. 1). The discharge rate of diesel fuel is determined with the help of electronic scales 1. Further from the fuel pump of high pressure 11 diesel fuel moves in the engine 12.

The engine gets warm up till the operational temperature using clean diesel fuel. Then, there is a measurement of effective indicators with the simultaneous analysis of the exhaust gases and their smokiness. Then by means of the valve 7 and the distribution valve 9 the engine 12 is taken over to a water-fuel emulsion which is delivered from capacity 4 via the filter 6, the valve 7 and the distribution valve 9 to the fuel pump of high pressure 11 and further to the engine 12. The consumption rate of a water-fuel emulsion is determined by means of electronic scales 2. Similar to work on diesel fuel, there is a measurement of effective indicators with the simultaneous analysis of the exhaust gases and their smokiness. Upon termination of an experiment for washing of a fuel system the engine 12 is taken to diesel fuel.

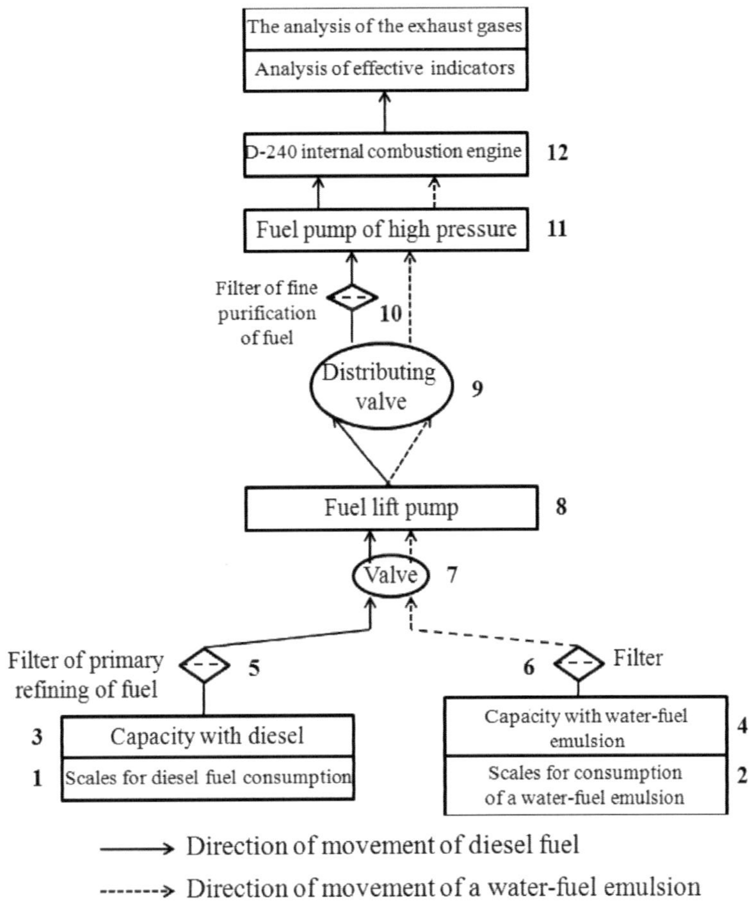

Figure 1 – Scheme of experimental installation

For additional treatment of a water-fuel emulsion in fuel system of diesel engines the device with the patent [11] held is developed. The general scheme of the device is presented in fig. 2. The fuel system of the diesel contains the fuel pump 1 of high pressure connected to a nozzle 5 by means of the fuel-supply line 2 of high pressures including a heating element 6. The heating element 6 is connected to the regulator of temperature of heating 3 and a preventer 4. The device consists of the fuel-supply line of high pressure 2 made in the form of a spiral, which is wrapped up by the isolating material 7 in which the heating element 6 included in an electric chain with the regulator of heating temperature 3 and a preventer 4 is located.

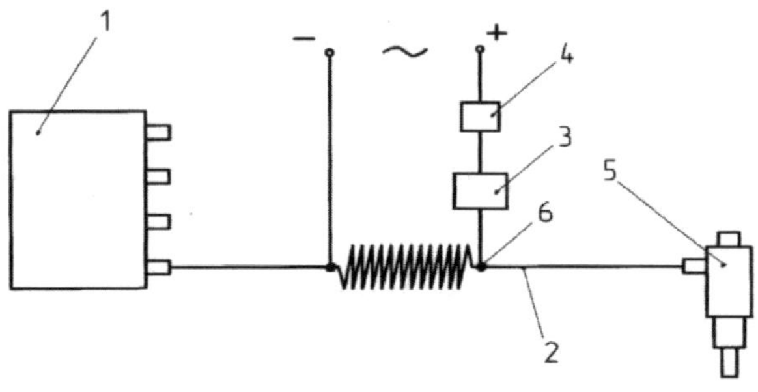

Figure 2 – The scheme of fuel system of the diesel with the developed device for processing of a water-fuel emulsion

Work of the device is carried out as follows. The water-fuel emulsion moves the fuel pump of high pressure 1 in a nozzle 5 via the pipeline of the high pressure 2 having the spiral form (the number of spiral turns depends on properties of a water-fuel emulsion). The water-fuel emulsion passing on a spiral is warmed up by a heating element 6 up to the required temperature at the exit and moves in a nozzle 5 (temperature of heating depends on properties of a water-fuel emulsion). In order to avoid an overheat of a water-fuel emulsion the electric chain with a heating element 6 has switched on the regulator of heating temperature 3 and a preventer 4.

3. Results and analysis

Results of tests of the diesel D-243 during the work using diesel fuel and a water-fuel emulsion of the developed composition with concentration of water 20 % are presented on graphic curve (fig. 3-6).

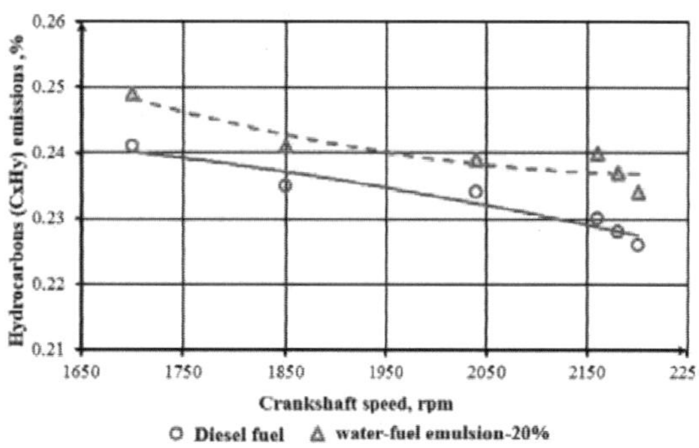

Figure 3. Change of level of hydrocarbons' emissions

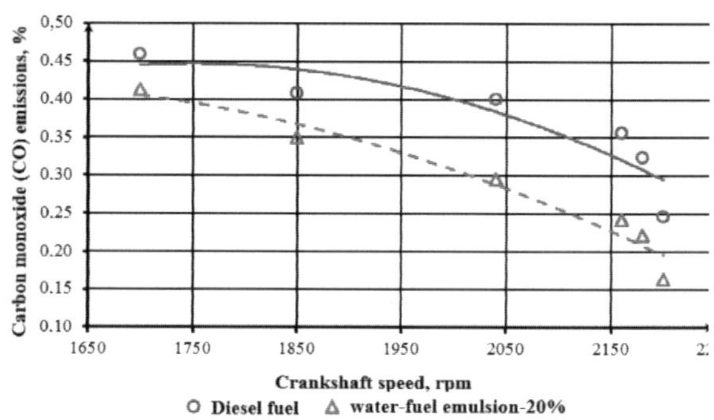

Figure 4. Change of level of carbon monoxides' emissions

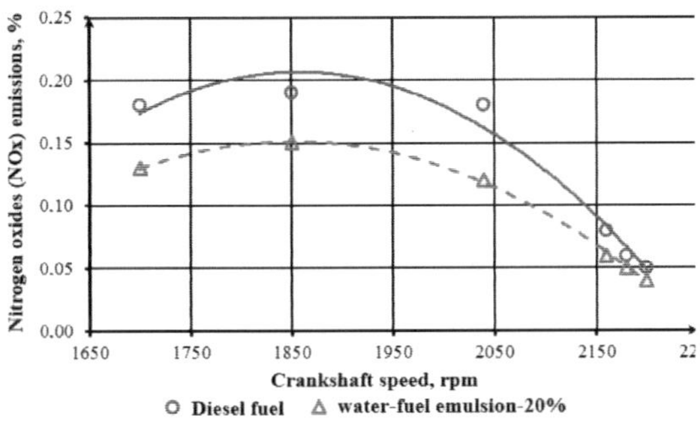

Figure 5. Change of level of nitrogen oxides' emissions

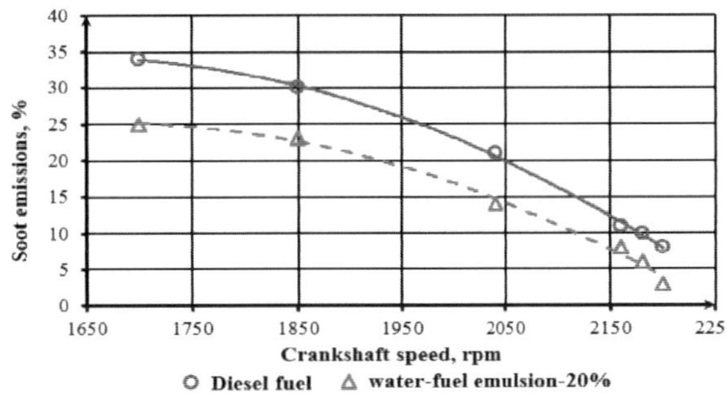

Figure 6. Change of level of the exhaust gases soot

In the combustion chamber after warming up of the sudden fracture of globules, splinters of sinking fractions (asphaltogenic), being moved with a high speed intensively interact with oxygen of an air charge, burn down almost completely, that is without formation of a varnish. The water phase which is in composition of the emulsified fuel, partially dissociates during fuel oxidation, then, in process of temperature increase, reaction of dissociation of water is sharply accelerated. The excess of atoms of hydrogen formed at dissociation reacts with atoms of oxygen and as a result, burning process is sharply accelerated.

In the course of burning of a water-fuel emulsion in a combustion chamber of diesels there is a conversion of the gas fractions formed as a result of phase transformations; the diesel fuel burning with a water vapor is enriched with hydrogen, and burns up with generation of additional energy almost completely with sharp reduction of hazardous emissions. At the same time the component wear of cylinder-piston group decreases significantly, and the group of the gas-removing channel is less exposed to abrasive wear, and, as a result, reliability of the fuel equipment as there is no coking of holes of injector spray nozzles increases.

The shift of the engine from diesel fuel to a water-fuel emulsion allows reducing considerably smokiness of the exhaust gases, emissions of nitrogen oxides and carbon. Emissions of hydrocarbons, on the contrary, are increased in all range of rotation frequencies of a crank shaft of the engine. Emissions of carbon oxides during the operation of the engine with heavy load are almost identical.

4. Conclusions

In the conclusion it should be noted that the water-fuel emulsion is an effective mean of improvement of carburetion quality and an intensification of burning process of fuel in diesels, especially in engines with unsatisfactory technical condition.

Application of a water-fuel emulsion allows improving ecological indicators of the diesel, reducing the content of harmful substances in the exhaust gases, thermal stress of details of cylinder-piston group, discoking of spraying nozzle holes, reducing waste of grease oil and temperature of the exhaust gases to improve resource indicators of the diesel.

The water fuel emulsion burns up quicker than diesel fuel. Processes of burning because of finely dispersed composition of an emulsion increase in a surface of burning of particles and good mixing with air are changed in a quality manner. Reduction of drops burning time of a water-fuel emulsion improves completeness of combustion of fuel and reduces carbon deposit (varnish) on working surfaces.

Partial dismantling of the engine after its operating time during 60 engine hours has shown that there is a removal of carbon deposition in cylinder-piston group which can be observed in the visual way.

As a result of the conducted researches it is found that for steady operation of the diesel and ensuring decrease in level of emission of hazardous substances the water-fuel emulsion with concentration of water in it 18-20% is optimum.

References

[1] Abu-Zaid M. Performance of single cylinder, direct injection Diesel engine using water fuel emulsions (2004) Energy Conversion and Management - Volume 45, Issue 5, pp. 697-705, doi: 10.1016/S0196-8904(03)00179-1.

[2] Buzdug S R, Sajin T. Aspects of mathematical modelling of heat and mass exchange at burning of droplet of water fuel oil emulsion (2013) Applied Mechanics and Materials - Volume 371, pp. 642-646, doi: 10.4028/www.scientific.net/AMM.371.642.

[3] Hountalas D T, Mavropoulos G C, Zannis T C, Mamalis S D. Use of water emulsion and intake water injection as NOx reduction techniques for heavy duty diesel engines (2006) SAE Technical Papers - SAE World Congress; Detroit, MI; United States; Code 90162, doi: 10.4271/2006-01-1414.

[4] Leng L. Yuan X, Zeng G, Wang H, Huang H, Chen X. The comparison of oxidative thermokinetics between emulsion and microemulsion diesel fuel (2015) Energy Conversion and Management - Volume 101, pp. 364-370, 2015, doi: 10.1016/j.enconman.2015.05.071.

[5] Lif A, Holmberg K. Water-in-diesel emulsions and related systems (2006) *Advances in Colloid and Interface Science* - Volume 123-126, Issue SPEC. ISS., pp. 231-239, doi: 10.1016/j.cis.2006.05.004.

[6] Misra M, Chen S, Grimes P. Water-fuel emulsions for energy application (2005) Proceedings of the Jan D. Miller Symposium - Innovations in Natural Resource Processing - Volume 2005, pp. 329-336.

[7] Wu Y-Y, Chen B-C, Hwang J-J, Chen C-Y. Performance and emissions of motorcycle engines using water-fuel emulsions (2009) International Journal of Vehicle Design - Volume 49, Issue 1-3, pp. 91-110.

[8] Lozhkin V N. Complex advancement of working processes of diesels for improvement of economic and ecological characteristics (2006) Engine-building – № 3. pp. 43-47.

[9] Isakov A Y. Formation of a competitive phase in case of overheats of water drops of a small amount (2006) Combustive-lubricating materials – № 9.

[10] Ivanov A S. Patent for the invention № 2349632 Russian Federation, MPK C10L 1/32. Method of preparation of a fuel emulsion/A.S. Ivanov. – 2007143006/04; Submitted 20.11.2007; Published 20.03.2009, Bulletin № 8.

[11] Ivanov A S., Smolin N. I., Startsev A. V., Morozov O. A. Patent for useful model № 79946 Russian Federation, MPK F02M 31/125. The device for treatment of a water fuel emulsion in fuel system of the diesel engine/A.S. Ivanov, N. I. Smolin, A. V. Startsev, O. A. Morozov. – 2007139391/22; Submitted 23.10.2007; Published 20.01.2009, Bulletin №2.

Analysis The Potential Implementation And Use Of The Renewable Energy Sources In The Murmansk Region

S M Chekardovskiy, A G Zakirzakov and E F Gordievskaya

Tyumen Industrial University
Volodarsky str. 38, Tyumen, Russian Federation, 625000

E-mail: Allbert-@mail.ru

Abstract. The paper presents the analysis of the geographical location of the Murmansk region and possibilities of using different renewable energy sources. Present the results of the analysis of using and practicality wind power to wind turbines to generate electricity. The paper presents of average annual consumption of electricity settlements in the Murmansk region, options for replacing the use of diesel fuel by wind energy and recommendations for possible areas of utilization of wind energy.

1. Introduction

The second half of the XXth century was marked by the beginning of the new energy sources search and development. The first energy conflict that occurred in 1979 provoked the beginning of the alternative and renewable energy sources development, including solar and wind energy, hydro and bio-energy, etc.

The reasons of the renewable energy sources development are not only the importance of energy security of the countries, but also ecological safety of citizens. Thus, according to the Tindell's Center of Climate Studies in Norwich (UK), it follows that in 2012 the volume of CO_2 emissions amounted to 35.6 billion tonnes, the share of China accounted for 28% with a specific volume of emissions per capita of 6.6 tons, the USA – 16% with a specific volume per capita of 17.2 tonnes and the European Union – 11% with a specific volume of emissions per capita of 7.3 tons [1].

The intensity of carbon dioxide emissions from the solid fuel consumption has not decreased until 2010. There is known a case, occurred in the 21st of October, 2013, when smog shrouded the majority of China cities and paralyzed the movement of vehicles and had a negative impact on the citizens' health. The level of harmful substances from the fuel and energy complex companies' electricity generation exceeded 40 times the acceptable level set by the World health organization. According to the World Bank rating, 16 of the 20 most polluted cities in the world are in China [2, 3].

Realizing the negative consequences of the fuel and energy sector development and harm to the citizens' health and environment, the governments of the majority countries have adopted the declaration, concerning the priority development of renewable energy sources. Starting with 2011, the investments in renewable energy development have had a positive trend and in 2015 they amounted to $ 329 billion [4,5].

Unfortunately, the Russian Federation considers the development of oil and gas industry as a priority one, however, in 2014-2015 the decline in world prices for hydrocarbon resources has prompted the government to consider the development of renewable energy in the country and export of the generated energy in the neighboring countries, in particular the Scandinavian Peninsula countries [6,7].

The purpose of this article is to analyze the potential implementation and use of the renewable energy sources in the Murmansk region.

2. Materials and methods

The analysis of the empirical material, consisting of 13 publications covering the period 2002 – 2016. The introduction and use of renewable energy is considered to be appropriate in the areas with the existing usage potential and lack of traditional hydrocarbon resources. Such areas include, for example, the Murmansk region, located in the European North of Russia and engaged in electricity generation primarily through imported fuel from the neighboring regions.

3. Results and Discussion

According to the Ministry of Architecture, Construction, Housing and Utilities of the Russian Federation there are 64 oil-fired boilers on the territory of the Murmansk region, where the purchase of oil products in the heating system makes up 1.2 million tons, and the consumption and production of local fuels in the total energy balance of the region is less than 1%.

Previously prospective use of hydro energy in the Murmansk region is currently experiencing a number of difficulties, caused by the development of almost all the possible locations of hydropower plants, the generation of which is not enough to cover the needs of the region [8,9].

Considering the energy potential of renewable energy sources it should be noted that the region has a low potential for the use of bioenergy, the basis of which is wood biomass, due to low volume of timber, weak infrastructure and the fact that a major part of the forests, located in the areas of consumption, refers to the protected forests. The use of solar energy is the least promising because the region is situated beyond the Arctic Circle, having borders with Finland and Norway that can be observed in figure 1.

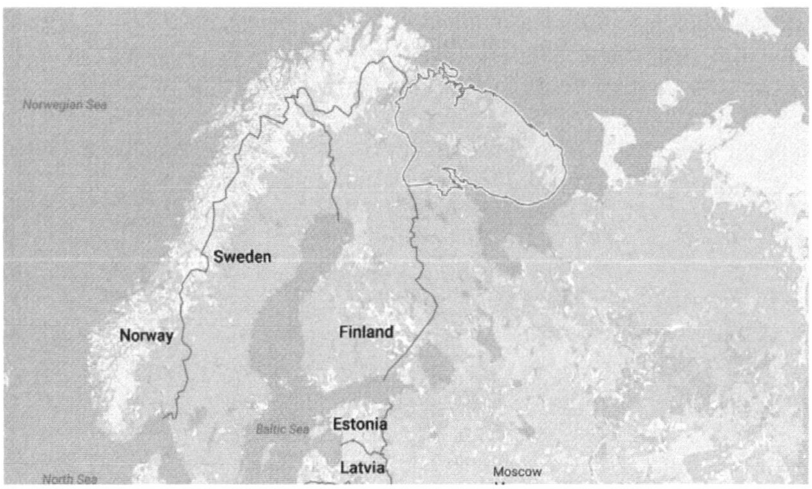

Figure 1 – The border of the Murmansk region with the neighboring countries and regions.

The analysis of solar radiation observations, conducted by the Centre of physicotechnical problems of power engineering of the North, has shown that the total solar radiation in the clear days ranges from 4500 to 5000 MJ/m^2, but the presence of high cloudiness and the harshness of Northern weather reduce the fraction of solar radiation that has reached the earth, almost by 75%. Taking into consideration this fact and the low number of sunny days during the year, the use of solar energy is impractical in this region [10].

The analysis of the scientific literature and region climate peculiarities has shown that the most promising sphere of renewable energy source use is the use of wind power. So the wind speed in the coastal areas is estimated to 10-12 m/s, as the distance moves away from the coastline, the speed is reduced to 5-6 m/s. The given speeds are the highest in the European part of the Russian Federation.

Taking into account the high wind energy potential, the high cost of electric and thermal energy in the region, as well as its successful experience of the wind farms implementation in the neighboring countries, it is considered appropriate to introduce and use the wind power plants, the competitiveness of which has been proven in the countries such as Germany, Denmark. Norway, etc [11, 12].

The available data on the long-term average wind speeds distribution at the height of 10 meters was marked on the map of the Kola Peninsula and is presented in figure 2 and table 1.

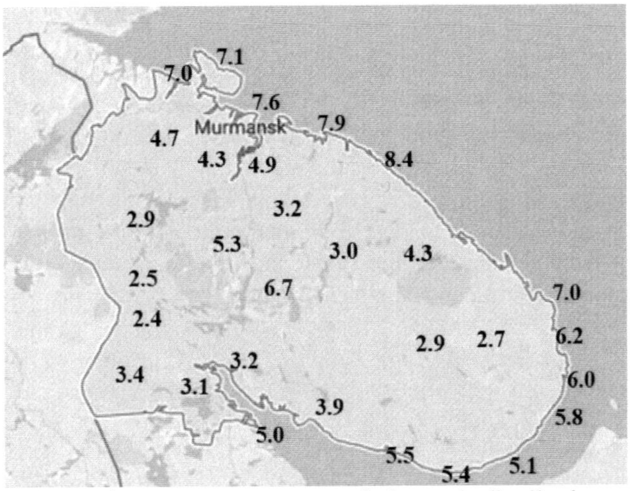

Figure 2. Long-term average wind speeds distribution.

Table 1 represents the characteristics of wind resources of the Kola Peninsula.

Table 1. The characteristics of the Kola Peninsula wind resources.

Characteristic	Zona 1	Zona 2	Zona 3
The average annual wind speed on 10 and 50 m., m/s	7.4 - 9.8	6.4 - 8.6	5.4 - 7.6
Specific wind energy, MW·h/m^2	10.6	7.7	5.2
The estimated wind speed, m/s	15.5	13.6	11.7

As it can be seen, at a height of 10 meters on the coastal areas the wind speed reaches 7-8 m/s, with the measurements height increasing the wind speed increases. Figure 3 shows the dependence of the wind growth on the height changing.

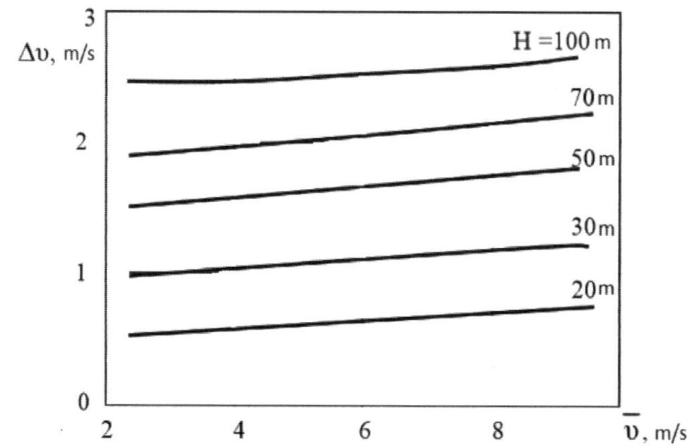

Figure 3 – The dependence of wind speed growth on the measurement height changing.

To assess the feasibility of the wind power generators use there has been carried out the analysis of speed and consumption in settlements, located in the remote areas of the southern part of the Murmansk region [13]. Annual measurement of air masses during every month took place in the following villages: Chapoma, Tetrino and Strelna. The measurement data are given in figure 4.

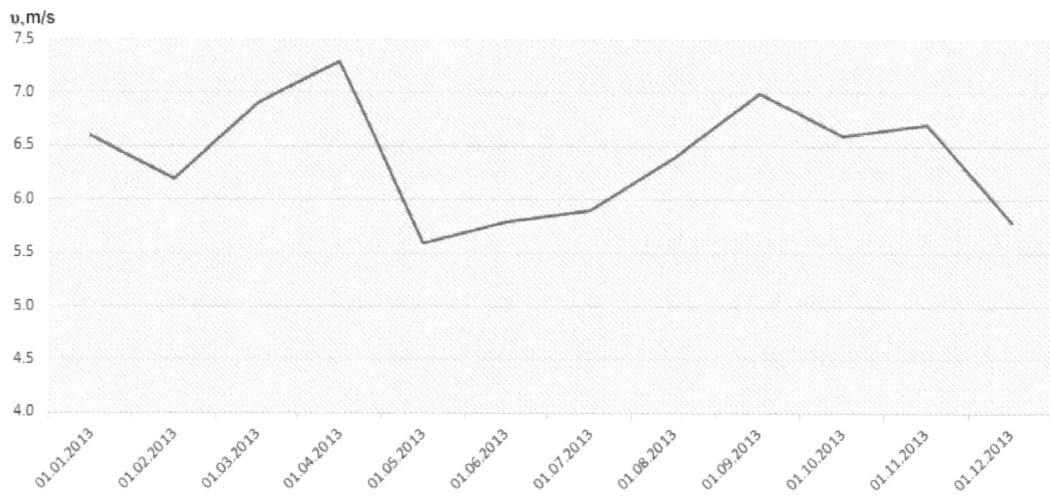

Figure 4 – Average monthly wind speeds.

An average annual wind speed is 6 m/s, which is a good indicator of the wind power generators introduction. Also, the graph of an average annual load has been evaluated and constructed and shown in figure 5.

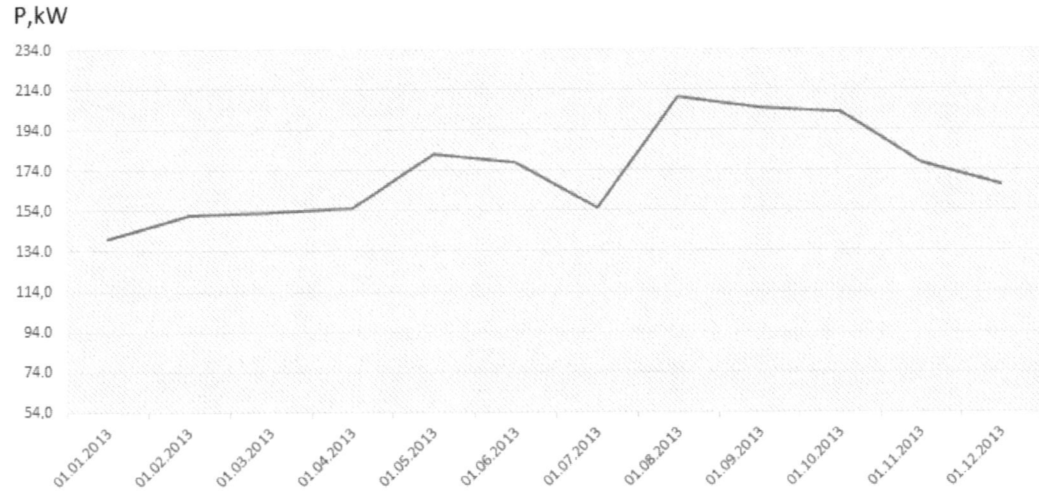

Figure 5 – Group villages' load graph.

The graph demonstrates that the annual average electricity load is 168 kW, maximum and minimum load value is 205 kW and 138 kW, respectively. The introduction of 4 units of wind power generators with a capacity of 50 kW will cover all the energy consumption in every village. The replacing of diesel fuel used for diesel power plant used, for example, in the village Chapoma by the wind power generators, the cost of 1 kW·h is reduced by 30%.

When you use the wind energy on the industrial scale on the territory of the Murmansk region, future export of the generated energy will be possible on the territory of Sweden, Finland and Norway.

4. Conclusion

The analysis of renewable energy sources implementation and use possibilities in the Murmansk region has shown that only wind power has the potential for implementation, covering domestic consumption and the generated electricity export to the neighboring countries.

The territory of the Barents Sea coast in the Murmansk region is one of the most windy sites and has a great potential of wind energy generating. The region has a favorable repeatability of the wind speeds, the steady prevailing wind with abnormally high levels of annual average speeds are observed here, the seasonal maximum of the wind coincides with the seasonal peak of energy consumption.

The main areas of possible wind energy use are: operation of large wind farms, included to the power grid; participation of the wind turbines in the electricity supply of autonomous consumers (joint work of diesel power plants and wind turbines), the participation of the wind power plants in the heat supply of consumers (joint work of boilers and wind turbines) and the use of wind energy in natural gas processing technologies.

References

[1] Dedecca J G, Rudi A. Hakvoort,. Roland Ortt J 2016 Market strategies for offshore wind in Europe: A development and diffusion perspective J. Renewable and Sustainable Energy Reviews 66 286–296

[2] Worrada Nookueaa, Pietro Elia Campanaa, Jinyue Yana 2016 Evaluation of Solar PV and Wind Alternatives for Self Renewable Energy Supply: Case Study of Shrimp Cultivation J. Energy Procedia 88 462–469

[3] Azad A K, Rasul M G, Rubayat I., Imrul R. Shishir 2015 Analysis of Wind Energy Prospect for Power Generation by Three Weibull Distribution Methods J. Energy Procedia 75 722-727

[4] Iov F, Ciobotaru M, Blaabjerg F 2009 Power Electronics Control of Wind Energy in Distributed Power Systems J. Renewable Energy 18 333-363

[5] Sharpe L 2002 Offshore generation looks set to take off J. IEE Review 48(3) 24 –25
[6] Sun Tao, Chen Z, Blaabjerg F 2005 Flicker Study on Variable Speed Wind Turbines with Doubly Fed Induction Generators Proc. IEEE ICEMS 2 919-924
[7] Akhmatov V 2003 Variable-speed Wind Turbines with Doubly-fed Induction Generators, Part III: Model with the Back-to-back Converters J. Wind Engineering 27(2) 79–91
[8] Slootweg J G, Kling W 2002 Modelling of Large Wind Farms in Power System Simulations J. IEEE Power Eng. Soc. Sum. Meeting 1 503–508
[9] Borisenko M M, Savarino M V, Zverev V G 1969 Vertical profiles of wind speed observations at the meteorological mast in Obninsk J. Proceedings of Voeikov Main Geophysical Observatory GGO 246 56-62
[10] Borisenko M M, Habarova E A, Zhiltsova E L 2008 Assessment of wind energy resources on the territory of Russia J. Proceedings of Voeikov Main Geophysical Observatory 557 53–66
[11] Minin V A, Bejan A V 2009 The heat supply of buildings with the wind turbines J. Energysaving and watertreatment 2 17-21
[12] Kopylov A E. 2008 Economics of renewable energy development in Russia J. Energy: Economics, technique, ecology 7 22-31
[13] Minin V A 2012 Prospects for development of renewable energy zones of the murmansk decentralized energy supply // J. of the Kola science centre of RAS 4(8) 110–122

Research Of The Influence Of Reftinskii SDPP`S Ash On The Processes Of Cement Stone`S Structure Forming

G A Zimakova, V A Solonina, M P Zelig

Tyumen Industrial University
Lunacharski str. 2, Tyumen, Russian Federation, 625001

E-mail: ksm@tgasu.ru

Abstract. The article describes the experimental research of cement stone. Cement stone forming involves highly dispersive mineral additive – ground ash. It is stated that the substitution of some part of cement with activated ash leaves cement strength high. This is possible due to the activity of ash in structure forming processes. Activation of ash provides the increase in its puzzolanic activity, complete hydration processes. it is stated that ash grinding leads to a selective crystallization hydrated neoformations. Their morthology is different on outer and inner surfaces of ash spheres. The usage of ash can provide cement economy on condition that rheological characteristics of concrete stay constant. Besides, the usage of ash will improve physical and mechanic characteristics of cement stone and concrete.

1. Introduction

The usage of coal as solid fuel by thermal power plants makes it inevitable to cumulate a great amount of products of combustion. According to the data of Russian Heat Engineering Scientific Research Institute, annual burning of solid fuel leads to creating 40-50 million tons of bottom ash, the largest part of which goes to ash dumps [3]. Namely, about 5 million tons of technogenic wastes are created at Riftinskii SDPP every year. More than 4.5 million tons of waste gets to ash dumps polluting air and water, and altering chemical and mineral structure of soil.

One of the most effective ways of ash waste recovery is the usage of wastes in producing building materials as the basic minerals substituting some part of cement binder.

The primary factor holding back the usage of some ashes is their non-homogeneity in a chemical composition as well as in the dispersiveness. Reftinskii SDPP regularly burns Ekibastuz coal. This allows to receive fly ash with constant characteristics. The chemical composition of Reftinskii ash includes oxides: SiO_2 – 60.0÷62.0%, Al_2O_3 – 29.0÷31.0%, Fe_2O_3 – 4.0÷5.0%, CaO+MgO – 1.5÷2.5%. Its structure contains glass particles of nearly ideal spherical shape [3]. The size of a granule is 30-80 mcm, 80-100 mcm until 25 %.

By now, researchers from all over the world have developed a lot of works connected with effective filling of cement matrix with technogenic waste flour [2÷10].

The conclusions made by the scientists cannot be doubted. They show that the characteristics of cement composites greatly depend on a disperse systems` structures which are the basis of their creation [2, 4, 5, 7]. The structural endurance of a disperse system, its rheological characteristics, a habit of action while flowing, the speed of destruction and reconstruction are closely connected [8, 10].

It is known that formation of a cement composite`s structure is a result of simultaneous actions of different factors and proceedings. The matter of interaction between products of cement stone hydration and mineral additives of ultra dispersive and nanometer range is the most complicated one while studying the process of structure forming.

Nano-additives are received through purposeful synthesis. They are referred to monomineral substances. Consequently, the view and depth of hydration processes will be rather similar. The usage of technogenic raw materials of polymineral nature will make it possible to provide the structure forming processes due to a range of chemical and physical processes. It is widely recognized that while using siliceous wastes, the interaction with hydration products takes place in a chemical way as a result of chemosorption of Portland cement by a silica additives.

2. Equipment and Instruments used in this research

Reftinskii ash grinding was executed in laboratory ball mill (LBM-7). Indices of dispersiveness were defined with the help of the meter SP-10. Cement composition samples have been kept in a water seal camera (WSC-1) at 20±2°C and relative humidity100% from the moment of their creation until testing. Compression strength of cube samples of dispersive cement stone 20x20x20 mm in size was defined under a hand hydraulic press (HHP-1).

Microphotographs of dispersive cement stone have been taken with an optical microscope of the inverted type GX-51.

3. Results and discussions

During the research work the tested ash has been ground for 1, 2 and 3 hours. The index of surface area of the particles in received probes is represented in Figure 1.

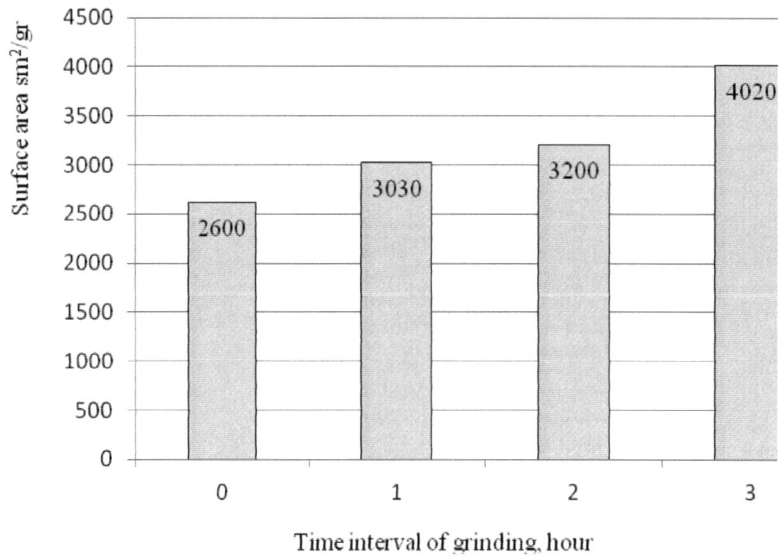

Figure 1 – Index of surface area of ground ash probes with different time intervals of grinding

The activity of the ground ash was defined according to the results of testing cement stone strength by compressing it. The amount of ash in the cement stone varied (the percent of ash added is indicated in Table 1). The samples were prepared from the paste of normal thickness, ground ash was added

IOP Conf. Series: Earth and Environmental Science **50** (2017) 012007 doi:10.1088/1755-1315/50/1/012007

while substituting some part of cement binder. Strength test by compressing cube samples 20x20x20 mm in size was executed at the age of hardening at 2, 7 and 28 days. Obtained results were compared with those of control composition of Portland cement CEM II/A-S 32.5 (Table 1, Figures 2-3).

Table 1 – Results of compression test of samples

Grinding time	Ash contents, %	Compression strength, kgs/sm^2			
		2 days	7 days	14 days	28 days
-	0	110	286	420	620
1 hour	10	149	339	382	480
	15	176	348	401	502
	20	107	328	390	440
2hours	10	120	317	422	501
	15	152	402	495	658
	20	132	230	382	490
3 hours	10	127	310	360	420
	15	148	305	365	410
	20	135	340	350	402

As the experiment shows, the strength of cement stone varies significantly under the influence of ash dispersiveness and the quantity of added ash.

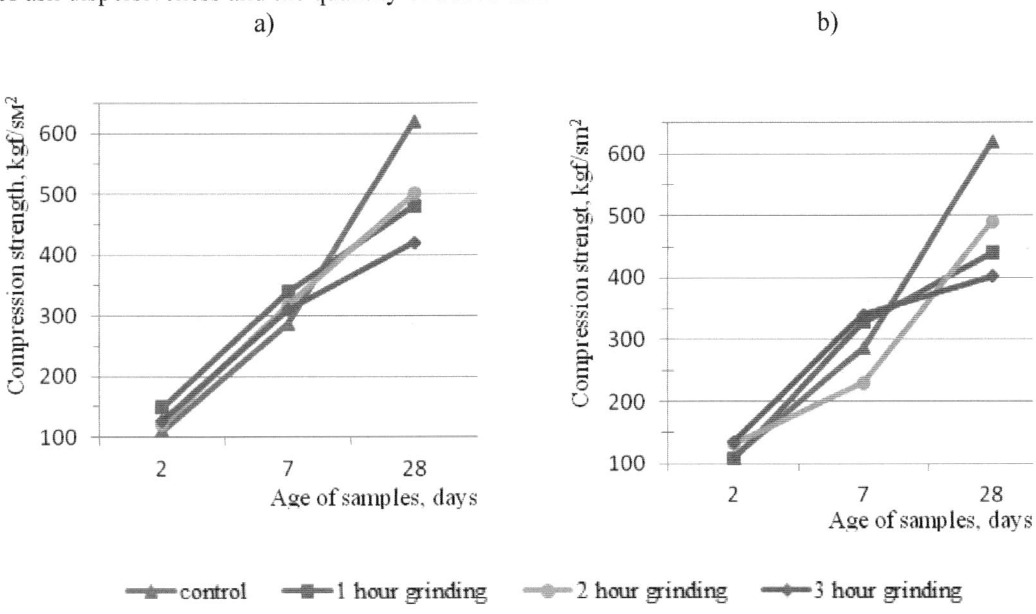

Figure 2 – Compression strength of samples with ash: a) 10 %; b) 20%

As it was found out in the experiment, increasing of the ash dispersiveness has a great influence upon cement growth kinetics. This phenomenon can be explained from the viewpoint of cluster formation as it is indicated by Solomatin [2, 8]. In case of 10% doze of ash the filling rate in the system is quite low. We can observe the decrease of strength because of lack of finely dispersed phase at packing the structure (Figure 2).

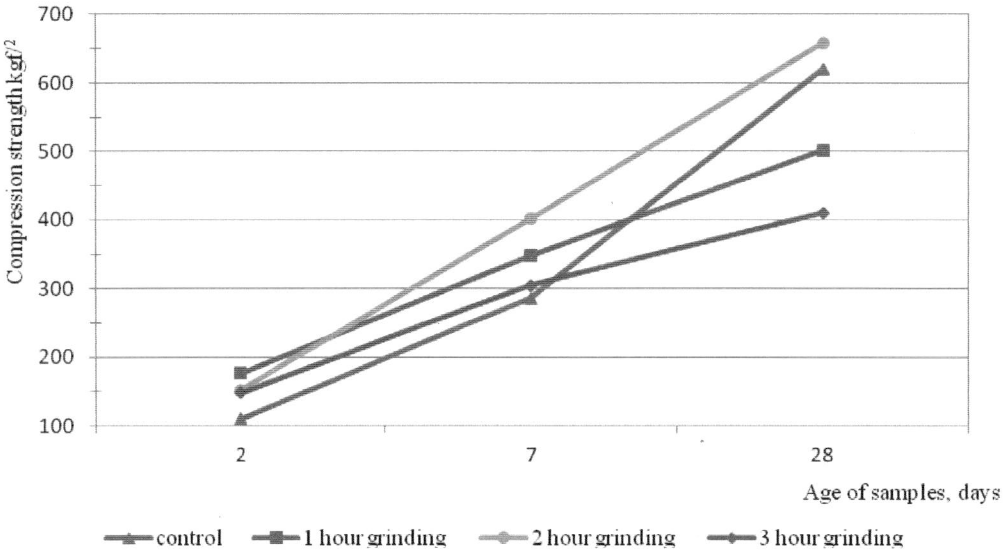

Figure 3 – Compression strength with ash contents 15%

According to the obtained indices of strength, the best grinding time is 2 hours. The further increase in the dispersiveness of ash (with surface area 4020 sm^2/gr) leads to some positive and negative effects. The positive one is pozzolanic intensity and microfilling effect. The negative one is the decrease in strength because of increase in water consuming quality of the system. At the age of 28 days the strength growth of cement composite is 6% on condition that the cement contains ground ash 15% of cement weight in amount, with surface area 3200 sm^2/gr (Figure 3).

In our opinion the process forming a monolithic structure of microfilled cement stone becomes possible while developing strong epitaxis contact interlocks with fragments of ash spheres and on the surface of unbroken spheres.

We developed a microscopic research of the samples of 90 days old. Their resistance was tested when they were 28 days old. This research was necessary for the analysis of the cement stone structure made from activate ash of 15% and cement (Figures 5-8).

It was stated that the dominating mechanisms of the interaction between cement, ash and water are physical and chemical processes: crystallization and epitaxis contacts.

Figure 5 shows the process of oriented growing and increasing of crystals on the surface of ash microsphere.

The process of crystal growing corresponds to chemoepitaxis and is divided into several stages: delivery of cations, anions, tetrahedral units, and products transforming into liquid in case of hydration of cement minerals; forming of layers on the parts where possible defects can be located; forming of near-surface centres of cryslallization; nucleation of the layer of their fusion until blanket layer formation.

The defects of ash microspheres (Figure 6) appearing in the process of mechanoactivation, are an energetically beneficial location for the nucleation and growth of the firm epitaxis layer.

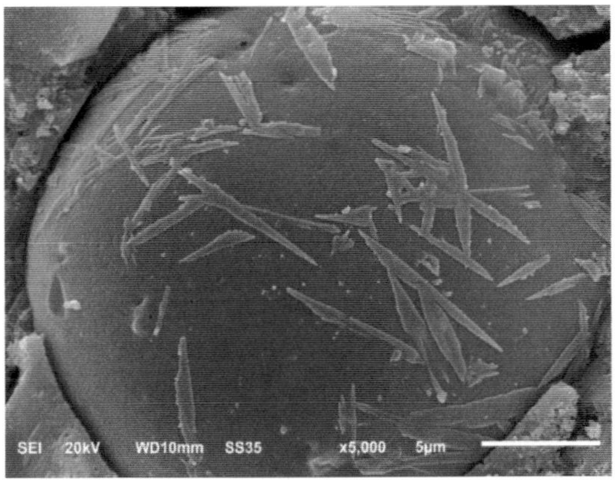

Figure 5 – A fragment of ash microsphere surface

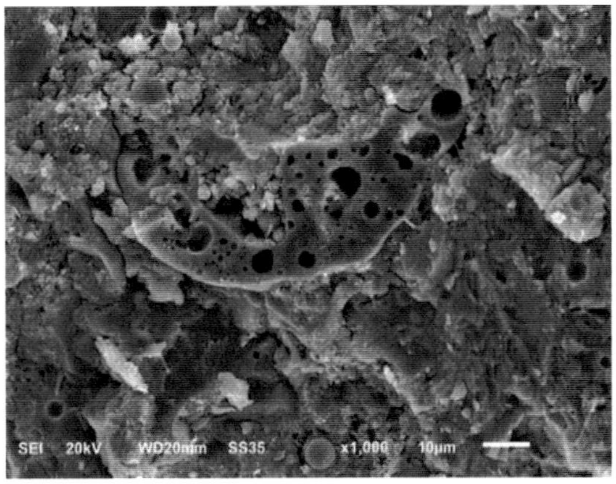

Figure 6 – A fragment of splintered ash microsphere

Figures 7, 8 demonstrate the process of forming the hydrates of filamentary morphology of C-S-H phase, platelike shape and six-sided calcium hydroaluminates and some quantity of hydrogarnets on the inner surface of ash microspheres. Regular joints of hydrosilicates and hydroaluminates in the form of dendrites of needle-shaped formations, platelike shape and six-sided crystals appear on the inner surface of the sphere.

As the results of the research state, agglomeration of ash microspheres (50 mcm in diameter) provides the formation of hydration phases in the zone of "free" space of the sphere. Neoformations are hydroaluminates of C_4AH_{19} type (a variation C_4AH_{13}), and some quantity of hydrohelenite. This is possible due to the penetrating qualities of the screen in relation to cations.

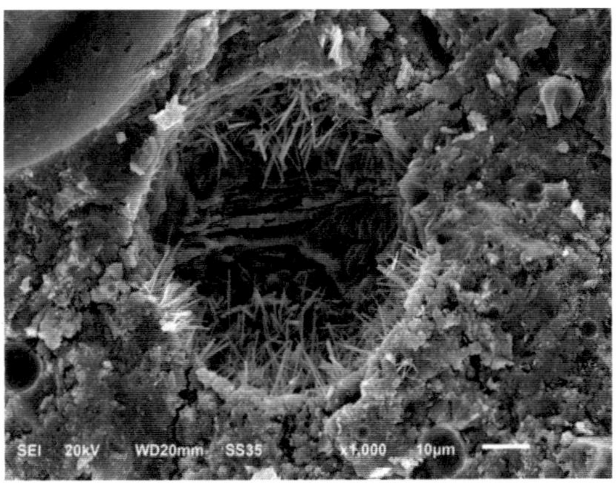

Figure 7 – Ash microsphere

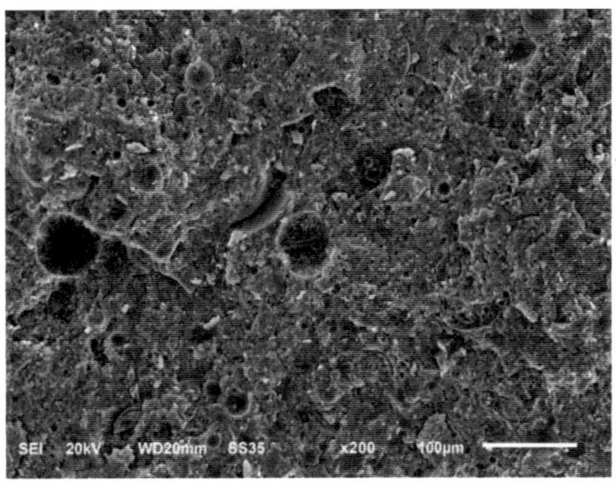

Figure 8 – Microstructure of cement stone with ash microspheres

We applied the slow-nucleation theory for the description of hydration process in the observed system containing ultradispersed ash. The theory states that the interaction of cement and water on the initial stage produce ions Ca^{2+}, OH^-, SiO_4^{4-}. Micro defects and destructions lead to the situation when the surface of the sphere begins to act as a screen through which Ca^{2+} penetrates first of all and participates in the synthesis of neoformations together with ash minerals and alumosilicate glass. The growth of crystals of dendrite-like shape is possible only in case of "free" space and ash activity. Some early scientific works [11, 12] do not contain any information about hydrate phases. They have not been defined.

By he age of 90 days crystals are develoed enough and provide more strength in cement stone, form mipor stone of high waterproofing qualities. Formation of such crystals can be possible only in case of mechanoactivation of ash.

The samples under test began to crack, chips appeared. This regularly happens with samples during compression testing. In microcracks (Figure 7) 60 days after testing one can notice neoformations and growth of crystal phases. Their morthological peculiarities are needle and tetrahedral shape, which are inherent for hydrosilicates. This can be explained by the "free" space in a microcrack and a long lasting reactive capacity of mechanoactivated ash. This capacity provides a self-healing process in cement stone for some period of time.

Conclusion

While studying the mechanisms of strength improvement of cement compositions containing mechanoactivated ash of some definite dispersiveness and granule forms one should pay a great attention to learning the processes of crystalforming which provide the strength of hardening systems . In the long run, it will be an undisputable factor in using ash in different types of cement composites. The structural typology of cement composite homogeneously mixed with microfilling (if its dispersiveness prevails that of cement binder) provides strength increase of contact zone due to the growth of junctions between particles, influence of capillary force and densification of a structure.

The active role of particles of ash microspheres in forming cement binding structure is proved by the results of electronic microscopial investigation.

Reftinskii SDPP experimental mechanoactivated ash with surface area 3200 sm^2/gr provides up to 15% of cement saving.

References

[1] Zyrianov V V. Fly ash – technogenic material / V.V. Zyrianov, D.V. Zyrianov. - M., 2009. 319 pp.

[2] Solomatov V I. Physical peculiarities of forming composite building materials` structure / Solomatov V I, Vyrovoi V N. // Building and Architecture. 1984 - № 8. - P. 59–64.

[3] Ramachandran V. Science about concrete: physic-chemical concrete study / Ramachandran V, Feldman R, Bodwen G; translated by Rosenberg T I, Ratinova Y B. - M. : 1986. - 280 pp.

[4] Kaprielov S S. The influence of cement stone structure with microsiliceous additives and superplasticizing admix on concrete qualities / Caprielov S S, Sheifeld A V, Crivoborodov Y R. // Concrete and reinforced concrete. - 1992. - № 7. - P. 4–7.

[5] Kalashnikov V I. The role of dispersive fillings in modern concrete / Calashnikov V I, Suzdaltsev O V, Drianin G P, Sekhsposian G P. //Building – 2014. – №7. P.11-21.

[6] Dvorkin L I. High strength concrete with fly ash additive / Dvorkin L I, Shabman I B, Chudnovski S M. [and others.] // Concrete and reinforced concrete. - 1993. - № 1. - P. 23–25.

[7] Bazhenov Y M. Modified high strength concrete / Bazhenov Y M, Demianova V S, Kalashnikov V I. - M., - 2006. - 368 pp.

[8] Solomatov V I. Intensive technology of concrete/ Solomatov V I, Takhirov M K. –M,, 1989. — 264 pp.

[9] Fenelonov V B. The properties of cenospheres and the mechanism of their formation during high-temperature coal combustion at thermal power plants / Fenelonov V B, Mel'gunov M S, Parmon V N. // KONA Powder and Particle J. -2010. -N 28. -P. 189-207.

[10] Dvorkin L I. concrete with mineral fillings / Dvorkin L I, Solomatov V I, Vorovoi V N, Chudnovski S M. -Kiev, 1991. -136 pp.

[11] Sagradian A A. The research of puzzolanic activity of ash microspheres / Sagradian A A, Zimakova G A. // Building. – 2012. – №2. P.43-47.
[12] Sagradian A A. The study of the qualities of heavy concrete modified by organomineral additive including ash microspheres / Sagradian A A, Zimakova G A. // Building. – 2012. – №4. P.26-31.

IOP Conf. Series: Earth and Environmental Science **50** (2017) 012008 doi:10.1088/1755-1315/50/1/012008

Natural Gas Use On Minibuses, Engaged In The Carriage Of Passengers And Baggage On The Regular Routes, As A Measure For Decrease In Harmful Environment Effects

E Chikishev, A Chikisheva, I Anisimov, D Chainikov

Motor transport operation, Industrial University of Tyumen
Volodarskogo Street 38, Tyumen, Russian Federation, 625000

E-mail: 4ainikov@mail.ru

Abstract. The paper deals with an option of increase the compressed natural gas use as a motor fuel for diesel minibuses to reduce emissions of harmful substances with exhaust gases. In terms of the Russian company LTD "WTC Automobilist", carrying passengers and baggage on regular routes by minibuses, the calculation of the park's natural gas needs is promoted. A mini CNG RS with optimal performance is suggested. The approximate payback period of the natural gas equipment installation on all buses of company and introduction into service period of mini CNG RS are calculated.

1. Introduction

Development of a street road network doesn't get on well at the majority of the countries behind rapid growth of the motor transport on roads. Congestions on the roads became a commonplace. This is particularly evident in cities with a population of over 500 thousand people. At the same time transport releases a huge amount of harmful substances into the atmosphere that causes an irreparable loss to the environment. Therefore many countries promptly expand the use of more environmental friendly alternative energy sources, reducing consumption of traditional oil motor fuels.

For Russia the most perspective is use of natural gas (NG) as the country possesses a great amount of the world's NG deposits (24.6%). It ranked second after the US for its production (18%). Today at state and regional levels taken a number of legislative initiatives, which will soon have to increase the consumption of natural gas by motor transport in compressed (CNG) and liquefied (LNG) states.

Countrywide construction of natural gas refuelling stations (CNG RSs) has to play a leading role in increase of this hydrocarbon consumption. In [1-6] it is noted that weak network of gas stations contains growth of the GBA fleet. In [3] it is noted that the rapid development of natural gas takes place in those countries where there are significant internal reserves in the given hydrocarbon.

At the same time the set of experiments carried out confirms that at combustion of NG in automobile engines less harmful substances emitted with the exhaust gases in comparison with both traditional gasoline and diesel fuel (DF) [7-20, etc]. Besides in all countries gaseous fuel is much cheaper in comparison with gasoline and diesel (as a rule, by 2-3 times). For example, in Russia this price proportion was regulated until recently by the Resolution of the Government of the Russian Federation of 15.01.93 № 31 "On urgent measures to expand the replacement of motor fuels with natural gas". However since 21.04.2015 this Resolution is repealed according to the Resolution of the Government of the Russian Federation of 10.04.2015 № 338 [21]. However, the upward trend in NG prices are not observed.

In the world, there are about 20 million natural gas vehicles (NGVs) (2% of the total global vehicle fleet) operating on NG. According to [22] the global NGVs fleet annually increases by 30%. A considerable part of them constitute the buses running on public transport routes. Therefore, research on the NG use by buses are relevant [15, 23, 24].

By the use of NG on motor transport in general and on buses in particular, Russia lags behind many countries. Recently, however, positive change in this direction are seen. Especially it is noticeable in such megalopolises as Yekaterinburg, Kazan, Moscow, Nizhny Novgorod and St. Petersburg. Nevertheless, even in these cities, the number of CNG RSs does not exceed 4 units. It is associated with some problems in the operation of natural gas buses (NGB), the main ones are significant zero runs to gas station and back and costs of time for them. One solution to this problem is to build a mini CNG RS in the territory of motor transport enterprise.

2. Methodology

2.1 Choice of the research object

In this paper the method of the efficiency increase of buses usage on the example of the entity of LTD "WTC Automobilist" (Tyumen) by transfer of its fleet to CNG and a construction of mini CNG RS in the territory of enterprise is offered. The choice of Tyumen is caused by the fact that since 2016, the city is included in the "Reduction of greenhouse gas emissions from motor transport in the cities of Russia" project. . The project is directed to decrease in emissions in the atmosphere of greenhouse gases by means of improvement of planning and management of an urban transportation by creation of effective systems of monitoring and promotion of environmentally friendly transport modes. Project implementation will be enabled at the expense of the Ministry of transport of the Russian Federation and the United Nations Development Programme with the assistance of the Global Environment Facility.

In Tyumen operates only one CNG RS, located in 12 km from the downtown. It is impractical to use this CNG RS bus fleet. Thus, today in the city a real alternative is development of the mini CNG RSs network which shall be located in close proximity or in the territory of the bus enterprises. At the same time one of the main installation criteria is availability of the gas pipeline as the mini CNG RS is connected to a gas distribution network.

Across the territory of the Tyumen region (autonomous regions and Khanty-Mansi Autonomous Dietrict (KMAD) and Yamalo-Nenets Autonomous Okrug (YNAD)) in general, and in particular the city of Tyumen there passes a huge kilometers amount of transmission and distribution pipelines. This is due to the fact that in YNAD produces 85% of Russian gas and practically in all settlements NG is used as fuel in many industries.

Within the borders of the city of Tyumen length of gas pipelines of various capacity is about 1000 km. Therefore installation of CNG RS is possible practically in any point of the can city. And pressure on the part from gas control points (GCP) to the consumer usually 0.05-0.3 MPas (0.5-3 bars) which is sufficient for the operation of mini CNG RS.

2.2 Transportation of passengers and baggage by bus of LTD "WTC Automobilist"

The kind of activity of LTD "WTC Automobilist" is rather extensive, however one of the main is to provide services in transportation of passengers and baggage on regular city routes. As of August, 2016 the organization has about 48 minibuses (15-26 places) are registered. These buses are mostly owned by individual carriers and the other entities which signed the contract for the organization and implementation of regular transportations of passengers and baggage on municipal Tyumen routes with LTD "WTC Automobilist". The bus fleet is constituted by the brands presented in the table 1.

Table 1 - Bus fleet of LTD "WTC Automobilist"

№	Bus brand	Busload, pearson	Year of manufacture
1	Fiat Ducato	15-18	2009-2011
2	Fiat FST613	22	2012
3	Peugeot Boxer (modifications)	22	2010-2013
4	GAZ 3302	22	2012
5	Citroen L4H2 and modifications	16-22	2013
6	Renault (modifications)	22-25	2013-2015
7	Ford (modifications)	25-26	2012-2014

All buses are equipped with diesel engines. They work on three regular routes of public transport:

1. Loop route № 52. Route: the square of Nemtsov – air-port Plekhanovo. Headway: on weekdays: 4-14 min., on weekends: 6-21 min. The length (distance) of a route in the direct direction: 5.94 km, in the return: 5.54 km. Quantity of stops in direct direction: 10, in the return: 9. On average the running time is from 06:20 till 21:39.

2. Loop route № 57. Route: the 1st residential district – the residential district "Moskovsky dvorik"). Headway: on weekdays: 1-20 min., on weekends: 5-14 min. The length (distance) of a route in the direct direction: 12.92 km, in the return: 15.25 km. Quantity of stops in direct direction: 27, in the return: 27. On average the running time is from 06:12 till 21:55.

3. Loop route № 68. Route: the Mys settlement– Bakinskikh Komissarov St.). Headway: on weekdays: 6-17 min., on weekends: 8-30 min. The length (distance) of a route in the direct direction: 14.04 km, in the return: 17.08 km. Quantity of stops in direct direction: 25, in the return: 33. On average the running time is from 06:11 till 21:48.

2.3 Calculation part

On brands of vehicles which are operated by the enterprise the gas equipment (for work on a gas-diesel cycle) and 3 ballons volume of 50 l everyone can be installed. Therefore, under standard operating conditions (air temperature of 20 °C, atmospheric pressure of 760 mm Hg) in these capacities, according to RD 03112194-1095-03 [25], 37,5 m^3 of methane can be filled (with a pressure in cylinders of 19,6 MPas). It should be noted that at vehicles operation on a gas-diesel cycle in country conditions fuel is spent in the ratio 70% of NG and 30% of DF. In city conditions this ratio is typically 50% to 50%. Consequently, the composition of the fuel-air mixture depends upon a number of factors (the driving style, engine wear, and road and climatic conditions, traffic intensity, frequency of acceleration and braking, etc.) and calculated by electronic vehicle control unit.

Let's determine what amount of natural gas it is necessary to fill all buses of the enterprise:

$$Q_{m^3} = A_b \times V \,, \tag{1}$$

Where Q_{m^3} – the number of the cubic meters of gas necessary for operation of all enterprise buses, m^3; A_b – the quantity of buss on duty, units; V – the total volume of gas ballons for one bus, m^3.

Let's say that the enterprise operates 48 buses which will be transferred to CNG and 3 ballons with total capacity of 37.5 m^3 will be established on all buses. Then, having added these data in a formula 1 we will obtain the following value:

$$Q_{m^3} = 48 \times 37.5 = 1800 \; m^3.$$

Further it is necessary to calculate the minimum daily need in NG of the buses fleet. For this purpose it is necessary to know:

1. Brands of buses and their fuel consumption. Let's consider on the example of the Peugeot L4H2M2 bus as they make a considerable part of the LTD "WTC Automobilist" fleet. About 16 such buses work at each of 3 routes. The basic fuel consumption norm of the considered bus in urban cycle

is 10.7 l per 100 km [26]. According to [27] let's accept that 1 m^3 of NG is equivalent to 1 l of liquid fuel.

2. The length of buses runs on routes. The length of one turnover of each route is equal:

№52	№57	№68
L_m = 5.94 + 5.54 = 11.48 km	L_m = 12.92 + 15.25 = 28.17 km.	L_m = 14.04 + 17.08 = 31.12 km.

Further we will determine run length for the working day. According to operations control of LTD "WTC Automobilist" on weekdays it is:

a. № 52 – Each bus carries out 16 turns a day. The total mileage of 1 bus (L_O) is:
$L_{O(№52)}$ = 11.48 x 16 = 183.68 km.

b. № 57 – Each bus carries out 7 turns a day. The total mileage of 1 bus is:
$L_{O(№57)}$ = 28.17 x 7 = 197.19km.

c. № 68 – Each bus carries out 6 turns a day. The total mileage of 1 bus is:
$L_{O(№68)}$ = 31.12 x 6 = 186.72 km.

Further we will determine the total daily mileage (L_{OM}) for all buses fleet by routes:

$L_{OM(№52)}$	$L_{OM (№57)}$	$L_{OM (№68)}$
183.68 x 16 = 2938.88 km.	197.19 x 16 = 3155.04 km.	186.72 x 16 = 2987.52 km.

The total daily mileage for all routes is:
L_{OM} = 2938.88 + 3155.04 + 2987.52 = 9081.44 km.

On the basis of the data of the total daily mileage by all routes we will determine day need of the fleet in fuel:

$$Q_O = 0.01 \times H_S \times L_{OM}, \tag{2}$$

Where H_s – basic fuel consumption norm for the vehicle milage, l/100 km; Q_O – day need of the fleet in fuel, l(m^3).

Having substituted values in a formula 2 we will receive the following:

$$Q_O = 0.01 \times 10.7 \times 9081.44 = 971.7 l(m^3).$$

As buses are operated within the city, a ratio of an CNG-DF consumption is accepted 50% to 50%. Let's determine, the day need in CNG ($Q_{O(CNG)}$) and DF ($Q_{O(DF)}$) during the work of buses on a gas-diesel cycle:

$$Q_{O(CNG)}\ 971.7 \times 50/100 = 485.85\ m^3 \qquad Q_{O(DF)}\ 971.7 \times 50/100 = 485.85\ l$$

Calculations are presented without the idle run and other factors causing increase in fuel consumption (climatic and road conditions, technical condition of the vehicle, driver skills etc.). The number of working days on one full CNG refueling (37.5 m^3), will determine by the following formula:

$$D_w = Q_{M^3} / Q_{O(CNG)}, \tag{3}$$

where D_w – The number of working days on one full CNG refueling, days.

On the basis of formula 3 we'll receive:

$$D_w = 1800/485.85 = 3.7\ days.$$

Knowing the day need in NG ($Q_{O(CNG)}$) within the given conditions, we will determine optimum, from the presented in the table 2, mini CNG RS for continual activity of the enterprise transport on gaseous fuel:

$$CNGRS = P_{CNGRS} \times T_R. \tag{4}$$

where P_{CNGRS} – one-hour efficiency of CNG RS, m^3/hours; T_R – the bus running time, h.

Table 2 – An example of some mini CNG RSs specifications and their cost

Model	Input pressure, bar	Output pressure, bar	Performance, nm³/hour	Cost (august 2016), thousand rub.
Coltri MCH-24	0.304	200/250	24	1223
Coltri MCH-30	0.304	200/250	30	1223
H8-11	2.53	200/250	40	2592
H6-11	4.053	200/250	45	2592
H8-15	3.04	200/250	50	2808
H6-15	6.08	200/250	60	2808

There are many companies In Russia that offer mini CNG RSs of various modifications: LTD "Elitgaz", LTD "Rademi", LTD "Broadway», «Fornovogas», LTD "Micrometan C" and others.

In view of the fact that 48 minibuses are opearted at the LTD "WTC Automobilist", it is necessary to calculate optimum compressor station for such number of vehicles.

Taking into account the working period of a public transport on regular routes in Tyumen from 6:30 till 22:00 (without time spent for a lunch and/or to a shift change) to satisfy the day needs in fuel of the considered enterprise the station capable in 15 hours to refuel about 486 m^3 of methane is necessary.

Let's find an optimum fuel station from the table 2.

$CNGRS_{Coltri\ MCH-24} = 24 \times 15 = 360\ m^3.$ $CNGRS_{Coltri\ MCH-30} = 30 \times 15 = 450\ m^3.$

$CNGRS_{H8-11} = 40 \times 15 = 600\ m^3.$ $CNGRS_{H6-11} = 45 \times 15 = 675\ m^3.$

$CNGRS_{H8-15} = 50 \times 15 = 750\ m^3.$ $CNGRS_{H6-15} = 60 \times 15 = 900\ m^3.$

As seen from the calculations mini CNG RSs H8-11, H6-11, H8-15, H6-15 satisfy specified requirements. A CNG RS Coltri MCH-30 has a little smaller performance than necessary. However, the difference in cost between Coltri MCH-30 and, for example, H8-11 is 1,369,000 rub. Therefore, it is economically reasonable to establish Coltri MCH-30.

Further we will calculate the purchase costs of the gas-ballon equipment (GBE). On the bus brands presented in table 1 usually all-metal ballon of the 1st type is established. By data [28] the cost of installation is about 177 thousand rubles (the GBE brand: Tamona Diesel Gas, 3 ballons of the 1st type). Let's calculate GBE purchase costs (C_P) for the enterprise:

$$C_P = 48 \times 177 = 8496\ thousand\ \ rub.$$

The cost of DF (C_{DF}) and CNG (C_{CNG}) in Tyumen as of the 2nd quarter 2016 is $C_{DF} = 37.1$ rub for 1 liter and $C_{CNG} = 13.3$ rub for 1 m^3.

Consequently, knowing the total day fuel requirement calculated according to the formula 2 let's define the overall cost of the bus fleet using diesel cycle (DC) (5) and gas-diesel cycle (GDC) (6), respectively:

$$CNGRS = P_{CNGRS} \times T_R. \tag{5}$$

$$C_{GDC} = Q_{O(CNG)} \times C_{CNG} + Q_{O(DF)} \times C_{DF}. \tag{6}$$

Where C_{DC}, C_{CNG} – total day expenditures during the operation on diesel cycle gas-diesel cycle respectively, rub.

Having substituted values in a formula 5 and 6 we will receive the following:

$$C_{DC} = 971.7 \times 37.1 = 36050.07\ rub.$$

$$C_{GDC} = 485.85 \times 13.3 + 485.85 \times 37.1 = 24486.8\ rub.$$

Day money economy (E_D) from use of a gas-diesel cycle in comparison with diesel are defined:

$$E_D = C_{DC} - C_{GDC}. \tag{7}$$

Having substituted values in a formula 7 we will receive the following:

$$E_D = 36050.07 - 24486.8 = 11563.27 \ rub.$$

To define the payback period of GBE it is used the following formula:

$$Pb_{GBE} = C_P / E_D , \quad (8)$$

where Pb_{GBE} – payback period from acquisition of the gas-ballon equipment, days.

Having substituted values in a formula 8 we will receive the following:

$$Pb_{GBE} = 8496000 / 11563.27 = 734.3 \ days.$$

Further we will calculate a payback period of mini CNG RS which we will determine by a formula:

$$Pb_{CNGRS} = C_{CNGRS} / E_D , \quad (9)$$

where Pb_{CNGRS} – payback period from acquisition of the mini CNG RS, days; C_{CNGRS} – cost of mini CNG RS, rub.

According to [29] cost of the Coltri MCH-30 CNG RS is 1223 thousand rubles.
Having substituted values in a formula 9 we will receive the following:

$$Pb_{Coltri\ MCH-30} = 1223000 / 11563.27 = 105.8 \ days.$$

Define total payback periods (Pb_{TOTAL}) of GBE and mini CNG RS according to the formula:

$$Pb_{TOTAL} = Pb_{GBE} + Pb_{CNGRS} . \quad (10)$$

Having substituted values in a formula 10 we will receive the following:

$$Pb_{TOTAL} = 734.3 + 105.8 = 840.1 \ days.$$

Thus, costs for re-equipment of the buses fleet for operation on a gas-diesel cycle and acquisition of the mini CNG RS of necessary performance pay off for 2 years and nearly 4 months.

3. Conclusions

Based on calculations it is possible to note that under the given conditions payback periods of transferring diesel minibuses on a gas-diesel cycle and construction the mini CNG RS are rather considerable. The reason is that in urban environments during engine operation on gas diesel cycle, due to the constantly changing speed regime, increasing the proportion of diesel fuel in the fuel-air mixture as compared to operation outside of urban. Therefore, fuel costs increase proportionally. Also the studied routes have rather low day mileage. That also affects the payback time.

Thus, enterprise economy from NG use can be reached in the medium term. However NG transport transport when working on gas-diesel cycle will emit less amount of harmful substanses with exhaust gases, cjmpared with diesel cycle. Therefore, to increase the use of NG on the transport and environmental improvement are invited to develop a plan of government (regional) enabling activities, such as:

1. To develop the plan for subsidizing of the bus enterprises. For example, in Tyumen to make changes to the resolution of Tyumen Administration of 29.08.2008 № 117-pk "On approval of the Order of subsidies provision for reimbursement of expenses related to the transport services provision within the organization of transport service for the population of Tyumen". To add point to grant a subsidy of 50% of the costs for the installation of GBE.
2. To reduce, cancel or temporarily not to levy a tax on vehicles using NG as fuel.
3. On the basis of the first two offers to change the requirements to the tender participants when carrying out tender on implementation of regular passengers and baggage transportations on municipal routes of Tyumen. To establish obligatory selection criterion in the form of the minimum quantity of

NG buses in the enterprise fleet to 30% (according to the order of the Government of the Russian Federation № 767-r of 13.05.2013. "On regulation of the relations to the sphere of gaseous motor fuel use, including natural gas as motor fuel").

References

[1] Chinese D, Patrizio P, Bonotto M A. Service station location model to explore prospects and policies for alternative transport fuels: A case of CNG distribution in Italy // Advances in Intelligent Systems and Computing - Volume 262, pp. 71-84, 2014, doi: 10.1007/978-3-319-04630-3_6.

[2] Chikishev E, Ivanov A, Anisimov I, Chainikov D. Prospects of and problems in using natural gas for motor transport in Russia // 2016 IOP Conf. Series: Materials Science and Engineering – Volume 142, 012110, 2016, doi: 10.1088/1757-899X/142/1/012110.

[3] Engerer H, Horn M. Natural gas vehicles: An option for Europe // Energy Policy - Volume 38, Issue 2, pp. 1017-1029, 2010, doi: 10.1016/j.enpol.2009.10.05.

[4] Han J, Li Y. Beijing necessity of development of natural gas buses // Advanced Materials Research / 2011 International Conference on Materials and Products Manufacturing Technology, ICMPMT - Volume 339, Issue 1, 28-30 Oct. 2011, Chengdu; China, pp. 509-516, doi: 10.4028/www.scientific.net/AMR.339.509.

[5] Jacqueline L Kirk, Abigail L Bristow, Alberto M. Zanni Exploring the market for Compressed Natural Gas light commercial vehicles in the United Kingdom // Transportation Research Part D: Transport and Environment - Volume 29, pp. 22-31. 2014, doi: 10.1016/j.trd.2014.03.004.

[6] Ma L, Geng J, Li W, Liu P, Li Z. The development of natural gas as an automotive fuel in China // Energy Policy - Volume 62, pp. 531-539, 2013, doi: 10.1016/j.enpol.2013.06.066.

[7] Anisimov I A, Chikishev E M. Studies of adaptability of vehicles run on natural gas and gasoline, to the actual air temperature // Gas industry – № 2(702), pp. 82-85, 2014.

[8] Bielaczyc P, Szczotka A, and Woodburn J. Regulated and Unregulated Exhaust Emissions from CNG Fueled Vehicles in Light of Euro 6 Regulations and the New WLTP/GTR 15 Test Procedure // SAE International Journal of Engines – Volume 8, Issue 3, pp. 1300-1312, 2015, doi: 10.4271/2015-01-1061.

[9] Bondarenko E V, Filippov A A. Evaluation of use of certain types of motor fuel for environmental safety criteria // Autogas filling complex + alternative fuel – 4 (52), pp. 31-35, 2010.

[10] Erokhov V I. Gas-diesel power supply system of new generation // Autogas filling complex + alternative fuel – №9(78), pp. 15-23, 2013.

[11] Gonçalves M, Jiménez-Guerrero P, Baldasano J M. Emissions variation in urban areas resulting from the introduction of natural gas vehicles: Application to Barcelona and Madrid Greater Areas (Spain) // Science of the Total Environment - Volume 407, Issue 10, pp. 3269-3281, 2009, doi: 10.1016/j.scitotenv.2009.01.039.

[12] Jahirul M I, Masjuki H H, Saidur R, Kalam M A, Jayed M H, Wazed M A. Comparative engine performance and emission analysis of CNG and gasoline in a retrofitted car engine // Applied Thermal Engineering - Volume 30, Issue 14-15, pp. 2219-2226, 2010, doi: 10.1016/j.applthermaleng.2010.05.037.

[13] Hassan M H, Kalam M A, Mahlia T M I, Aris I, Nizam M K, Abdullah S, Ali Y. Experimental test of a new compressed natural gas direct injection engine // Energy and Fuels - Volume 23, Issue 10, pp. 4981-4987, 2009, doi: 10.1021/ef8011382.

[14] Kumar P, Gupta N C. Commuter exposure to inhalable, thoracic and alveolic particles in various transportation modes in Delhi // Science of the Total Environment - Volume 541, pp. 535-541, 2016, doi: 10.1016/j.scitotenv.2015.09.076.

[15] Lim C, Kim D, Song C, Kim J, Han J, Cha J-S. Performance and emission characteristics of a vehicle fueled with enriched biogas and natural gases // Applied Energy - Volume 139, pp. 17-29, 2015, doi: 10.1016/j.apenergy.2014.10.084.

[16] Liu J, Yang F, Wang H, Ouyang M, Hao S. Effects of pilot fuel quantity on the emissions characteristics of a CNG/diesel dual fuel engine with optimized pilot injection timing // Applied Energy - Volume 110, pp. 201-206, 2013, doi: 10.1016/j.apenergy.2013.03.024.

[17] Oprešnik S R., Seljak T, Bizjan F, Katrašnik T. Exhaust emissions and fuel consumption of a triple-fuel spark-ignition engine powered passenger car // Transportation Research Part D: Transport and Environment - Volume 17, Issue 3, 2012, pp. 221-227, doi: 10.1016/j.trd.2011.08.002.

[18] Pastorello C, Dilara P, Martini G. Effect of a change towards compressed natural gas vehicles on the emissions of the Milan waste collection fleet // Transportation Research Part D: Transport and Environment - Volume 16, Issue 2, pp. 121-128, 2011, doi: 10.1016/j.trd.2010.09.002.

[19] Sahoo B B, Sahoo N, Saha U K. Effect of engine parameters and type of gaseous fuel on the performance of dual-fuel gas diesel engines-A critical review // Renewable and Sustainable Energy Reviews - Volume 13, Issue 6-7, pp. 1151-1184, 2009, doi: 10.1016/j.rser.2008.08.003.

[20] Rose L, Hussain M, Ahmed S, Malek K, Costanzo R, Kjeang E. A comparative life cycle assessment of diesel and compressed natural gas powered refuse collection vehicles in a Canadian city // Energy Policy - Volume 52, pp. 453-461, 2013, doi: 10.1016/j.enpol.2012.09.064.

[21] The resolution of the Government of the Russian Federation of April 10, 2015 № 338 " On the Annulment of the decision of the Russian Government dated 15 January 1993 № 31". – Moscow, 2015. – pp. 1.

[22] The world fleet of the vehicles using gas fuel annually increases by a third [Electronic resource]: http://www.uralavtogaz.ru/about/news/2014/04/14/77/ (accessed September 8, 2016).

[23] Wang R, Wu Y, Ke W, Zhang S, Zhou B, Hao J. Can propulsion and fuel diversity for the bus fleet achieve the win-win strategy of energy conservation and environmental protection? // Applied Energy - Volume 147, pp. 92-103, 2015, doi: 10.1016/j.apenergy.2015.01.107.

[24] Zhang, S, Wu Y, Hu J, Huang R, Zhou Y, Bao X, Fu L, Hao J. Can Euro V heavy-duty diesel engines, diesel hybrid and alternative fuel technologies mitigate NOX emissions? New evidence from on-road tests of buses in China // Applied Energy - Volume 132, pp. 118-126, 2014, doi: 10.1016/j.apenergy.2014.07.008.

[25] Guidance document 3112199-1095-03. Guidelines on the organization and operation of gas-cylinder vehicles that run on compressed natural gas. – Ministry of Transport, Road Department, JSC SaRIoMT - 2002. 59 p.

[26] Peugeot Boxer [Electronic resource]: http://www.professional.peugeot.ru/media/deliacms/media/68/6843-939cfe.pdf (accessed September 8, 2016).

[27] The Resolution of the Ministry of Transport of the Russian Federation № AM-23-r (an edition of 14.07.2015 № NA-80-r No.) "About enforcement of methodical recommendations of "Fuels and lubricants consumption norms on a motor transport". – M.: Ministry of Transport of the Russian Federation, 2015. – 215 pages.

[28] GBO installation cost calculator [Electronic resource]: http://elitegas.ru/gbo/calculators/ gbo-price-calculator/ (accessed September 10, 2016).

[29] Methane compressors (NGV RI, NGV RS) [Electronic resource]: http://elitegas.ru/gbo-methane/cng-compressors/ (accessed September 10, 2016).

Coal Mining Machinery Development As An Ecological Factor Of Progressive Technologies Implementation

A B Efremenkov[1,a], A A Khoreshok[2,b], S A Zhironkin[3,c], A V Myaskov[4,d]

[1] Yurga Technological Institute (Branch) of National Research Tomsk Polytechnic University
Leningradskaya st. 26, Yurga, Russian Federation, 652055
[2] T.F. Gorbachev Kuzbass State Technical University
Vesennyaya st. 28, Kemerovo, Russian Federation, 650000
[3] National Research Tomsk Polytechnic University
Lenina av. 30, Tomsk, Russian Federation, 634050
[4] National University of Science and Technology MISiS, College of Mining
Leninsky av. 4, Moscow, Russian Federation, 119049

E-mail: ytitpu@tpu.ru[a], haa.omit@kuzstu.ru[b], zhironkin@inbox.ru[c], mgi@misis.ru[d]

Abstract. At present, a significant amount of energy spent for the work of mining machines and coal mining equipment on coal mines and open pits goes to the coal grinding in the process of its extraction in mining faces. Meanwhile, the increase of small fractions in mined coal does not only reduce the profitability of its production, but also causes a further negative impact on the environment and degrades labor conditions for miners. The countermeasure to the specified processes is possible with the help of coal mining equipment development. However, against the background of the technological decrease of coal mine equipment applied in Russia the negative impact on the environment is getting reinforced.

1. Introduction

For the last decade the intensive development of mechanic methods of coal mining has led to a significant deterioration of the coal quality. According to statistics even in 1975 the mechanization level (for coal seams with the dip angle up to 35 degrees) in major coal regions of Russia was 73.7-90.0%. Now its indicator has reached 100% [1-2].

Currently, the main part of the underground coal mining is carried out by narrowly gripping coal-mining machines [3-5].

With the growth of their power capacity there is a tendency to increase the degradation of extracted coal. Over-degraded mined coal badly affects the economic performance of the mines. Most of coal mining enterprises losses are associated with constantly deteriorating quality of the coal.

Increased varietal composition requirements are mainly implemented to energy coal. Excessive degradation in size of such coal complicates their separation on classes, transportation and leads to a significant decline of coal selling prices.

Significant part of the extracted coal, and in Kuzbass region (Western Siberia, Russia) it reaches up to 60%, comes to the coking [6]. All of coking coal is subjected to upgrading operations. Costs of small classes of coal upgrading by washing methods are in 4-8 times higher than for large classes.

2. Material and Method

In recent years, the quality of coal products is associated with an increase of the ash content of mined coal, and less attention is paid to its varietal composition. But these factors are interconnected with each other. So, with the increase in volumes of coal upgrading, the growth of crushed coal leads to the output of much larger number of un-upgradable wastage. Its processing and inclusion in the material production require additional costs.

Thus, from an economic point of view, the advisability of reducing the degradation in size of extracted coal is undeniable.

Coal over-degradation in size in mechanized extraction also leads to significant dust emissions. In this specific dust emission exceeds the permissible limits (MPC). High air dust content in coal mine complicates the work of the staff, increases the risk of work operations and reduces productivity. Therefore there is a need to stop coal-mining machine, to apply additional measures of dust suppression which complicate the organization of the working process, reduce productivity of mining faces.

In the structure of occupational diseases of miners the leading place is taken by dust pathology of the respiratory system. Especially dangerous for miner's health is the dust of coking coal.

High saturation of Kuzbass with coal companies, chemical plants, power stations significantly degrades the environment. Rapid development of new mines and open pits is encouraged by the growth of mining machinery specific productivity. That increases not only the risk of the air pollution but the water contamination [7-8].

Currently, total emissions into the air of Kuzbass region from stationary sources are on average 1.5 million tons for the year, including 0.18 million tons from coal enterprises (11.7%). Coal industry enterprises emit up to 9.3 thousand tons of suspended solid contaminants annually polluting residential areas of Kuzbass miners' towns [9-10]. The given social and environmental aspects of the coal degradation in size problem show urgent necessity to reduce mined coal crushing.

Deterioration of technical and economic parameters of coal enterprises, including the quality of extracted coal has been caused by poor supplying the industry with modern equipment, materials and by the narrow use of modern coal mining technologies, by increasing trend of immunity of coal production of the science and technology achievements, by the complexity of geological conditions of coal mining.

The enterprises of coal industry operate 30-35% of physically deteriorated and depreciated equipment, and only 10-15% of used mining equipment complies with modern technical level widely accepted in the world [11]. In addition, the most of used mining equipment is operated in inappropriate geological and technical conditions.

We also observe clearly insufficient efforts of domestic researchers involved in the development of new coal mining equipment and improvement of existing machinery.

From 1963 to 1993, Russian coal industry produced about 80 new and modified pieces of coal mining machinery [11].

For their comparison using the quality criterion (the factor of grade) the perfection coefficient k and the fineness index km were calculated. These indicators uniquely characterize the cutting modes from the grinding point of view. Obtained index km = 0.015-0.02 corresponds to the effective cutting tools and methods and progressive cutting regimes.

So, today the situation at coal industry enterprises shows that the latest machinery science achievements cannot find a proper application.

Thus, for example, to equip narrowly gripping coal-mining machines, which carry out the bulk of the underground coal mining, we developed screw actuators with different types of cutters and disc cutters.

They provide a large increase in the output of large coal grades by 6-14%. Area of application of the screws with such cutters is only 20-25% of developing coal seams and screws with disc cutters do not even come out of the prototype stage, although there are some technical specifications for these screws and disc cutters. The reason for the limited use of these screws is that the extra cost of using more expensive and less reliable cutters screws is referred to the segments of coal mines, and the economic effect is achieved by the whole mine only after coal sorting and upgrading.

Thus, the actual coal miners (segment, mine) are not economically stimulated by improving the quality of the extracted coal and consequently, by implementing new technology. In addition, shortcomings in the existing coal pricing in Russia today give a small difference in the price for concentrate and raw coal, which results in a small economic effect of improving coal varietal composition.

3. Results and Discussion

To ensure the improvement of the coal grade in the whole industry it is required to plan and standardize the varietal composition of the coal produced at all stages of the mining process. For operational analysis the varietal composition techniques it is necessary to assess the contribution of each operation of mining process in the coal crushing and, above all, the contribution of destruction as the basic operation.

For this purpose, in recent 15 years sufficient work of researches around the world has been done to improve the working bodies of coal-mining and tunneling machinery [12-19].

Analysis of the researches in this area shows that the special attention should be paid to the following directions:

- the study and improvement of coal transportation and loading process by screws and other working bodies;
- selecting the optimal parameters of cutting diagrams and arrangement of cutting tools application area;
- the study of coal crushing and improving the efficiency of coal-mining machines according to the factor of mined coal's grade;
- the study of regime and parameters of destruction;
- improving the reliability and durability of coal-mining machine working bodies and determining the rational area of their application.

Thus a large amount of theoretical and experimental studies conducted in the last decade and aimed at improving the working bodies of the narrowly gripping coal-mining machines, revealed qualitative and quantitative dependences between the loading capacity of the screws and their constructive and regime parameters. Based on these we developed a number of recommendations for their improvement.

The research in the field of rational choice of cutting schemes and cutters' placement were carried out in the following aspects: obtaining better-quality coal composition and the least dust emission, effective work in seams with rock layers and inclusions, and reducing the dynamic loads on the drive.

The methodology of choice of rational cutting schemes and cutters' placement on the screw comprised the basic regularities of coal cutting process by a single instrument. The main conclusion was that the specific energy consumption and output of degradation in size coal decrease hyperbolically with the increase in average cutting depth values hcp and cross-section of the chips Scp. The choice of rational values of hcp and Scp is recommended to make in the area of stabilization of specific energy consumption and grade.

The sufficient amount of researches is devoted to the problems of coal grinding and increase in the output of high-quality coal [20-24].

It should be noted that the studies of coal grinding and a choice of rational parameters of destruction and schemes of tools are closely connected and, as a rule, were carried out comprehensively. However, considering the great economic importance in energy, social and economic aspects, we will focus on this issue, as it's no doubt that the growth of mechanized coal extraction and power capacity of coal-mining machine lead to a decrease in the output of high-quality coal.

The physical nature of the coal grinding and brittle rock by cutting tools can be explained by the mechanism of destruction by cutters. It determines the presence of the continuous series of coal pieces of various sizes, ranging from fine (dust) fractions in the products of coal destruction.

The main factors determining varietal composition of coal are the following: coal destructibility, destruction scheme parameters and cutting conditions, the methods of mining face processing, working body design.

With the increase the share of narrowly gripping coal-mining machines in the volume of coal production and their growing power available in Kuzbass and other coal basins we observe the tendency to increase the volumes of degraded in size mined coal (Fig. 1).

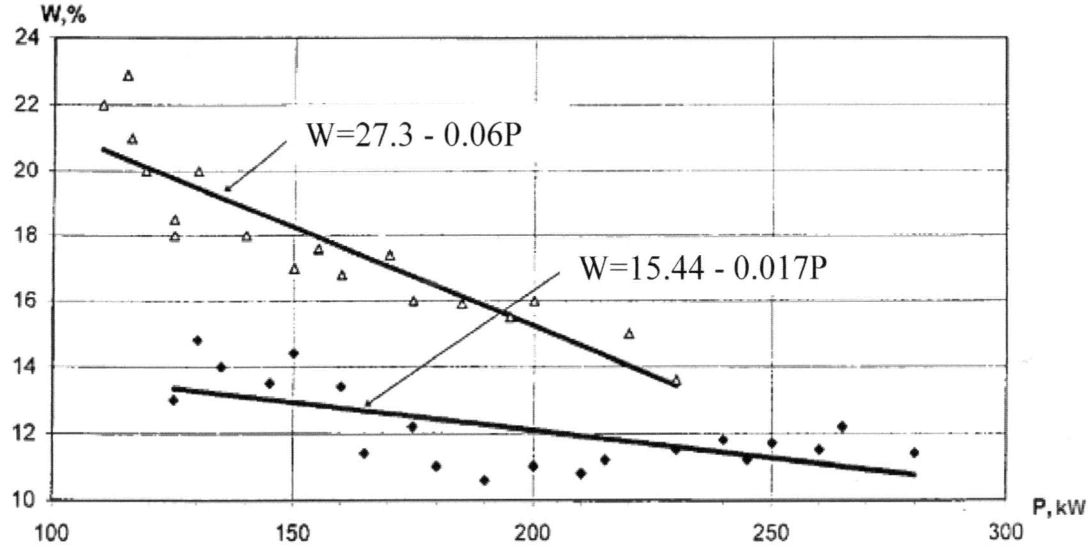

Figure 1 – The dependence between output of the coal of large and middle fractions (W, %) and coal-mining machines' power capacity (P, kW): Δ - for coal company "Kuzbass Coal Inc."; ▲- for the whole Kuzbass

Fig. 1 shows the variation of the dependence between the output and large-medium grades coal and power capacity of coal-mining machinery.

Analysis of these dependences shows a systematic decrease in the output of large and middle classes coal, an increase in the output of coal fines, and the relationship of these indicators with power capacity of coal-mining machines. Naturally, overgrinding of mined coal cannot positively affect the economic indicators of coal mines.

It is known that high requirements regarding varietal composition of coal are presented mainly to energy coals.

Their excessive grinding together with the increasing complexity of dividing the coal into classes (sorting) and transportation, is accompanied by a significant decline in selling prices. For example, a change in the actual wholesale price of energy coal, taking into account its quality, reaches up to 180%, i.e. selling price of coal, depending on its sort, changes in 1.5-3 times.

The main part of coal (about 60%) produced in Kuzbass comes to coking.

In recent years the amount of coal production in Kuzbass having been constantly increasing but the share of coal upgrading is declining, against the raise of output of small and middle coal fractions.

Crushing ability for coal from various seams is quite different.

Therefore, even using the same type of coal-mining machines with established modes of operation can lead to varietal composition of mined coal. It was proved by the sieve analyses of samples taken on "Leninskugol Inc." enterprise directly in the mining faces of a number of mines. The culm output (0-6 mm grade) was from 32 to 60% and the output of coal with class less than 13 mm did not exceed 40-50%.

4. Conclusion

The foregoing leads to the conclusion that the reduction in degradation in size of mined coal is of great economic importance for the coal industry, not only in Russia but also in other coal producing countries.

Coal overgrinding during cutting does not only reduce its commercial value, but also provides a significant dust generation, which is the starting point of high dust emissions of operating coal-mining machines in working faces.

The high dust content complicates the work of the staff, increases occurrence rate of industrial injuries and reduces productivity of mining faces. For example, in mechanized faces with the largest dust emissions it is almost impossible to perform working operations on the outgoing air stream.

The analyzed data testify that the work of modern coal-mining machines in a significant number of mining faces is characterized by a low grade of produced coal and very large dust emission. It has a great importance for environment-friendly use of natural resources, which will be the key point of interest in coming years [25].

5. Acknowledgement

The research was made in T.F. Gorbachev Kuzbass State Technical University under State Assignment N10.782.2014 K "Development of high-performance process for complex processing of low-grade coals and waste coal with producing a low ash coal-and-oil concentrate, composite fuels, rare earth and trace elements".

References

[1] Khoreshok A A, Zhironkin S A and Tyulenev M A et al. 2016 Innovative technics of managing engineers' global competencies IOP Conference Series: Materials Science and Engineering 142 (1) 012122

[2] Tyulenev M A, Lesin Yu, Vik S and Zhironkin S 2016 Methodological Bases of Advanced Geo-ecological Problems Resolving in Neo-industrial Clusters Proceedings of the 8th Russian-Chinese Symposium "Coal in the 21st Century" pp 333-336

[3] Ryzhkov YA, Gogolin VA and Karpenko NV 1992 Modelling the structure of solid masses of lump and granular materials (plane problem). Journal of Mining Science 28(1) pp 6-12

[4] Khoreshok A A, Buyankin P V, Vorobiev A V, Dronov A A 2016 Simulation of Stress-Strain State of Shovel Rotary Support Kingpin IOP Conference Series: Materials Science and Engineering 127 012014

[5] Ryzhkov YA, Lesin YV, Gogolin VA, Karpenko NV 1996 Modeling the structure of fragmented and granular material: Three-dimensional problem. Journal of Mining Science 32(3) pp 188-191

[6] Tyulenev M, Zhironkin S, Kolotov K and Garina E 2016 Background of innovative platform for substitution of quarry water purifying technology Pollution Research 35 (2) pp 221-226

[7] Tyulenev M A, Gvozdkova T N and Zhironkin S A et al. 2016 Justification of Open Pit Mining Technology for Flat Coal Strata Processing in Relation to the Stratigraphic Positioning Rate Geotechnical and Geological Engineering 34 (6) doi:10.1007/s10706-016-0098-3

[8] Tyulenev M A and Lesin Y V 2014 Justification complex purification technology open-pit mines wastewater Symposium of the Taishan academic forum – Project on mine disaster prevention and control pp 441-444

[9] Tyulenev M, Zhironkin S and Litvin O 2015 The low-cost technology of quarry water purifying using the artificial filters of overburden rock Pollution Research 34 (4) pp 825-830

[10] Lesin Y V, Luk'yanova S Y and Tyulenev MA 2015 Formation of the composition and proper-

ties of dumps on the open-pit mines of Kuzbass IOP Conference Series: Materials Science and Engineering 91 (1) 012093

[11] Tyulenev M A, Zhironkin S A, Garina E A 2016 The method of coal losses reducing at mining by shovels International Journal of Mining and Mineral Engineering 7 (4) DOI: 10.1504/IJMME.2016.10000781

[12] Andreev G E 1991 A review of the Brazilian test for rock tensile strength determination. Part I: calculation formula Mining Science and Technology 12 pp 445-456

[13] Peide S 1991 Geothermal prediction of country rock in deep mines Mining Science and Technology 10 pp 433-438

[14] Shixiong Zhang and Guangxu Tong 1991 Influence of block boundary weakening on the caving process Mining Science and Technology 10 pp 157–166

[15] Yuejin Li and Singh R N 1991 An overview of condition monitoring and an expert system for longwall mining machinery Mining Science and Technology 12 pp 279-290

[16] Nazarova L A, Freidin A M and Neverov A A 2005 Chamber Mining with Roof Caving at the Nikolaevsk Mine Journal of Mining Science 41(4) pp 342–349

[17] Neverov S A 2012 Types of the Orebodies on the Basis of the Occurrence Depth and the Stress State. Part II: Orebody Tectonotypes and Geomedium Models Journal of Mining Science 48(3) pp 421–428

[18] Neverov S A and Neverov A A 2013 Geomechanical Assessment of Ore Drawpoint Stability in Mining with Caving Journal of Mining Science 49 (2) pp 265–272

[19] Zhurilo A G, Zhurilo D Yu and. Rudnev A V 2015 Brief review of the development of cutting tools Ferrous metals 1 pp78-86

[20] Tanaino A S and Botvinnik A A 1999 Three-dimensional solution of mining-geometric problems in a graphic dialog regime in planning opencasts Journal of Mining Science 35 (6) pp 640-653

[21] Cushtar C, Matti M and Veron S. 2011 Industrial coal demand in China: a provincial analysis J. Resource and Energy Economics 33 (1) pp12–35

[22] Demirel N 2011 Effects of the rock mass parameters on the dragline excavation performance Journal of Mining Science 47 (4) pp 441-449

[23] Oparin V N, Cheskidov V I, Bobyl'sky A S and Reznik A V 2012 The sound subsoil management in surface coal mining in terms of the Kansk-Achinsk coal basin Journal of Mining Science 48 (3) pp 585-594

[24] Mattis A R, Cheskidov V I. and Llabutin V N 2012 Choice of the hard rock surface mining machinery in Russia Journal of Mining Science 48 (2) pp 329-338

[25] Sherer D and Attig T 1983 Ethics and the Environment (New Jersey: Prentice-Hall, Inc.) 554 p

Rational Use of Land Resource During the Implementation of Transportless System of Coal Strata Surface Mining

T Gvozdkova[1, a], M Tyulenev[2, b], S Zhironkin[3, c], V A Trifonov[4, d], Yu M Osipov[4, f]

[1] Mezhdurechensk Branch Kuzbass State Technical University,
Stroitelei av. 36, Mezhdurechensk, Russian Federation, 652881
[2] T.F. Gorbachev Kuzbass State Technical University
Vesennyaya st., 28, Kemerovo, Russian Federation, 650000
[3] National Research Tomsk Polytechnic University
Lenin Av. 30, Tomsk, Russian Federation, 634034
[4] Yurga Technological Institute (branch) of National Research Tomsk Polytechnic University Leningradskaya st. 26, Yurga, Russian Federation, 652055

E-mail: [a] kuzstu@inbox.ru, [b] tma.geolog@kuzstu.ru, [c] zhironkin@inbox.ru, [d]v.trifonof@rambler.ru, [f]umo1943@yandex.ru

Abstract. Surface mining and open pits engineering affect the environment in a very negative way. Among other pollutions that open pits make during mineral deposits exploiting, particular problem is the landscape changing. Along with converting the land into pits, surface mining is connected with pilling dumps that occupy large ground. The article describes an analysis of transportless methods of several coal seams strata surface mining, applied for open pits of South Kuzbass coal enterprises (Western Siberia, Russia). To improve land-use management of open pit mining enterprises, the characteristics of transportless technological schemes for several coal seams strata surface mining are highlighted and observed. These characteristics help to systematize transportless open mining technologies using common criteria that characterize structure of the bottom part of a strata and internal dumping schemes. The schemes of transportless systems of coal strata surface mining implemented in South Kuzbass are given.

1. Introduction

For the last decade coal mining in Kuzbass increase by 6-10 million tons annually. As far as the main part of Kuzbass' coal in mined by open pit enterprises (65%), the increase of coal production causes landscape destruction by 35-40 square kilometers every year [1-4]. The transportless system of surface mining may significantly decrease the earth consumption but requires more complicated technological schemes of coal deposit exploiting [5-9]. Special difficulties appear when transportless mining system is implemented for the strata of several coal seams development [10-14]. Currently transportless methods allow efficient developing the several flat or inclined coal seams strata with a total capacity of 50 m or more, along with forming three-layers dumps of overburden rock. Often there is an excess of transportless quarry face over transport one that leads to leads to a reduction in the volume of rock developing with transportless technology [15-19]. The capacity of internal dump is not fully used as potentially possible. In order to avoid this limitation, it is necessary to increase the volume of overburden rock, piled in internal dump. This requires solving a number of techno-economic issues related to the placing additional volumes of overburden in the internal dumps and to the calculation of economically proved cutback overburden layer height [20-21].

For optimization of parameters of surface mining transportless technology it is necessary to satisfy conditions of mining efficiency target indicators preservation and to decrease the general costs of overburden [22-24]. This article describes transportless technological schemes of surface mining and generalizes their common characteristics [25-26]. Accordingly the method of overburden distribution (with preparation of additional capacity in earlier piled dump layer and without preparation), two types of transportless technology are analyzed.

2. Material and Methods

In Tomusinsky and Mrassky geological and economic areas of Kuzbass (Western Siberia, Russia) coal deposits are characterized by succession bedding of coal seams in the rocks (Fig. 1).

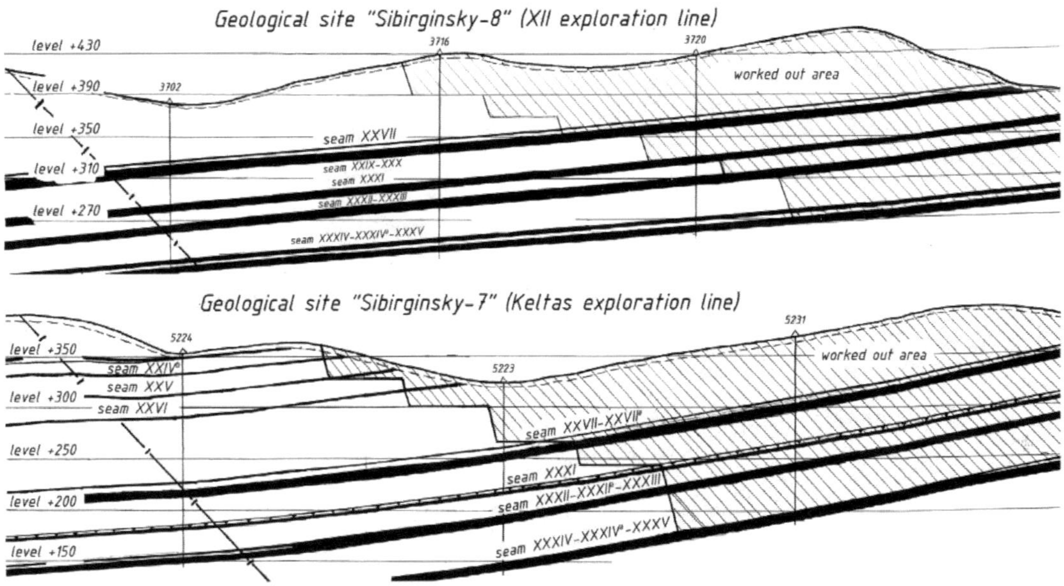

Figure 1 – Typical geological sections on exploration lines for "Krasnogorsky" open pit.

The coal fields are exploited by modern large open pits: "Krasnogorsky", "Sibirginsky", "Mezhdurechensky", "Olzherassky". The overburden rocks are siltstone and mudstone [Tyulenev & Lesin, 2014; Lesin *et al.*, 2015] with Protodyakonov scale of hardness of 5-7, as well as large-sandstone with hardness coefficient (*f*-rate) of 8-10.

The exploiting of open pit fields is made by mixed deepening-and-continuous longitudinal semi-locating mining method. Accordingly this, the upper part, up to 30-70 m, with one or two seams is developing by transport technology with carrying of rocks to the outer and inner dump; the lower part, with a total capacity of rock strata 49-52 m, and including one or two coal seams, is developed by transportless technology with transshipment of rocks into the internal dump. Analysis of the parameters of coal seams bedding in strata showed that according to the conditions of seams bedding in the strata the structures of geological sites can be divided in terms of τ, which characterizes the type of structure of the lower part of strata to a height of two seams and their interbed.

$$\tau = H_2/H_1, \tag{1}$$

where: τ – the rate characterizing location of the middle seam in a strata; H_1 – the height of the lower interbed, m; H_2 – the height of the upper interbed (between the second and the third seams), m.

We defined the average τ rate for "Krasnogorskiy" open pit the following way:

$$\tau = 13/33.6 = 0.38 \tag{2}$$

for "Mezhdurechenskiy" open pit (as shown on Fig. 2):

$$\tau = 30/25 = 1.2 \tag{3}$$

On Fig. 2 *Ws* is a stope width.

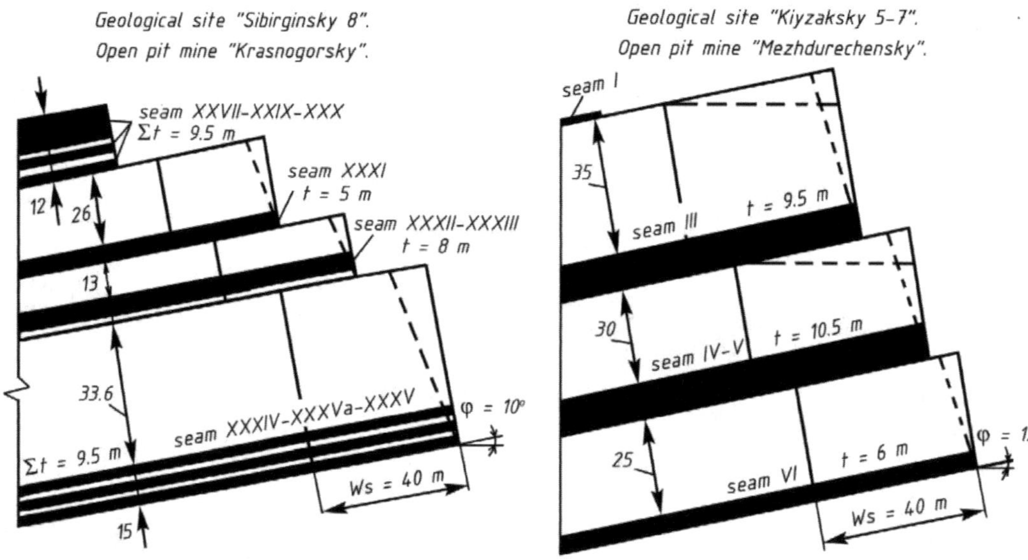

Figure 2 – Average parameters of coal seams bedding on geological sites of "Krasnogorsky" and "Mezhdurechensky" open pits

The introduction of this indicator lets us consider rock strata from the soil of the upper (third) bank to the roof of the lower seam as the mountain massif of the total thickness of 49-52 meters, divided by a median seam on interbeds of the same or different capacities. Based on this, we can exercise a single methodological approach to the analysis of the development of all geological sites.

Describing the bedding conditions of seams in geological areas as a whole, the significant differences between them, as indicated by the index τ, should be noted.

The schemes of excavation of flat coal seams strata for the development by transportless technology have a number of common features, which we defined having analyzed some papers. These features include: the number of seams and interbeds exploiting by transportless technology; the method of strata of two seams exploiting; the number of piled layers of inner dump; allocation scheme of coming into the dump overburden rock by piled layers. The equipment used and its parameters are not applied to the generalizing features because in each case its choice is based solely on the parameters of the seams bedding [27-32]. In some researches in surface mining technological schemes and methods there were several characteristics of coal strata development described. They include number of exploiting seams and interbeds, simultaneous or consequent exploiting, number of piled layers of internal (and, in some cases, external dump).

Considering all characteristics of surface mining technological schemes and methods for inclined coal strata development, we defined special features for transportless system for the conditions of

South Kuzbass geological area described in Fig. 1 and Fig. 2. These special features of transportless system include the following:

1. The priority number of exploiting seams and interbeds is two. One seam and one interbed are developed on two sites (administrative section №4 on the geological site "Sibirginsky 8" and the geological site "Kiyzaksky 5-7"). In the geological section "Sibirginsky 1-3" by transportless technology strata of two seams and two interbeds is developed, but the top high interbed (42-49 m) is developed by one bench by transportless technology, and the bottom, with low height (4-7 m) by transport.

2. To develop the strata of two seams two ways are used: with the simultaneous (parallel) development of two interbeds and two layers; with a sequence in descending order working of interbeds and layers. The simultaneous method is used in those areas where the rate $\tau = 0.3 \div 1$; consequent – when $\tau > 1$.

3. The number of piled layers of internal dump basically equals three, which is determined by the height of working out rock strata (49-52 m), placed in a dumping site by using draglines with boom length of 85-90 m. In the development of rock strata with less height a two-layer dump is piled.

In exceptional cases, four-layer dumps are piled. It's possible in mining rock strata with height of more than 52 m in a specific profile of dump layer [33]. In addition, in piling of three-layer dumps in case of exceeding the volume of the rock their stacking capacity the preparations for the additional capacity for the third layer are made [34]. Removed rock in this case is placed in the fourth layer.

The distribution of entering to the dump rock in dump layers is carried out by two known schemes.

The first scheme. All incoming to the dump rock is located in the first and second layers, and to increase the capacity of the second layer in the previously piled dump the additional content for the second layer is prepared, and extracted rock is piled in the third layer. This scheme is applied at the simultaneous development of two seams and two interbeds.

The second scheme. All incoming to the dump rock is distributed in three layers without preparing any additional content. The scheme is applied to both simultaneous and consequent ways: with the first and second strokes of dragline extracted rock is being piled in the first and second layers of the dump, being intermediate bulk which requires recasting [35]. With special stroke dragline, being set at the top of the second layer moves the intermediate bulk to the third layer.

The described schemes differ in dragline's movement along the front-line of work in the development of rock banks.

3. Results and Discussion

We generalized and organized transportless systems of several coal seams strata open pit mining in terms of τ rate and proposed the dump schemes of piled layers (Fig. 3, Fig. 4).

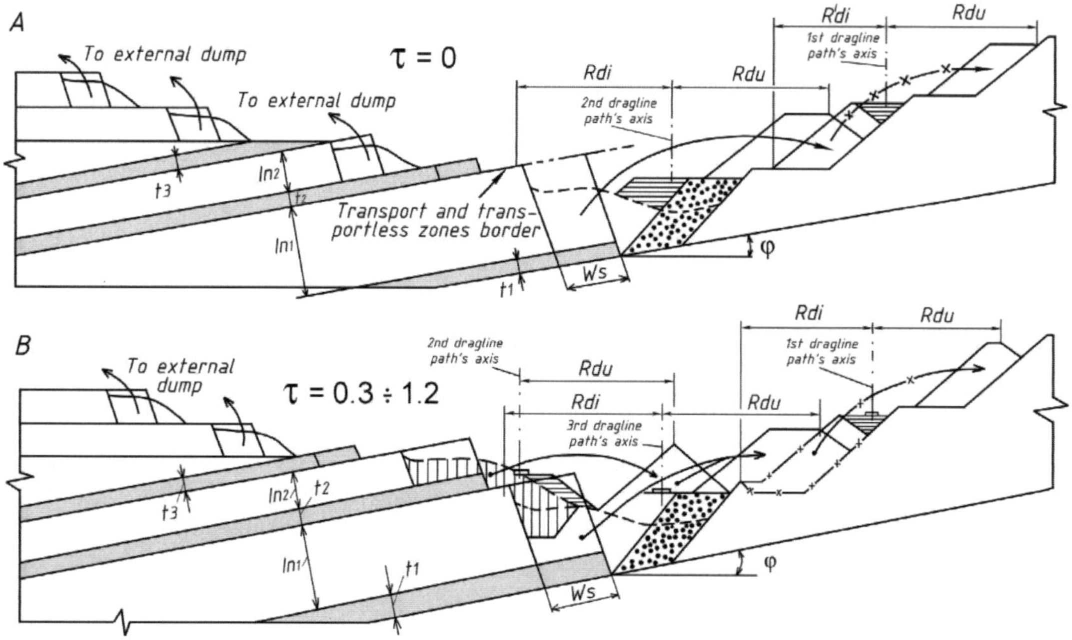

Figure 3 – Transportless schemes of excavation for conditions of South Kuzbass open pits fields A – exploiting of one interbed (τ = 0); B – simultaneous exploiting of two interbeds (τ = 0.3-1.2)

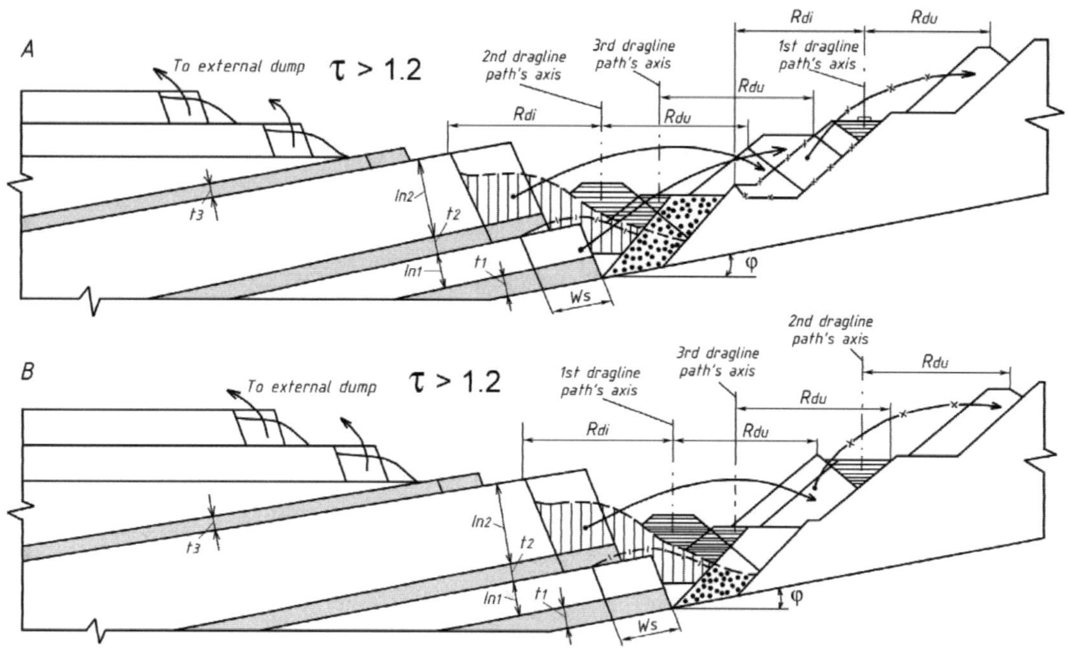

Figure 4 – Transportless schemes of excavation for conditions of South Kuzbass open pits fields with consequential exploiting of two interbeds (τ > 1.2): A – the scheme with preparing of primary capacity in the third layer; B – the scheme without preparations

On these figures: t_1, t_2, t_3 – thickness of 1st, 2nd and 3rd seams respectively, m; In_1, In_2 – thickness of 1st (lower) and 2nd (upper) interbeds, m; *Ws* – stope width, m; *Rdi* – dragline's digging radius, m; *Rdu* – dragline's dumping radius, m; φ – bedding angle, degrees.

All four technological schemes of transportless system of coal strata surface mining were implemented at different districts of "Krasnogorsky" and "Mezhdurechensky" open pits. The result of pilling internal pumps in the mined-out space of these quarries was saving 3 km^2 of the land suitable for agriculture in 2012, 4 km^2 in 2013 and 3 in 2014 (assumed for all six open pits of South Kuzbass).

Systematization of transportless technological schemes was made using τ rate which characterizes location of the middle seam in strata of three flat or inclined seams. Introduction of this indicator allows considering the strata as a solid array with a general thickness of 49-52 m (for conditions of Southern Kuzbass). On this basis we developed common methodical approach to analysis of development of geological segments with similar conditions of bedding of coal seams. It was defined that at values of τ rate less than 1.5 it is effectually to use consequent exploiting system and with τ rate from 1.5 to 3 – simultaneous exploiting system. If τ rate is more than 3 than both exploiting system are effective. Specificity of defining the most preferable transportless scheme of excavation and dumping permits mining enterprises to safe large land resources transferring maximum volumes of overburden in internal dumps [36]. In general high perspectives of transportless system implementation at surface mining are expected to expand the limits of using mining machines [37-39].

4. Conclusions

Analysis of transportless technological schemes used in existing open pits showed that basically two inclined seams are developed, more rarely one seam. The generalization of transportless technological schemes is made using some selected features: the order of overburden bench at the coal-face side (simultaneously or sequentially), the scheme of piling of internal three-layer dump (the preparation of additional container in the second layer or without preparation). As a result four currently used typical schemes of excavation and internal dumping were emphasized.

5. Acknowledgment

The authors would like to thank to prof. Alexey Khoreshok, prof. Yury Lesin for their discussions during the study. We thank anonymous referees for constructive and useful comments on the papers sent via e-mail.

References

[1] Tyulenev M A, Gvozdkova T N and Zhironkin S A et al. 2016 Justification of Open Pit Mining Technology for Flat Coal Strata Processing in Relation to the Stratigraphic Positioning Rate Geotechnical and Geological Engineering 34 (6) doi:10.1007/s10706-016-0098-3

[2] Tyulenev M A, Zhironkin S A, Garina E A 2016 The method of coal losses reducing at mining by shovels International Journal of Mining and Mineral Engineering 7 (4) DOI: 10.1504/IJMME.2016.10000781

[3] Lesin Y V, Luk'yanova S Y and Tyulenev M A 2015 Formation of the composition and properties of dumps on the open-pit mines of Kuzbass IOP Conference Series: Materials Science and Engineering 91 (1) 012093

[4] Tyulenev M A, Khoreshok A A, Garina E A, Danilov S and Zhironkin S 2016 Adaptive technology of using backhoes for full coal extraction Proceedings of the 8th Russian-Chinese Symposium "Coal in the 21st Century: Mining, Processing, Safety" pp 111-115.

[5] Lokhanov B N, Zakharov Yu A, Bereznyak M M and Kalinin A V 1967 Open-cut mines in the Kuzbass: Progress and prospects Soviet Mining 3 (5) pp 523-527

[6] Bereznyak M M, Kalinin A V and Pronoza V G 1970 A method of calculating the width of the caved rubble in the transportless system of working a series of sloping beds Soviet Mining 6 (6) pp 638-643

[7] Matushenko V M 1975 A method of comparing excavation equipment Soviet Mining 11 (5) pp 576-578
[8] Kuznetsov V I, Mattis A R, Tashkinov A S, Vasil'ev E I and Zaytsev G D 1997 Efficiency of excavation of overburden rock at quarries with the use of blast-free technology Journal of Mining Science 33 (5) pp 471-477
[9] Oparin V N, Cheskidov V I, Bobyl'sky A S and Reznik A V 2012 The sound subsoil management in surface coal mining in terms of the Kansk-Achinsk coal basin Journal of Mining Science 48 (3) pp 585-594
[10] Molotilov S G, Cheskidov V I, Norri V K and Botvinnik A A 2009 Methodical principles for planning the mining and loading equipment capacity for open cast mining with the use of dumpers. Part II: Engineering capacity calculation Journal of Mining Science 45 (1) pp 43-58
[11] Scott B, Ranjith P G, Choi S K and Manoj K 2010 A review on existing opencast coal mining methods within Australia Journal of Mining Science 46 (3) pp 280-297
[12] Demirel N 2011 Effects of the rock mass parameters on the dragline excavation performance Journal of Mining Science 47 (4) pp 441-449
[13] Hummel M 2012 Comparison of existing open coal mining methods in some countries over the world and in Europe Journal of Mining Science 48 (1) pp 146-153
[14] Prakash A, Mallika V, Murthy S R and Singh K B 2013 Rock excavation using surface miners: An overview of some design and operational aspects International Journal of Mining Science and Technology 23 (1) pp 33-40
[15] Tanaino A S and Botvinnik A A 1999 Three-dimensional solution of mining-geometric problems in a graphic dialog regime in planning opencasts Journal of Mining Science 35 (6) pp 640-653
[16] Tanaino A S and Cheskidov V I 1999 Substantiation of the sequence for opencast mining of a series of flat and inclined strata using the mined-out space for internal dumps Journal of Mining Science 35 (3) pp 304-313
[17] Nazarov I V 2011 Numerical modeling of overburden rehandling with draglines Journal of Mining Science 48 (1) pp 55-61
[18] Cushtar C, Matti M and Veron S. 2011 Industrial coal demand in China: a provincial analysis J. Resource and Energy Economics 33 (1) pp12–35
[19] Mattis A R, Cheskidov V I and Llabutin V N 2012 Choice of the hard rock surface mining machinery in Russia Journal of Mining Science 48 (2) pp 329-338
[20] Cheskidov V I, Norri V K and Sakantsev G G 2014 Diversification of open pit coal mining with draglining Journal of Mining Science 50 (4) pp 690-695
[21] Vukotic I, Kecojevic V, Zhang W, Cai Q and Chen S 2013 Optimization of transport passage with dragline system in thick overburden open pit mine International Journal of Mining Science and Technology 23 (6) pp 901-906
[22] Levenson S Ya and Gendlina L I 2014 Safe dumping equipment Journal of Mining Science 50(5) pp 938-942
[23] Lintukangas M, Suihkonen A, Salomäki P and Selonen O 2012 Post-mining solutions for natural stone quarries Journal of Mining Science 48(1) pp 123-134
[24] Tyulenev M, Zhironkin S, Kolotov K and Garina E 2016 Background of innovative platform for substitution of quarry water purifying technology Pollution Research 35(2) 221-226
[25] Tyulenev M, Zhironkin S and Litvin O 2015 The low-cost technology of quarry water purifying using the artificial filters of overburden rock Pollution Research 34 (4) pp 825-830
[26] Zhironkin S A, Khoreshok A A, Tyulenev M A et al. 2016 Economic and Technological Role of Kuzbass Industry in the Implementation of National Energy Strategy of Russian Federation IOP Conference Series: Materials Science and Engineering 142 (1) 012127
[27] Khoreshok A A 2002 On side cutting bit when operating at sheerer drums Ugol' 7 pp 10-11

[28] Lekontsev Yu M, Sazhin P V, Temiryaeva O A, Khoreshok A A and Ushakov S Yu 2013 Two-side sealer operation Journal of Mining Science 49(5) pp 757-762
[29] Aksenov V V, Khoreshok A A and Beglyakov V Y 2013 Justification of creation of an external propulsor for multipurpose shield-type heading machine - GEO-WALKER Applied Mechanics and Materials 379 pp 20-23
[30] Kovalev V A, Gerike B L, Khoreshok A A and Gerike P B 2014 Preventive maintenance of mining equipment based on identification of its actual technical state Symposium of the Taishan academic forum – Project on mine disaster prevention and control pp 184-189
[31] Ryzhkov Y A, Gogolin V A, and Karpenko N V 1992 Modelling the structure of solid masses of lump and granular materials (plane problem). Journal of Mining Science 28(1) pp 6-12
[32] Ryzhkov Y A, Lesin Y V, Gogolin V A, Karpenko N V 1996 Modeling the structure of fragmented and granular material: Three-dimensional problem. Journal of Mining Science 32(3) pp 188-191
[33] Khoreshok A A, Buyankin P V, Vorobiev A V, Dronov A A 2016 Simulation of Stress-Strain State of Shovel Rotary Support Kingpin IOP Conference Series: Materials Science and Engineering 127 012014
[34] Tyulenev M A, Lesin Yu, Vik S and Zhironkin S 2016 Methodological Bases of Advanced Geo-ecological Problems Resolving in Neo-industrial Clusters Proceedings of the 8th Russian-Chinese Symposium "Coal in the 21st Century" pp 333-336
[35] Tyulenev M A and Lesin Y V 2014 Justification complex purification technology open-pit mines wastewater Taishan Academic Forum – Project on Mine Disaster Prevention and Control pp 441-444
[36] Khoreshok A, Tyulenev M and Vöth S 2016 Conditions for Minimum Dynamic Loading of Multi-brake Hoists Proceedings of the 8th Russian-Chinese Symposium "Coal in the 21st Century" pp 239-245
[37] Khoreshok A A, Zhironkin S A and Tyulenev M A et al. 2016 Innovative technics of managing engineers' global competencies IOP Conference Series: Materials Science and Engineering 142 (1) 012122
[38] Efremenkov A B 2011 Forming the subterranean space by means of a new tool (geohod) Proceedings of the 6th International Forum on Strategic Technology IFOST 6021037
[39] Efremenkov A B and Timofeev V Y 2012 Determination of necessary forces for geohod movement Proceedings - 2012 7th International Forum on Strategic Technology IFOST 6357729
[40] Efremenkov A B and Aksenov V V et al. 2012 Force parameters of geohod transmission with hydraulic drive in various movement phases Proceedings - 2012 7th International Forum on Strategic Technology IFOST 6357716

NBIC-Convergence of Machinery and Basic Technologies as the Ecological Factor of Wellbeing

S A Zhironkin[1, a], K A Kolotov[2, b], A E Genin [3, c], F V Agafonov[4, d], S A Kovalevsky[5, e]

[1] Yurga Technological Institute (branch) of National Research Tomsk Polytechnic University, Leningradskaya 26, Yurga, Russian Federation, 652055
[2] National Research Tomsk Polytechnic University Lenina av. 30, Tomsk, Russian Federation, 634034
[3] Branch of JSC "SO EES" Siberian ODU Kuzbasskaya st., 29, Kemerovo, Russian Federation, 650000
[4] Kemerovo State Agricultural Institute Markovzeva st. 5, Kemerovo, Russian Federation, 650056
[5] T.F. Gorbachev Kuzbass State Technical University, Vesennyaya st. 28, Kemerovo, Russian Federation, 650000

E-mail: [a] zhironkin@inbox.ru, [b]594950@bk.ru, [c]geninanton@gmail.com, [d]kemshi-ise@mail.ru, [e]koval71@mail.ru

Abstract. The development of science and technology in the XXI century is on the way of the extensive use of natural resources. As a result, the scientific and technical progress does not initiate the reduction of environmental damage but encourages its cumulative growth. So the environmental problems have become a serious threat to social wellbeing of all mankind. The development of the core technologies doesn't allow creating the situation in which the new knowledge would generate positive changes in the environment. Therefore, among the issues that define the future of advanced environmental technologies, and potentially significant for the understanding of the humanities, there is the phenomenon of technological convergence and, in particular, the predicted convergence of the nano-, bio-, information and cognitive technologies (NBIC). The consequence of NBIC-convergence may be the transformation of the humanity into a single global mind, which can be linked into a solid set of technologies, establishing united standards for social wellbeing and environmental technologies.

1. Introduction

Earlier stages of technological convergence, starting from combining scientific knowledge with the technical activities at the beginning of the industrial revolution, and ending with the most important technological convergence of the XX century - the emergence of information and communication networks - have created miracle expectations in the minds of scientists, a breakthrough that has incalculable social consequences. However, despite the widespread diffusion of the new technologies in human life and a significant increase of social wellbeing in the 20th century, its eco-factor is still determined by the boundaries of the application of the basic technologies. In other words, technological convergence hasn't allowed overcoming the limitations of nature yet other than reinforcing a devastating effect on the environment by the industry - the main achievement of modern civilization. Under such conditions, social wellbeing, raised by industrial technologies on previously unattainable level, cannot be truly continuous, since it depends on the balance of nature-destructive and environment-saving technologies. As until now this balance is disturbed in favor of irreparable damage to the environment, the social wellbeing, even during the human's life, cannot be continuous. One of the possible ways out of the "vicious circle" of technological, environmental and social problems is the development of NBIC-convergence technologies.

2. Material and Method

Many scientists say that nature is "wrong" too often, so we could not unconditionally rely on it [1-10]. Features of this approach can be found in technological convergence studies. For example, one of the reports of the US National Science Foundation was named "Converged technologies for improving human performance" [11]. Its authors suggested that the use of appropriate technologies will contribute to the world peace and the development of fruitful cooperation between people and machines, which will result in the reduction of environmental damage brining human wellbeing growth to a minimum.

Nevertheless, the impact of technological development on the continuous wellbeing of mankind is positively evaluated by not all scientists. So, after a successful start of the process of convergence, when Richard Feynman [12] as early as in 1959 described possibility of manipulation of matter at the molecular level, and Swiss researchers Heinrich Rohrer and Gerd Binnig created a scanning tunneling microscope that could show individual atoms and move them [13], a number of authors stated the danger that the new technology could bring to the world. For example, Eric Drexler warned of the apocalyptic scenario of progress in nano-technology, thanks to which humans can lose ecological niche [14], and R. Smalley, John. Whitesides, R. Kurzweil initiated the discussion about the prospects of the impact of new molecular electronics, nanobiology, algorithmization of the human mind on ecology of the human environment [15].

However, the significant number of scientists, mainly in the fields of humanitarian knowledge was the proponents of the use of new technologies for the growth of social wellbeing.

A number of researchers analyzing wellbeing emphasize the possibilities of new technologies in provision of individuals with wealth and maintenance of social and economic stability of society [16-19]. Other authors give priority to the perception by the individual, community, and even a certain nation of objective living conditions [20-25].

The problem of continuous social wellbeing was closely studied by the authors who observe it as a set of external conditions, economic opportunity, technological development and professional skills making it possible to receive certain benefits over the lifetime period [26-29]. Some researches connect continuous wellbeing with the "lifetime employment" in the high-tech Japanese corporations [30], others – with the social paternalism in the socialist countriesё as an attempt to approve the continuous wellbeing [31].

Despite the importance of the wellbeing achieved with the basic technologies, we believe it to be "non-continuous", as it lasts as long as its two main conditions are fulfilled.

The first condition of «non-continuous» social wellbeing is to achieve a certain level of environment preservation in general.

The second condition is the existence of an individual satisfactory state of physical and mental health, sufficient for successful business or labor. And it also depends upon the environmental problems.

Therefore, as soon as one of these conditions is violated, mainly due to the environment harmful technologies, «non-continuous» wellbeing can be lost. At the same time, millions of people constantly face the choice of working in harmful and stressful environment. Accordingly, there is no unity in the research of social scientists devoted to the dilemma between using basic technologies to improve material wellness and keeping nature green (they both, as it was mentioned above, are the basic of continuous wellbeing).

Thus, the proponents of "welfarism" point out that the pursuit of maximum material wealth has some kind of affectation, under the influence of which a person for some time can forget his desire for continuous wellbeing and neglect some threats for his health or put up with severe stress [32]. Using harmful technologies for achieving momentary happiness from the growth of material wellness is a sign of «non-continuous» wellbeing in case people have been deliberately neglecting their health preservation for many years. This can be illustrated by the following examples [33]. People often think that a new car will make them happy, but after two weeks its utility growth is reduced. They think that the life and work in bad ecological conditions give them some extra money and bring them happiness, but it turns out that feeling unwell and destroyed health have much more intensive and long-term negative impact on their wellbeing.

In turn, the supporters of the liberal model of wellbeing assessment offer to take into account people' own preferences in the analysis of various factors of social wellbeing [34]. This depreciates the role of ecological factors in achieving continuous wellbeing and increases the role of other factors, mainly, the profit from selling raw materials and energy (for example, the possibility of getting natural resource rent). This also can explain fairly high stress level, various diseases among the workers of mines, quarries, chemical plants etc.

At the same time, a gradual change in attitude towards ecological factor as a key point of continuous wellbeing illustrates «Easterlin's paradox» [35]. It consists in the fact that in the developed countries the average level of wellbeing, subjectively evaluated by a large number of respondents, is relatively insensitive to changes in income. That is, people are increasingly aware that they tend to material achievements that have little impact on social wellbeing, in particular continuous.

However, in an effort to bring their wellbeing in line with modern ecological standards, people increasingly appreciate an opportunity to live in ecologically favorable environment and to take effective control over their health.

The lifetime health preservation problem makes people look widely at the essence of ecology and nature-friendly technologies. In addition to air and water pollution reduction, avoidance of harmful for environment technologies, the risk of technogenic accidents, continuous wellbeing involves reducing the risk of chronic and fatal diseases leading to premature mortality. It can be achieved with NBIC-converged technologies.

3. Results and discussion

Taking into account the opinion of above mentioned authors, we consider ecological factor of social wellbeing as a condition of human existence, which is stipulated by actual performance of new nature-saving technologies. Long-term, not limited by old technologies, stay in such a space we consider to be continuous wellbeing.

It sharpens the problem of "embedding" the modern ecological technologies associated with resource saving, water, air and soil purifying into the continuous wellbeing model. We note that the feature of present stage of technological changes is convergence of basic technologies, which is called NBIC-convergence [11]. NBIC is an acronym for the following types of technologies: N -nano; B - bio; I -info; C - cogno. With regard to the continuous wellbeing, it means the connection of ecological, chemical and economic (marketing, investment) technologies within a new industry – nature purifying.

A key role in the ecological basis of continuous wellbeing we give to chemistry and biotechnologies, which, together with nano-technology, are one of the most important and the fastest "horizontal" technological areas. The intensive development of biotechnologies is stipulated by not only the success of chemistry and molecular biology, but also the crisis of traditional technologies (especially against the background of new trends in medicine and ecology), the need to make purifying facilities, to preserve natural resources, to reduce emissions of pollutants. The presence of serious scientific groundwork and experimental development in the field of continuous wellbeing convergent technologies will make it possible in the coming years, to develop affordable purifying systems [36-38], to expand significantly the scope of preventive harmful influence on the environment [39-43], to involve business firms and governments to ecological projects investment. Except this direct effect on the process of achieving the continuous wellbeing, the convergence of NBIC technologies will also provide the achievement of a number of goals, indirectly determining the conditions for continuous human wellbeing:

- counter the spread of various types of contamination [44-45];
- producing chemicals for broad application in purifying facilities and in producing renewable materials intended to replace existing primary products;
- conversion into a routine advance development of nature-saving and purifying convergent technologies.

Distribution of NBIC-convergence in the field of wellbeing changes the question about the value of the development and implementation of innovative chemical and biochemical technologies into purifying equipment. If referring to «non-continuous» wellbeing new material things and services have individual value, than when applied to the continuous wellbeing they are valuable for the whole society. Therefore, proclamation of continuous wellbeing as the imperative of ecological development of modern society will undoubtedly require a significant price reduction of nature-saving equipment and technologies. And this price reduction can be achieved only with transferring NBIC innovations into large-scaled industrial producing.

The current development of purifying and nature-saving equipment market is characterized by the following features:

1. The increased volume of this production for ecological purposes (as for the last decade the annual increase of output was up to 10%).
2. An increasing number of customers – large industrial firms and corporations, mainly in the USA, Western Europe, Japan (more than 250 specialized facilities are annually being installed). In the coming years it is expected to increase their number in 2 times).
3. Higher-than-anticipated growth of consumption of the ecologically clean production.
4. Extension of the areas of purifying facilities applications.
5. Increased requirements to the technologies of industrial production (economic cost-effectiveness and environmental safety).

The global market of nature-saving production for ecological purpose has a strong annual growth tendency up to 15%. This trend is expected to continue during the next 15-20 years. Growth in production of nature-saving products and equipment is provided both by expanding the areas of their application and by the development of innovative technologies in this field greatly reducing the cost of their mass application.

So it is impossible to consider the modern discourse of social wellbeing to be complete due to the absence of a single paradigm that integrates objective factors of its dynamics and the subjective perception of the totality accessible tangible and intangible values by the individual. Up to now philosophers, sociologists, economists, ecologists cannot accurately match human wellbeing with various forms of life - material, human, spiritual and social [46]. Therefore, the social scientific analysis does not allow achieving the required identity of factors, criteria, evaluations of individual and social significance of the various ecological factors. As a consequence, it is difficult to search for consistent and effective ways of improving both the environment and social wellbeing.

This significantly improves the relevance of converged ecological and social wellbeing studies, becoming of its diversified methodology, and opens up the prospect for a new discourse – the continuous wellbeing, based on NBIC-convergence.

Modern methodology of social wellbeing tends to the economic and sociological research. Accordingly, new technological possibilities and innovations in the medicine field, bio-chemistry, and microbiology opening the access to the new values of environment conditions, without taking into account the future opportunities. Within the framework of the existing methodology it is difficult to identify ecological factor of continuous wellbeing that accompanies a life of all society, as during this period the availability of the achievements of progress and their distribution radically expands.

Therefore, the key issue in understanding of the continuous wellbeing nature is: "Can prosperity entirely be measured by tangible achievements over the lifetime period?" This issue has become particularly relevant in the last two decades, when the breakthrough innovations have accelerated the pace of progress but have not helped to reduce the negative impact on the nature. As a result the increased environmental risks leads to asymmetry of social wellbeing in different countries, shifts emphasis from possession of material goods towards individuals' access to nature-saving technology.

Institutional environment of formation of continuous social wellbeing throughout a person's life is also undergoing major changes. Along with protection from the destructive consequences of external shocks associated with the risks of inflation, devaluation, cyclical downturn and unemployment social protection mechanisms generated by the state are increasingly focused on stimulating technological innovation.

The increase of innovation ecological factor of continuous social wellbeing is also stipulated by the cyclical nature of material goods consumption and its high dependence on the institutional maturity of a society. In such circumstances, the main vector of the convergence of innovative technologies determining the continuous wellbeing is nature preservation and provision of longevity in the process of reducing mortality from dangerous diseases caused by harmful environment. The synchronicity of material benefits provision and preservation of health, natural resources in the face of increasing instability in ecology becomes the main measurable instrument of continuous social wellbeing.

4. Conclusion

In summary, we emphasize the important role of the convergence of innovative technologies in the fields of medicine and biochemistry in the formation of the continuous social wellbeing. This continuousness means a lifelong access to both material goods and the latest technologies, preservation of health and prevention of dangerous diseases. Even today, these technologies are becoming the dominant component of the continuous wellbeing, determining its standards in developed countries and directly affecting the consumption of high-tech goods. New diagnostic tools, vaccines and medicinal drugs can be obtained with the help of convergent medical and biotechnologies.

The peculiarity of the development of convergent biomedical technologies in the XXI century is not only in their rapid growth in the applied aspect, but also rapid expansion in human daily life. But for the continuous social wellbeing the development of innovative technologies gives another, more significantly prolonged, effect. It is an exceptional opportunity for sustainable development of the society in which the growth of economic activity of people will be accompanied by the extension of their physical wellbeing in the process of preventive diagnosis and disposal from dangerous diseases.

5. Acknowledgement

The research was made in T.F. Gorbachev Kuzbass State Technical University under State Assignment N10.782.2014 K "Development of high-performance process for complex processing of low-grade coals and waste coal with producing a low ash coal-and-oil concentrate, composite fuels, rare earth and trace elements".

References

[1] Anastas P T and Kirchhoff M M 2002 Origins, current status, and future challenges of green chemistry Acc. Chem. Res 22 pp 686-694

[2] Lesin Y V, Luk'yanova S Y and Tyulenev M A 2015 Formation of the composition and properties of dumps on the open-pit mines of Kuzbass IOP Conference Series: Materials Science and Engineering 91 (1) 012093.

[3] Tyulenev M A, Khoreshok A A, Garina E A, Danilov S and Zhironkin S 2016 Adaptive technology of using backhoes for full coal extraction Proceedings of the 8th Russian-Chinese Symposium "Coal in the 21st Century: Mining, Processing, Safety" pp 111-115

[4] Tyulenev M A, Lesin Yu, Vik S and Zhironkin S 2016 Methodological Bases of Advanced Geo-ecological Problems Resolving in Neo-industrial Clusters Proceedings of the 8th Russian-Chinese Symposium "Coal in the 21st Century" pp 333-336

[5] Aksenov V V, Khoreshok A A and Beglyakov V Y 2013 Justification of creation of an external propulsor for multipurpose shield-type heading machine - GEO-WALKER Applied Mechanics and Materials 379 pp 20-23

[6] Lesin Y V, Luk'yanova S and Tyulenev M 2010 Mass transfer of dispersed particles in water filtration in macro-grained media J. Journal of Mining Science 46 (1) pp 78-81

[7] LesinY V, Luk'yanova S Y and Tyulenev M A 2015 Formation of the composition and properties of dumps on the open-pit mines of Kuzbass IOP Conference Series: Materials Science and Engineering 91 (1) 012093

[8] Tyulenev M A and Lesin Y V 2014 Justification complex purification technology open-pit mines wastewater Symposium of the Taishan academic forum – Project on mine disaster prevention and control pp 441-444

[9] Khoreshok A A, Zhironkin S A and Tyulenev M A et al. 2016 Innovative technics of managing engineers' global competencies IOP Conference Series: Materials Science and Engineering 142 (1) 012122

[10] Roco M C and Bainbridge W S 2003 Converging Technologies for Improving Human Performance. Nanotechnology, Biotechnology, Information Technology and Cognitive Science (Dordrecht: Kluwer Academic Publishers) 482 p.

[11] Feynman R 1962 Quantum Mechanics and Path Integrals (Oxford: Oxford Univ. Pub.) 224 p.

[12] Binnig G and Rohrer H 1979 Gerät zur rasterartigen Oberflächenuntersuchung unter Ausnutzung des Vakuum-Tunneleffekts bei kryogenischen Temperaturen Europäische Patentanmeldung 20 (9) pp 79

[13] Drexler E 1992 Nanosystems: Molecular Machinery Manufacturing and Computation (Los Altos: Wiley Pub.) 376 p

[14] Byrd D and Yavelow C 1986 The Kurzweil 250 Digital Synthesizer Computer Music Journal 10 (1) pp122-138

[15] Bartel A P and Lichtenberg F R 1991 Technical Change, Learning, and Wage (Washington: National Bureau of Economic Research) 2732 p

[16] Brehm J and Rahn W 1997 Individual-Level Evidence for the Causes and Consequences of Social Capital American Journal of Political Science 41 322-326

[17] Berkman L, Glass T, Brissette I and Seeman T 2000 From social integration to health: Durkheim in the new millennium Social Science and Medicine 51 (6) pp 843-857

[18] Fields G S, Yoo G 2000 Falling labour income inequality in Korea's economic growth: Patterns and Underlying Causes Review of Income and Wealth 46 (2) 139-160

[19] Gallup G H, Hill E 1960 The secrets of long life (New York: Geis Associates, Random House) 366 p

[20] Max-Neef M 1995 Economic Growth and Quality of Life: A Threshold Hypothesis Ecological Economics 15(2) pp 115-118

[21] Clark A E and Oswald A J 2002 A Simple Statistical Method for Measuring How Life Events Affect Happiness International Journal of Epidemiology 31(6) pp 1139-1144

[22] Melchior M, Niedhammer I, Berkman L, Goldberg M 2003 Do psychosocial work factors and social relations exert independent effects on sickness absence? A six year prospective study of the GAZEL cohort J Epidemiol Community Health 57(4) pp 285–293

[23] Diener E 2005 Guidelines for national indicators of subjective wellbeing and ill-being Social Indicators Network News 84 pp 4-6

[24] Deaton A 2008 Income, health, and wellbeing around the world: Evidence from the Gallup World Poll Journal of Economic Perspectives 22(2) pp 53-72

[25] Barysheva G A and Novoselova E G 2014 Methodology of Application of the Structural Shift Mechanism for Regulation of the National Economic Management System J. Applied Mechanics and Materials 682 pp 550-554
[26] Roco M C, Bainbridge W S 2002 Converging Technologies for Improving Human Performance nanotechnology, biotechnology, information technology and cognitive science: Report, June URL: http://www.wtec.org/ConvergingTechnologies/1/NBIC_report
[27] Harter J and Gurley V 2008 Measuring wellbeing in the United States Assoc. Psichol. Sci. 12 pp 23-26
[28] Kahneman D and Riis J 2005 Living, and thinking about it: Two perspectives on life (Oxford: Oxford University Press) 482 p
[29] Koshiro K 2012 Lifetime employment in Japan: three models of the concept. Foreign Labor Developments Bureau of Labor Statistics 21 August pp 266-281
[30] Coons C, Weber M 2013 Paternalism: Theory and Practice (Cambridge: Cambridge University Press) 458 p
[31] Burns J H 2005 Happiness and utility: Jeremy Bentham's equation Utilitas 17 pp 46–61
[32] Fleurbaey M 2009 Beyond GDP: The Quest for a Measure of Social Welfare Journal of Economic Literature 47 (4) pp 1055-1075
[33] Sen A 1985 Commodities and Capabilities (Amsterdam: Elsevier) 422 p
[34] Easterlin R A 1995 Will Raising the Incomes of All Increase the Happiness of All? Journal of Economic Behavior and Organization 27 (1) pp 35-47
[35] Tyulenev M, Zhironkin S and Litvin O 2015 The low-cost technology of quarry water purifying using the artificial filters of overburden rock Pollution Research 34 (4) pp 825-830
[36] Tyulenev M A, Gvozdkova T N and Zhironkin S A et al. 2016 Justification of Open Pit Mining Technology for Flat Coal Strata Processing in Relation to the Stratigraphic Positioning Rate Geotechnical and Geological Engineering 34 (6) doi:10.1007/s10706-016-0098-3
[37] Zhironkin S A, Khoreshok A A, Tyulenev M A et al. 2016 Economic and Technological Role of Kuzbass Industry in the Implementation of National Energy Strategy of Russian Federation IOP Conference Series: Materials Science and Engineering 142 (1) 012127
[38] Aksenov V V, Khoreshok A A and Beglyakov V Y 2013 Justification of creation of an external propulsor for multipurpose shield-type heading machine - GEO-WALKER Applied Mechanics and Materials 379 pp 20-23
[39] Efremenkov A B and Timofeev V Y 2012 Determination of necessary forces for geohod movement Proceedings - 2012 7th International Forum on Strategic Technology IFOST 6357729
[40] Efremenkov A B 2011 Forming the subterranean space by means of a new tool (geohod) Proceedings of the 6th International Forum on Strategic Technology IFOST 6021037
[41] Khoreshok A A, Mametyev L E, Borisov A Y, Vorobyev A V 2016 Influence of the Rigid Connection between Discs in the Tetrahedral Prisms on Equivalent Stresses When Cutting Work Faces IOP Conference Series: Materials Science and Engineering 127 012039
[42] Khoreshok A A, Buyankin P V, Vorobiev A V, Dronov A A 2016 Simulation of Stress-Strain State of Shovel Rotary Support Kingpin IOP Conference Series: Materials Science and Engineering 127 012014
[43] Efremenkov A B and Aksenov V V et al. 2012 Force parameters of geohod transmission with hydraulic drive in various movement phases Proceedings - 2012 7th International Forum on Strategic Technology IFOST 6357716
[44] Lesin Y V, Luk'yanova S and Tyulenev M 2010 Mass transfer of dispersed particles in water filtration in macro-grained media J. Journal of Mining Science 46 (1) pp 78-81
[45] LesinY V, Luk'yanova S Y and Tyulenev M A 2015 Formation of the composition and properties of dumps on the open-pit mines of Kuzbass IOP Conference Series: Materials Science and

Engineering 91 (1) 012093
[46] Fakhrislamova E I, Dolgikh T V, Pevneva I V and Formulevich I V 2015 Trends and Prospects of Higher Education Development in Russia by the Example of Training of Specialists in the Field of Economics Indian Journal of Science and Technology 8 (S10) pp 1-9

Environmental Improvement Of Opencast Mining

S Prokopenko [1,2,a], A Sushko[1], Yu Filatov[2], M Kislyakov[3] and I Kislyakov[3]

[1] Yurginsky Technological Institute (Branch), Tomsk Polytechnic University 652055, Kemerovo Region, Yurga, Leningradskaya St., 26, Russia;
[2] JSC «NC VostNII», 650003, Kemerovo, Institutskaya str., 3, Russia;
[3]Kemerovo Institute of Russia Economic University named after G.V. Plekhanov 650992, Kemerovo, avenue Kuznetskiy, 39 Russia;

[a]E-mail: sibgp@mail.ru

Abstract. Existing classifications of waste dumps in the quarries are given and their phenomenological nature is clarified. The need to identify the essence of the term "dump" is shown as well as the idea of "dump" as an artificial formation with everted and mixed rocks distanced from the quarry. Essential classification of man-made rock formations in quarries is developed. Characteristic of variations of man-made waste formations in quarries is developed. To reduce harmful effects of open-pit mining, dumps should be substituted with strat-lays - man-made structures relevant to natural stratification of litho-substances. Construction of strat-lays would improve ecological and technological culture of open cast mining.

Introduction

Kuznetsk coal basin is characterized by a variety of geological and mining conditions of coal seams exploitation. Currently, in the Kemerovo region, there are 60 active and 5 coal pits under construction with total production capacity more than 150 million tons of coal per year. In 2015, almost 140 million tons of coal were mined opencast and 76 million tons - underground; that is, more than 400 million m^3 of overburden and country rock were displaced into waist-rock disposals (ettles). In recent years, waste rocks are displaced by road (over 50%), railway (20%), excavator dumping (20%) and hydraulic transport (5%).

Kuzbass open pits and mines are winning coal at depths of 0-500 m, bringing to the surface all characteristic lithological varieties: loose overburden (clay, loam, sandy loam), half-rock (siltstone, mudstone, coal, burnt rock) and bedrock (sandstone, pyrite). Since coal deposits mining hundreds of waste dumps have been formed in Kuzbass region. The dumps total volume overtops 25 billion m^3.

The largest dumps are formed by open-cast mining. The dumps' heights vary in a wide range: single-stage dumps, piled at slope of response, have a height of 15 to 90 m; multistage dumps have heights of 40 to 160 m with the resulting angles of slopes 16-36 degrees. The region has 50 hydraulic-mine dumps with a total volume of clay soils about 1 billion m^3. Hydraulic-mine dump heights vary from 4 to 73 m, areas - from 6 to 736 m, capacities - from 0.6 to 100 million m^3 [1].

Rock mining dumps are the most large-scale technosphere objects of the Earth. As a result of the rocks loosening, their total volume exceeds quarry production itself. Waste rocks occupy vast areas of the Earth's surface, contaminate surrounding agricultural and forest lands with fugitive fine particles, change conditions and movement of ground- and surface waters and damage the biosphere [2].

Magnitude and hazard of waist grounds and rocks cause attention of scientists and experts. As a result of investigation they developed versatile classification of dumps:

1. According to a construction technology: bulldozed, excavated, aggraded, and combined.
2. According to a construction method: areal, peripheral, combined.
3. According to position relative to the quarry: external and internal.
4. According to rock transporting: road, rail, hydraulic dumping and combined.
5. According to a number of dumped stages: single-stage and multistage.
6. According to configuration: conical, ridge-like, truncated pyramid.

These classifications are phenomenological considering dumps as phenomena. Classification of dumps' structural composition organizes them into two classes of composition (one-component and composite) and into 6 structural groups and 16 types [3]. However, none of classifications concerns the nature of the dump, its idea and conceptual content. Consequently, variant structures may be called "dumps"; that may confuse mining experts and bring errors in design and construction documents. To eliminate possible inconsistencies we should clarify the nature of the concept "dump", define range of application of the term, identify its varieties, define their distinctive characteristics.

Materials and research methods

The researches were conducted in coal pits of southern, central and northern Kuzbass. More than 60 dumps of various configurations were examined: truncated pyramid, ridge-like, conical and technogenic hydraulic-mine dumps, constructed on various technologies [4-6]. Results of scientific investigations on construction methods, structure and properties of dumps, obtained by other authors for conditions of the Kuzbass, the Donetsk Basin, Kazakhstan, Yakutia, South Africa, Australia, have been used [7-16]. Revealing of the nature of dumps was carried out with methods of dialectical cognition and etymological analysis. Methods of scientific analysis, generalization and systematization were used to determine distinguishing characteristics and classification of man-made dumps.

Results and discussion

Updating the idea of the dump requires a deeper knowledge. Knowledge, philosophically, is a form, a relationship, extracted from the thing and indicated by a sign or text. In the process of scientific investigation one should "transpose... a form (idea) immersed in the thing "into the human mind" [17]. These cognitive abilities are determined by the skills to use already acquired knowledge and habits, to critically assess their own abilities and weaknesses, to be able to strive for personal and professional development, the ability to concentrate and be immersed in the object under study [18]. Let us clarify the concept "otval" (dump). Etymological analysis of the word shows that it contains two components: the prefix "ot-" (out/ from/away) and the root "val" (rampart/wall). The word "val" (wall) has been used since ancient times, when people shielded their homes with a ditch and a rampart (wall). The ditch "tore" and lowered the Earth's surface, the rampart increased it. This large height difference prevented penetration of animals or enemies. In the Middle Ages the ditch and the rampart were widely used in fortification of cities and fortresses in Europe and Asia.

The word "val" in the Russian language means not only "piled up", but also "brought down", "upside down". The rampart was formed on the edge the ditch while digging by stacking surface soil, then the middle soil. At the top of the rampart (point B) the soil from the ditch bottom (point A) was laid (Figure 1).

Thus, the natural soil layers in the rampart were everted "upside down". Stratification of a man-made rampart/dump was opposite to the natural litho-structure. With the emergence of powerful excavation and transport equipment and increasing of depth of excavation there emerged a necessity to erect the dump/rampart distant from the quarry (trench/ditch). At this time, there appears the concept "ot-val" – a dump distanced, moved away. If the "rampart" (val) is adjacent to the ditch with bottom edge (point B), the "dump" (otval) is already formed in the distance, beyond the possible collapse of the pit (the ditch) wall, reducing risk of ground fall. Thus, the "dump" means an artificial formation with everted, mixed grounds and distanced from the quarry.

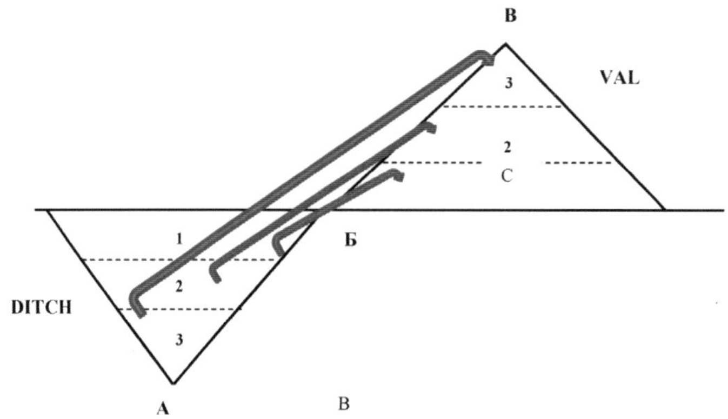

Figure 1 - Formation of val (rampart) from ditch ground

Excavated rock/soil layers 1, 2, 3, 4, 5 are displaced to the dump, wherein, in order to reduce costs, the rock/soil is usually dumped downward from points A, B, C, D, etc.; that mix rock material is taken from different working levels (Figure 2).

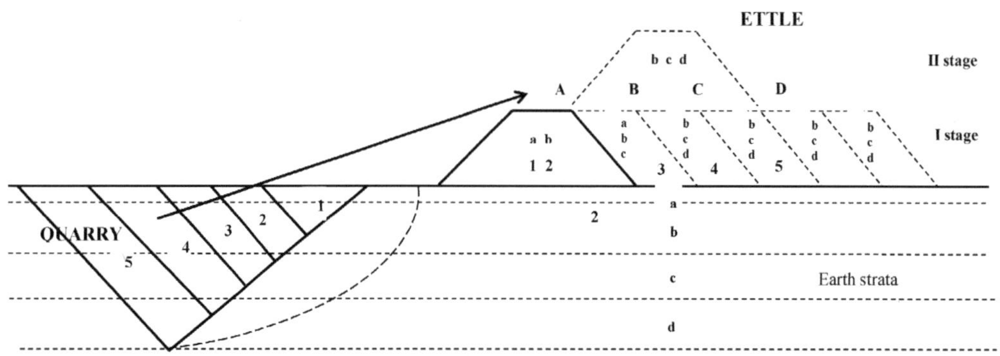

Figure 2 - Stratigraphy of natural and man-made rock mass in quarry

Mining operations are accompanied by cutting new working levels in the thickness of the earth's crust underlying strata. As a consequence, the underlying rocks are laid into the dump, bringing to the surface "dead" metamorphic and igneous rocks. Wherein, quaternary deposits "a" and sediments "b" occur at the bottom, under dump thickness. Deformation of natural lithosphere stratification sequence damages ecological well-being of mining areas, because the deeper into earth crust interior, the more lifeless substance is. All natural environments suffer: water, air, soil, plants and animals. Mine reclamations are not able to restore natural well-being, as they are chronically falling behind deformations. Long-term effects of unfavorable transformation of the earth's crust is currently difficult to predict without the relevant studies, however, it is obvious that in the future adaptive capacity of the Earth can not cope with anthropogenic impacts of this kind and magnitude.

The essence of the concept "otval" (dump) makes evident inadequacy of using such term for naming anthropogenic soil massifs, dumped in the quarry stripped area; since no "distancing" or "displacing" from the quarry does not occur. Word combination "vnutrennii otval" (internal dump) is an oxymoron; more representative is the term "zaval" (fill/heap), which contains the concept of "zavalivat -*verb*" (filling/heaping). Other word combination "vneshnii otval" (external dump) is a tautology.

It seems appropriate to replace a general term "otval" with a number of more precise terms. So a ridge-like construction, formed with dragline, shovel or hydraulic shovel along the working trench, we propose to call "val" (wall). Rocks, transported from the quarry or dumped with the excavator, form a "otval" (dump) distanced, disposal, moved away. When moving soil with directed explosion, as well as formation of dragline spoil banks or motor-dumps, a "zaval" (heap/fill) occurs - a massif of mixed soils/rocks inside quarry. It differs significantly from the dump in shape, hydrological properties and environmental friendliness [19]

Cultural development of current and future mining determines the need for a new approach to construction of waste rock dumps. The dump, as a place of waste disposal, should be considered from the point of view of Stratigraphy. Millions of cubic meters of overburden, extracted from the depths to the surface (only in Kuzbass almost 1.0 billion m^3 every two years), actually form a new lithosphere stratum. When forming these rock strata, it's essential to restore natural formation of lithological varieties for preservation living conditions on the surface.

This change in mining is feasible technologically, organizationally and economically. When adjusting values, and the economic side of the question solvable [20]. Anthropogenic construction preserving the Earth's stratification may be called a new composite term "strasyp"(strat+layer=strat-lay) from "stra-" (stratum) and "sypat - *verb*" (to layer) i.e to lay material orderly under control. "Strasyp" (strat-lay) should replace the current "dump" - a stratigraphically irregular and disordered construction, in the near future to reduce the negative impact on the environment (Figure 3).

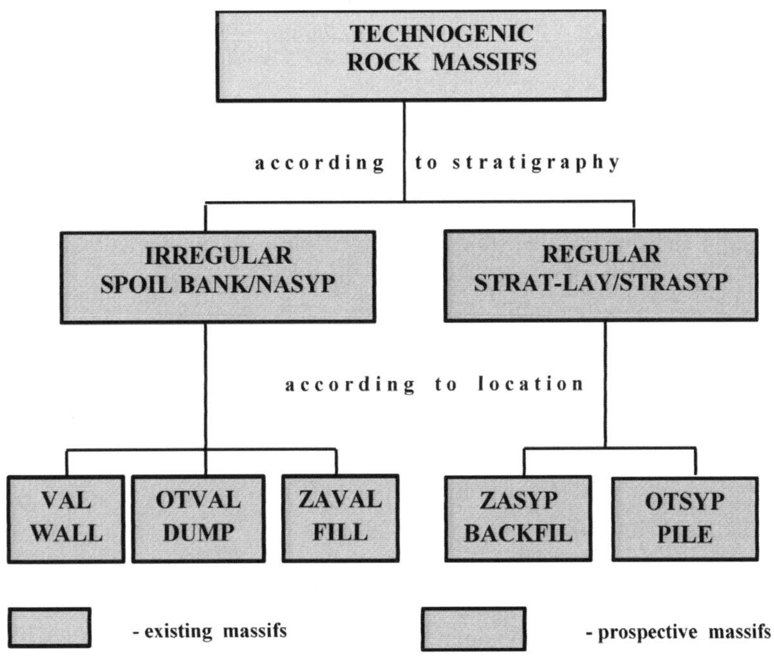

Figure 3 - Essential classification of man-made massifs in quarries

In the pits with continuous mining (Krasnoyarsk Territory of Russia, Ukraine, Australia, USA, etc) "heaps" (zaval), constructed in the stripped area, are formed as "strat-lays" since rocks are displaced from work areas to dump areas level-by-level. In such massifs natural sequence of rock placement is maintained; that minimizes the adverse impacts. It is advisable to name such formations "zasyp" (back-fill) as it is a filling formation with regular stratification of the bulk material (Table. 1).

This experience should be applied in pits with dragline mining; layered dumping in compliance with natural interposition of rock layers will improve man-made rock formations. There appear rock formations, named "otsyp" (pile); that is, distanced rocks layered in accordance with natural bedding. Change-over from "ot-val"(dump) to "ot-syp"(pile) and "za-syp"(back-fill) marks a new stage of technological and ecological culture of open cast mining.

Table 1 Essential characteristic of man-made rock formations in quarries

Essential characteristic	Man-made rock formations				
	SPOIL BANK			STRAT-LAY	
	WALL	DUMP	FILL	BACKFILL	PILE
Location relative to the quarry	In contact	At a distance	Inside	Inside	At a distance
Stratigraphy	Reversed order	Absent. Irregular	Absent. Irregular	Complies with natural	Complies with natural
Location relative to the earth's surface	Above	Above	Below	Below	Above
Spatial form	Ridge-like	Truncated cone or pyramid	Fills quarry space	Fills quarry space	Truncated cone or pyramid
Rock laying technology	Stratified	Irregular	Irregular	Stratified	Stratified

Conclusion

The study results allow revising the idea of rock man-made dumps and define the scope of the term. The essential classification of man-made rock formations helps to clarify specific terminology, thereby improving ecological and technological culture of mining, and construction of man-made formations preserving the natural stratification of litho-substances. Construction of strat-lays reduces waste formation and increases environmental friendliness of open pit coal mining; that reduces negative consequences of violating the integrity of the Earth's crust.

References

[1] Zharikov V P, Ermoshkin V V and Kleimenov R G 2012 Sustainable land management in the formation of dumps and gidrootval on cuts of Kuzbass The Mountain of inform.-analit.newsletter. 2 pp 78–83.
[2] Mikhal'chenko V V and Prokopenko S A 1992 Clean technology - the future of the open coal production in the Kuzbass Coal 1 pp 11–14.
[3] Prokopenko S A and Ludzish V S 1991 Classification of structures dump arrays mines Kuzbass Labour Safety in industry 10 pp 35–36.
[4] Prokopenko S A 2005 To coal power - high technologies oal' 7 pp 55–57.
[5] Prokopenko S A, Ludzish V S, Kurzina I A and Sushko A V 2015 Alternative source of energy in operation of power-driven machines in coal mines Gornyi zhurnal/ Mining Journal 1 pp 75–77.
[6] Mikhailov A A, Lesin Yu V and Prokopenko S A 1989 Improvement of dumping to improve the effectiveness of remediation on the mines of Kuzbass Coal 10 pp 41–42.
[7] Kryachko O Yu 1980 Management of waste dumps of open cast mining Moscow, Nedra, 180 p.
[8] Maleev N G and Kurowski M N 2003 The construction plans and the formation of multilevel dumps Modern technologies of mineral resources development. SB scientific papers 1 Krasnoyarsk: Publishing house Khusmith pp 67–69.
[9] Kutepov Yu I and Kutepova N A 2008 Regularities of formation pore pressure while hydrochlorothiazie and dumping dry tailings The Mountain of inform.-analit. newsletter 11 pp 54–57.

[10] Instruction to prevent spontaneous combustion, fire and dismantling dumps Approved. by order of Rostekhnadzor 738 from 23.12.2011 Available at: URL: http://docs.cntd.ru/document/902324483 (Last accessed date 04.06.2016).

[11] Bahaeva S P Prostov S M 2011 Integrated monitoring of technogenic soil massifs of coal mines Safety in industry 4 pp 34–36

[12] Cheskidov V V 2011 Complex sensing alluvial deposits gidrootval No. 3 cut "Kedrovsky" Mining industry 6 pp 25-29[13] Kalashnik A I Zaporozhets D V Dyakov A Yu and Demagin A Yu 2014 GPR sounding of snow-rock dumps of JSC "Apatite" Bulletin of the Kola science centre of RAS 3(18) pp 65–67.

[13] Gaponov Yu S 2015 Geomechanical substantiation of stability of dumps taking into account the influence of the characteristics of the destroyed rock mass/ abstract Dis c and. tech. sciences S.-Petersburg 212 p.

[14] Macarisina Yu I 2016 Improvement of environmental safety of waste dumps of coal mines of Donbass with the help of physical and phyto-techniques Dis. cand. of agricultural sciences Lugansk pp 235.

[15] Lesin Yu V, Luk'yanova S Yu and Tyulenev M A 2015 Formation of the composition and properties of dumps on the open-pit mines of Kuzbass IOP Conference Series: Materials Science and Engineering 6 Article number 012093.

[16] Nevelev A B and Khudyakova N L 2015 Philosophy of cognition: proc. manual Chelyabinsk: publishing house of Chelyabinsk. state University press 123 p.

[17] Lizunkov V G, Marchuk V I and Podzorova E A 2015 Identification of Criteria, Features and Levels of Economic and Managerial Competencies Development for Bachelors in Mechanical Engineering Procedia - Social and Behavioral Sciences 206 pp 388–393.

[18] Mikhal'chenko V V, Prokopenko S A, Orlov V G and Koksin A V 1991 Semessersesy the technology for thick sloping and steep deposits Coal 5 pp 44–46.

[19] Prokopenko S A and Ludzish V S 2014 Problems of innovative development of the mining Enterprises of Russia Gornyi zhurnal/ Mining Journal 1 pp 47–49.

The Modelling Of Basing Holes Machining Of Automatically Replaceable Cubical Units For Reconfigurable Manufacturing Systems With Low-Waste Production

N M Bobrovskij [a], D G Levashkin[b], I N Bobrovskij[c], P A Melnikov[d], A A Lukyanov[e]

Togliatti State University
Belorusskaya str.14, Togliatti, Russian Federation, 445667

E-mail: [a] Bobrnm@yandex.ru, [b]LevashkinD@gmail.com, [c]Bobri@yandex.ru, [d]Topavel@mail.ru, [e]A.lukyanov@tehnomasch.ru

Abstract. Article is devoted the decision of basing holes machining accuracy problems of automatically replaceable cubical units (carriers) for reconfigurable manufacturing systems with low-waste production (RMS). Results of automatically replaceable units basing holes machining modeling on the basis of the dimensional chains analysis are presented. Influence of machining parameters processing on accuracy spacings on centers between basing apertures is shown. The mathematical model of carriers basing holes machining accuracy is offered.

Introduction

The present stage of engineering industry development is characterized by a heading automatically replaceable machine systems and units, for solving of the nomenclature expansion, of released products, abbreviation of a time and resources on manufacture preparation problems. To overcome a complex of industrial contradictions it is necessary at the expense of theoretical researches conducting, working out of new designs. So continuation of configurations and designing machine systems, automatic transfer lines development theory is the theory of configurations of automatic machine systems of variable configuration and structure (reconfigurable manufacturing systems - RMS). Reconfigurable it is directed configuration and architecture RMS changes. Along with known solutions it was possibly to provide with application in the capacity of basing, fastening and details transportation units - automatically replaceable cubical units (carriers) [1-5].

This engineering solution improves the accuracy and performance of the parts. The most complete processing of one part is implemented in a separate working position. Implementation of low-waste production based on RMS is ensured by reducing metal consumption in machine tool systems, reducing extent of the production process, reducing the number of working positions, required areas and multiple reduce of production waste [5-14]. A heading of carriers is connected with some questions of their manufacturing accuracy maintenance.

Formulation of the problem

It is possible to present a design of the carrier in the form of a cube (Figure 1) where on each basil-plane is available basing holes T_n, $T_{(n+1)}$, $T_{(n+2)}$, T_m. On two non-adjacent holes, for example, $(T_n, T_{(n+2)})$ and $(T_{(n+1)}, T_m)$ there is a carrier basing on a working position, under the two non-adjacent holes and a basil-plane (Figure 2).

For raise of efficiency of the automated machining the carrier with the details installed on lateral basil-plane is resurface on a working position. Thus there is an interleaving of a complete set of basing holes and a basil with conservation of the theoretical scheme of basing [1]. Resurfacing the carrier provides change of attitude of a detail. In *XOY*-system group of holes T_n, $T_{(n+1)}$, $T_{(n+2)}$, T_m dimensional communications define: $A_{ij}^{н}$ - not recut spacings on centers between holes $(i \neq j = 1,2,3,4, m)$, A_{ij}^{n} - recut spacings on centers. Each processed holes T_{ij} possesses a set of parameters $A_{ij}^{н}$, A_{ij}^{n}, L_{xij}, L_{yij}, B_{ij} (Figure 2) which define an arrangement of each hole concerning a basil of carrier *G*. The parameters influencing accuracy of a mutual bracing of holes are spacings on centers $A_{ij}^{н}$, A_{ij}^{n}, the set tolerance of hole - $T_{\Delta m}^{i}$.

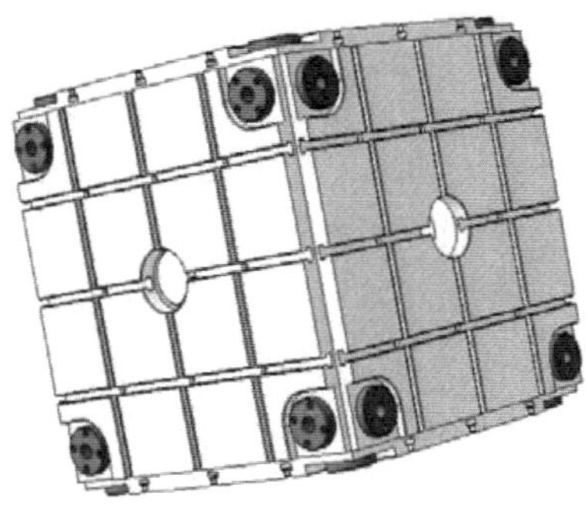

Figure 1 - Design of the carrier

So accuracy of hole T_n machining is defined by sizes L_n in *OX, OY* - directions (Figure 3). In an *OX* axis direction the arrangement of hole T_n is characterized: nominal size L_n, size $L_{(n)\,p} = = L_n + \Delta L_n$, where ΔL_n - a deviation of size L_n. The data set of sizes forms a dimensional chain of hole T_n machining. The closing chain link is size $L_{(n)\,p}$, increasing a deviation ΔL_n, reducing - face value L_n.

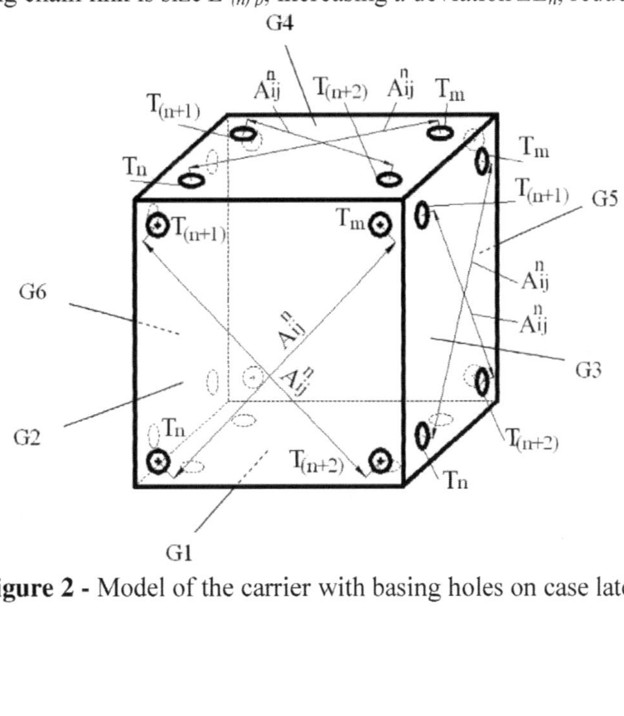

Figure 2 - Model of the carrier with basing holes on case lateral faces

At calculation of link $L_{(n+1)}$ accuracy for holes T_{n+1} will observe a chain of sizes $B_{n\,(n+1)}$, $A_{n\,(n+1)}$, L_n, $L_{(n+1)}$ (Figure 4). Sizes L_n, $B_{n\,(n+1)}$ - have the set rating values and limits deviations ΔL_n, $\Delta B_{n\,(n+1)}$. Link $A_{n\,(n+1)}$ - defined (face value is not known), is known its limiting deviation $\Delta A_{n\,(n+1)}$.

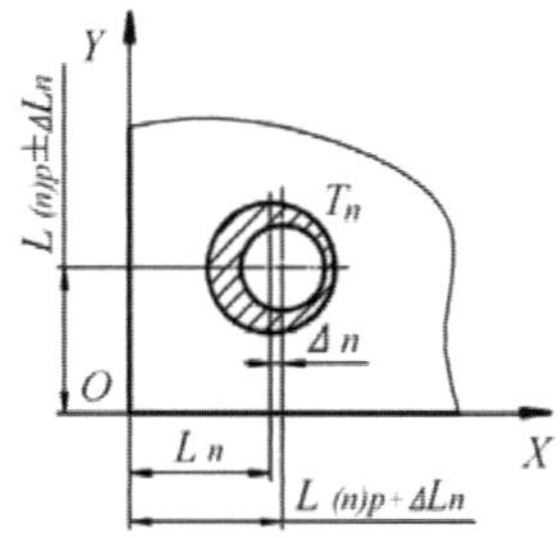

Figure 3 - Dimensional chain

At closing link $L_{(n+1)}$ theoretical values and limiting deviations $t_{L\,(n+1)}$ are set. It is necessary to count magnitude of required link $A_{n\,(n+1)}$ face value. At calculation of link $L_{(n+2)}$ accuracy holes T_{n+2} will observe a chain from sizes $B_{n\,(n+2)}$, $A_{n\,(n+2)}$, L_n, $L_{(n+2)}$, and also a chain which is switching on sizes $A_{n\,(n+1)}$, $A_{n\,(n+2)}$, $A_{(n+1)\,(n+2)}$. Link $A_{n\,(n+2)}$ - defined (face value is not known), is known its limiting deviation $\Delta A_{n\,(n+2)}$.

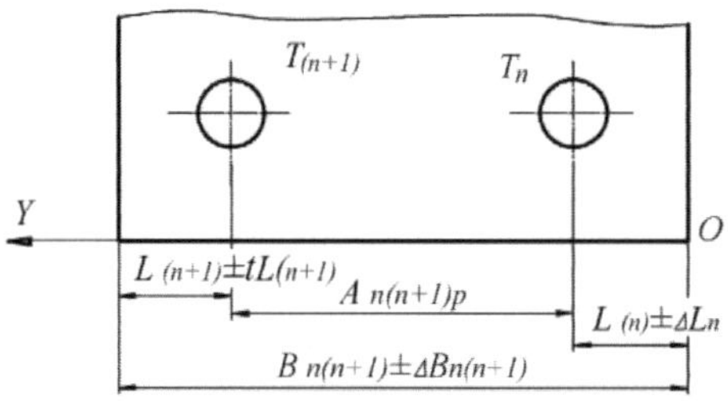

Figure 4 - Dimensional chain hole T_{n+1} machining

At closing link $L_{(n+2)}$ theoretical values and limiting deviations $t_{L\,(n+2)}$ are set. It is necessary to count magnitude of required link $A_{n\,(n+2)}$. face value. For spacing on centres size $A_{(n+1)\,(n+2)}$ it is necessary to count magnitude of face value, and to check, whether there are settlement limiting values of link $A_{(n+1)\,(n+2)}$ for its regulated values.

Accuracy of each holes machining is characterised by several closing links of dimensional chains. For hole T_n it is sizes L_n in *OX, OY* - directions. For hole T_{n+1} it is sizes $A_{n\,(n+1)}$, $L_{(n+1)}$. For hole T_{n+2} it is sizes $A_{n\,(n+2)}$, $L_{(n+2)}$, $A_{(n+1)\,(n+2)}$. Accuracy of hole T_m machining is defined by accuracy of following links chains, in a *OX* - direction: $B_{n\,(n+1)}$, $A_{(n+1)\,m}$, L_m, $L_{(n+1)}$, in *OY* - direction: $B_{n\,(n+2)}$, $A_{(n+2)\,m}$, L_m, $L_{(n+2)}$. Also independent chains form the diagonal crossing sizes, in *OX, OY* - directions: $A_{n\,(n+2)}$, $A_{(n+2)\,m}$, $A_{(nm)}$; $A_{n\,(n+1)}$, $A_{(n+1)\,m}$, $A_{(nm)}$ (Figure 5).

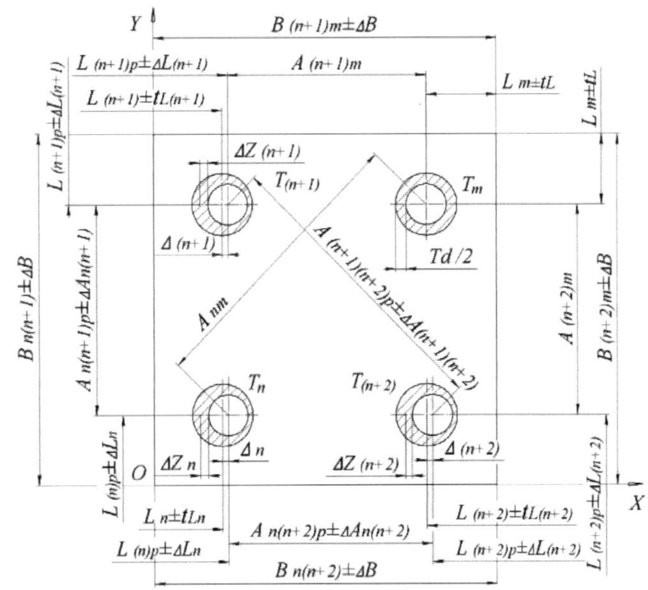

Figure 5 - The settlement of basing holes machining dimensional chains

Thus, for maintenance of carrier basing accuracy of basing holes T_n, $T_{(n+1)}$, $T_{(n+2)}$, T_m machining it is necessary to stand equal in limits of maximum deviations recut spacing on centres sizes $A_{(n+1)(n+2)}$, $A_{(nm)}$, that is possible if equality in tolerance limits of sizes L_{ij} and spacings on centres A_{ij} between basing holes is stood. For hole T_m it is sizes $A_{(n+2)\,m}$, $A_{(n+1)\,m}$, $A_{(nm)}$, L_m.

Method of investigation

Let's make an auxiliary dimensional chain concerning each closing link and we will write down the additional equation defining rating value of an unknown making link. Also considering, that machining of holes begins, for example, with aperture T_n, sequence of holes T_n, $T_{(n+1)}$, $T_{(n+2)}$, T_m machining accuracy calculation can look like: 1) machining of hole T_n, machining of hole T_{n+1}, machining of hole T_m; 2) machining of hole T_n, machining of hole T_{n+2}, machining of hole T_m (Figure 5). The simplified record of dimensional chains for each alternative sequence of machining can look like: 1) T_n, $T_{n+1} \rightarrow T_m$; 2) T_n, $T_{n+2} \rightarrow T_m$. In both cases hole T_m is closing. It allows, using results of calculations to define alternative of hole T_m machining (a chain 1 or 2), providing the highest accuracy of its positioning concerning next holes T_n, $T_{(n+1)}$, $T_{(n+2)}$ in each basil G.

The arrangement of hole T_m of relative holes T_n, T_{n+2} is defined by rating values of closing links $A_{(n+2)\,m}$, $A_{(nm)}$ and sizes $L_{(n)}$, $L_{(n+2)}$ in OX, OY - directions:

$$L_{(n)P}=L_{(n)}\pm(\Delta_{(n)}+tL), \tag{1}$$

$$L_{(n+2)P}=L_{(n+2)}\pm(\Delta_{(n+2)}+tL). \tag{2}$$

where $L_{(n)}$ - rating value of distance of hole $T_{(n)}$ to edge of a basil of the carrying agent, $\Delta_{(n)}$ - value of a deviation of hole $T_{(n)}$ taking into account elastic deformations of the tool and a carrier, tL - a limit deviation on manufacturing of holes $T_{(n)}$ and $T_{(n+2)}$, $L_{(n+2)}$ - rating value of distance of hole $T_{(n+2)}$ to edge of a basil of the carrier, $\Delta_{(n+2)}$ - value of a deviation of hole $T_{(n+2)}$ taking into account elastic deformations of the tool and a carrier basil, tL - a limit deviation on manufacturing of hole $T_{(n+2)}$.

The interconnection of sizes $A_{(n+2)m}$ and $A_{(nm)}$ is being out through a system gear ratio $\xi=\sqrt{2}$ equal-arrangement holes of settlement chain T_n, $T_{n+2} \rightarrow T_m$. Deviation closing links $A_{(n+2)\,m}$ and $A_{(nm)}$ are defined by relationships:

$$\Delta A_{(n+2)m} = \Delta B+(tL_{(n+2)}+\Delta_{(n+2)})+tL. \tag{3}$$

$$\Delta A_{nm} = \xi\,(\Delta B+(tL_n+\Delta_n)+tL) \qquad (4)$$

Let's mark out ΔA_{nm} - a deviation of a closing size $A_{(nm)}$, $\Delta A_{(n+2)\ m}$ - a deviation of a closing link $A_{(n+2)\,m}$, ΔB - a deviation of carrier basil size B_{ij}, tL - a maximum deviation on machining of hole T_m, $tL_{(n+2)}$ - a maximum deviation on machining of hole $T_{(n+2)}$, tL_n - a maximum deviation on machining of $T_{(n)}$.

Let's define displacement (correction) ***e*** a design value of closing links $A_{((n+2)\ m)\ P}$ and $A_{(nm)\ P}$ concerning their face values $A_{(n+2)\ m}$ and A_{nm}:

$$e_{(n+2)m} = (A_{min((n+2)m)P} - A_{min((n+2)m)}) + tL + (\Delta B+(tL_{(n+2)}+\Delta_{(n+2)}); \qquad (5)$$

$$e_{(nm)} = (A_{min((nm)P} - A_{min((nm)}) + (tL(\xi)^{-1}) + (\Delta B+(tL_{(n)}+\Delta_{(n)})\xi. \qquad (6)$$

Where $A_{min\ ((n+2)\ m)\ P}$ and $A_{min\ ((nm)\ P}$ - the minimum design values of links $A_{(n+2)\ m}$, A_{nm}; $A_{min\ ((n+2)\ m)}$ and $A_{min\ ((nm)}$ - the minimum values of links $And_{\ (n+2)\ m}$ and A_{nm}; $e_{\ (n+2)\ m}$ - correction of a closing link $A_{(n+2)\ m}$ design value; $e_{\ (nm)}$ - correction of a closing link design value A_{nm}.

Using expressions (5,6) is being out calculation of links $A_{(n+2)\ m}$ and A_{nm} correction on values $e_{(n+2)\ m}$ and $e_{\ (nm)}$, taking into account deviations of links $L_{(n)}$, $L_{(n+2)}$. On design values of closing links $A_{((n+2)m)P}$ and $A_{(nm)\ P}$ accuracy of machining in *XOY* - system (Figure 5) hole T_m of settlement chain T_n, $T_{n+2} \rightarrow T_m$ in *OX*, *OY* - directions is provided.

For settlement chain T_n, $T_{n+1} \rightarrow T_m$ the rule of hole T_m of relative holes T_n, T_{n+1} is defined by rating values of closing links $A_{(n+1)\ m}$ and $A_{(nm)}$, the rule of holes T_n, T_{n+2} is defined by sizes $L_{\ (n)}$, $L_{\ (n+1)}$ in *OX*, *OY* - directions. Expressions for calculation of sizes rating values $L_{\ (n)\ P}$, $L_{\ (n+2)\ P}$ on an auxiliary dimensional chain (Figure 5) look like:

$$L_{(n)P}=L_{(n)}\pm(\Delta_{(n)}+tL), \qquad (7)$$

$$L_{(n+1)P}=L_{(n+1)}\pm(\Delta_{(n+1)}+tL). \qquad (8)$$

$L_{\ (n)}$ - rating value of distance of hole $T_{(n)}$ to edge of a carrier basil; $\Delta_{\ (n)}$ - value of a deviation of hole $T_{(n)}$ taking into account elastic deformations of the tool and a carrier basil; tL - a maximum deviation on manufacturing of holes $T_{(n)}$ and $T_{(n+1)}$; $L_{\ (n+1)}$ - rating value of distance of hole $T_{\ (n+1)}$ to edge of a carrier basil; $\Delta_{\ (n+1)}$ - value of a deviation of hole $T_{\ (n+1)}$ taking into account elastic deformations of the tool and a carrier basil; tL - a deviation on manufacturing of hole $T_{\ (n+1)}$.

Expressions (13) represent dependence of nominal sizes $L_{\ (n)}$, $L_{\ (n+1)}$ from machining parametres, the tolerances of holes in diameters $D_{\ (n)}$, $D_{\ (n+1)}$. Deviations of closing links $A_{(n+1)\ m,}$ $A_{(nm)}$ are defined according to:

$$\Delta A_{(n+1)m} = \Delta B+(tL_{(n+1)}+\Delta_{(n+1)})+tL. \qquad (9)$$

$$\Delta A_{nm} = \xi\,(\Delta B+(tL_n+\Delta_n)+tL). \qquad (10)$$

Let's mark out ΔA_{nm} - a deviation of a closing size $A_{(nm)}$; $\Delta A_{(n+1)\ m}$ - a deviation of a closing link $A_{(n+1)\ m}$; ΔB - a deviation of size B_{ij} of a carrier basil; tL - a maximum deviation on machining of aperture T_m; $tL_{\ (n+1)}$ - a maximum deviation on machining of hole $T_{\ (n+1)}$; tL_n - a maximum deviation of hole $T_{(n)}$.

Let's define displacement (correction) ***e*** a design value of closing links $A_{((n+1)\ m)\ P}$ and $A_{(nm)\ P}$ concerning their face values $A_{(n+1)\ m}$ and A_{nm}:

$$e_{(n+1)m} = (A_{min((n+1)m)P} - A_{min((n+1)m)}) + tL + (\Delta B+(tL_{(n+1)}+\Delta_{(n+1)}); \qquad (11)$$

$$e_{(nm)} = (A_{min((nm)P} - A_{min((nm)}) + (tL(\xi)^{-1}) + (\Delta B+(tL_{(n)}+\Delta_{(n)})\xi. \qquad (12)$$

$A_{min((n+1\ m\ P}$ and $A_{min((nm)P}$ - the minimum design values of links $A_{(n+1)\ m}$ and A_{nm}; $A_{min\ ((n+1)\ m)}$ and $A_{min\ ((nm)}$ - the minimum values of links $A_{(n+1)\ m}$ and A_{nm}; $e_{(n+1)\ m}$ - correction of a design value of a closing link $A_{(n+1)\ m}$; $e_{(nm)}$ - correction of a design value of closing link A_{nm}.

Modeling solutions

Using expressions (11, 12) is being out calculation of links $And_{\ (n+2)\ m}$ and A_{nm} correction on values $e_{\ (n+2)\ m}$ and $e_{\ (nm)}$, taking into account deviations of links $L_{\ (n)}$, $L_{\ (n+2)}$. Generally for machining of basing hole T_m it is necessary to execute preliminary with given machining accuracy of basing holes T_n, T_{n+1}, T_{n+2}. Thus in calculations dimensional chains $T_n \rightarrow T_{n+1}$, T_n, $T_{n+1} \rightarrow T_{n+2}$ (Figure 5). For calculation of equal-arrangement holes (T_n, $T_{(n+1)}$, $T_{(n+2)}$, ... T_m) machining accuracy following equations are gained. For rating value of size $A_{n\,(n+1)}$:

$$e'_{n(n+1)} = A_{(n(n+1))P} - (A_{max(n(n+1))} + A_{min(n(n+1))})/2; \quad (13)$$

$$e''_{n(n+1)} = (A_{min(n(n+1))P} - A_{min(n(n+1))}) + tL_{(n+1)} + (\Delta B + \Delta L_{(n)}); \quad (14)$$

$$e'''_{n(n+1)} = (A_{min(n(n+1))P} - A_{min(n(n+1))}) + tL_{(n+1)} + (\Delta B + (tL_{(n)} + \Delta_{(n)})); \quad (15)$$

$$e'_{n(n+1)} = e''_{n(n+1)} = e'''_{n(n+1)} \quad (16)$$

$A_{min\,(n\,(n+1))\,P}$ - the minimum design value of link $A_{n\,(n+1)}$; $A_{min\,(n\,(n+1))}$ - the minimum value of link $A_{n\,(n+1)}$; $e_{n\,(n+1)}$ - correction of a closing link design value $A_{n\,(n+1)}$; $\Delta L_{(n)}$ - a deviation at machining of hole T_n; ΔB - a deviation of a carrier basil size B_{ij} ; $tL_{(n+1)}$ - a maximum deviation on machining of hole $T_{(n+1)}$; tL_n - a maximum deviation on machining of hole $T_{(n)}$, $\Delta_{(n)}$ - a deviation on machining of hole $T_{(n)}$ taking into account elastic deformations of the tool and a carrier basil. For maintenance of given machining accuracy is necessary performance of equality (16) according to a condition for link A_{nm} ($e_{nm}=0$). For rating value of size $A_{n\,(n+2)}$:

$$e'_{n(n+2)} = A_{(n(n+2))P} - (A_{max(n(n+2))} + A_{min(n(n+2))})/2; \quad (17)$$

$$e''_{n(n+2)} = (A_{min(n(n+2))P} - A_{min(n(n+2))}) + tL_{(n+2)} + (\Delta B + \Delta L_{(n)}); \quad (18)$$

$$e'''_{n(n+2)} = (A_{min(n(n+2))P} - A_{min(n(n+2))}) + tL_{(n+2)} + (\Delta B + (tL_{(n)} + \Delta_{(n)})); \quad (19)$$

$$e'_{n(n+2)} = e''_{n(n+2)} = e'''_{n(n+2)} \quad (20)$$

$A_{min\,(n\,(n+2))\,P}$ - the minimum design value of link $A_{n\,(n+2)}$; $A_{min\,(n\,(n+2))}$ - the minimum value of link $A_{n\,(n+2)}$; $e_{n\,(n+2)}$ - correction of a closing link design value $A_{n\,(n+2)}$; $\Delta L_{(n)}$ - a deviation at machining of hole T_n; ΔB - a deviation of a carrier basil size B_{ij}; $tL_{(n+2)}$ - a maximum deviation on machining of hole $T_{(n+2)}$; tL_n - a deviation on machining of hole $T_{(n)}$; $\Delta_{(n)}$ - a deviation on machining of hole $T_{(n)}$ taking into account elastic deformations of the tool and a carrier basil. For rating value of a size $A_{(n+1)\,(n+2)}$:

$$e'_{(n+1)(n+2)} = A_{(n+1)(n+2)P} - (A_{max(n+1)(n+2)} + A_{min(n+1)(n+1)})/2; \quad (21)$$

$$e''_{(n+1)(n+2)} = (A_{min(n+1)(n+2)P} - A_{min(n+1)(n+2)}) + tL_{(n+2)}\,(\xi)^{-1} + (\Delta B + \Delta L_{(n+1)})\xi; \quad (22)$$

$$e'''_{(n+1)(n+2)} = (A_{min(n+1)(n+2)P} - A_{min(n+1)(n+2)}) + tL_{(n+2)}\,(\xi)^{-1} + (\Delta B + (tL_{(n+1)} + \Delta_{(n+1)})\xi); \quad (23)$$

$$e'_{(n+1)(n+2)} = e''_{(n+1)(n+2)} = e'''_{(n+1)(n+2)}\ . \quad (24)$$

$A_{min\,(n+1)\,(n+2))\,P}$ - the minimum design value of a link $A_{(n+1)\,(n+2)}$; $A_{min\,(n+1)\,(n+2))}$ - the minimum value of a link $A_{(n+1)\,(n+2)}$; $e_{(n+1)\,(n+2)}$ - correction of a closing link design value $A_{(n+1)\,(n+2)}$; $\Delta L_{(n+1)}$ - a deviation at machining of hole T_{n+1}, ΔB - a deviation of a carrier basil size B_{ij}; $tL_{(n+2)}$ - a maximum deviation on machining of hole $T_{(n+2)}$; $tL_{(n+1)}$ a - maximum deviation on machining of hole $T_{(n+1)}$; $\Delta_{(n+1)}$ - a deviation on machining of hole $T_{(n+1)}$ taking into account elastic deformations of the tool and a carrier basil. For rating value of a size $A_{(n+2)\,m}$:

$$e'_{(n+2)m} = A_{((n+2)m)P} - (A_{max(n+2)m} + A_{min(n+2)m})/2; \quad (25)$$

$$e''_{(n+2)m} = (A_{min((n+2)m)P} - A_{min(n+2)m}) + tL + (\Delta B + \Delta L_{(n+2)}); \quad (26)$$

$$e'''_{(n+2)m} = (A_{min((n+2)m)P} - A_{min(n+2)m}) + tL + (\Delta B + (tL_{(n+2)} + \Delta_{(n+2)})); \quad (27)$$

$$e'_{(n+2)m} = e''_{(n+2)m} = e'''_{(n+2)m}\ . \quad (28)$$

$A_{min\,((n+2)\,m)\,P}$ - the minimum design value of a link $A_{(n+2)\,m}$; $A_{min\,((n+2)\,m)}$ - the minimum value of a link $A_{(n+2)\,m}$; $e_{(n+2)\,m}$ - correction of a closing link design value $A_{(n+2)\,m}$; $\Delta L_{(n+2)}$ - a deviation at machining of hole $T_{(n+2)}$; ΔB - a deviation of a carrier basil size B_{ij}; $tL_{(n+2)}$ - a maximum deviation on machining of hole $T_{(n+2)}$; tL - a maximum deviation on machining of hole $T_{(m)}$; $\Delta_{(n+2)}$ - a deviation on machining of hole $T_{(n+2)}$ taking into account elastic deformations of the tool and a carrier basil. For calculation of rating value of a size $A_{(n+1)\,m}$:

$$e'_{(n+1)m} = A_{((n+1)m)P} - (A_{max(n+1)m} + A_{min(n+1)m})/2; \quad (29)$$

$${}''_{(n+1)m} = (A_{min((n+1)m)P} - A_{min(n+1)m}) + tL + (\Delta B + \Delta L_{(n+1)}); \quad (30)$$

$$e'''_{(n+1)m} = (A_{min((n+1)m)P} - A_{min(n+1)m}) + tL + (\Delta B + (tL_{(n+1)} + \Delta_{(n+1)})); \quad (31)$$

$$e'_{(n+1)m} = e''_{(n+1)m} = e'''_{(n+1)m} \quad (32)$$

$A_{min\,((n+1)\,m)\,P}$ - the minimum design value of a link $A_{(n+1)\,m}$; $A_{min\,((n+1)\,m)}$ - the minimum value of a link $A_{(n+1)\,m}$; $e_{(n+1)\,m}$ - correction of a closing link design value $A_{(n+1)\,m}$; $\Delta L_{(n+1)}$ - a deviation at machining of hole $T_{(n+1)}$, ΔB - a deviation of a carrier basil size B_{ij}; $tL_{(n+1)}$ - a maximum deviation on machining of hole $T_{(n+1)}$; tL - a maximum deviation on machining of hole $T_{(m)}$; $\Delta_{(n+1)}$ - a deviation on machining of hole $T_{(n+1)}$ taking into account elastic deformations of the tool and a carrier basil.

Concerning closing hole T_m machining can be spended out in OX, OY - directions (Figure 5), according to calculations of nominal sizes of chains T_n, $T_{n+1} \rightarrow T_m$ and T_n, $T_{n+2} \rightarrow T_m$ results. Direction sampling is defined by calculations corrections $e_{(nm)}$ on each settlement chain results:

$$e_{(nm)}' = A_{(nm)P} - (A_{max(nm)} + A_{min(nm)})/2; \quad (33)$$

$$e_{(nm)}'' = (A_{min(nm)P} - A_{min(nm)}) + tL(\xi)^{-1} + (\Delta B + \Delta L_{(n)})\xi; \quad (34)$$

$$e_{(nm)}''' = (A_{min(nm)P} - A_{min(nm)}) + tL(\xi)^{-1} + (\Delta B + (tL_{(n)} + \Delta_{(n)})\xi); \quad (35)$$

$$e_{(nm)}' = e_{(nm)}'' = e_{(nm)}''' \quad (36)$$

$A_{min\,(nm)\,P}$ - the minimum design value of a link $A_{(nm)}$; $A_{min\,(nm)}$ - the minimum value of a link $A_{(nm)}$; $e_{(nm)}$ - correction of a closing link design value $A_{(nm)}$, $\Delta L_{(n)}$ - a deviation at machining of hole T_n, ΔB - a deviation of a carrier basil size B_{ij}; $tL_{(n)}$ - a maximum deviation on machining of hole $T_{(n)}$; tL- maximum deviation on machining of hole $T_{(m)}$; $\Delta_{(n)}$ - a deviation on machining of hole $T_{(n)}$ taking into account elastic deformations of the tool and a carrier basil.

Conclusions

In production conditions of low-waste RMS carrier is constructive conferred a basing redundancy property under the condition of variable spatial position in the work area in order to enable the simultaneous advance of the tool to multiple parts disposed on the lateral faces of its body G_i. This allows:

1. Reducing metal consumption of machine tool systems and obtaining a number of technical effects that simplify the organization of the production process.
2. Reducing waste of the production concerning to traditional production forms of organization.
3. In order to enable the accuracy of spatial arrangement of parts on the RMS carrier in this study were obtained dependences of basing parameters on its manufacturing precision parameters of basing holes of its body. Thus, on carrier basing accuracy the essential agency renders mutual bracing accuracy of basing holes on each basil. Oscillations of basing holes mutual bracing accuracy are caused by alternativeness of machining concerning each carrier basil dimensional chains.
4. The carrier machining algorithmic sequence where each subsequent hole is machining taking into account a lapse of the previous hole machining and maximum deviations of spacings on centres between them is offered.
5. The machining process mathematical model, allowing to count accuracy of basing holes spacings on centres taking into account at stage of their manufacturing lapse of machining process is offered.
6. Results of modelling are confirmed by calculations of carrier basing accuracy and results of its one basil basing holes T_n, T_{n+1}, T_{n+2}, T_m, machining.

References

[1] Cao S K, Yang S W, Chen X S, Wang G C, Song W W, Milling machine design based on open CNC system, Applied Mechanics and Materials, 52-54 (2011) 887-892. Information on https://www.scopus.com/inward/record.uri?eid=2-s2.0-79954456243&partnerID=40&md5=223dfcd1b59608abbb73c25b694d4716 DOI: 10.4028/www.scientific.net/AMM.52-54.887

[2] Chu C-L, Tai T-Y, Liu Y-H, Lu C-T, Chuang C-H, Liao H-W, Development of high-precision micro CNC machine with three-dimensional measurement system, International Journal of Automation and Smart Technology, 2(2) (2012) 95-101. Information on https://www.scopus.com/inward/record.uri?eid=2-s2.0-84905676733&partnerID=40&md5=7612a5d68d06aaf5c460023bf63b7b63 DOI: 10.5875/ausmt.v2i2.138

[3] Zakharov O V, Bobrovskij I N, Kochetkov A V, Analysis of Methods for Estimation of Machine Workpiece Roundness, Procedia Engineering, 150 (2016) 963-968. Information on http://www.sciencedirect.com/science/article/pii/S1877705816313807 DOI: 10.1016/j.proeng.2016.07.071

[4] Mehrabi M G, Ulsoy A G, Koren Y, Reconfigurable manufacturing system and their enabling technologies, International Journal of Manufacturing Technology & Management, Proquest ABI/INFORM, 1 (2000) 114-131. Information on http://www.inderscience.com/info/inarticle.php?artid=1330 DOI: 10.1504/IJMTM.2000.001330

[5] Mustapha N, Daoud A K, Wassy I S, Availability modeling and optimization of reconfigurable manufacturing systems, Journal of Quality in Maintenance Engineering, Emerald Group Publishing Limited, 9(3) (2003) 284-302.

[6] Bobrovskij N M, Melnikov P A, Grigoriev S N, Bobrovskij I N, Aspects of thermal field by wide burnishing, IOP Conf. Series: Materials Science and Engineering, 91 (2015) 012035. Information on http://iopscience.iop.org/article/10.1088/1757-899X/91/1/012035 DOI:10.1088/1757-899X/91/1/012035

[7] Pérez R, Dávila O, Molina A, Ramírez-Cadena M, Reconfigurable micro-machine tool design for desktop machining micro-factories, 7th IFAC conference on manufacturing modelling, management, and control, MIM 2013, (2013) 1417-1422 Information on http://wwwsciencedirectcom/science/article/pii/S1474667016344913 DOI: 103182/20130619-3-RU-301800031

[8] Sung-Yong S, Tava L O, Derek Y H, An approach to scalability and line balancing for reconfigurable manufacturing systems, Integrated manufacturing systems, 12(7) (2001), 500-511 Information on http://pubdbwiwiuni-jenade/publications/publication_view/3464 DOI: 101108/09576060110407815

[9] Koren Y, Shpitalni M, Design of reconfigurable manufacturing systems, Journal of manufacturing systems, Elsevier Science Publishing Company, (2011) ISSN: 0278-6125 Information on http://www-personalumichedu/~ykoren/uploads/Design_of_Reconfigurable_Manufacturing_Systemspdf DOI: 101016/jjmsy201101001

[10] Patil, S Van J, Vyatkin V, Pang C, Patil S, On composition of mechatronic components enabled by interoperability and portability provisions of IEC 61499: A case study, 2013 IEEE 18TH International conference on emerging technologies and factory automation, ETFA 2013, IEEE Industrial Electronics Society, University of Cagliari, 2013 ISSN: 1946-0740 Information on: http://ieeexploreieeeorg/document/6648136/?reload=true DOI: 101109/ETFA20136648136

[11] Abrishambaf R, Hashemipour M, Bal M, Structural modeling of industrial wireless sensor and actuator networks for reconfigurable mechatronic systems, The International Journal of Advanced Manufacturing Technology, Springer-Verlag London Ltd, 64(5) (2013) 793-811 ISSN: 0268-3768 Information on http://linkspringercom/article/101007/s00170-012-4070-y DOI: 101007/s00170-012-4070-y

[12] Plitea N, Lese D, Pisla D, Vaida C, Structural design and kinematics of a new parallel reconfigurable robot, Robotics and Computer-integrated Manufacturing, Elsevier Science Publishing Company, Inc, 29(1) (2013), 219-235 ISSN: 0736-5845 Information on http://wwwsciencedirectcom/science/article/pii/S0736584512000786 DOI: 101016/jrcim201206001

[13] Bobrovskij N M, Melnikov P A, Grigoriev S N, Bobrovskij I N, Simulation of thermal fields using different types of wide burnishing, IOP Conf Series: Materials Science and Engineering, 91 (2015) 012034 Information on http://iopscienceioporg/article/101088/1757-899X/91/1/012034 DOI: 101088/1757-899X/91/1/012034

[14] Abdi M R, Labib A W, Performance evaluation of reconfigurable manufacturing systems via holonic architecture and the analytic network process, International Journal of Production Research, Taylor & Francis Group, 49(5) (2011) 1319-1335 ISSN: 0020-7543 Information on http://wwwtandfonlinecom/doi/abs/101080/002075432010520989?mobileUi=0&journalCode=tprs20 DOI: 101080/002075432010520989

Integrated Development Of Noise-Dust Woodworking Machines At The Design Stage

A N Chukarin [a],Y I Buligin [b], L N Alexeenko [c], V A Romanov

Don State Technical University
Sq. Gagarina 1, Rostov-on-Don Russian Federation, 344010

Email: [a] opm@rgups.ru, [b]bulyur_rostov@mail.ru, [c]puchinka78@mail.ru

Abstract. The article deals the problem of creating of integrated security systems from the effects of hazardous and harmful factors of woodworking machinery on the design stage. Proposed the machine device design, which provides noise-dust protection of the operator fulfill the criterion of maximum permissible levels of noise and dust concentrations.

Introduction

Model woodworking machines are widespread and intensively exploited not only in woodworking industry, but in machine-building enterprises too, in particular, furniture shops and foundry areas. Existing modeling tools have technical characteristics, such as precision machining, performance, the reliability of the relevant international standards rigs, but for security duty have poor performance, because creating a working zone operators increased noise levels and concentrations of dustiness, much higher than health standards. Therefore the problem of noise and reduce dustiness model woodworking machines станков is relevant for mechanical engineering and has a great scientific and technical and socio-economic importance*.

Experimental studies of noise and vibration model machines

Experimental studies of noise spectra and model machines vibration performed under conditions foundry model areas JSC " Rostvertol " and LTD PC NEVZ. Measuring levels of vibration and noise were carried with acoustic meter "Ecophysics".

Research of formation regularities of noise spectra and model machines vibration conducted idling and cutting in various wood species and under typical processing conditions. When measuring noise levels microphone installed at the operator's workplace. When measuring vibrations piezosensor with a magnet fastened to the basic elements of machine tools, and to the blanks stucked with special mastic. It should be noted, that through measuring octave vibration levels fixed values not in normability workplace particle frequency, but in the particle frequency 31.5 - 8000 Hz.

Comparison of noise and vibration spectra allows qualitatively assess contribution of the individual sources into forming sound field in the working zone and identify dominant sources of noise. Idling at maximum rotation speeds sound pressure levels already exceed the maximum allowable value (fig.1).

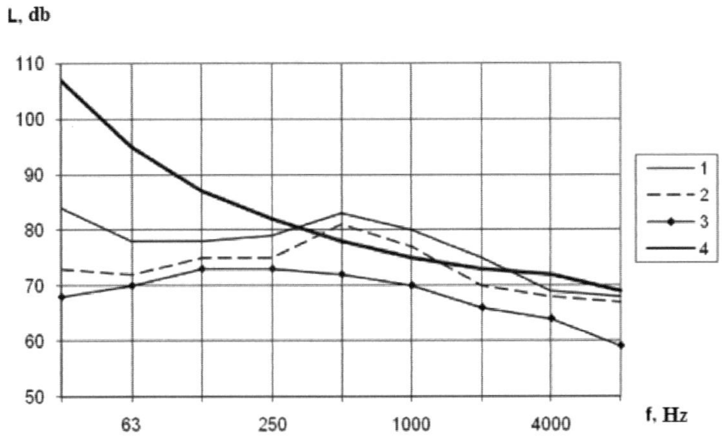

Figure 1 – The spectra of noise in the working area medium modelular machine idling: 1 - at rotation speed 6000 rpm; 2 - at rotation speed 4500 rpm; 3 - at rotation speed 1500 rpm; 4 - limit spectrum.

At maximum rotation speed 4500 и 6000 rpm sound pressure levels of the machine already exceed the maximum allowable value. For example, at rotation speed 4500 rpm sound pressure levels exceed the maximum permissible in the fifth and sixth octaves with center frequencies 500 и 1000 Hz (respectively) for 3-4 dB. By increasing rotation speed to 6000 rpm excess of sound pressure levels over the maximum admissible is 5dB in the fifth octave with center frequencies 500 Hz, 4 dB in the sixth octave with center frequencies 1000 Hz, 2 dB in the seventh octave with center frequencies 2000 Hz.

When cutting processing noise spectra in the working area changes significantly. In the noise spectra dominatштп medium and (especially) high-frequency part of the spectrum. For example, for light model machine most intensive components of noise spectra are located in the fifth to tenth octaves (fig.2). Excess of sound pressure levels observed at a spindle rotation speed 3000 rpm and amounts 2-5 dB (processing pine) in the frequency range 1000-8000 Hz. Increasing spindle rotation speed, the intensity of the sound emitting increases too.

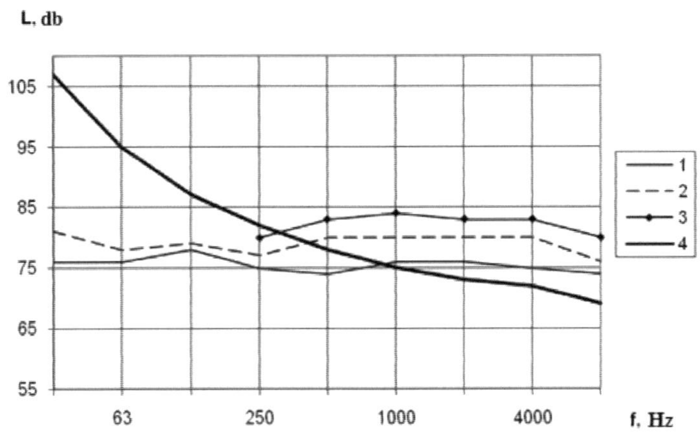

Figure 2 – The spectra of light model machine noise: 1 - processing pine (n=3000 rpm); 2 - processing pine (n=6000 rpm); 3 - processing oak (n=6000 rpm); 4 - limit spectrum.

Expands the active range of the spectrum, which has excess of sound pressure levels over the maximum allowable values. In particular, excessing noise levels by processing pine is from 2 to 8 dB in the frequency range 500-8000 Hz. When processing hardwood sound pressure levels increasing. For example, when processing oak blanks (n=6000 rpm) sound pressure levels in comparison with when processing pine blanks increasing by 3-4 dB, which coincides accurately with theoretical data (estimated value of 4 dB). Excess of sound pressure levels over the maximum permissible amounts of 3-10 dB. It should be noted the nature of identity of noise spectrum for processing different types of wood. Similar results obtained on average model machine FMS. Increasing sound pressure levels in comparison with a light model machine is generally about 2-3 dB in the mid-part of the spectrum 500-1000 Hz it could be assumed, that this fact is explained by increasing cutting power (the theoretical value of 3.2 dB). Increasing sound pressure levels over the maximum allowable values reaches 8-11 dB in medium- and high-frequency part of the spectrum 500-8000 Hz.

To clarify the contribution of noise sources forming a sound field in model machines working zone the measurements of vibration speed on the supporting elements of the system and processing blanks were provided [1].

The highest levels of vibration speed were fixed on buildings spindle headstocks. The levels of vibration speed of the motor body at low frequencies (unto 250 Hz) 8-15 dB are lower, then of body headstock. However directly in medium and high frequency part of the spectrum 500-8000 Hz levels of vibration speed of the motor body are practically the same then of the body headstock. Vibration spectra of frame, table and, especially, blanks,

are pronounced low-frequency character. The main sources, determining the excess of sound pressure levels over the maximum allowable values in the fifth-sixth octaves are: body headstock and cutting tool, firstly, and so the main drive motor.

These data are the foundation to develop technical solutions to ensure sanitary standards noise in the working area of model tools.

Experimental research of model machines dustiness and CRC efficiency

Disperse composition of dust we investigated by sieve analysis and microscopy by standard methods using apparatus for particle size distribution research, including standard set of sieves, vibration system and body with electrical enclosure, and MIN-8 microscope. As a result of preliminary investigations established, that the particle size distribution of all used in the experiments kinds of dust material obeys a log-normal distribution. Fig.3 shows dependence between frequency speed of model machine and size of resulting from processing materials of wood dust particles.

From the obtained data we can conclude, that wood dust and sample taken for analysis is coarsely and has property of polydispersity. Considering, that considered wood dust is coarse, advantage may be given to inertial cyclone duct-collectors, rather than textile filters.

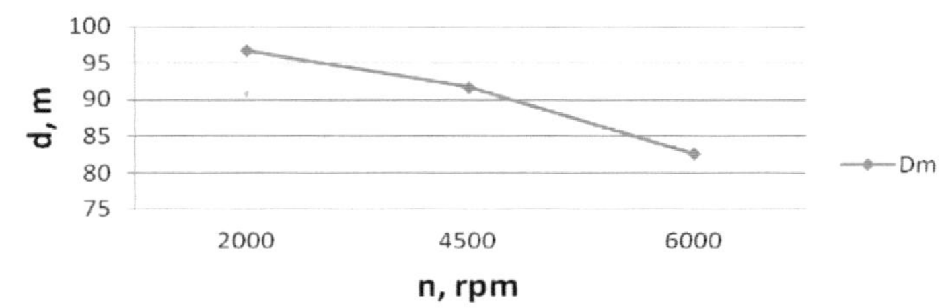

Figure 3 – The dependence of the average wood dust median diameter from the operation mode of model woodworking machine

Besides, increasing wood processing speed on model machines at high rotation speed reduces the average median particle size and decreases capture efficiency of conventional cylindrical cyclones (lower then 92.4%).

Therefore the choice of cleaning vehicles is based on the use of cyclones with advanced conical part and adjustable design parameters for maximum efficiency. CRC, due to the conical profile, uses increased centrifugal force, acting on a dust particle, and the ability to control the immersion depth of the exhaust pipe into cyclone body reduces dust removal by secondary eddy currents of air. Regulation of relations between the volume of working part of the CRC and hopper allows to capture dust of different density with more efficiency [2,3].

Experimental data led to the conclusion, that the taken dust samples are coarse disperse, as well as polydisperse. Further, fixed times, when released a blowout of bulk material from the outlet of cyclone with the same meaning of feed air flow Q= 25 m^3/h (V=4.5 m/s). Then was measured the mass of sawdust in the hopper using exact electronic scales and dust cleaning rate was calculated:

$$\eta = \frac{m_{ex}}{m_{en}} \cdot 100\%, \qquad (1)$$

where - mass of dust after treatment, g; - mass of dust before treatment, g.

Research of the efficiency of wood dust collecting were made for different amounts of hoppers from 5 l to 60 l in the particle frequency 40-315 micrometer at a rate of 27 m^3/h. The efficiency of dust cleaning is higher in the conical cyclone in contrast with cylindrical cyclones and increases in the case of increasing the ratio of the volume of the hopper to the working volume of the cyclonea. At the fig. 4 shown efficiency dust collecting of wood dust with conical cyclone depending on its disperse composition, competition of hopper volume to the working volume of the cyclone with optimum ratio of the outlet pipe height to the height of the cyclone Hp/Hc=0,6. Ratio of hopper volume the working volume of the cyclone changed in the course of research and respectively were equal: for hopper capacity of 19 l: Vh/Vcc=0.7; for 30 l: Vh/Vcc=1.11, for 60 l: Vh/Vcc=2.22. As can be seen from the experiments, dust cleaning efficiency higher for the hopper volume of 60 l and ratio Vh/Vcc=2.22.

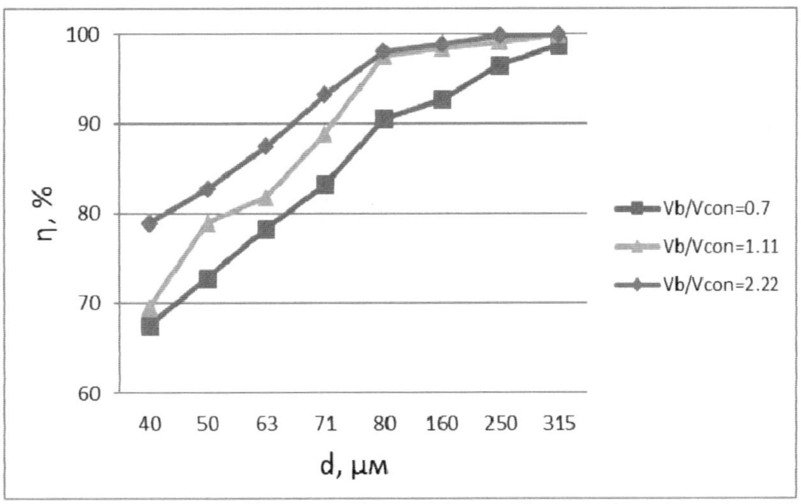

Figure 4 – Dust collecting efficiency of wood dust with conical cyclone depending on the dispersed composition, attitude of hopper volume to the working volume of cyclone with optimum ratio Hp/Hc=0.6

Experimental studies of the impact of exhaust pipe immersion depth on dust collection efficiency by cyclones with a hopper volume of 30 l in the particle frequency of wood dust from 40 to 315 micrometer (fig. 5,6).

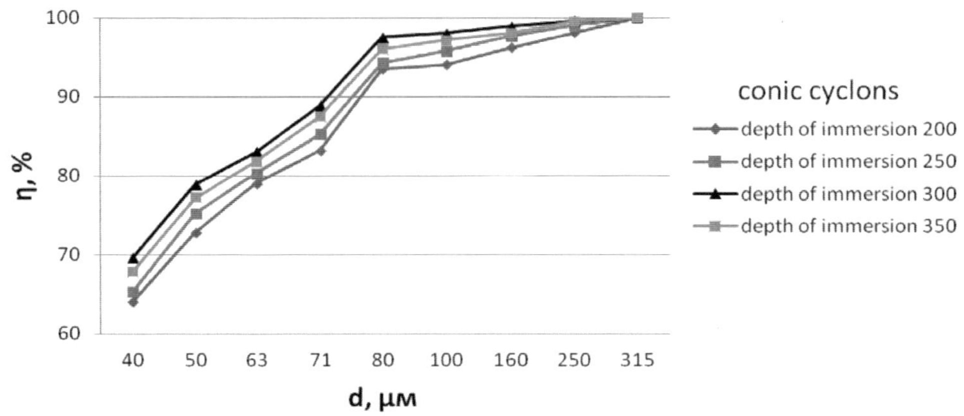

Figure 5 – Dust collecting efficiency of wood dust with conical cyclone depending on the pipe immersion depth of exhaust pipe

From the results of experimental studies it can be made the following conclusions:

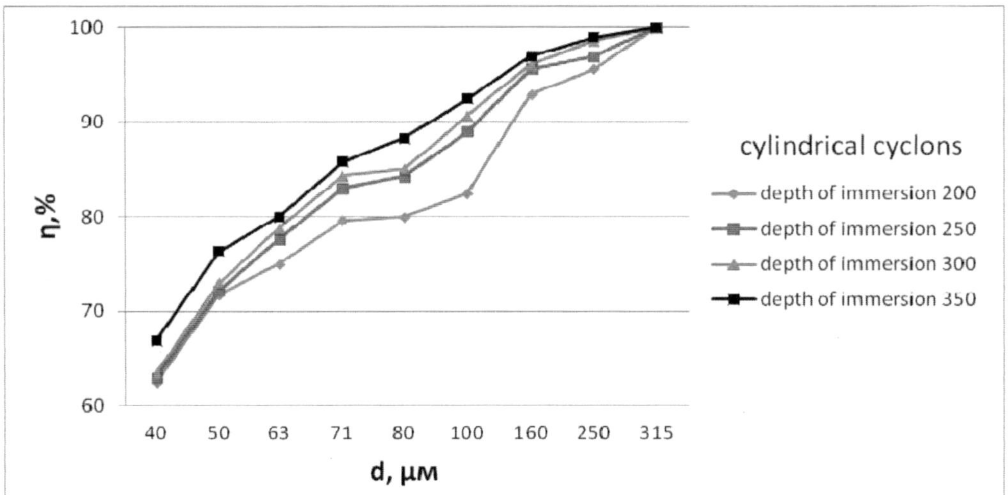

Figure 6 – Dust collecting efficiency of wood dust with cylindrical cyclone depending on the pipe immersion depth of exhaust pipe.

1. With increasing depth of immersion exhaust pipe, at the same values of input speeds of dusty air, dust collection efficiency values of the conical device is much higher than that of cylindrical, most clearly seen this effect at speeds 15-20 m/s.

2. When the immersion depth of the outlet pipe is h=300mm the resistance of the conical cyclone unit takes its maximum value, and dust collection efficiency reaches its maximum value 97-99 %. Conversely, with a further increase of the outlet pipe immersion depth ещ h=350 mm the efficiency of the conical cyclone falls.

3. Cylindrical cyclone has maximum resistance at a depth of immersion of exhaust pipe to h=350 mm. At this depth experimentally established the greatest efficiency of dust collecting too, which, however, much lower, than the same for conical device (88-92%).

It was established, that the construction of dust collectors of non-cylindrical shape (ball cyclone or a cyclone with a reverse cone (CRC)) with adjustable geometric parameters enable carry out the cleaning dust-air flow process more effective in comparison with the conventional cylindrical cyclones. Have been founded the optimal attitude of volumes of the working part of the CRC and the hopper, and so attitude of immersion depth of the exhaust pipe to the height of the working part of the machine, in which the dust cleaning efficiency increases to 99 % [3].

Design of noise-dust protection system for woodworking machines.
Below there is the structure of comprehensive protection of operators from the noise and dust impacts (fig.7) [4-10]. The dominant role in the formation of noise spectra in the working zone of the cutting tool sound emitting of the model machines at high frequencies and headstock body at medium frequencies allows to restrict with the noise protection system only the cutting zone.

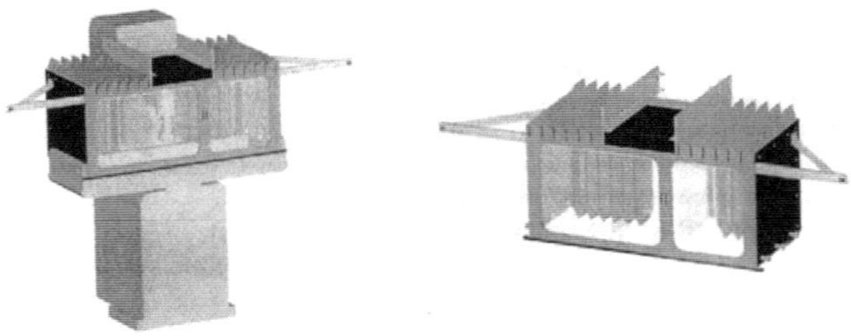

Figure 7 – Noise protective fencing and aspiration cover wood (model) machines

Processing zones protections have a sufficiently high sealability, which increases the concentration of dust under protective device. Therefore, based on the concentration of dust in the cutting zone, calculated concentration of dust under enclosing construction and compared with the experimental values.

Installation of cutting area sound dampening enclosure of light and medium model machines and acoustic screens of medium machines engines provided performance of sanitary norms of noise in the operator's work places. The use of this air purification system from dust reduced the concentration of wood dust in the working zone to 3.5-4 mg/m^3, which is 2-2.5 mg/m^3 lower than maximum allowable concentration.

Conclusions
Analysis of working conditions of operator's work places of model milling machines showed, that disparity sanitary conditions observed for only two dangerous and harmful production factors – dust and noise. The dust concentration exceeds the maximum permissible concentration 6 times, and the sound levels of maximum permissible limit – to 15 dBA.

Experimental studies of vibration and noise spectra, conducted in conditions of foundry model areas revealed the dominating in in excess of sanitary norms noise sources and in fact determined the choice of maximum permissible sound pressure level execution methods.

Developed the construction of noise-dust protection system for model machines. Calculating of design parameters produced by the criterion of performing of maximum permissible levels of noise and dust concentrations.

* The work was performed within the Federal Target Program within the government order Ministry of Education and Science of the Russion Federation to conduct fundamental research work in 2014-2016 on this topic: "Development of bases of designing technology complex systems and operator protection from the effects of hazardous and harmful factors".

Reference
[1] Romanov V A, Chukarin A N, Fleck B M. Refinement of the vibration spindle headstocks calculation methods for milling and drilling woodworking machines DSTU Reporter. – Rostov-on-Don, 2013. - № 1/2 (70/71), p.86-92.

[2] Meskhi B Ch , A N Mikhailov, Buligin Y I, Alexeenko L N, Denisov O V, Panchenko O S. Dust collector Patent number 2506880 of 20 02 2014.

[3] Buligin Y I, Panchenko O S, Romanov V A, Denisob O V. Increased air dedusting efficiency of metalworking and woodworking industries working zones/ // DSTU Reporter – Rostov-on-Don, 2013 – № 7/8(75), p 49-57.

[4] Meskhi B Ch, Ryabyh G Y. Vibroakustic characteristics and dustiness of woodworking machinery Monography - Rostov-on-Don: DSTU Publishing center, 2007 – 153 p.

[5] Meskhi B.Ch., A.G. Li., Tsvetkov V.M. Mathematical models of noise generation process with interrupted cuts // Pub. IU AP. № 1, 2004:-p. 3-12.

[6] Meskhi B.Ch. Improving the working conditions of operators of machine tools by reducing the noise in the work area (theory and practice). Rostov-on-Don: Publishing center DSTU, 2003. — 131 p.

[7] Meskhi B.Ch. Noise characteristics of grinding machines, The Second All-Russian School Seminar with international participation «New in theoretical and applied acoustics», SPb, 17-18 october 2002: Proceedings / edieted by. N.I. Ivanova, p. 55-56.

[8] Noise abatement in manufacturing: Directory / Edieted by. Y.Y.Yudina. -M.: Engineering, 1985. -400 p.

[9] Meskhi B.Ch. Noise characteristics of grinding machines, The Second All-Russian School Seminar with international participation «New in theoretical and applied acoustics», SPb, 17-18 october 2002: Proceedings / edieted by. N.I. Ivanova. BSTU.: 2002. p. 55-56.

[10] The method of reduction of industrial noise. Naturliche Gerausche // Production. 1999. - № 47. - p. 46.

IOP Conf. Series: Earth and Environmental Science **50** (2017) 012015 doi:10.1088/1755-1315/50/1/012015

Environmental Aspects Of The Green Surface Plastic Deformation Technology Of Car Parts

S N Grigoriev [1, a], N M Bobrovskij [2, b], I N Bobrovskij [2, c], P A Melnikov [2, d], A A Lukyanov [2, e]

[1] Moscow State Technological University Stankin, Vadkovskij per. 1, Moscow, Russian Federation, 127055
[2] Togliatti State University, Belorusskaya st. 14, Togliatti, Russian Federation, 445667

E-mail: [a]Rector@stankin.ru, [b]Bobrnm@yandex.ru, [c]Bobri@yandex.ru, [d]Topavel@mail.ru, [e]A.lukyanov@tehnomasch.ru

Abstract. Foreign and domestic experience in development of dry processing technologies are considered. The results of the introduction of dry processing technologies (cutting, boring, milling, drilling) on the industrial companies in Germany are given. The negative impact on the environment and human health is shown. The possible ways of leakage of lubricoolant components in the atmosphere and soil are considered. Lubricoolants are considered as a required permanent component. Three main tasks for lubricoolant: cooling, lubricating and chip disposal are discribed.

Introduction

Modern engineering plant consumes ten thousand tons of lubricoolants required for machining processes of machine parts. In existing machining technologies and particularly in surface plastic deformation processing (SPD), processing combined with supply of lubricoolants which have a negative impact on the environment and health of production personnel [1]. Some research's shows that energy costs of using lubricoolants (supply, maintenance and cleaning in the circulation system) reach up to 80% of the total energy costs.

The present stage of development of mechanical engineering is characterized by statement and solution of extreme problems to find the optimal conditions in the processing under the changed significance criteria of parameters of machine parts, the change to so-called "green" technologies. This increase of research interest in such complex tasks associated with limit of natural, material, human resources and need to save energy and materials.

There are three main ways of green machining technologies development:

1. Dry processing;
2. Lubricoolant replacement for environmentally friendly materials (vegetable-based and organic lubricants);
3. Using micro supply of lubricoolant during processing.

In 2012 import quota in the Russian Federation of machine and tool production was 91.2%; machines – 93.3%, tools – 88.1%, abrasive production – 83.4%. The value of exports of machine and tool production is high - 58.1%; machines - 41.9%, for the tools - 61.0%, abrasive production - 77.6%.

Therefore, engineering solutions in this area may contribute to not only import substitution, but forthcoming of export-dependent production. The leading countries of the engineering sector (USA, Japan, Germany) have not such a complex problems as in Russian Federation. Nowadays, active steps are carried out in the solution of the problem. "The Presidential Commission for Modernization and Technological Development of Russia's Economy" was established.

Europe experience: refusal of lubricoolant

Since the 90s of the last century many European companies in collaboration with institutions have been performing developments in the theoretical and practical methods of machining without lubricoolants for Environmental Security of production and expected economic effect. The major project "Dry processing" is carried out in Germany in the spring of 1994 and 2002. It covered a fundamental scientific research, development and optimization of technologies for tools, equipment and processes for machining different materials. The project was implemented under the Federal Ministry of Education, Science, Research and Technology. The project involved 24 companies (DaimlerChrysler AG, Gühring, Huller Hille, Mapal, etc.), institutes and innovation technology centers which was organized into 5 groups with representative industrial companies leaders. In general, dry processing technologies (dry cutting, turning operation, milling, drilling) except SPD processing were investigated.

Two project groups have implemented developments into mass production in December 1996: aluminium milling on the company "Dasa" in Augsburg and dry drilling on company "HDM". Huller Hille company has developed a machining center for dry machining of engine parts (cylinder head).

At the factory of "BMW" company in Steyr on a new automated line in a dry processing was carried out boring of holes of 6-cylinder aluminum block. Cost analysis showed an economy of 17% due to decrease of investment and operating costs. Deep boring is implemented in the company "Daimler-Benz". Introduction of dry machining in Bosch company was planned in 1999-2000, in Nuremberg is production of body of throttle valve (1100 parts / day). In Homburg is boring and milling of body of hydraulic valves from casting (28,000 parts / year). In Feuerbach is the process of deep boring in steel (4000 parts / day).

The BMW, Daimler-Benz companies and ISF Institute of Dortmund's University developed a list of requirements for the machine, operating on dry processing technologies in mass production. These requirements are implemented on the milling machine "Specht 500-t" in Huller Hille company.

Russian Federation: refusal relevance of lubricoolant

The mechanical assembly production of JSC "AVTOVAZ" operates 356 units of metal-working equipment that consumes 2 thousand tons of fire hazard lubricoolants per year which 60% are irreversible losses. These liquids have a low ignition temperature; it creates the risk of fire; deteriotates the working conditions which cause occupational diseases; leads to environmental damage.

Researches of Russian scientists confirm the fact that petroleum oils vapors can lead to the body's defeat up to the lipoid pneumonia. Hydrocarbon vapors have a narcotic effect, triethanolamine causes dermatitis, sodium nitrite contributes to gas exchange abnormalities in humans body and damage to the heart muscle, the chlorinated additives can cause damage to the liver and kidneys, trichloroethane is a source of phosgene liberation, etc.

Air pollution occurs not only during operation, but also as a result of evaporation and combustion of used lubricoolants. However, the main danger it is their leak into the soil and get into the subterranean water.

Despite of constant improvement of machining technologies, tools and equipment, lubricoolants are considered as a required permanent component. Lubricoolant performs three main tasks for the steady processing: cooling, lubricating and chip disposal. Total rejection of lubricoolant can solve many problems.

In opinion of the public and consumers the corporate policy of modern industrial company is not acceptable without her efforts to preserve ecological balance. The constant raising the legal requirements for environment protection leads to the search for environmentally friendly processes in mechanical processing.

Experience of processing without lubricoolants technology

Between 2001 and 2012 by scientist of the Togliatti State University (TSU) have been developed: scientific bases of surface plastic deformation processing without lubricoolants of car's parts; process mathematical models; algorithms for calculation of process conditions (confirmed by a certificates of algorithm registration); technological limitations of the process; processing techniques; tools and devices designs implemented in production and patented. The results of work published in more than 100 articles, 50 domestic and International proceedings, were granted 11 patents and 3 certificates of algorithm registration, prepared for defense 1 doctoral thesis and 3 candidate theses.

Environmental compatibility as perspective of developed technology

Components of lubricoolants, bactericides and fungicides, reaction products as well as foreign substance can cause a series of occupational diseases. The German Agency for Environmental Protection also announced that up to 30% of all occupational diseases of engineering companies employees associated with exposure to lubricoolants in labor activity process. Evaporation and combustion of exhaust lubricoolants pollute the atmosphere. Toxic components (sulfur dioxide, organic chlorine and heavy metal compounds) with clouds spread across the planet. Moreover, combustion of synthetic oil based lubricoolants has unpredictable effects.

The leakage of lubricoolants into the soil and infiltration into the subterranean water leads to the formation of "oil lens". The speed of oil propagation is 10.2-10.5 m/s in width and depth. Oil migrates with subterranean water and contaminates it with components: polycyclic aromatic hydrocarbons, polychlorinated biphenyls, sulfur- and chlorine-containing additives, biocides, nitrites, organic metal contents (lead, barium, antimony, zinc). Especially low degree (10-30%) of biodegradation of these substances leads to ecological disbalance: intensive reproduction and mutation of microorganisms assimilating mineral oil.

In this way, processing with total rejection of lubricoolants significantly improves the environmental performance of production and reduces the stress on the region's ecosystem and there is no need to dispose of exhaust lubricoolants.

Technology implementation

The collective of authors propose a finish surface plastic deformation processing technology of surface of revolution without lubricoolants.

It is worth nothing that surface plastic deformation processing does not withdraw the metal. Surface irregularities, remaining from the previous operation, are partially or completely smoothed during processing, and the surface acquires brightness [2-14]. Thus, there is no problem of chip removal from the treatment region as in cutting or abrasion processing, though in SPD method the wear fragments are accumulated on tool's working surface.

Consequently, SPD processing depends on detergency value of lubricoolant which is determined by degree of treatment region purification of sludge containing the fine chip particles of workpiece and tool material (carbides and non-metallic inclusions). Mechanical and electrostatic van der Waals forces retain the sludge on surface of solid's.

The smaller sludge particles easier penetrate into the tool and workpiece and more impacted on the solid surface degrading processing technological parameters. In case of improper lubricoolant purification the sludge are deposited on the surface of workpiece directly from the liquid. The visual control of contaminated workpiece's surface is complicated.

To resolve this problem authors have developed a device for SPD processing of parts without lubricoolants [15]. The device is directed to solve the technical problem of working surface cleaning during processing.

IOP Conf. Series: Earth and Environmental Science **50** (2017) 012015 doi:10.1088/1755-1315/50/1/012015

The technical result is provided by the device for SPD processing of revolution surfaces with lubricoolant which remains on workpiece surface after previous operation. It contains the drive holder and tool holder and processing tools (indenters) equipped with rectilinear edge scrapers. The edges are located lengthwise the workpiece surface generatrix, and scrapers contact the workpiece surface at least with one edge.

Figure 1 shows the proposed device. Indenters 1 are pressed against the processing surface with the operating force and fixed in the holder 2 with screws 3.

Fine-fibre felt scrapers 4 are secured on instruments 1. Scrapers are fixed by the inner salient in the hole to the rear end to the indenter or by assembling with a tightness, as shown in the figure 1. The hole in scraper performed with diameter of approximately 1 mm less than the diameter of the tool.

The device operates as follows:

Workpiece starts to rotate. Indenters 1 are contacted with workpiece's surface towards the arrow on figure 1 and pressed against it with the operating force. After that, tools are feeded lengthwise the axis of workpiece, wherein scraper 4 contacts the processing surface with edges "a". Edges of scraper "a" captures sludge particles remaining on the workpiece's surface or occured during processing.

Commonly, the amount of sludge is relatively small. The experience of device showed that the processing of such workpiece as rear axle shaft with two-shift operation mode and the productivity of the machine up to 180 parts/hour requires to replace scraper at most once every two weeks.

The application of this device provides for operator-setter the possibility of permanent equipment visual inspection immediately after the part out of the treatment zone, and also provides the required surface quality and tool's durability.

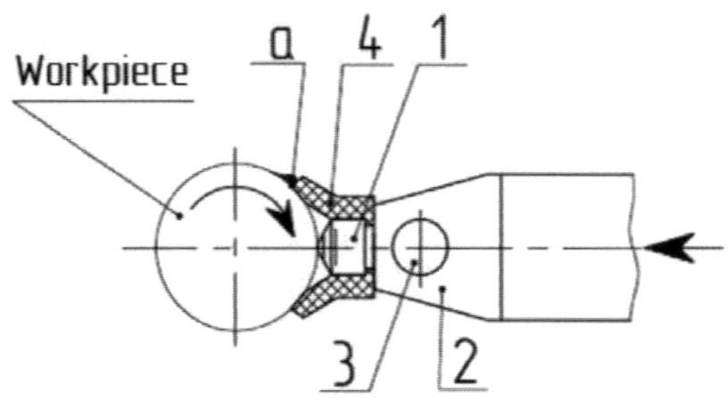

Figure 1 – Engineering solutions implemented for SPD processing without lubricoolants

Practical application

The generalization of the results of theoretical and experimental studies allowed the development and implementation of high productive SPD processing with wide indenter without lubricoolants in mass production. In this processing method the deformation area width matches the width of processing surface [16-18].

At present, with use of SPD technology without lubricoolants produced more than 15 million parts. The research works are carried out to expand the range of parts processed by the proposed technology. Figure 2 shows the machine for processing by SPD without lubricoolants.

Figure 2 – Machine for processing by SPD without lubricoolants

The application of this developed technology of car part's gland necks processing reduced the number of defects "leaking of gland" 3 times during the warranty period. After polishing by traditional technology (Figure 3) surface receives the required microrelief in connection with gland in a long time period in contradistinction to SPD method (Figure 4).

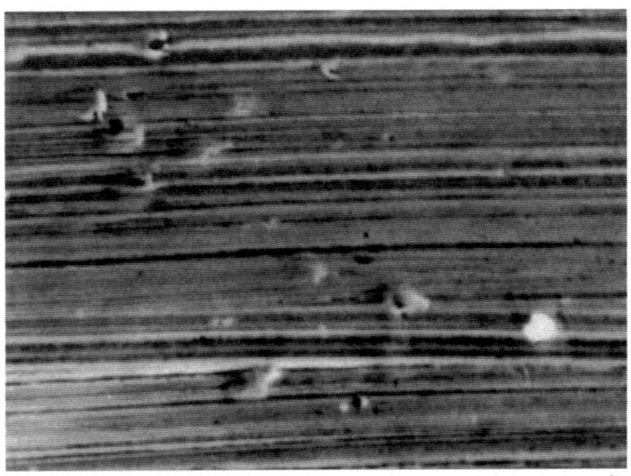

Figure 3 – Surface after polishing processed by "Nagel" machine

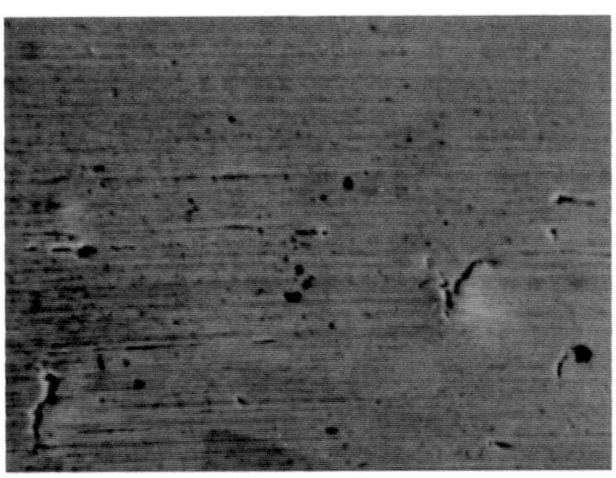

Figure 4 – Surface after SPD processing without lubricoolants

Results and Discussion

Innovative technology of surface plastic deformation processing without lubricoolants has no application analogues in mass production in Russian Federation and abroad. That is the only processing without lubricoolant which uses native tools and technology that is implemented in mass production in Russian Federation.

The total economic effect from the technology implementation in mass production more than 1.5 billion rubles.

The proposed engineering solutions are exclusively based on domestic technology and tools, thus expensive imported equipment (costs up to EUR 1 million per unit) are eliminated and the load of engineering enterprises of the Russian Federation are increased. Processing technology without lubricoolants eliminates the fire hazard, environmental pollution and harm to human health.

Conclusion

The SPD processing technology without lubricoolants has shown high quality results in production of automotive components of LADA cars. Developed technology used for finishing processing of gland neck of primary shaft of LADA car and has the following parameters [19]:

- Processing method: Surface plastic deformation;
- Processing surface: gland neck of primary shaft D 20…40 mm;
- Processing length: 10…20 mm;
- Initial surface roughness: less than Ra 0.8;
- Number of concurrent instruments: 2;
- Nominal force on the tool: adjustable, 150…250 N;
- Nominal spindle speed: 500 rev/min (adjustable by continuously frequency converter);
- Nominal feed: 0.12 mm/rev (0.06 mm/rev for each tool);
- Drive of the tool: hydraulic;
- Drive of feed (slide movement): hydraulic;
- Nominal cycle time: 28.8 sec;
- Productivity with the efficiency factor 0.8: - 100 parts / hour.

Thus, the proposed technology can be applied to any engineering enterprises in processing of the surface of revolution of the parts working in terms of friction, for example: gland neck of shafts, directing gland neck of shafts, surfaces of rolls.

References

[1] Kamble P S, Jadhav V S, Experimental study of roller burnishing process on plain carrier of planetary type gearbox, International journal of modern engineering research, 2(5) (2012) 3379-3383.

[2] Mayank K P, Manish D P, Optimization of input process parameters for vibratory ball burnishing process using RSM, International journal of research in engineering & applied sciences, 2(2) (2012) 372-381. Information on: http://euroasiapub.org/wp-content/uploads/2016/09/39-3.pdf.Information on: http://citeseerx.ist.psu.edu/viewdoc/download?doi=10.1.1.417.243&rep=rep1&type=pdf.

[3] Bobrovskij N M, Melnikov P A, Grigoriev S N, Bobrovskij I N, Aspects of thermal field by wide burnishing, IOP Conf. Series: Materials Science and Engineering, 91 (2015) 012035. Information on http://iopscience.iop.org/article/10.1088/1757-899X/91/1/012035 DOI:10.1088/1757-899X/91/1/012035

[4] Bobrovskij N M, Melnikov P A, Grigoriev S N, Bobrovskij I N, Simulation of thermal fields

using different types of wide burnishing, IOP Conf. Series: Materials Science and Engineering, 91 (2015) 012034. Information on http://iopscience.iop.org/article/10.1088/1757-899X/91/1/012034 DOI: 10.1088/1757-899X/91/1/012034.

[5] Kuznetsov V P, Smolin I Y, Dmitriev A I, Konovalov D A, Makarov A V, Kiryakov A E, Yurovskikh A S, Finite element simulation of nanostructuring burnishing, Physical Mesomechanics, 16 (1) (2013) 62-72. Information on: https://www.scopus.com/inward/record.uri?eid=2-s2.0-84876512923&partnerID=40&md5=00ee115f8291c87727443f4baab07b3e DOI: 10.1134/S1029959913010074.

[6] Kuznetsov V P, Skorobogatov A S, Gorgots V G, Yurovskikh A S, The analysis of speed increase perspectives of nanostructuring burnishing with heat removal from the tool, IOP Conference Series: Materials Science and Engineering, 124(1) (2016) 012127. Information on: https://www.scopus.com/inward/record.uri?eid=2-s2.0-84975745777&partnerID=40&md5=d172fb1942804f2147dedaf58b28ecaa DOI: 10.1088/1757-899X/124/1/012127.

[7] Kuznetsov V P, Dmitriev A I, Anisimova G S, Semenova Yu V, Optimization of nanostructuring burnishing technological parameters by Taguchi method, IOP Conference Series: Materials Science and Engineering, 124(1) (2016) 012022. Information on: https://www.scopus.com/inward/record.uri?eid=2-s2.0-84975706896&partnerID=40&md5=c9fb4cd34a1ffcdaa0d9521d84324eb6 DOI: 10.1088/1757-899X/124/1/012022.

[8] Popova M A, Kuznetsov V P, Lesnikov V P, Popov N A, Konakova I P, The structure and mechanical properties of single-crystal nickel alloys with Re and Ru after high-temperature holds, Materials Science and Engineering A, 642 (2015) 304-308. Information on: https://www.scopus.com/inward/record.uri?eid=2-s2.0-84936999787&partnerID=40&md5=c89b23ed44459075b7af40abc6a02c9b DOI: 10.1016/j.msea.2015.06.094.

[9] Gulhane U D, Mishra S B, Mishra P K, Enhancement of surface roughness of 316L stainless steel and TI-6AL-4V using low plasticity burnishing: DOE approach, International journal of mechanical engineering and technology, 3(1) (2012) 150-160. Information on: http://www.iaeme.com/MasterAdmin/UploadFolder/STAINLESS%20STEEL.pdf.

[10] Babu P R, Ankamma K, Prasad T S, Raju A V S, Prasad N E, Optimization of burnishing parameters and determination of select surface characteristics in engineering materials, Optimisation of burnishing of engineering materials, 37(4) (2012) 503-520.

[11] Afazov S M, Becker A A, Hyde T H, Mathematical modeling and implementation of residual stress mapping from microscale to macroscale finite element models, Journal of manufacturing science and engineering, 134(2) (2012). Information on: http://manufacturingscience.asmedigitalcollection.asme.org/mobile/article.aspx?articleID=1461038 DOI: 10.1115/1.4006090.

[12] Zhou R, Cao J, Ehmann K, Xu C, An investigation on deformation-based surface texturing, Journal of manufacturing science and engineering, 133(6) 2011. Information on: http://manufacturingscience.asmedigitalcollection.asme.org/article.aspx?articleid=1460597 DOI: 10.1115/1.4005459.

[13] Dyl T, Dolny K, The influence of the burnishing on technological quality of elements of part shipping machines, Journal of KONES powertrain and transport, 17(2) (2010) 89-95.

[14] Dabeer P S, Purohit G K, Effect of ball burnishing parameters on surface roughness using surface roughness methodology, Advances in production engineering & management, 5(2) 2010 111-116. Information on: http://apem-journal.org/Archives/2010/APEM5-2_111-116.pdf

[15] Bobrovskij N M, Bobrovskij I N, Gomelskij M V, Ezhelev A V, Melnikov P A, Lukyanov A A, The device for processing surfaces of revolution by burnishing. Patent, Russian Federation, no 2482953, 2011.

[16] Bobrovskij N M, Melnikov P A, Grigoriev S N, Bobrovskij I N. Modeling of Surface Topography After Burnishing Processing // Applied Mechanics and Materials. 2015. T. 770. p. 397-401. URL: http://www.scientific.net/AMM.770.397

[17] Melnikov P A, Bobrovskij I N, Grigoriev S N, Bobrovskij N M. Burnishing Tool Face Optical

Control Method // Applied Mechanics and Materials. 2015. T. 770. p. 248-252. URL: http://www.scientific.net/AMM.770.248 (дата обращения: 23.08.2016).
[18] Bobrovskij I N, Grigoriev S N, Melnikov P A, Bobrovskij N M. Research Of Hardalloyed Burnishing Tool Durability With Coatings By Ion-Plasmous Sputtering Method // Applied Mechanics and Materials. 2015. T. 770. p. 274-278.
[19] Small innovative enterprise "Tehnomasch+". Information on: http://en.tehnomasch.ru.

Research Of Polytropic Exponent Changing For Influence Evaluation Of Actual Mixture Composition On Hydrocarbons Concentration Decreasing On Deep Throttling Operation

N M Smolenskaya[a], **V V Smolenskii** [b], **I Bobrovskij** [c]

Togliatti State University,
Belorusskaya st. 14, Togliatti, Russian Federation, 445667

E-mail: [a]Nata_smolenskaya@mail.ru, [b]Biktor.cm@mail.ru, [c]Bobri@yandex.ru

Abstract: The purpose of this article is to present study of polytropic exponent as rating of thermodynamic process in internal combustion motor operating to deep throttling in a subcase of idle running. It is necessary to consider the influence of hydrocarbon part in exhaust gases in a process of development a new internal combustion engines especially on deep throttling operation: on combustion procedure, on irregularity of exhaust gases composition.

Introduction

Deep throttling operating mode attributes are negatives: low-pressure on bump, working and exhaust strokes, reverse emissions exhaust gases cylinders and even in intake system in the phase of valve overlap; increasing of residual gas ratio and rate of inertia; oil absorption in combustion chamber through interstice in tube rings and guide bushings of the intake valves. All this deteriorate the ignition and combustion, and lead to misfire, incomplete combustion, and to increasing of CH concentration in exhaust gases [1].

On deep throttling operation present high combustion period instability of sequential periods, this defines by changing of residual gases inertia, and so, incoming charge reactivity. Usually, periods with most comprehensive combustion follows by periods with misfire and combustion degradation. Actual working substance mixture composition is non-stable and changes from period to period, that changes thermodynamic processes of compression, burning and expansion. Hydrogen reforms hydrocarbon fuel burning conditions which is important for deep throttling operations [2, 3, 4].

Polytropic exponent

One of the thermodynamics process integral characteristic is polytropic exponent which indicates changing of heat flow direction and intensity [5, 6]. On figure 1 represented schematic changing of polytropic exponent and marked thermodynamic system main curve cusps. Point 1 indicates beginning of compression on curve piece to piece where compression polytropic exponent equivalent adiabatic exponent, which results in heating of combustible mixture from engine cylinder walls, which degrades and heat flow becomes equivalent 0 in point 1. On curve piece from point 1 to point 2 happens heat sink from contracting combustible mixture through cylinder walls. Point 2, where polytropic and adiabatic exponents matches again, and generated heat quantity becomes equal heat quantity, degrades through cylinder walls [7].

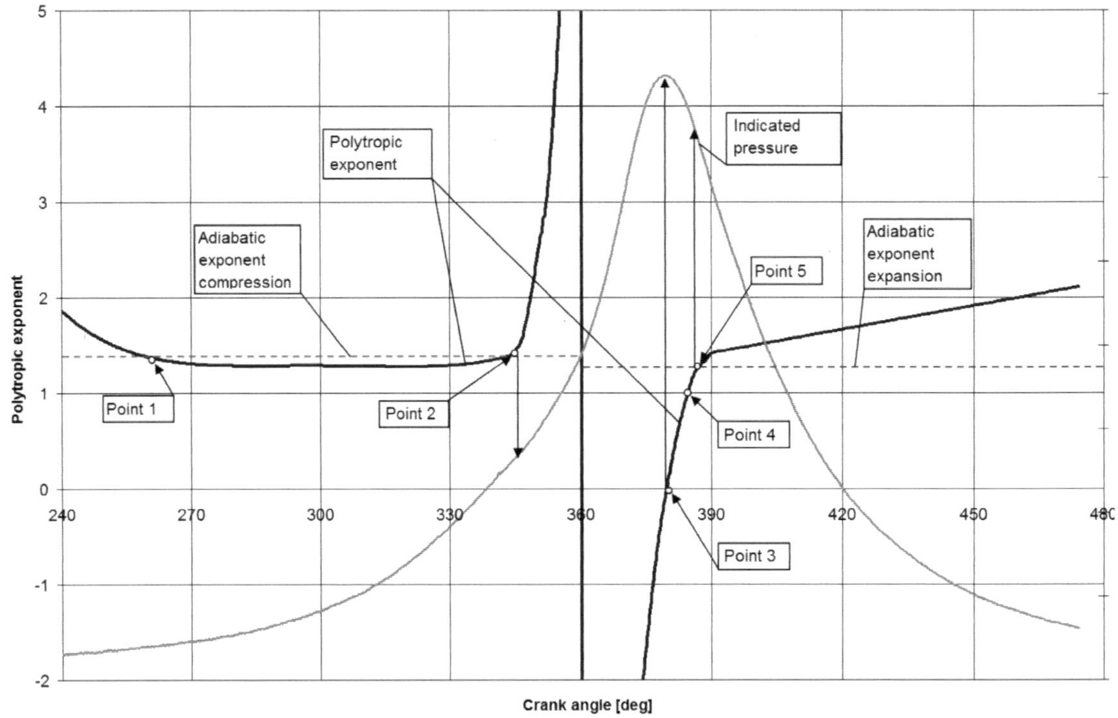

Figure 1 - Polytropic exponent characteristic curve actual cycle explosion engine with spark ignition

On curve piece from point 2 to top dead point (TDP), equivalent 360 crank angle; combustion process on compression stroke and polytropic exponent tends to infinity i.e. isochoric process. On curve piece from TDP to point 3 polytropic exponent takes a negative value, relevant active heat emission on expansion stroke. Point 3, where n = 0 responds to maximum cylinder pressure. On curve piece from point 3 to point 5, polytropic exponent becomes equality to adiabatic exponent, occurs afterburning on expansion stroke. In point 4 polytropic exponent equals one (isothermal process means that quantity of heat returning through cylinder walls equal to quantity of heating power). On curve piece from point 5 exhaust stroke describable «graded» zone, where combustion procedure ends, and linear increase zone, where heat sink increasing because of area of the cooling surface.

Therefore, changing thermodynamics process integral characteristic research on deep throttling operating mode is actual to indicate ability to increase fuel effectiveness [1, 4, 7, 8].

Experimental technique

Experimental investigation was held on engine VAZ-2111 (Figure 2), upgraded with special pressure plate (thickness is 4 mm), compression ratio decrease from 9.9 to 7.5.

Indexing system was used to analyze radiant quantities of thermodynamics and combustion processes, and additional equipment was used to determine air and fuel output. META Avtotest was used to determine exhaust gas composition. Tensiometer of pressure DMVG (produced by MVG company) with proper signal multiplication. This sensor does not need cooling; it is thermostable, fast response is possible because of high natural frequency – 20 kHz. Optical sensor was used as crankshaft position sensor, and data was registrated with system L-Card L-783M: all-in-one PCI card, allowing to add a signals of megahertz range digital signals access.

Figure 2 – VAZ-2111 engine with sequential multiport fuel injection

Figure 3 – Special pressure plate

Results and Discussion

Acquired data of pressure history changing shows that on idling mode there is irregularity of operation. Figure 4 represents an example of pressure changing oscillograph record with mixture composition (α) 0.988.

On main procedure on idling mode there is periods, which can be described as misfire periods, perfect combustion and partial combustion (Figure 4a). So end gas composition, and composition of air-fuel mixture depends on how went combustion on previous cycle. In the work of V.F. Kamenev [1] end gas composition in sequential cycles is not equivalent. After first cycle of perfect combustion end gases becomes inactive, and on next cycle they ballasting air-fuel mixture, so occurs partial combustion and stable flame front cannot be created. That means that in next cycle combustion will take place on all combustion space, but it's often when behind flame front established considerable quantity of suboxide combustion residue. In next cycle they with end gases are ideal activating agents, which leads to perfect combustion.

Hydrogen as gasoline additive can lead to increasing of formation stable heat source in ignition [9], that leads to decreasing of misfire (Figure 4b).

IOP Conf. Series: Earth and Environmental Science **50** (2017) 012016 doi:10.1088/1755-1315/50/1/012016

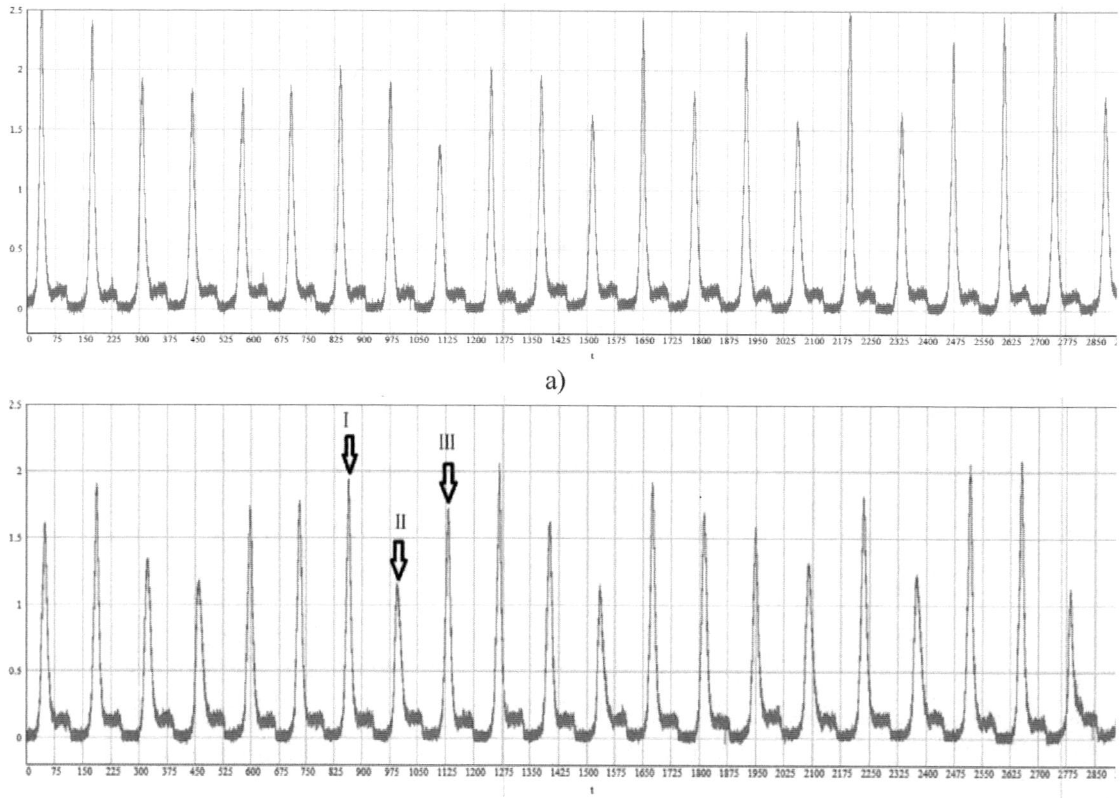

a)

b)

Figure 4 – Pressure oscillograms a) gasoline, α=0.988; b) gasoline with 6% hydrogen, α=0.98

Polytropic exponent changing analysis for 3 successive cycles shows us: perfect combustion cycle than misfire cycle and partial combustion cycle with considerable quantity of suboxide combustion residue. Figure 5 represents equitation of pressure changing (Figure 5a) and polytropic exponent in compression, combustion and expansion cycles (Figure 5b) of 3 successive cycles on idle.

Evaluation changing can be started from polytropic exponent in start of compression (part from 220 to 250 crank angle), polytropic exponent maximum in II cycle, which success I cycle (perfect combustion), which maximize uses of combustion procedure heat, reducing end gas temperature. In III cycle, polytropic exponent is minimum, that means that heat flow from cylinder walls will be much less. Evaluation polytropic exponent in main part of compression (part from 250 to 320 crank angle) shows, that polytropic exponent decreasing more in II cycle. In this cycle end gases mainly consist of perfect combustion products (from teracidic gases), and in I and III end gases consist of partial combustion products, which leads to higher polytropic exponent. Polytropic exponent averages on part from 220 to 320 crank angle as follows: in I cycle 2.066; in II cycle 2.205; in III cycle 1.894. It is notable that in I cycle sequential cycle with partial combustion end gases consist of more light atoms than in III cycle after misfire in II cycle. That can allow to expedite process of flame front establishing, indicates by fast increasing polytropic exponent after ignition.

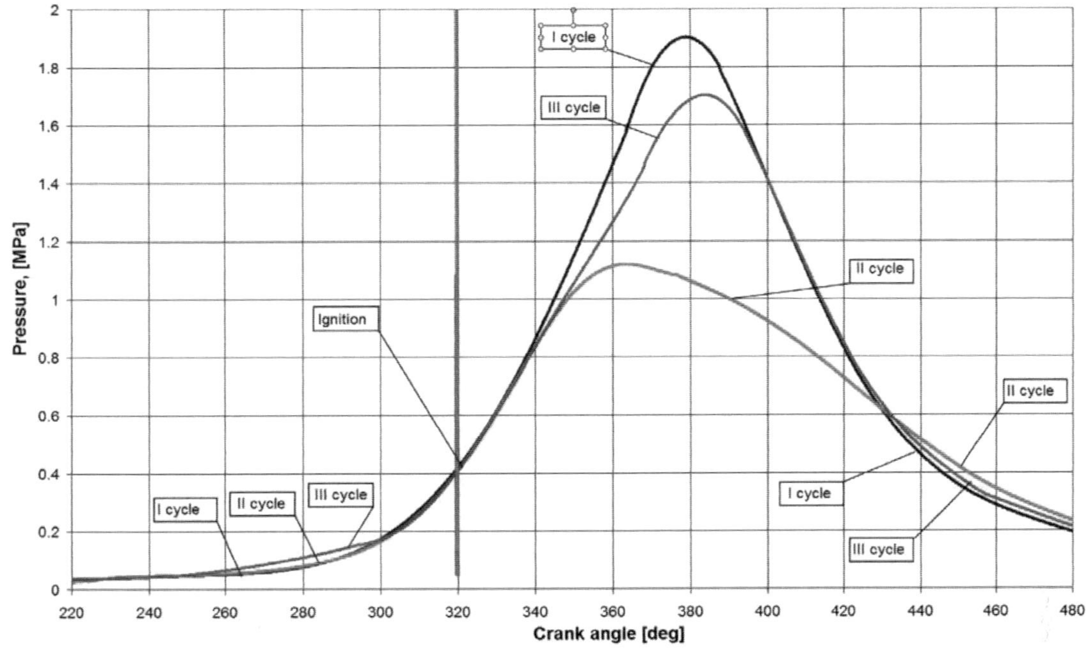

a)

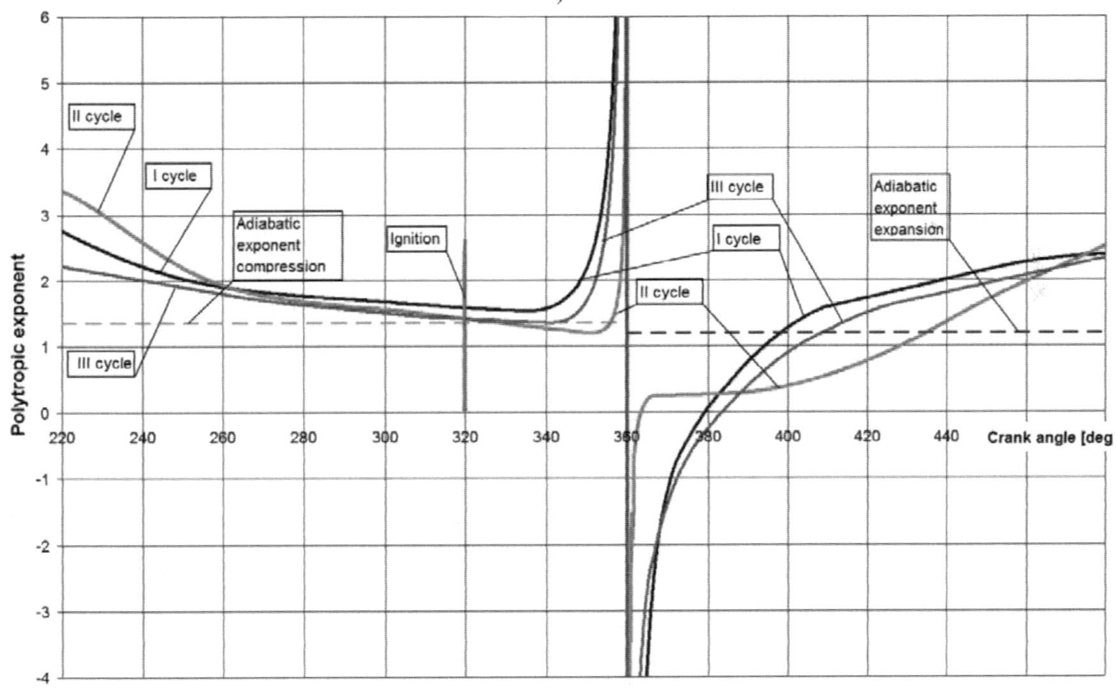

b)

Figure 5 – Changing of pressure (a) and polytropic exponent (b) in process compression, combustion and expansion cycles of three successive cycles on idle

Polytropic exponent changing analysis in the end of combustion cycle and on expansion in II cycle with misfire is long zone shows values lesser than 1. This means low-intensity combustion process and in I, III cycle process end faster and polytropic exponent value shows burning process completeness.

In I cycle there is heat sink in cylinder walls and in III cycle still partial combustion product is presented which decreases polytropic exponent. Average values of polytropic exponent in expansion on part from 400 to 480 crank angle as follows: in I cycle 1.713; in II cycle 1.548; in III cycle 1.635.

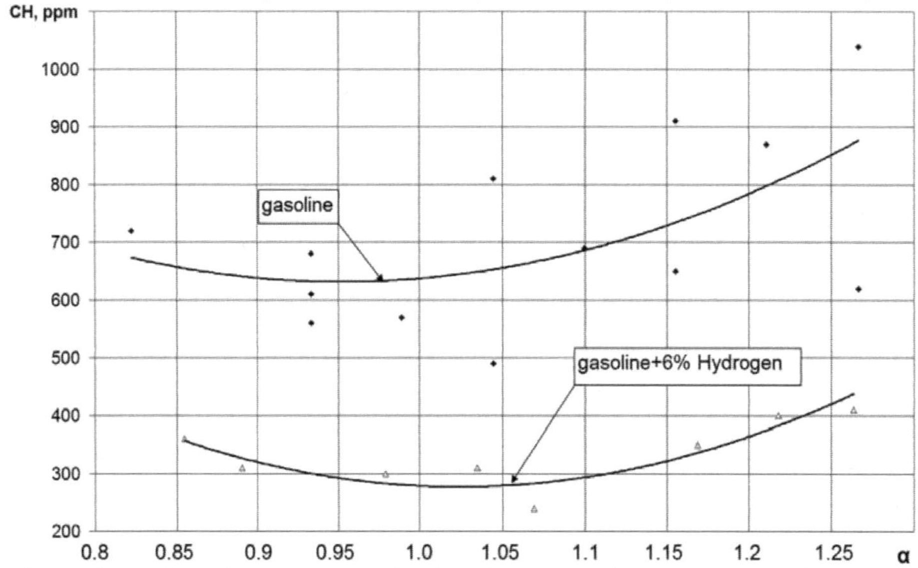

Figure 6 – Hydrocarbons concentration in exhaust gases in VAZ-2111 engine on idle

On figure 6 represented results on problem of decreasing of hydrocarbons concentration in exhaust gases, illustrates that hydrogen addition increases stability and burning rate. Reduction of misfire of gasoline with 6% hydrogen addition [2, 3, 9, 10] provides in average emission reduction on hydrocarbons in two times. Hydrocarbons concentration in exhaust gases in leaning more than α = 1.15 explained worsening combustion terms, increasing cycle irregularity and misfire rates in proportion for gasoline with and without 6% hydrogen.

Conclusion

1. Polytropic exponent research showed connection between air-fuel mixture burning conditions and end gas composition and combustion irregularity on deep throttling operation. Polytropic exponent analysis in compression, burning and expansion allows to deeper understand different factors on subject thermodynamic system and quantify composition and qualities of fuel-air mixture in considerable conditions of irregularity.
2. Hydrocarbons concentration in exhaust gases by hydrogen addition decreases number of misfires and increases cycle stability on idle.

Acknowledgments

This work was financially supported by the Russian Foundation for Basic Research (project no. 16-38-00331).

References

[1] Kamenev VF, Scientific bases and ways to improve the toxic characteristics of automobile engines with spark ignition, doctoral thesis 05.04.02 NAMI, Moscow, (1996) 454.

[2] Gortyshov Y F, Gureev V M, Misbakhov R S, Gumerov I F, Shaikin A P., Influence of fuel hydrogen additives on the characteristics of a gas-piston engine under changes of an ignition advance angle, Russian Aeronautics, 52(4) (2009) 488-490. Information on https://www.scopus.com/inward/record.uri?eid=2-s2.0-76449108200&partnerID=40&md5=7d22fcb28bdacfabed914ef05922e463 DOI: 10.3103/S1068799809040199.

[3] Bortnikov L N., Pavlov D A., Rusakov M M., Shaikin A P, The composition of combustion products formed from gasoline-hydrogen-air mixtures in a constant-volume spherical chamber, Russian Journal of Physical Chemistry B, 5(1) (2011) 75-83. Information on https://www.scopus.com/inward/record.uri?eid=2-s2.0-79956326816&partnerID=40&md5=8010b46bf50435298d67999d72a65398 DOI: 10.1134/S1990793111010039.

[4] Nemati A, Fathi V, Barzegar R, Khalilarya S, Numerical investigation of the effect of injection timing under various equivalence ratios on energy and exergy terms in a direct injection SI hydrogen fueled engine, International Journal of Hydrogen Energy, 38(2) (2013) 1189-1199. Information on: https://www.researchgate.net/profile/Arash_Nemati3/publication/257174777_Numerical_investigation_of_the_effect_of_injection_timing_under_various_equivalence_ratios_on_energy_and_exergy_terms_in_a_direct_injection_SI_hydrogen_fueled_engine/links/0deec536f02e74e6d3000000.pdf DOI: 10.1016/j.ijhydene.2012.10.083.

[5] Teh K.-Y, Miller S L , Edwards C F, Thermodynamic requirements for maximum internal combustion engine cycle efficiency. Part 1: Optimal combustion strategy, International Journal of Engine Research, 9(6) (2008) 449-465. Information on https://www.researchgate.net/publication/245395230_Thermodynamic_requirements_for_maximum_internal_combustion_engine_cycle_efficiency_Part_1_Optimal_combustion_strategy DOI: 10.1243/14680874JER01508.

[6] Chintala V, Subramanian K A, Assessment of maximum available work of a hydrogen fueled compression ignition engine using exergy analysis, Energy, 67 (2014) 162-175. Information on https://www.researchgate.net/publication/260211945_Assessment_of_maximum_available_work_of_a_hydrogen_fueled_compression_ignition_engine_using_exergy_analysis DOI: 10.1016/j.energy.2014.01.094.

[7] Li Y, Jia M, Chang Y, Kokjohn S L, Reitz R D, Thermodynamic energy and exergy analysis of three different engine combustion regimes, Applied Energy, 180 (2016) 849-858. Information on: http://www.sciencedirect.com/science/article/pii/S0306261916311254 DOI: 10.1016/j.apenergy.2016.08.038.

[8] Zhao Z, Wang S, Zhang S, Zhang F, Thermodynamic and energy saving benefits of hydraulic free-piston engines, Energy, 102 (2016) 650-659. Information on: https://www.researchgate.net/publication/299420570_Thermodynamic_and_energy_saving_benefits_of_hydraulic_free-piston_engines DOI: 10.1016/j.energy.2016.02.018.

[9] Shaikin A P., Galiev I R., On the effect of temperature and the width of the turbulent combustion zone on the ionization detector readings, Technical Physics, 61(8) (2016) 1206-1208. Information on https://www.scopus.com/inward/record.uri?eid=2-s2.0-84981489611&partnerID=40&md5=2a171a70052f225423d5f1082d29a746 DOI: 10.1134/S1063784216080247.

[10] Wang B, Wang Z, Shuai S, Xu H, Combustion and emission characteristics of Multiple Premixed Compression Ignition (MPCI) mode fuelled with different low octane gasolines, Applied Energy, 160 (2015) 769-776. Information on: http://www.sciencedirect.com/science/article/pii/S030626191500152X DOI: 10.1016/j.apenergy.2015.01.115.

Automated System Of Monitoring Of The Physical Condition Of The Staff Of The Enterprise

A Pilipenko

Orel State University named after I.S. Turgenev
Str. Komsomolskaya, 41, Orel, Russian Federation,302001

E-mail: a@pilipenko.info

Abstract. In the work the author solves an important applied problem of increasing of safety of engineering procedures and production using technologies of monitoring of a condition of employees. The author offers a work algorithm, structural and basic electric schemes of system of collection of data of employee's condition of the enterprise and some parameters of the surrounding environment. In the article the author offers an approach to increasing of efficiency of acceptance of management decisions at the enterprise at the expense of the prompt analysis of information about employee's condition and productivity of his work and also about various parameters influencing these factors.

1.Introduction

In modern machine-building or tool production the working personnel are succeeded by expensive and high-precision machines, still they are managed by people. Only training of the specialist in work at one of models of machines costs at least 200 000 rubles. Increase in demand for specialists controlling dangerous and expensive equipment leads to the intention to secure the employee and to increase the quality of his working space and life. Poisoning with toxic evaporations and gases, reduction of attentiveness and increasing of mistakes, a nervous stress and aggressive behavior are characteristic signs of any large production enterprise which do not only have a negative effect on quality of engineering procedures, but also attract danger to life of employees. The current state of information technologies allows developing telemedicine technologies, and also to move to a formulation of the question about automation of collection of data on psychophysical conditions of a person and further about forecasting of medical conditions [1]. Further development of technologies of monitoring of a condition of a person and forecasting of psychophysical conditions will allow to realize adaptive intellectual management systems which work algorithms directly depend on the operator and his indicators of health at given time.

It is well-known that as signals of a functional condition of a person indicators of skin and galvanic reaction, speech activities, arterial pressure, a tone of vessels, etc. are used. Based on the data obtained by means of these signals it is possible to determine a current condition of a person. Having a sufficient set of reliable data on a condition of a person and surrounding environment, it is possible to diagnose and predict his medical conditions starting with emotional changes and finishing with diseases of specific organs. And the chief engineer and the workshop manager will obtain exact and prompt information on a condition of employees of the workshop and will be able to develop engineering procedures more effectively and to optimize work of the workshop in general that will allow to increase product quality and amounts of its release thanks to the correct working schedule and distribution of employees on suitable technological transactions; to change working conditions by release from unskilled, monotonous, hard and harmful labor, improvement of safety conditions, decrease in losses of working hours from an industrial traumatism and professional diseases; economy and release of labor power.

In this connection it seems reasonable to develop new methods of getting information signals of a condition of a person and algorithms of data collection, and also to create combined dynamic methods of collecting information. Getting information in real time from devices of data collection generates super big data sets [2,3], and their handling requires accomplishment of optimization of algorithms of quick search. All this confirms relevance of developments of multiple parameter devices of collection of data on condition of an operator and reflects possibilities of development of the project.

Work purpose: increasing of safety of production and labor productivity of personnel.

2.Results and Discussion

The object of monitoring is personnel and their workplace and the object of automation is the equipment of personnel. With the purposes of increasing of safety of engineering procedures and labor productivity of working personnel the following requirements were imposed to the system:

1. Collection of data on a biometric condition and location of a person (monitoring of the electrocardiogram and skin and galvanic reaction, determination of location on GPS).
2. Collection of data on a surrounding situation: temperature, moisture content, concentration of dangerous gases in the air (methane, carbon dioxide, compounds of nitrogen).
3. Data transmission from the portable device, located on the worker's body, to the safety operator by means of Bluetooth transceivers.
4. Handling of collected information, reduction of information to some general form.
5. Comparison of collected information with reference values of biometric parameters of a condition of a person and determination of deviations.
6. The alarm system on an automated workplace of the safety operator about extraordinary deviations of parameters with indication of an employee's name, all obtained data at the given moment and his location for rendering the urgent help to him.
7. The analysis of the reasons of deviations of the compared parameters depending on a surrounding situation.
8. In the long term: Ensuring issue of negative results of monitoring for the emergency services and forecasting of medical conditions of the person.

Proceeding from the stated problem we carried out the choice of equipment necessary for creation of a prototype of the automated hardware and software system of monitoring of a condition of personnel, then the block diagram of a complex was constituted, it is represented in the figure 1.

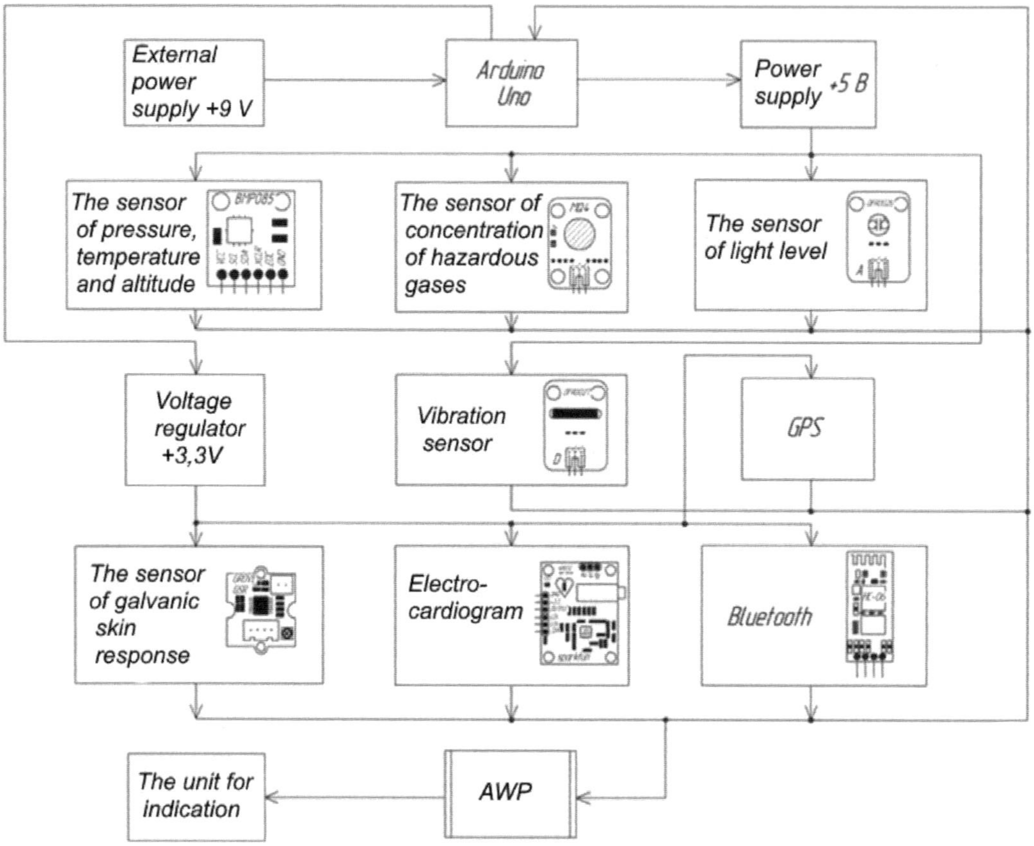

Figure 1- Block diagram of the system

As we can see from the figure, a power supply of +9 V is brought to the hardware computing platform Arduino, from Arduino a power supply of +5 V moves to the sensors (the light sensor, the sensor of concentration of hazardous gases, the sensor of pressure, temperature and altitude and the sensor of vibration), and via the built-in voltage regulator +3.3V moves to other sensors (the sensor of the skin and galvanic response, electrocardiogram, GPS module for determination of location), and also to the module of wireless data transfer Bluetooth HC-06 transmitting a signal to an automated workplace of the operator (personal computer) with the installed program where data on each user of the device are displayed. The data contain information about user's parameters and a light signaling in case of an output of any parameter out of limits of an admissible norm.

Based on the block diagram the circuit diagram is constructed. It is given in figure 2.

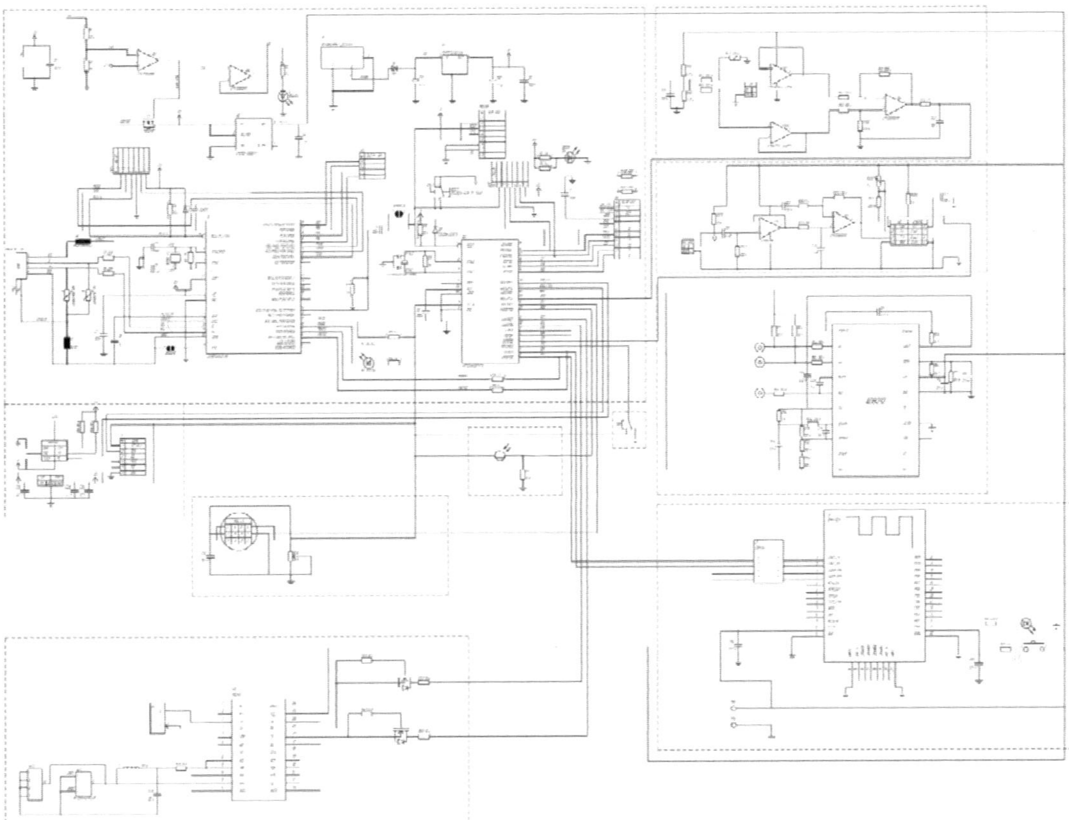

Figure 2- Circuit diagram of the system

For creation of a prototype of the system we chose the hardware computing platform ArduinoUno due to the simplicity of its connection to the selected sensors and programming. For convenient programming of the interface and data output from analog sensors the Processing language was selected [4]. It allows to obtain data from Arduino and to output necessary information in real time easily. The Processing language is based on Java, so the code created in it doesn't require compilation and is executed line by line. As a development environment we selected PDE (ProcessingDevelopmentEvironment) which is similar to Arduino IDE. After final debugging of the program the page html-code with Java applet is created. It can be posted on the website and provides remote system management.

The algorithm of system operation is provided in figure 3. After execution of standard procedures of connection of libraries and initialization of variables the program realizes primary data collection from sensors. Further it is necessary to make calibration of the device. It is necessary as the user can have normal values of the skin and galvanic response and a heart rhythm different from standard. For calibration the device takes off data from sensors three times, at the same time the user shall confirm consent to calibration by means of the button to provide protection against false operation. So, for example, each person has his own physiological level of skin resistance. The individual normal range of the signal levels for a person is set and we can judge stressful influence according to a deviation from this range. Thus it becomes possible to reveal events which had stressful or psychotraumatic character for this person [5].

Figure 3 - Algorithm of work of the system

After accomplishment of calibration the cycle in which the obtained data are compared to limits of normal values is carried out. In case of detection of extreme indications the secondary measurement of indicators is performed to exclude a mistake, then the obtained data are compared to reference values again and if they go beyond admissible limits again, then the device displays the message on a request for sending the message on emergency situation.

In figure 4 we provide the picture of display of parameters of the employee in a window of the operator realized in IDE Processing [6]. To display the graph we will require the following part of a program code in a cycle with a precondition:

```
for (int x=150; x <width-1; x="" ++)="" <br="">{ check (x, 255,0,0);
line ((width) - x, 6+ (height/6) *0)+ ((height-1-getY (valuesA[x-150]))/6),
(width) - 1-x, 6+ (height/6) *0)+ ((height-1-getY (valuesA[x-149]))/6));
check (x, 0,255,0);
line ((width) - x, 4+ (height/6) *1)+ ((height-1-getY (valuesB[x-150]))/6),
(width) - 1-x, 4+ (height/6) *1)+ ((height-1-getY (valuesB[x-149]))/6));
check (x, 0,0,255);
line ((width) - x, 2+ (height/6) *2)+ ((height-1-getY (valuesC[x-150]))/6),
(width) - 1-x, 2+ (height/6) *2)+ ((height-1-getY (valuesC[x-149]))/6));
check (x, 255,255,0);
line ((width) - x, 0+ (height/6) *3)+ ((height-1-getY (valuesD[x-150]))/6),
(width) - 1-x, 0+ (height/6) *3)+ ((height-1-getY (valuesD[x-149]))/6));
check (x, 0,255,255);
line ((width) - x,-2+ (height/6) *4)+ ((height-1-getY (valuesE[x-150]))/6),
(width) - 1-x,-2+ (height/6) *4)+ ((height-1-getY (valuesE[x-149]))/6));
check (x, 255,0,255);
line ((width) - x,-4+ (height/6) *5)+ ((height-1-getY (valuesF[x-150]))/6),
(width) - 1-x,-4+ (height/6) *5)+ ((height-1-getY (valuesF[x-149]))/6));}
```

Having analyzed the data it is possible to draw the conclusions:

- the third from above graph displays a maximum permissible air pollution in 452 c.u. in case of the maximum harmless level of 454 c.u.
- the fourth graph SGR (skin and galvanic reaction) specifies that the employee experiences the feeling of hunger.

The indications of the sensors and the system are checked at the control stand for aero gas control shown in figure 5. It includes high-precision sensors with sensitive elements from the Honeywell [7] and Lamtec [8] companies with a permissible error no more than 2%.

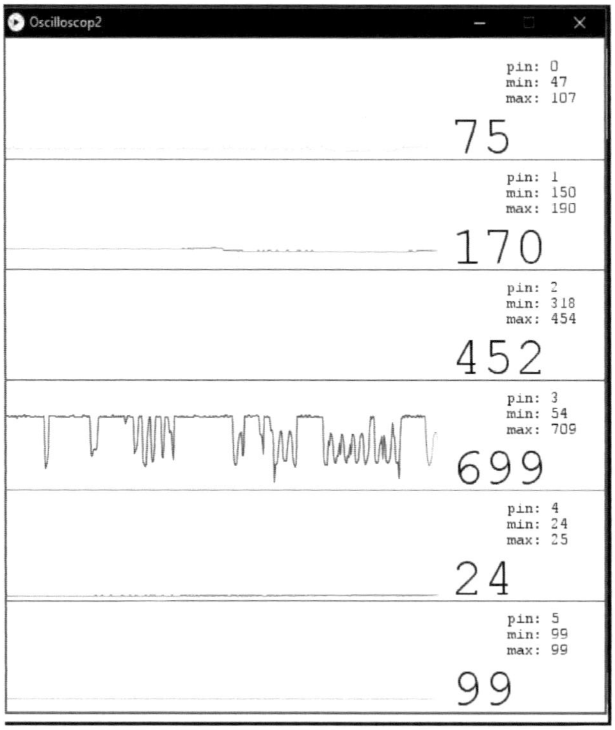

Figure 4 - A picture of the screen of the operator in Processing IDE.

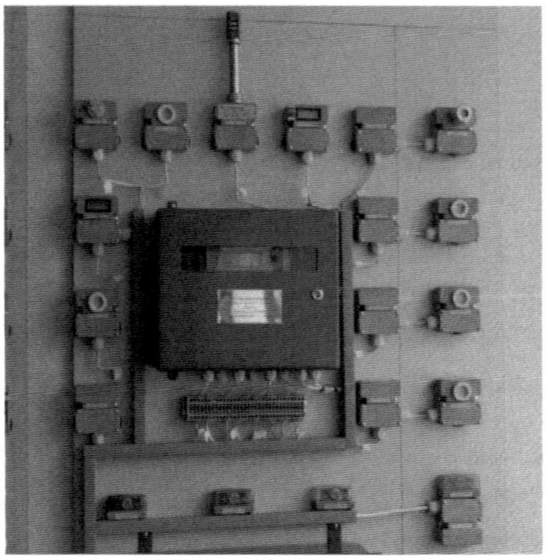

Figure 5 - Checking stand

3. Conclusion

As a result of the conducted applied researches the automated system of monitoring of a condition of working personnel is developed. It represents a portable individual block allowing to realize monitoring of movements of employees in a workshop, deviations in a psychophysiological condition of a person and deviations in parameters of aerogasdynamic control. The received prototype allowed to research a condition of working personnel on different industrial enterprises in various working conditions. For example, in case of excess of content of carbon dioxide [9] in the air to 9%

productivity decreases by 12%, and in case of feeling of hunger productivity of workers decreases by 18%. The developed system can become a part of large system of monitoring of safety of industrial enterprises, and also be used in municipal economy for monitoring of a condition of city resources and processes that will allow to improve quality of human life at the urbanized territory[10]. Similar systems do not only allow to reduce time of response to emergency, but also to optimize engineering procedures and an enterprise management system in general. All developments are conducted with the assistance of a grant of the Russian President and are regularly registered in Ruspatent that confirms their innovation and relevance.

References

[1] Abashin V, Pilipenko O (2013) Urbanosfera. Personal'naya urbanosfera [Urbanospherum. Personal urbanospherum]. Biospheric compatibility: person, region, technologies, no 2, pp. 29-34. (in Russian)

[2] Pilipenko A, Pilipenko A (2015) Podxody' k razrabotke sistemy' sbora informacii o sostoyanii gorodskoj sredy' [Approaches to development of the system of collection of information about condition of the urban environment]. The collection of scientific works following the results of the international scientific and practical conference "Technical science: tendencies, prospects and technologies of development", no 2, Volgograd, 228 p. (in Russian)

[3] Saigusa N et al. Dataset of CarboEastAsia and uncertainties in the CO_2 budget evaluation caused by different data processing //Journal of forest research. 2013. Vol. 18. 1. pp. 41-48.

[4] D'Ausilio A Arduino: A low-cost multipurpose lab equipment //Behavior research methods. 2012. T. 44. 2. p. 305-313.

[5] Programmnoe obespechenie dlya avtomatizacii pomeshhenij na platforme Arduino (2015) [The software for automation of rooms on the Arduino platform]. Certificate of registration Software for the Computer. RF. no. 2015662105; applic. 28.09.2015; published 17.11.2015 (in Russian)

[6] Yi I et al. Communication between Physical and Virtual Models in Designing Smart Buildings //International Conference on Cooperative Design, Visualization and Engineering. Springer Berlin Heidelberg, 2012. pp. 126-133.

[7] Arkilic E B, Schmidt M A, Breuer K S Sub-nanomol per second flow measurement near atmospheric pressure. Experiments in fluids. 1998. Vol. 25. 1. pp. 37-41.

[8] Guth U, Zosel J, Shuk P. Combustion Control Sensors, Electrochemical //Encyclopedia of Applied Electrochemistry. Springer New York, 2014. pp. 230-237.

[9] Billings W D et al. Arctic tundra: A source or sink for atmospheric carbon dioxide in a changing environment? //Oecologia. 1982. Vol. 53. 1. pp. 7-11.

[10] Chen H et al. The inspection of the triggering mechanism for a hazardous mudflow in an urbanized territory. Environmental Geology. 2004. Vol. 45. 7. pp. 899-906.

Algorithm Research Exposure Dust Emissions Enterprises Of Building Production On The Environment

E V Omelchenko, E A Trushkova, M V Sidelnikov, S L Pushenko, E V Staseva

Architectural and construction academy of the Don State technical university
st. Socialisticheskaya, 162, Rostov-on-Don, Russian Federation, 344000

E-mail: ayurvedaom@mail.ru

Abstract: This paper describes the algorithm research level of exposure to dust emissions, the organization and layout of the sanitary protection zone of the enterprise of building production on the basis of statistical data on emissions share in Russia in the Rostov region, the issues of the impact of harmful substances on the human body and developed measures to reduce air emissions.

Introduction

All the Rostov region, is subject to strong air pollution. Adversely impact on human health and its performance determines the need for additional measures for air protection, taking into account the specific characteristics of enterprises in the region. For the testing methods used in the algorithm research level of exposure to dust emissions of enterprises of building production as objects of study chosen by the company - "Sebryakovcement", "Kavkazcement" holding "Eurocement Group", "Chiri-Yurt cement factory" - representatives of the basic cement of the construction industry.

While in Russia is currently operating 49 (working - 45), with a total annual capacity of enterprises – 73.7 million tons of cement - the degree of capacity utilization - only 50%. The largest share in total Russian production occupy Northwest, Central Black Earth, Volga, North Caucasus, Urals, which produce cement, not only for their own needs, but also deliver it to the neighboring regions. Southern Federal District of Russia most of all full of cement plants.

The main environmental problems of all plants cement production are emissions of solids (dust). The main sources of emissions: the explosions in the quarry with extraction of raw materials, crushers and hammer mills, wet/dry grinding mills, clinker kiln, trucks for transportation of raw materials and finished products.

Technique

To identify the level of exposure to cement construction dust, we used the results of certification of workplaces on working conditions and instrumental measurements made during production monitoring. In 2015, the laboratory ambient air studies in the Rostov region grew by 13.8%. The proportion of samples exceeding 5 LAC (Limit Admissible Concentration) increased and amounted to 0.029%. In 2015, the analysis of air quality of the city revealed with atmospheric pollution levels above LAC (ranked 5 places): Rostov-on-Don, Azov, Shakhti, Bataysk, Taganrog. [1-10].

With limited funding organizations there is a need to develop cost-effective spending.

We propose an algorithm to study the level of exposure to dust emissions of enterprises of building production on the environment, providing specific characteristics of enterprises and developing a system to prevent dust production areas, comprising the steps of:

1. Analysis of the climate, the local geographic and meteorological features of the location of enterprises of the construction industry: the use of schematic maps of the city and its surroundings with landmarks; applying to the industrial site map diagram with the main emission sources, sanitary protection zone (SPZ), territory, weather stations, points of observation for atmospheric pollution; definition of the diurnal variation of the intensity and frequency of surface and elevated temperature inversions, different speeds and wind directions, calm, rain, fog; consideration of local topographic features at atmosphere pollution calculation and setting LAE (Limit Admissible Emission);

2. Assessment of environmental hazard building enterprise (EHE): using dimensionless ratio K_{EHE} determined by the ratio of the parameter P to the consumption of contaminated gases G_g, m^3/s [1, 2]:

$$K_{EHE} = P_{v\,i} / 10^3\, G_g. \qquad (1)$$

The variable parameter P_v for each pollutant is calculated by a known formula, with the origin taken to cut the source of pollution

$$P_v = \Sigma\, T \cdot P \cdot B_{ji} \cdot R_{ji} \qquad (2)$$

where required for the dilution air $T \cdot P \cdot B_{ji}$, m^3, determined according to the formula:

$$T \cdot P \cdot B_{ji} = 10^3\, M_{ji} / C_{LACi} \qquad (3)$$

parameter R_{ji} is defined by the formula:

$$R_{ji} = m_{ij} / (H_j + D_j) \cdot C_{MAXi} / C_{LACi}, \qquad (4)$$

m_{ij} - the number of sources in the company, the same type of emitting pollutants, pcs .; C_{MAXi} - the maximum concentration of the pollutant in the mouth of the source mg / m^3; C_{LACi} - limit admissible concentration in mg / m^3; D_j - the diameter of the mouth of the source of m; H_j - height j-th source, m.

The essence of the introduction of an algorithm for determining K_{EHE} reduced to the calculation of the dimensionless C_i concentrations of pollutants at the exit of the chimney and change of dimensionless coordinates η, ξ, H (respectively, the dimensionless transverse to the longitudinal and vertical coordinates) assigned to the characteristic linear dimension of the sanitary protection zone L_{SPZ}.

In accordance with the algorithm determined risk category 4 company depending on the known criteria P_v (formula (2)):

$$P_v \geq 10^8 \text{ - 1st category,}$$
$$10^8 > R \geq 10^6 \text{ - 2nd category,}$$
$$10^6 > P \geq 5 \cdot 10^6 \text{ - 3rd category;}$$
$$P < 5 \cdot 10^4 \text{ - category 4.}$$

The boundaries of the sanitary protection zone can also be determined by a change in the dimensionless ratio of conditions $K_{EHE} \leq 2$.

Studies on the average performance of the plants showed that when the amount of SPZ: $L_{SPZ} < 25$m area for production of building materials dates back to the 2nd category of environmental risk, at 25 $< L_{SPZ} < 50$m - to the 3rd category of danger, 50 $< L_{SPZ} < 75$m - to 4 th hazard category. In the calculations of the mass emissions approximated ingredients and is related to the total time of the plants for the season. You can select SPZ equal $L_{SPZ} = 75$m, for which pollution below the LAC. From experiments it follows that condition $K_{EHE} \leq 2$ is fulfilled when the amount of sanitary protection zone $L_{SPZ} > 75$m.

3. Conduct hazard identification in the workplace: an analysis of dust emitted substances and their combinations with the sums the harmful effect, their quantitative characteristics, the identification of the actual value; characteristic of applied technologies, pollution control equipment installed chimneys, the quantitative characteristics of emissions from the rationale for the measurement data and balance sheets, the type of pollution control equipment, the composition of raw materials, fuel, a survey of similar productions, literature data; stationary characteristics, route and supervision under a torch for air pollution, current scientific and technical forecast of its change, the rate of excess MPC and hazard classes of substances [12-14].

4. Evaluation of the hygienic conditions of the population living in the sanitary protection zones and areas to be included in the sanitary protection zones; analysis of data on the impact of pollution of the atmosphere with dust on human health and the environment: forests, parks, wildlife, historical and cultural monuments, buildings, power lines, metal products (corrosion); analysis of actual field comparison of the results of maximum one-time concentrations of substances in the atmosphere and design of the field according to the actual release of the parameters (on individual substances); inventory assessment of concentrations of emission sources parameters and observation substances in the atmosphere using accepted when calculating the values of the background concentrations; development of card schemes to the calculation results of air pollution at the adverse weather conditions and the release of all the substances and combinations of substances; development of measures to reduce emissions, up to partial or total stop of the enterprise under adverse weather conditions in short periods of air pollution.

5. Assessment of the impact of levels of dust emissions building production plants, animals and humans: observation of the change in the chemical composition of plants and snow cover; changes in plant species composition, anthropogenic transformation of plant communities along the factor of contamination: as a result found that the cement dust from the animal causes damage to the skin integument, the development of inflammation of the mucous membranes, affects the respiratory tract, settling in the bronchi and lungs, and in humans cement dust causes a number of serious diseases of the respiratory system - pneumoconiosis caused by fibrogenic mineral dust (silicosis, anthracosilicosis, asbestosis) and silicotuberculosis, provided that silicosis is a determining cause disability or death; alkaline cement base and a high allergenicity chromate lead to serious diseases of the skin, respiratory tract and mucous membranes of the oral cavity and the nasopharynx.

6. Development of greening project (dendrology project and estimates for the implementation of the works) for the enterprises of the construction industry in view of the formation of schemes and shelterbelts location, alleys, taking into account the differentiated range selection for each zone area (depending on the gas concentration) on species of plants, shrubs and trees in the buffer zone, as they are an effective way to protect residential areas from the harmful effects of industrial emissions enterprises, the value of which is determined by the current legislation in accordance with the hazard class enterprise [15].

The main focus for the construction industry enterprises is to choose rocks when landscaping with high dust binding properties. When you create a project of landscaping the sanitary protection zone and carrying out landscaping work, all healthy and valuable in - sanitary hygienically existing plantations must be saved.

In this case at least 50% of the total number of trees planted should occupy the main tree species, which has the highest sanitary efficiency, viability in these soil and climatic conditions, and gas resistance in relation to the emissions of the industrial enterprises [7].

All kinds of industrial plants on its territory must have plantations, passing the strip around the perimeter for decorating unsightly industrial buildings, commercial and warehouse premises and territories occupied by containers, parts, products finished goods, raw materials and other materials, for example, from trees with high crown in two or three with a number chess planting. The second part of the plantations are obscuring hiking path runs from shop to shop, to the warehouse of finished products, to the entrance and the dining room. For such landings are recommended plants with dense canopy of high altitude, high-growth, providing maximum microclimate effect, in the shortest possible time and with the cost of care minimum: Ash green, Ailanthus altissima, Elm plumose branch y etc. Third part of the plantations are forming recreation areas workers at lunchtime. They are created when large enterprises next to each shop or department for a small part of the territory of the whole work. These areas need to be addressed as a quiet rest areas shaded resting places, with watering sites, vertical gardening and ornamental plantings pleasing the eye (colors, beautiful flowering shrubs, lawns). The fourth part of the plantations are to ensure that in the conditions of the Rostov area shading buildings and plants from excessive solar radiation. Moreover, depending on the nature of the production process of features solve the problem of intensity and density of planting along the fronts so as to not degrade the shading operating conditions. For landings along the shops on the narrow strip of land is better suited to other pyramidal and Turkestan poplar and vines (grapes and wild fiveleaves its shape).The green dress industrial site always reflects the peculiarities of production. If a chemical or metallurgical enterprise, it is very sharply limited range of species, due to high demands on the gas resistance. If this enterprise construction industry (ferroconcrete plants, cement plants), combined heat and power plants or fertilizer companies, the focus is on the choice of landscaping rocks at high dust binding properties.

If necessary, absorb and dissipate odors industrial complexes, selected plants, the most efficient in gas exchange, such as the Canadian poplar, ash, maple [8-12].

For the preparation of the project of landscaping the territory sanitary protection zone needs the following material:

- Long-term plan of building area of sanitary protection zone, tallies with the general plan of the city or industrial area;
- The scheme of transport and pedestrian roads, as well as the scheme of underground utilities;
- Vertical planning of territory and cartogram excavation;
- Micro and macro-climatic characteristics of the area;
- Quantitative and qualitative data of air pollution and the radius of the spread of harmful emissions;
- Layout and inventory data on existing green spaces sanitary protection zone;
- Design assignment.

The project landscaping sanitary protection zone are: dendrology project, explanatory note and the estimates for the production of greening works. Dendrology project consists of a scheme of placing of green spaces of different types, scale 1:1000; dendrology plan a scale of 1:500 with an indication of the structure, size, array (profiles and details in scale 1:200 or 1:100), the range of plants for landscaping sanitary protection zone; alignment of the drawing at a scale of 1:500.

Explanatory note to the landscaping project of sanitary protection zone proves:

- Selection of plants layout of various types in the direction of the smoke dependence with existing objects and sites, released during the demolition and withdrawal from the area of residential development;
- Structure of individual protection zones, arrays, arrays of optimal size, the extent of landscaping;
- Select the range of trees, shrubs, flower lawn, plants for landscaping sanitary protection zone, depending on the climatic region, the nature of industrial production, the effectiveness of this breed in clean air, its gas resistance;
- Agro-technical recommendations for the creation and maintenance of green spaces in the area and measures to improve the gas resistance of plants;
- A feasibility study of this option.

Result

Thus, the algorithm research level of exposure to dust emissions of enterprises of building production reveals the basic problem of the construction industry - the imperfection of the process, since the Russian cement is produced by high-cost, outdated technologies; most atmospheric particles held in the air for a long time, do not precipitate at the ejection locations and dry atmosphere flows fall into regional and global contaminants.

Conclusions

The received algorithm allows to identify ecologically fragile enterprise and to actively influence the parameters of the processes for the stabilization of the ecological situation at the plant and within residential areas. In addition, taking into account the analytical relationships obtained during mathematical simulation model is formed by environmental hazard assessment based on the famous K_{EHE} criterion [1]. It allows you to predict air pollution as a function of meteorological parameters of anthropogenic dispersal migration of dust aerosols and develop an automated complex of measures on greening of industrial areas.

References

[1] Kanishchev A N. Scientific and actual bases of decrease in environmental pollution by toxic gas and thermal emissions of an automobile and road complex: abstract yew. Cand.Tech.Sci. Volgograd: VGASA, - 2002.

[2] Omelchenko E V. Thesis. On competition of an academic degree of Candidate of Technical Sciences "Decrease in technogenic environmental pollution by organized dust emissions of productions of road-building materials": Rostov N / Д: RGSU, - 2002.

[3] Durov V V. Protection of atmospheric air in the cement industry / V. V. Durov//Cement and its application.-1998. - No. 6.

[4] Haustov A P. Management of environmental management / Haustov A P, Redina M M. - M.: The higher school, - 2006.

[5] Demyanova V S. The disperse filled cements on the basis of lithotripsy. Ecological aspects of production / Demyanova V S, Kazin G N. - M.: The Young Guard, - 2006.

[6] Kannel W B. Left ventricular hyperthrophy as a risk factor: the Framingham experience.// J. of hypertension , 1991,9 (Suppl 2): S3-S9.

[7] Resnik L M. Ionic basis of hypertension, insulin resistance, vascular disease and related disorders // Am.J Hypertens.- 1993.- Vol. 6.-P.S123-S134.

[8] Andreeva E E. K ocenke riskov dlja zdorov'ja rabotnikov pri klassifikacii ob#ektov sanitarno-jepidemiologicheskogo nadzora i planirovanii proverok (na primere g. moskvy) / E.E. Andreeva // Analiz riska zdorov'ju. - 2016. - № 2 (14). - S. 84-92.

[9] Artem'eva A A. Ocenka vlijanija faktorov proizvodstvennoj sredy na sostojanie uslovij truda i professional'nuju zabolevaemost' rabotnikov // A.A. Artem'eva // Federal'noe agentstvo zheleznodorozhnogo transporta, FGBOU VPO «Moskovskij gosudarstvennyj universitet putej soobshhenija (MIIT). - 2014. - S. 15-18.

[10] Denisov Je I. Problema real'noj jeffektivnosti individual'noj zashhity i privnosimyj risk dlja zdorov'ja rabotnikov / Je.I. Denisov, T.V. Morozova, E.E. Adeninskaja, N.N. Kur'erov // Medicina truda i promyshlennaja jekologija. - 2013. - № 4. - S. 18-25.

[11] Zaharenkov V V. Ocenka professional'nogo riska dlja zdorov'ja rabotnikov promyshlennyh predprijatij na osnove medicinskoj tehnologii / V.V. Zaharenkov, A.M. Oleshhenko, I.P. Danilov, D.V. Surzhikov, V.V. Kislicyna, T.G. Korsakova // Akademicheskij zhurnal Zapadnoj Sibiri. - 2013. - T. 9. - № 2 (45). - S. 8.

[12] Zaharenkov V V. Primenenie novoj medicinskoj tehnologii dlja ocenki professional'nogo riska zdorov'ju rabotnikov promyshlennyh predprijatij / Zaharenkov V V, Oleshhenko A M, Danilov I P, Surzhikov D V, Kislicyna V V, Korsakova T G, Motuz I Ju // Vestnik Kuzbasskogo nauchnogo centra. - 2013. - № 17. - S. 28-30.

[13] Zeljaeva N V. Ocenka zabolevaemosti rabotnikov promyshlennyh predprijatij i puti ee snizhenija (na primere predprijatij otrasli mashinostroenija g. Nizhnego Novgoroda) / Zeljaeva N V // Fundamental'nye issledovanija. - 2014. - № 7-4. - S. 698-705.

[14] Kislicyna O A. Nezashhishhennost' v sfere truda rabotnikov v Rossii kak faktor riska plohogo zdorov'ja i povedenija s riskom dlja zdorov'ja: gendernyj aspekt / O.A. Kislicyna // Upravlenie zdravoohraneniem. - 2015. - № 1 (43). - S. 66-77.

[15] Habibulina V M. Zdravoohranitel'naja politika: pragmatichnost' na osnove otkrytyh dannyh medicinskoj statistiki / V.M. Habibulina // Materialy Mezhdunarodnoj mezhdisciplinarnoj konferencii. - 2015. - S. 380-390.

Evaluation Of Rotation Frequency Gas-Diesel Engines When Using Automatic Control System

A Zhilenkov, A Efremov

Peter the Great Saint-Petersburg Polytechnic University
Polytechnicheskaya, 29, St.Petersburg, Russian Federation, 195251

E mail: zhilenkovanton@gmail.com

Abstract. A possibility of quality improvement of stabilization of rotation frequency of the gas-diesels used as prime mover of generator set in the multigenerator units working for abruptly variable load of large power is considered. An evaluation is made on condition of fuzzy controller use developed and described by the authors in a number of articles. An evaluation has shown that theoretically, the revolution range of gas-diesel engine may be reduced at 25-30 times at optimal settings of the controller in all the power range. The results of modeling showing a considerable quality improvement of transient processes in the investigated system at a sharp change of loading are presented in this article.

1. Introduction

An introduction of gas-diesel engines as prime movers of generating sets is one of the most perspective trends of multigenerating autonomous grids of a number of transport objects and an industry. First of all it concerns self-contained drilling rigs where associated gas can be used as fuel of gas-diesel engines that increases their efficiency, profitability, reliability and reduces the number of hazardous emissions. The first condition directly depends on ensuring stability of a shaft rotation frequency of gas-diesel primary engine of the generating set (GS). The reasons of low stability of the gas-diesel engine revealed on the basis of the analysis of specialized literature and our own full-scale studies have been described by authors in a number of works [1-4]. Analysis of published data and the problem statement.

2. The improved dynamic model of GDGS

Authors have offered a specified model of GDGS describing the generating set both in a statics, and in dynamics for a research of dynamic characteristics. Authors have shown [4] that the dynamic model of the gas-diesel engine at account of an effect of disturbances on loading can't be adequately described by the mathematical models offered in literature which aren't including an account of a turbo-charging. Its block diagram represented in fig. 1.

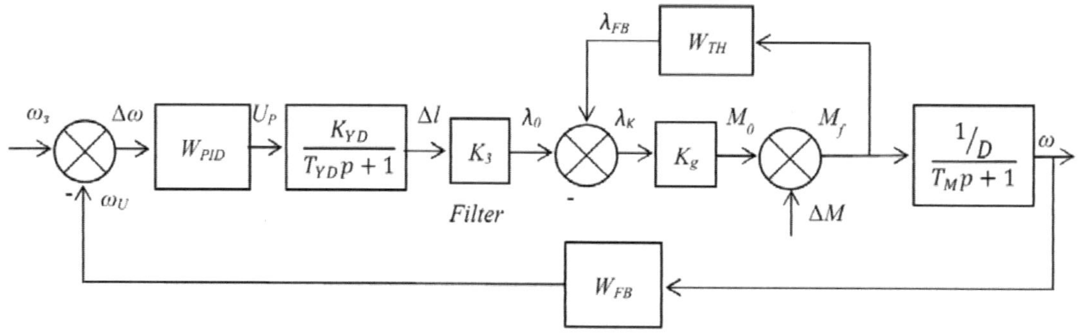

Figure 1 – A block diagram of GDGS: W_{PID} –a block with a transmissibility of PID controller; D – coefficient allowing self-regulating properties of prime mover; T_M – equivalent engine time constant allowing lag of flyweight masses of engine and a rotor of a generator; K_g – elememnt gain displaying conversion of flow of gas-air mixture while shaft torque; M – shaft torque; $\omega_з$ – preset speed; ω_u – true speed; $\Delta\omega_з$ – speed error; $U_П$ – output voltage of PID controller; K_{YD} – block gain, describing travel of fuel rack; T_{YD} – time constant of block, describing travel of fuel rack; λ_0 – quantity of gas-air mixture; λ_{OC} –gas-air mixture being returned trough turbocharger system; λ_K – quantity of gas-air mixture, being taken for establishing of shaft torque; M_0 – torque, being established by gas-diesel engine; ΔM –resistance torque about the shaft, established by adjusent gas-deisel generator; M_f – net moment about the shaft; ω – shaft rotation speed; W_{TH} – turbocharger block.

This model is described in detail in [1-4].

Adaptive nonlinear control system of GDGS revolutions. Authors have suggested the adaptive regulator which is carrying out automatic identification of GDE parameters and adaptive correction of the PID-regulator settings of GDE [2-4] for increase in efficiency stabilization systems of rotation frequency of gas-diesel engines (GDE) in gas-diesel-generator set (GDGS) of the autonomous grid (AG) operating with abruptly variable loads.

The block diagram of the investigated system including the controller (C) with the tuning unit (TU) of parameters of the controller is represented on fig. 2.

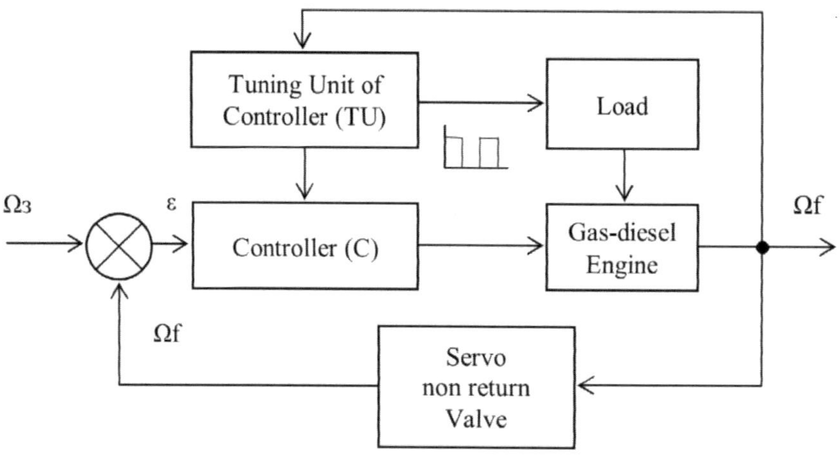

Figure 2 – A block diagram of adaptive controller GDE with automatic identification of parameters: $\Omega_з$ – assigned rotation frequency; Ω_f –actual rotation frequency; ε – error signal

3. Evaluation of theoretical efficiency of stabilization of GDE revolutions at adaptive control

As it is found[2], really functioning GDGS can be considered as the perfect machine which is effected by the normal-mode interference with a certain [4] physical nature and parameters. It is also may be assumed that the suggested control algorithm [2] will allow lifting significantly stability of revolutions of GDE in all power range of load.

The block diagram of the GDE is represented in the figure3. The normal-mode interference in the form of the destabilizing function is summed up with an output signal – the rotation speed of a GDE shaft.

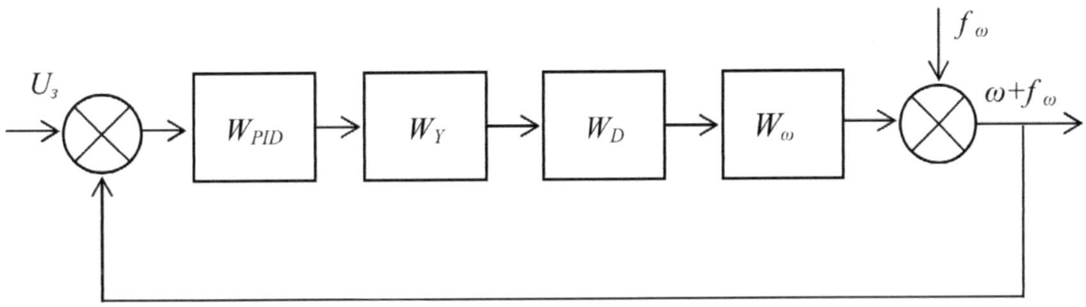

Figure 3 – The block diagram of the Gas diesel engine W_{PID} – PID controller of GDE rotation frequency; W_Y - output signal amplifier of PID controller; W_D – DC machine(flap controller); Wω – block describing dependence of revolutions of gas-diesl engine on rotation angle of flap; U_3 - output signal of speed setter; f_ω – normal-mode interference; ω –rotation frequency of GDE.

Let's present the description of blocks of the scheme fig. 5:

$$\left.\begin{aligned} &W_{\mathrm{PID}} = \frac{T_p T_g p^2 + K_p T_i p + 1}{T_i p}; W_y = \frac{K_y}{T_y p + 1}; \\ &W_{\mathrm{D}} = \frac{(T_{TH} p + 1)}{\dfrac{T_{TH}}{1 - K_D K_{TH}} p + 1} \cdot \frac{K_D}{1 - K_D K_{TH}}; W_\omega = \frac{1/D}{\dfrac{T_E}{D} p + 1} \end{aligned}\right\} \qquad (1)$$

Assuming $W_K = W_{PID} \cdot W_Y \cdot W_{\mathrm{D}} \cdot W_\omega$,

calculate interference level at the output of closed loop system $f_{\omega 3}$:

$$f_{\omega 3} = \frac{f_\omega}{1 + W_K} = f_\omega \cdot \frac{1}{1 + W_K} = f_\omega \cdot W_3 \qquad (2)$$

Interference power spectral density at the output of closed loop

$$S_{\omega 3}(\omega) = S_\omega \cdot \left|W_3(j\omega)\right|^2 = S_\omega \cdot \left|\frac{1}{1 + W_K(j\omega)}\right|^2 \qquad (3)$$

From the given formula, taking into account a character of linear amplitude frequency characteristic (LAFC) stable system for the area of effective interference cancellation where $|W_K| >> 1$ is:

$$S_{\omega 3}(\omega) \approx S_\omega \cdot \frac{1}{\left|W_K(j\omega)\right|^2} \qquad (4)$$

Let's consider now LAFC of the closed loop system more detail.

The calculated values of coefficients a_i and b_i and for all load range of GDGS are tabulated in table 1.

Table 1

The calculated values for the coefficients a_i and b_i the load P

P,% a_i,b_i	0	20	40	60	80	100
a_3	0	16.3	12.1	8.8	6.9	5.5
a_2	0	54.6	41	27.76	23.5	18.9
a_1	0	1.1	1.86	2.22	2.0	1.7
b_3	0	0.1	0.19	0.4	0.5	0.3
b_2	0	2.2	4.0	7.8	10.3	6.3
b_1	0	4.5	5.0	6.4	6.6	5.6

According results of calculations of polynomial coefficients $A(a_i, p)$ and $B(b_i, p)$ it has been revealed that they actually compensate each other. At the same time the error doesn't exceed in the narrow frequency range of 10 dB. For this reason equation for evaluation of efficiency of interference cancellation f_ω in diesel revolutions W_k may be present as:

$$W_K \approx \frac{K_y K_D}{D(1-K_D T_{TH})} \cdot \frac{1}{T_i p} = \frac{K_E}{T_i p} \tag{5}$$

Transmissibility of closed loop system has the form of:

$$W_3(p) = \frac{1}{1+W_K(p)} = \frac{T_i p}{T_i p + K_E} = \frac{\left(T_i / K_E\right)}{\frac{T_i}{K_E} p + 1} \tag{6}$$

Interference power spectral density in output revolutions of engine.

$$S_{\omega Z}(\omega) = S(\omega) \cdot \frac{\left(\frac{T_u}{K_E} j\omega\right)^2}{1+\left(\frac{T_1}{K_E}\right)^2 \omega^2} \tag{7}$$

Variance of output signal – is calculated from equation used for calculation of correlation function:

$$R_{out}(\tau) = \int_{-\infty}^{+\infty} S_{\omega Z}(\omega) e^{-j\omega\tau} d\omega \tag{8}$$

which for $\tau = 0$ is rearranged to the form:

$$R_{out}(0) = \sigma^2_{out} = \int_{-\infty}^{+\infty} S_{\omega Z}(\omega) d\omega \tag{9}$$

Since at values ω_i, for which $\left(\frac{T_1}{K_E}\right)^2 \omega^2 >> 1$, $W_Z(j\omega) = 1$, then spectrum of output parameter is influenced only by low frequency band for which

$$W_Z(j\omega)=\left(\frac{T_1}{K_E}\right)^2\omega^2 \quad (10)$$

In view of stated variance of output signal - with consideration of value $S(\omega)$ may be calculated by means of:

$$R_{out}(0)=\sigma_{out}^2=\frac{1}{2\pi}\left(\frac{T_I}{K_E}\right)^2\int_{-\infty}^{+\infty}\frac{\sigma_{in}^2\alpha}{\omega^2+\alpha^2}\cdot\frac{d\omega}{\omega^2}=\left(\frac{T_1\cdot\sigma_{in}}{\sqrt{2}\cdot K_E\cdot\alpha}\right)^2 \quad (11)$$

Thus it has been revealed that it will be possible to reduce a range of gas diesel engine revolutions for the whole power range at optimum tuning of PID controller 25-30 times.

4. Quality investigation of frequency regulation of rotation and power on computer model

The general block diagram of GDGS connected in parallel is provided in fig. 4. Allocation of active power between GU of automated power system (APS) depends on the accuracy of regulation of turns of GDGA. The general block diagram of GDGU in parallel is represented in fig. 4. active power allocation between GU of APS depends on the control accuracy of GDGU's revolutions [5-9].

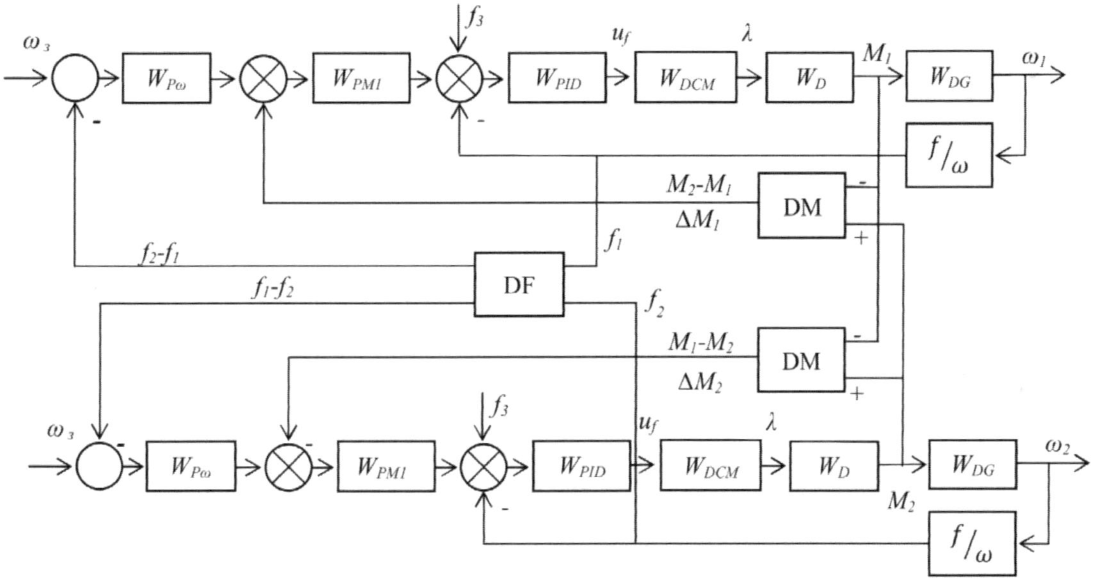

Figure 4 – The block diagram of parallel connection of two GDGUs: W_{PO} – frequency rotation control block of GDE; W_{PM1} и W_{PM2} – power control block of 1st and 2nd GDE; DCH- frequency sensor; DM – power sensor ; M - torque; f – rotation frequency

It consists of two GDGUs operating for active and inductive load of different power (0.2 MW and 0.3 MW at the power of each GDGU 0.67mvt). GDGU incorporates the following blocks: model of the gas-diesel engine, excitation systems of the generator, switchboards, measuring units. Control of settings of regulators is carried out via virtual COM ports from the external fuzzy controller.

Results of a research of frequency stabilization process of GU's rotation G1 (generator 1) and G2 (generator 2) are presented in fig. 8, while abrupt change in loading. It is visible from diagrams that the use of the fuzzy controller allows improving qualitatively a stabilization process of revolutions.

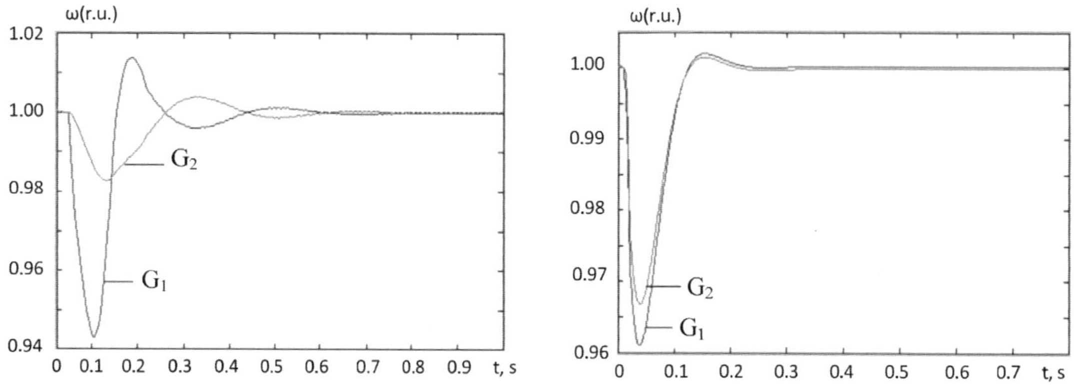

a –PID-controller only b – fuzzy controller

Figure 5 – Stabilization of GDGU's revolutions.

Results of a transient process research of active power control generated by generator units operating in parallel G1 and G2 are given in fig. 6. Transient power process is caused by oscillations of GDGU's revolutions at abrupt change in load. It is visible from diagrams that use of the fuzzy controller allows to improve qualitatively the accuracy of distribution of active power between GDGUs operating in parallel.

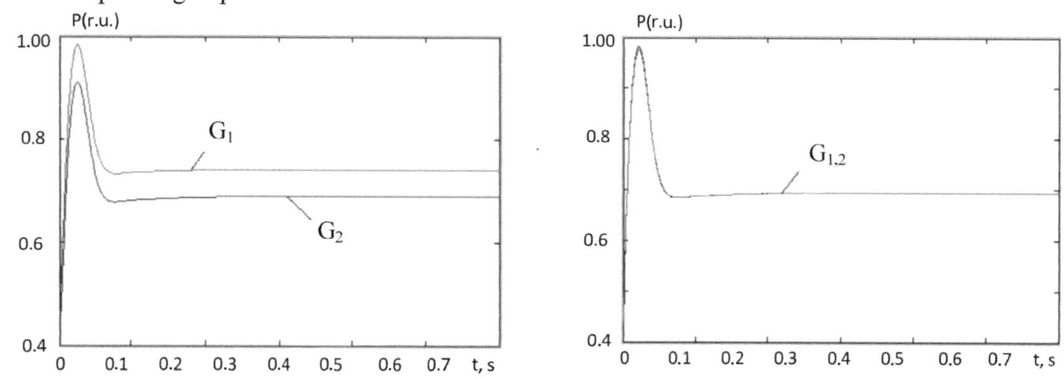

Figure 6 – The analysis of the adaptive controller stabilizes expected value of frequency oscillations at the preset value of 50 Hz

Irregularity in power distribution when static and dynamic made no more than 5% that, by a long short is in limits which don't exceed standards, being specified by classification societies.

Frequency deviations of the voltage being generated didn't exceed values of +/-0.2 Hz being admissed by the Register.

Time of transient process didn't exceed 3s, overshoot hasn't exceeded 2%, maximum deviation of revolutions in transient processes didn't exceed 4%.

5. Conclusion

The analysis of results shows that use of the adaptive regulator of rotating speed of gas-diesel engine including a fuzzy controller provides reducing of oscillations GDGU's revolutions theoretically about 25-30 times.

The mathematical analysis and computer modeling showed that worked out methods and devices of automatic control and regulation of GDGUs operating in parallel increase stability of their revolutions and accuracy of proportional allocation of the power being returned by them, providing a possibility of effective and safe operation of GDGU in a parallel as a part of APS.

References

[1] Nyrkov A, Sokolov S, Zhilenkov A, Chernyi S 2016 Complex modeling of power fluctuations stabilization digital control system for parallel operation of gas-diesel generators. 2016 IEEE NW Russia Young Researchers in Electrical and Electronic Engineering Conference (EIConRusNW),. 636-640. doi: 10.1109/EIConRusNW.2016.7448264

[2] Zhilenkov A, Chernyi S 2015 Investigation performance of marine equipment with specialized information technology, Procedia Engineering 100 1247–1252

[3] Zhilenkov A 2015 Improving the efficiency of automatic control systems of power quality of stand-alone power systems. EasternEuropean Journal of Enterprise Technologies 6 (9) 10–16

[4] Chernyi S and Zhilenkov A 2015 Modeling of complex structures for the ship's power complex using XILINX system. Transport and Telecommunication 16 (1) 73–82

[5] Chernyi S 2016 Use of Information Intelligent Components for the Analysis of Complex Processes of Marine Energy Systems. Transport and Telecommunication Journal, 17 (3) 202–211

[6] Kutateladze S S 1979 Fundamentals of the heat transfer theory Atomizdat 415

[7] Delaur M C, Chan V S, Murray D 2003 B A simultaneous PIV and heat transfer study of bubble interaction with free convection flow Experimental Thermal and Fluid Science 27 911–926

[8] Kitagawa A, Uchida K, Hagiwara Y 2009 Effects of bubble size on heat transfer enhancement by sub-millimeter bubbles for laminar natural convection along a vertical plate International Journal of Heat and Fluid Flow 30 778–788

[9] Nyrkov A, Budnik V, Sokolov S, Chernyi S 2016 The Algorithm of Development the World Ocean Mining of the Industry During the Global Crisis. IOP Conference Series: Materials Science and Engineering 142 012121

IOP Conf. Series: Earth and Environmental Science **50** (2017) 012020 doi:10.1088/1755-1315/50/1/012020

Application Biosphere Compatibility Concept To Evaluate The Quality Of Urban Environment By Bioindication Methods

S Vorobyov

Oryol State Universiti,
Komsomolskya Street 95, Oryol, Russian Federation, 302026

E-mail: vorser323@rambler.ru

Abstract. This article focuses on the application of methods biondisation different types of urban green areas to assess the quality of urban environment from the standpoint of compatibility biosphere concept. To assess urban environmental quality, we used a variety of areas of the city of Orel with different levels of human impact.

1. Introduction

In modern conditions, when urban development is not possible without focusing on a limited area of population, industry and transport, pollution reaches such proportions as the current system of environmental regulation quality of urban ecosystems can not fully perform its function of maintaining the balance of the biosphere and the technosphere [1,2,3]. The problem of preservation of the environment is based on the provisions of the concept of sustainable development [4,5], which served as the basis for developing strategies for the protection of the environment in most countries [2,5], including Russia. Despite the enormous positive role of sustainable development in international cooperation on the protection of the environment (agreement on limiting greenhouse gas emissions) and urban development, still not existed approaches to transition to biosphere compatibility city [6].

In this regard, of particular interest is the concept of biosphere compatibility offered by Vice President of Russian Academy of Architecture and Construction Science [7]. The concept of biosphere compatibility is means the balance of biosphere and technosphere, for evaluation of this balance we can use as indicators of environmental pollution traditional values of industrial emissions compared to maximum permissible concentration, green areas, the intensity of a cleaner environment by recreational areas and other indicators of the anthropogenic impact on the biosphere. The general concept of hierarchy and relationships represented in the matrix of the transformation of the city in biosphere compatibility's city (Table 1) [7].

Table 1 - Matrix development of urban areas based on the balance of homeostasis Biotechnosphere

1. The harmonious coexictence of urban areas and the natural environment	4. Regalatory support maintaining the Balance of Biotechnosphere	7. Functions of the city, providing development of urban areas
2. Analysis of the interaction of the components of urban ecosystemes (emissions technosphere, biological objects, people)	5. Knowledge management as a basis biosphere compatibility processes in urban ecosystems	8. Reliability – program for the development of urban settlements based on the concept of biosphere compatibility
3. Balance Biotechnosphere – Triple drmographic, technosphiric biological indicators of urban ecosystems	6. Progress, a comparison of the planned parameters of urban ecosystems with actual	9. Knowledge – new approaches to the urban environment and the development of urban areas

Implementation of the second cell of the matrix is not possible without the current assessment of environmental quality of the urban environment as a starting point for predicting the development of urban areas and for development of environmental protection measures. As one of the promising methods for assessing the quality of the environment are the methods of bioindication. Methods compare favorably to traditional laboratory studies the major media (air, soil, water) simplicity, speed of execution and low cost [7]. The use of bioindication methods does not require special education and allows to evaluate the quality of environment directly by city planner engineer.

Engineer designer can evaluate the impact on the environment by assess the state of greenery. To do this, he can use the methods of bio-indication [8]. These methods differ in speed and ease of implementation (such as fluctuating asymmetry coefficient). They allow you to partition the same green planting on the degree of adsorption of contaminants. This is an important question because now in the concept of biosphere compatibility is considered as a homogeneous formation of uniformly absorb pollution.

In this approach, the numerical value of the biospheric compatibility is defined as follows [7]:

$$\eta = \sum_{n} \sum_{i} (D_{in} \cdot \xi^{*}_{in} \cdot \Theta_{in} - A_{in} \cdot \gamma_{in} \cdot m_{in}) \qquad (1)$$

where the right side of this formula is a relative index of pure environment; left - a relative indicator of pollution from the technosphere with maximum concentrations that allow the development of urban area.

D_{in} - The ratio of the required area of the biosphere wich necessary to neutralize the pollution from the technosphere is based on the i-th source polution in the n-th function of the city; ξ^{*}_{in}- Uniformity coefficient of the biosphere, taking into account the different intensities of pollutants; Θ_{in} - the number of sources, pollution from which can be absorbed by the biosphere, in the zone of influence of the i-th source in the implementation of the n-th function of city; A_{in}- The amount of pollution from the i-th source in the implementation of the n-th function of the city for the territory of distribution of pollution; γ_{in} - The coefficient of reduction of pollution parameters from each source; m_{in} - The relative number of sources, pollution from which must be absorbed by the biosphere in the zone of influence of the i-th source in the implementation of the n-th function of the city.

For calculation of the ratio of the area of the biosphere in the region or district of the village needed to neutralize contaminants from (D_{in}) we had the formula [7]:

$$D_{in} = \{(V_{in} / \Theta_{in}) / k_{in}\} / S_{all} \quad (2)$$

where V_{in} - volume of contaminants from the i-th source for implementing the n-th function of city kg/year; k_{in} - the volume one pollutant wich absorb by 1 m2 biosphere, kg/year; S_{all} - total area (m^2).

The value of parameter pollution from the i-th source in the implementation of the n-th function of the city (A_{in}) is calculated by the formula [7]:

$$A_{in} = (S_{con} / \Theta_{in}) / S_{all} \quad (3)$$

where S_{con} - the area of contamination from the i-th pollutant in the implementation of the n-th function of the city, m^2.

The conception of expanded reproduction of the main productive forces can be written as [7]:

$$\eta > 0; \quad (4)$$

$$\eta < \quad (5)$$

With a positive balance of bio-and technosphere (see formula 4) provided the main productive force growth and natural population growth, otherwise (see formula 5) holds regressive development of territory [7].

2. Results and discussions

As an example, we have analyzed the land territory Orel city as a typical regional center of Central Federal District of Russia. To detect anthropogenic impacts on urban ecosystems we have identified ecotopes - land urban areas with similar parameters of the biosphere (the area and species composition of green territories), the technosphere (polluters), demographic (number of inhabitants) and urban development (type of building, the proximity of industrial enterprises, roads) factors. As these areas we have chosen the following areas: the area adjacent to the housing number 16 (Moskovskay street 77) Oryol State University (area 1), the park near main building Oryol State Universitry (Komsomolskay street 95) (area 2), area adjacent to the Museum of Oryol writers (area 3). All of these areas are characterized by the similar buildings (five-storey buildings, the lack of high-rise buildings, flat terrain), the characteristics of human impact (prevalence of vehicle emissions from major highways), similar areas (about 4 ha), the same composition of green spaces (the main tree species are the English oak, Norway maple and Heart- linden, plants of the same age and size), the same number of inhabitants (about 14000 people). All of these areas are characterized by the same set of factors of anthropogenic impact and varying degrees of quantitative manifestation of these factors. As a control area that not feel the impact of these factors, we have chosen the area of Medvedevsky forest (area 4).

Figure 1 – Location of Area 1

Figure 2 – Location of Area 2

Figure 3 – Location of Area 3

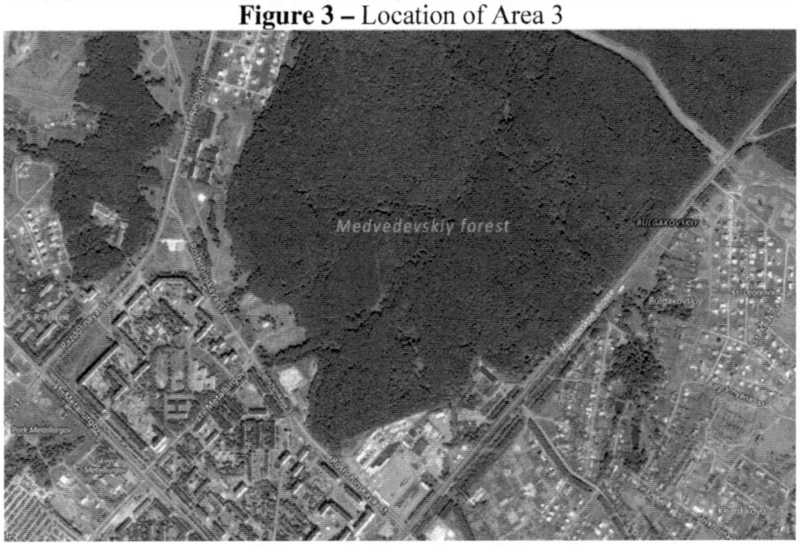

Figure 4 – Location of Area 4

The main source of pollution at all this sections was selected vehicles, as it accounts for a significant part of the emissions in atmosphere Oryol. To determine the number of vehicles for the subsequent calculation of emissions, at study sites were counted in the rush hour, within a year. Noted the number of cars in 30 minutes, followed by recalculation of 1 hour. Car traffic was divided into categories: passenger road transport; cargo, with a diesel engine; truck, with a carbureted engine; bus, diesel engine; with carburetor engine bus; trolley bus; minibuses. Separation was made to more accurately assess the environmental burden, as each of these categories is characterized by a specific set of emissions. Traffic is amounted 2067, 1528, 1231 cars in hour, for areas 1, 2 and 3 respectively.

The main source of contamination at all study sites were selected vehicles, as it accounts for a significant portion of the emissions in the city of Oyoll. To determine the amount of pollutants from vehicle emissions has been used a technique developed by A. Voeikov [9].

To assess the quality of the environment, we used the method of calculating the index fluctuating asymmetry, is, the variance based on the difference between the sides are not zero, and from some differences between the parties taking place in the given sample and calculated as follows:

$$\sigma^2 = \frac{\Sigma(d_{i-r}M_{d)}}{n-1} \quad (6)$$

where d_{i-r} - the average difference between the parties; M_d - difference of characteristic values. These indices are calculated by the following formulas:

$$M_d = \frac{\Sigma d_{i-r}}{r} \quad (7)$$

$$d_{i-r} = \frac{2(d_i - d_r)}{d_i + d_r} \quad (8)$$

where d_i - value of characteristic on the left side; d_r - value of characteristic on the right side; n - sample size.

Processing plant sample in determining the coefficient of fluctuating asymmetry is measuring the lengths of the veins on the leaf blade right and left.

We investigated the main tree species in Oryol (oak (Quercus robur L.), maple (Acer negundo L.) and linden (Tilia cordata Mill)) which accounted for 80% of all green spaces[10]. Plants were chosen the same age and size. Foliage urban green space was selected according to accepted procedures in the lower part of the crown at arm's length with the side facing the roadway. The leaves were collected one part of the medium sized type.

We obtained the following results on the emissions of vehicles on the study areas (Table 2):

Table 2 - Distribution of total and maximum single emission investigated areas

Investigated area	Indicator	
	Total emissio, t/year	Maximum single ejection, g/s
Area 1	13454	15.2789
Area 2	10274	1.0471
Area 3	9338	0.8971

To clarify the mutual influence of vehicle emissions on the investigated areas in Oryol, we have analyzed the meteorological data, summarized in Table 3. These data represent the average number of prevailing winds for 2015 year.

Table 3 - Interference investigated areas in the context of wind direction

Contaminated areas	Polluting areas		
	Area 1	Area 2	Area 3
Area 1	N, N-E, W, S-W, S, S-W	S-E	S-W
Area 2	N-W	N, N-E, E, S-W, S, S-W, W, N-W	W
Area 3	N-E	E	N, N-E, E, S-E, S, S-W, W, N-W

Then calculations were made using the procedure described in [9]. Results are summarized in Table 4.

Table 4 - Formation of pollutants at investigated areas in view of the air transport

Area	Emission, t/year
1	19619
2	12419
3	9153

The analysis of the above table shows that, given the volume of the air transport of pollutants in the area 1 rose by 45%, on a area 2 increased by 21%, and in the area 3 decreased by 2%. This is due to the predominance of winds western and southern direction of the geographical location of the Oryol.

When we analysed bioindication methods on the investigated areas we obtained the results (Table 5):

Table 5 - Area of leafs of green spaces.

Area	Tree leaf's area of urban green spaces, cm^2		
	Oak	Maple	Linden
Area 1	15.2	20.7	14.5
Area 2	19.7	25.3	16.1
Area 3	26.3	36.7	20.3
Area 4 (control)	27.5	44.3	24.2

Common to all of those species examined trends relates the fact that the area of leafs with decreasing the number of vehicles is increasing and is the highest in the control group of plants. The biggest area of leafs for all three species (after the control group) is marked on the area 3, then area 2 and the smallest in the area 1. Thus, we have the effect of vehicle emissions on the studied plants, which is higher in those areas where more number of cars. But each breed has their differences.

Leaf area of maples (Acer negundo L.) is the highest among the studied species of urban green space, which is related to species characteristics of the breed. The largest area of leafs is in the control group (area 4), then by area 3, area 2 and the minimum area is marked to plants area 1. The difference between the highest and lowest value was 2 times.

Thus, the observed correlation between the number of vehicles and the amount of leaf area on the test plants. This dependence confirms the correlation coefficient constituting r = -0,92.

Average values of leaf area of oak less than maple, but more then linden, is also associated with species characteristics. The highest values of leaf area also fall on the plants in the control group (area 4). The difference between the highest and lowest values is also 2 times.

The correlation coefficient values of the leaf area of oak and vehicles on study sections is r = -0,96 and indicates a pronounced inverse relationship.

Average values of leaf area linden is lowest among the studied species of urban green spaces. The highest values of leaf area in the breed also fall on the plants in the control group (area 4) and the whole dynamics of the distribution of leaf area linden on tested areas looks the same as in the previous two cases. Differences between the linden leafs on the areas are not as big as the other species studied urban green space this fact indicates the greater stability of this breed compared to other breeds.

To assess the impact of pollutants we used the method of study of fluctuating asymmetry coefficient, the values of which are given in Table 6.

Table 6 - Values of the coefficient of fluctuating asymmetry

Area	Fluctuating asymmetry coefficient values for plants		
	Oak	Maple	Linden
Area 1	0.025	0.020	0.017
Area 2	0.023	0.017	0.015
Area 3	0.020	0.014	0.010
Area 4 (control)	0.006	0.005	0.005

On the storage capacity of green space influenced by many factors, from weather conditions to species composition, because different types of urban green spaces are characterized by unequal resistance to absorbed pollutants. Thus, the biosphere compatibility can be calculated by the formula:

$$\xi^* = 1 - (S_{gs} \cdot \sigma_c) \tag{9}$$

Where S_{gs} - the total area of green spaces ; σ_c - coefficient of fluctuating asymmetry.
Using the above formulas were obtained the following results shown in Table 7.

Table 7 - Values of the coefficient of uniformity of the biosphere on the studied sections

Indicator	Studied area			
	Area 1	Area 2	Area 3	Area 4
ξ^*	0.15	0.19	0.23	0.79

Substituting the values obtained in the formula (1) we received updated information (Table 8).
Table 8 - Values of the coefficient biosphere compatibility on the studied sections

Indicator	Studied area			
	Area 1	Area 2	Area 3	Area 4
η	0.24	0.32	0.41	0.97

From the obtained values of the biosphere compatibility can be concluded that the current state of ecosystems at all studied areas indicates the presence of degradation processes due to anthropogenic impacts that the biosphere cannot fully accumulate pollution and regenerate the city ecosystem.

3. Conclusions

The highest values of the coefficient fluctuating asymmetry observed in plants maple on all three points, followed by plant oak and linden lowest observed. In plants of all three species investigated urban green spaces greatest values of the index of asymmetry observed at the area №1, then area №2, №3 and the lowest in the area №4.

In control group the coefficient asymmetry values is smaller than in the samples at investigated areas, and their differences are not so obvious that due to the fact that in a relatively favorable from an environmental point of view the conditions the plants are not exposed to stress like in urban environments. In an urban setting green spaces show better ability to adapt. That's why the difference in biological indicators are more obvious. Distribution of values of the coefficient fluctuating asymmetry coincides with the dynamics of the distribution of the emissions in the investigated areas.

There is a direct correlation of coefficient asymmetry of green space and emissions at study sites, as indicated by high values of the correlation coefficient of this factor (0.97). The influence of meteorological factors, which manifests itself in the predominance of winds west and south directions, is increased human impact on area №1 by the air pollution from areas №3 and №2. Our future investigations will be associated with the improvement of methods of assessing the quality of the environment, based on the concept of biosphere compatibility with the influence of the prevailing direction of air flow in an urban environment as an important factor in enhancing anthropogenic impact.

References

[1] Kolloros G A. model applicable in coses of road impacts on habitate environment // G. Kolloros, A. Athanasopulou. – Jornal of International Scientific Publications: Ecology&Safety. V. 10. – 2016 – p. 85-94.
[2] Yiu N A. taxonomic review of the application of safety management system in constraction.// N. Yiu, D. Chan. - Jornal of International Scientific Publications: Ecology&Safety. V. 19. – 2016 – p. 394-408.
[3] Grem T. Insurance value of Green Infrastructure in and Araund cities. // T. Grem, J. Kroneneberg, E. Andersson. - Ecosystems. V. 19. – 2016 – p. 1051-1063.
[4] Heydinger J. Reinforcing the ecosystem Services Perspective: The Temporal Component// Ecosystems. V. 19. – 2016 – p. 661-673.
[5] Bertram J. Combinig machanism and drift in community ecology: a novel statistical machanics approach. – J. Bertram, R. Dewar. – Theoretical Ecology. – V. 8 – 2015, p. 419-435.
[6] Ilyichev V. About the dynamic formation of the urban livelihood system compatible with the biosphere. // V. Ilyichev, S. Emelyanov, V. Kolchunov., N. Bakaeva. – Applied mechanics and materials. V. 725-726, 2015, p. 1224-1230.
[7] Ilyichev V. Estimation of indicators of ecological safety in civil engineering.// PROCEDIA ENGINEERING. – V. 117 – 2015, - p. 126-131.
[8] Vorobev S. Evaluation of urban environment by bioindication methods from position the concept of urban settlements biosphere compatibility // S. Vorobev, D. Kozlov. The strategies of Modern Science Development: Proceedings of the V International scientific-practical conference (Yelm, WA, USA, 26-27 December 2013). – Yelm, WA, USA: Science Book Publishing House, 2013. – p.149-152.
[9] Vorobev S. The influnce of meteorological factors on the pollution emission by vehicles in the city Orel // S. Vorobev, D. Kozlov - VI International scientific conference «21 century: fundamental science and technology» (North Charleston, SC, USA, 20-21 Aprylr 2015) Charleston, SC, USA: Science Book Publishing House, 2015. – 11-14 p.
[10] Vorobev S. Using bioindication methods for assessing the environmental quality of urban settlements from the perspective of the concept of biosphere compatibility on an example of the Oryol city // Vorobev S, Kozlov D. - Lambert academic publishihg: Saarbruken, Deutschland, 2014. – p. 54.

IOP Conf. Series: Earth and Environmental Science **50** (2017) 012021 doi:10.1088/1755-1315/50/1/012021

A Thermo-Physical Model Of Destruction Of Contaminants By Means Of A Water-Ice-Jet Cleaning Technology

Yu Stepanov, M Burnashov, E Stepanova

Orel State University named after I.S. Turgenev
Komsomolskaya st.95, Orel, Russian Federation, 302026

E-mail: yury057@yandex.ru

Abstract. The reader will achieve a benchmark understanding of the essence of cleaning for the removal of contaminants from machine elements by means of cryo jet\ water-ice jet with particles prepared beforehand. This paper represents the classification of the most common contaminants appearing on the surfaces of machine elements after a long-term service. The conceptual contribution of the paper is to represent a thermo-physical model of contaminant removal by means of a water ice jet. In conclusion, it is evident that this study has shown the dependencies between the friction force of an ice particle with an obstacle (contamination), a dimensional change of an ice particle in the cleaning process and the quantity of heat transmitted to an ice particle.

Introduction

The hydro jet technologies are widely used in industry. They are necessary for cutting materials, cleaning contaminants off the surfaces of machine elements, varnish removal, blasting of thin bores etc. [1 - 20].

Problem description

During an operation of machine elements contaminants (which differ in kind, composition, properties and the strength of the adhesion to the surfaces of machine elements) are being formed on their outer and inner surfaces. These contaminants reduce their durability, service life and resistance to corrosion.

When repairs are carried out, contaminants undermine labour productivity, accuracy of monitoring and defectation of machine elements, reduce repair quality and resource of repaired machines and their elements.

The most widespread causes of contaminative coating formation include an emulsive and oil slick, an introduction of the contaminants from the environment, thermal decomposition of oils, an oxydation of metal surfaces, burnt-on (sticking) sand, oil remains, scale etc.

Contaminants on the objects under repairs are divided into the following types due to their chemical composition:

- organic (oil and fat deposit, varnish coating films, lubricants);
- non-organic (scale, road mud, corrosion products);
- mixed (carbon deposit, varnish, grease, industrial contaminants).

Contamination of plant items, assembly units and machine elements include:

- outer deposit,
- remains of lubricative materials,
- carbon deposit (building-up),
- corrosion products,
- scale,
- remains of old paint coatings.

The most common and widespread contaminants are remains of fuel-lubricative materials and products of their transformation. Drastic changes of lubricative materials, occuring during use of machines, are caused by the processes of 'aging'-oxydation and polymerization. The following types may be singled out:

1. an incomplete combustion of fuel,
2. oxydation products,
3. products of destruction of hydrocarbons,
4. polymerization products,
5. products of condensation and coagulation of hydrocarbon and heteroorganic compound,
6. corrosion products and biodamage of metals in the medium of remains of fuel-lubricative materials.

According to up-to-date state of the field under discussion the industrial processes must be environment-friendly, energy-efficient and meet industry and evolving governmental regulations. These demands are also obligatory for cleaning for the removal of contaminants from surfaces.

Water jet methods are considered to be the most highly promising and universal methods of cleaning among existing ones.

At the same time, a hydroabrasive cleaning with sand and similar materials, performing the function of an abrasive, has a wide range of drawbacks. Let's mention some of them:

- the abrasive complicates cleaning of inner and inaccessible surfaces;
- roughening of clean surfaces;
- the necessity of utilization of the abrasive, contaminated during the blasting process;
- a high cost of an abrasive material and its delivery.

Therefore, renewal and replacement of the abrasive material is absolutely necessary for a wide introduction of hydroabrasive technology of cleaning of machine elements.

Moreover, water-ice jet cleaning technology is not only an ecologically clean but also a cost-saving method.

Ice particles possess properties of hard particles, which allow them to remove a great deal of contaminants during accelerating to higher speeds and avoid causing damage to a base of surfaces, undergoing cleaning. Use of water and ice makes regeneration of clean media easier.

In comparison with other methods of cleaning, water-ice jet technology obtains some advantages:

1 minimization of harmful impact upon environment;
2 a low-cost cleaning materials;
3 an absence of the necessity to transport and keep large amounts of the abrasive material;
4 a possibility of a closed and waste-free cycle;
5 a reduction in an abrasive influence on material;
6 the abrasive does not block up peep-holes of machine elements joints;
7 an absence of dust in the cleaning process;
8. a high durability of an instrument.

A water-ice jet formation for cleaning machine elements surfaces is run in the following way: a highly pressured stream, having a diametre ranging from 0,15 to 0,3 mm, is being mixed with entrained stream of ice particles according to an ejection principle. As a result, a two-phased structure, called a water-ice jet, is formed. Use of ice particles, prepared beforehand, is preferable and optimal for this process. The size of particles varies from 1.5 to 2 mm (Figure 1).

Figure 1 – The picture of an ice particle

Mechanism of coating destruction, caused by water-ice jet, is determined by a multiphase character of highspeed flow (water-ice particles) (Figure 2). Having spent part of its energy on an acceleration of ice particles, water jet creates voltage, equal to its strength, in the material, which is being destroyed. At the same time, ice particles are making a full impact on the material, causing microcracks. They concentrate stresses, which enhances cleaning efficiency in general.

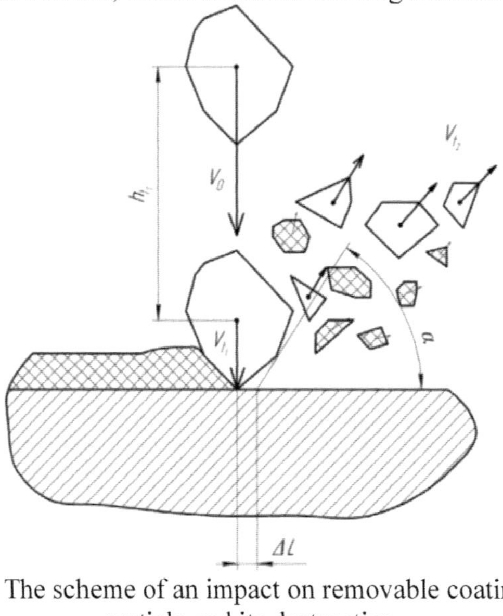

Figure 2 – The scheme of an impact on removable coating of an ice particle and its destruction.

Economical efficiency is achieved in the process of deriving granulated ice by means of crushing ice lumps in industrial ice generators. During further mechanic crushing necessary fraction is separated from ice crumb. After production of working granulated ice with a help of this method, the model of an ice particle was created in the programming environment Compass-3D (Figure 3).

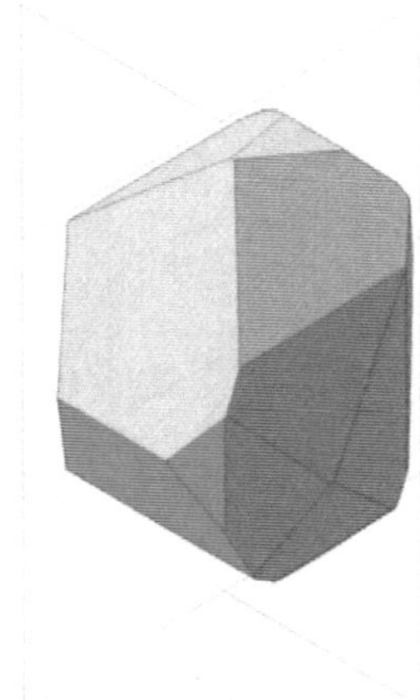

Figure 3 – The scene of the model of an ice particle (an isometric projection)

Thermo-physical model of destruction

While being formed, water-ice jet may undergo the following phases:

1. A highly pressured water jet bursts from a stream-forming checkerwork, creating an area of low pressure in a plenum (mixing) chamber, where ice particles are being entrained from the supplying canal.
2. Water jet ice particles are entrained by a water flow and mix with it, causing an intensive heat exchange.
3. Mixture of water and ice particles is being accelerated in a collimator.

Energy is generated in the closed system 'an ice particle-a destructed material-machine elements' by means of impulsive force and interaction and is being transformed into a heat stream and a work of cutting. The definition of a fraction of energy, spent on heating an ice particle, accompanied by creation of local zones of reversed phase transfer (transmission) (RPT) of ice and heating of material, drums up a particular interest.

Contaminating coatings have a low capacity for heat transfer, therefore, temperature of a contact surface 'an ice particle-material' will determine mechanic and thermal physical properties of material of coating.

The scheme of an interaction between the surface and an ice particle is represented in picture 4.

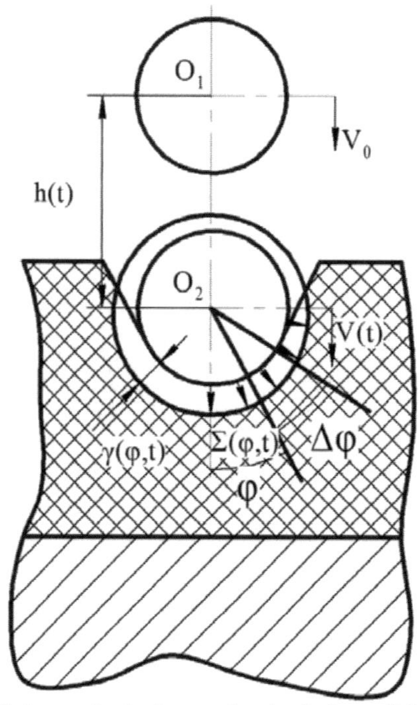

Figure 4 – Schematic design and calculation of the process of interaction of a water jet and ice particles with coating.

An equation of an ice particle motion in the material prepared to be removed may be written in the following form:

$$m(t)\frac{\mathrm{dV}}{\mathrm{dt}} = -\frac{1}{2}C_x\rho_c(T)S(h)V^2(t) - \sigma_{\mathrm{res}}(t)S(h) - F_{\mathrm{fr}}(t,\phi); \quad (1)$$

where *m(t)* is a current value of the mass of an ice particle, taking into consideration RPT on the contact surface; *V(t)* is a current value of the speed of penetration; C_X is a drag coefficient of an ice particle; *S(h,φ)* is a current value of a projection of section area of an ice particle, taking into account depth of penetration and presence of RPT on the elementary area within the solid angle $\Delta\varphi$:

$$S(h,\phi) = \pi R^2(t,\phi)\cdot\left[\frac{h(t)}{R(t,\phi)}H(R-h) + H(h-R)\right]; \quad (2)$$

where *h(t), R(t)* are the functions of depth of penetration and the current radius of an ice particle, taking into consideration RPT respectively; *H(...)* is the Heaviside step function.

The first component of the right part of an expression (1) characterizes the loss of kinetic energy of an ice particle, when interacting with the material. The second component reflects the resistance of material to destructive impact, and the third component is the friction force, influencing the particle, while penetrating into the removable material.

Friction force will depend on availability of liquid phase, which is relevant to the scheme of motion of an ice particle in the material with phase transitions on the contact surface:

$$\begin{aligned} F_{\mathrm{Tp}}(t,\phi) &= f_V(t)\sigma_r(t,\phi)S(h,\phi)\cdot H[\mathrm{dA}(t,\phi) - (q + C_V\Delta T_1)\mathrm{dm}(t,\phi)] + \\ &+2\mu_1\frac{V(t,\phi)}{\gamma(t,\phi)}S(h,\phi)\cdot H[(q + C_V\Delta T_1)\mathrm{dm}(t,\phi) + \mathrm{dA}(t,\phi)]; \end{aligned} \quad (3)$$

$$\begin{aligned} \sigma_x(t,\phi) &= \frac{1}{2}\rho_c(t)V^2(t,\phi); \\ \sigma_r(t,\phi) &= \frac{\upsilon_{\mathrm{ice}}}{1-\upsilon\mathrm{ice}}\sigma_x(t,\phi); \end{aligned} \quad (4)$$

where $(q + C_V\Delta T_1)\mathrm{dm}(t,\phi)$ is a value of the energy necessary for the transformation of the solid phase of the spherical layer by the mass dm in liquid state; μ1 is a dynamic viscosity coefficient of liquid phase; γ is a size of a gap between ice particle surface and a profile of a cavity in an obstacle.

The value of γ in bounds of a discrete angular coordinate $\Delta\varphi$ is defined by the mass of the solid phase, limited by the surface in the form of a spherical layer, 'lost' in each particular discrete moment of time. $f_V(T)$ is a function of a speed coefficient of friction; $\sigma_X(t,\varphi)$ is a value of an axial tension on the contact surface of an ice particle with an obstacle, generally equal to a pressure of a velocity (dynamic, kinetic-energy) head; $\sigma_r(t,\varphi)$ is a value of radial voltage on the contact surface of an ice particle with an obstacle.

The work of the resistance forces, determining the intensity of the heat of the contact area, is represented with a help of the following sum:

$$A(t,\varphi)= A1(t,\varphi)+ A2(t,\varphi)+ A3(t,\varphi), \quad (5)$$

where $A_1(t,\varphi)$, $A_2(t,\varphi)$, $A_3(t,\varphi)$ is a work of inertial constituent of force of resistance, work of a friction force and work of forces of plastic resistance to the deformation of coating respectively.

Change of the size of an ice particle, when cleaning within an angular coordinate φ, is given according to the following dependency:

$$d(t,\phi) = 2\cdot\sqrt[3]{\frac{3}{4\pi\rho_{\text{ice}}}\left(m(t) - \int_0^{\frac{\pi}{2}} m_{\text{ice}}(\phi)d\phi\right)}; \quad (6)$$

where $\rho_л$ is a density of an ice particle; mл(φ) is a mass of an ice particle, «lost»,

$$A(t,\varphi)=Q(t,\varphi) \quad (7)$$

where $Q(t,\varphi)$ is a the thermal energy of the system of the contact of bodies.

In accordance with the condition of heat transfer through boundary surface of two bodies in contact in case of an ideal thermal contact [4] and the law of conservation of energy, an amount of ice, having been transmitted into an ice particle, and a respective change of its mass may be determined in a following way:

$$\begin{gathered} dQ_1(t,\phi) = \frac{v_m}{v_{\text{ice}}} dA(t,\phi); \\ m(t) = m(t-\tau) - \int_0^{\phi_K} \frac{Q_1(t,\phi)}{q + C_V^m \Delta T_1} d\phi, \end{gathered} \quad (8)$$

where ΔT_1 is a range of temperature change of the contact layer of an ice particle, determining the RPT moment, ($\Delta T_1 = T_{RPT} - T_{01}$; T_{RPT} is a temperature of RPT; T_{01} is an initial temperature of an ice particle; m_0 is an initial mass of an ice particle.

Taking into consideration the fact that a (described in this paper) single act of an interaction of an ice particle with the material of coating, suffering damage, is similar to others, given a particular similarity coefficient, it becomes possible to calculate the percentage of energy, spent on heating of an ice particle, which is accompanied by appearance of local areas of a reverse phase tranfer (transmission) (RPT) of ice and on heating of the material.

Conclusion

A received thermo-physical model allows to develop and, having been proved scientifically, to appoint the modes of water-ice cleaning:

- water jet pressure;
- a diameter of a jet forming nozzle;
- the dimensions of ice particles;
- the temperature of storage of ice pellets;
- the distance from the nozzle to the surface;
- feeding speed of a nozzle apparatus during processing.

References

[1] Kurco M C, Chadwick R F. Werkstoffbearbeitung mittels Hochdruck // Technika (Suisse). - 1972. -21. -N 3. -P. 2079 - 2083.

[2] B-L Liu, L-H Liu. Research on the preparation of the ice jet and its cleaning parameters // Papers presented at the 14th Internationals on Jetting Technology. Held in Brugge, Belgium, 21-23 September. P. 203-210.

[3] Hashish M, Dunsky C M. The formation of cryogenic and abrasive-cryogenic jets // Papers presented at the 14th Internationals on Jetting Technology. Held in Brugge, Belgium, 21-23 September. P. 329-343.
[4] Kiyohashi H, Handa K. "A study of Production of Ice Particles by the Heat of Vaporization of Cryogenic Liquefied Fuels and Their Applications in Ice Jets", Proceedings of the International Symposium on New Applications of Waterjet Technology, Ishinomaki, Japan, pp. 51-60, 1999.
[5] Liu H, t. Butler A vanishing abrasive cryogenic jet for airframe repainting// Processing of the new applications of water jet technology. Isinomaki, Japan, 1999. – pp. 51 – 60.
[6] Water jet cutting: a production tool // Machining and Production Engeneering.-1983.-Vol. 141, N3631.- P. 18-19.
[7] Hegland D E. Supersonic water jet slashes through tough materials // Automation. - 22. - N 8. - 1975. - P. 40 - 44.
[8] Hensen K F, La-Brush E C. Material removal by high-pressure liquid jet at ten kilolars // Trans. ASME. - B. 97. - N 3. - 1975. - P. 1067 - 1073.
[9] Vatter och polymer-nytt skarwertyg // Nord Emball. - 1975. - 41. - N 2. - P. 25-26.
[10] Pushkarev A Y, Golovin K A, Zabin A B, Yeruhimivich Y I. Simulation of Hydro-Abrasive Cutting Process//Proceeding of the International Symposium on New Applications of Water Jet Technology. Oktober 19-21, 1999. Ishinomaki, Japan. Ishinomaki Senshu Universiti.- P. 421-425.
[11] Louis T J. Fluid Jet Tehnology Fundamentals and Applications, 5th American Waterjet Conference, Toronto, Canada, August, 1989, pp. 145 - 168.
[12] Hessling M. Recent Examination Relating to the Effects of the Abrasive Material, Operating Parameters and Rock Properties on the Depth of Cut Obtainable with Abrasive High Presser Water Jets when Cutting Rock, 9th International Symposium on Jet Cutting Technology, Sendai, Japan, October, 1988, paper G3, pp. 357 - 376.
[13] On the modelling of abrasive waterjet cutting / El-Domiaty A A, Shabara M A, Abdel-Rahman A A, Al-Sabeeh A K // Int. J. Adv. Manuf. Technol. - 1996.-12. -№4. -C.255-265.
[14] The liquid cleaver / Etchells Paul // Aerosp. Compos, and Mater. - 1991. – 3, №1 - C 40 - 43.
[15] Techniques to increase water pressure for improved water - jet – assisted cutting / Kovscek P D, Taylor C D, Thimons E D // Rept. Invest / BurMines US Dep. Inter. -1988 - № 9201 - C1-10.
[16] Abrasive waterjet sound power measurement / Merchant Howard C, Chalupuik James D. //Noise Countr. Eng. J. -1987 - 29. - № 3 - C. 85-89.
[17] Anon, High Pressure Water Jet Systems – Part 2, No. 4, June, 1993, pp. 20 –23.
[18] Yazici, Sina, Abrasive Jet Cutting and Drilling Rock, Ph.D. Dissertation in Mining Engineering, University of Missouri - Rolla, Rolla, Missouri, 1989, 203 p.
[19] Faber K and Oweinah H. Influence of Process Parameters on Blasting Performance with the Abrasive Jet, paper 25, 10th International Symposium on Jet Cutting Technology, Amsterdam, October, 1990, pp. 365 - 384.
[20] Chalmers E J. Effect of Parameter Selection on Abrasive Waterjet Performance, 6th American Water Jet Conference, Houston, Texas, August , 1991, pp. 345 - 354.

Eco-analytical Methodology in Environmental Problems Monitoring

M I Agienko[1, a], E P Bondareva[1, b], G V Chistyakova[1, c], O V Zhironkina[1, d], O I Kalinina[2, e]

[1] Kemerovo Institute (branch) «Plekhanov Russian University of Economics» Kuznetsk av., 39, Kemerovo, Russian Federation, 650992
[2] T.F. Gorbachev Kuzbass State Technical University
Vesennyaya st., 28, Kemerovo, Russian Federation, 650000

E-mail: [a]marinaivanovnaag@mail.ru, [b]bondarevaep@list.ru, [c]galvik06@mail.ru, [d]o-zhironkina@mail.ru, [e]koi2008@list.ru

Abstract. Among the problems common to all mankind, which solutions influence the prospects of civilization, the problem of ecological situation monitoring takes very important place. Solution of this problem requires specific methodology based on eco-analytical comprehension of global issues. Eco-analytical methodology should help searching for the optimum balance between environmental problems and accelerating scientific and technical progress. The fact that Governments, corporations, scientists and nations focus on the production and consumption of material goods cause great damage to environment. As a result, the activity of environmentalists is developing quite spontaneously, as a complement to productive activities. Therefore, the challenge posed by the environmental problems for the science is the formation of geo-analytical reasoning and the monitoring of global problems common for the whole humanity. So it is expected to find the optimal trajectory of industrial development to prevent irreversible problems in the biosphere that could stop progress of civilization.

1. Introduction

Currently, the idea of rethinking of the relations between Man and Nature, their harmonious interaction and a caring natural resource usage have become the most topical subject for the discussion by philosophers. The necessity of changing the role of a human being in the physical world, forming environmental friendly consciousness is caused, on the one hand, by significant deterioration of the ecological situation in the world and the increasing number of apocalyptic forecasts of the mankind development [1-2]. Floods, fires, sudden temperature fluctuations, drought, tsunamis – these are the phenomena and processes that have had an especially negative impact on human society and natural systems recently. On the other hand, humanity has faced the new environmental challenges that were not comprehended before. The threats to the environment, such as the risk of global warfare, mass migration, climate change, are beyond the means accumulated in the progressive states [3].

Consequently, the main issue that the environment puts before philosophers is to find an answer to new challenges that ultimately will determine the future of civilization [4]. For that to happen, philosophy and ecology must interact [5]. Philosophy can help to solve environmental problems in various areas, as it stimulates the formation of new social consciousness, focused on the necessity to overcome environmental contradictions related to cultural attitudes of the past, it contributes overcoming narrow-mindedness of some academic positions, one-sidedness of spiritual and practical orientation of Man in his relationship with nature, fragmentation of opinions [6-10].

That is philosophical consideration of current environmental situation that allows scheduling such solutions to environmental problems that are positive for society and the environment. Having become a driving force of society values reorientation, philosophical principles of interaction between nature and society, thereby, perform the role of one of the main factors of self-organization of the system "Man – Environment". In the future, while selecting a strategy for improving an environmental situation, one will need to take many factors into account. And the longer activities to protect the environment from human being will be devoid of a philosophical conception, the more factors will arrive.

2. Material and method

Some achievements of scientific and philosophical thought deserve especially high praise, as they have a direct and significant impact both on people's minds, and their environmental activities [11-15]. V.I. Vernadsky and his followers' doctrine about the noosphere proposed a way of achieving the harmony between Man, Society and Nature and thus solving environmental problems [16]. In this doctrine the major factor in harmonizing relations between society and nature is called a rational activity of a social human being. This activity can give stability and rational forms of existence and development to his living environment.

Also an important achievement of philosophy is the conception of an ethical relationship to nature. The fundamental positions of this conception were formulated by Albert Schweitzer. He formulated a well-known bioethics principle of "Reverence for Life" [17]. In modern philosophical and ethical thought this principle was further developed and modified within the principles of environmental ethics, bioethics, including bioethics in medicine. This is currently one of the promising areas of ethics and general philosophical thought.

The ideas of N.N. Moiseev on the environmental imperative and the coevolution of nature and society, that reflected the current state of the environmental situation in the world and approaches to its stabilization and improvement, are also significant [18].

The conception of "sustainable environmental development", collectively elaborated and proposed to the world community by a number of UN institutions on environmental issues and included in many regulatory documents adopted at international, regional and national levels, plays an important role. For the first time this conception was adopted as a guideline to follow at the UN Conference on the Environment in Rio de Janeiro in 1992.

At the same time, important issues determining ecological consciousness of modern society stay without proper attention of philosophers. How should philosophical thought react to environmental problems and challenges? What is the essence of the philosophy of ecology? What is the content of the philosophical aspect in the approaches aiming to solve environmental problems?

It is important that it is philosophy that traditionally considers problems of interaction of Man, Society and Nature. Philosophy has always reflected the problems of human and nature coexistence, seeking to impart a certain harmony of their interaction through a spiritual understanding of oneself and the World of Nature, and thus spiritual activity aimed at the transformation of Nature.

Philosophy has made significant progress in understanding and clarifying the environmental problems that exist in modern world:

- Philosophers give generalized ideological interpretation of environmental problems;
- Philosophy has developed and categorized the conceptual definitions on the problems of interaction between Man, Society and Nature, and, consequently, ecological form of social consciousness;

- Philosophy along with science and other types of intellectual activities is actively engaged in the process of searching the most effective ways of solving environmental problems.

At present the need for harmonization of relations between society and nature is quite realized. Besides, there is the understanding of the fact that the mankind should follow in one direction as a single whole while seeking for solutions of environmental problems. In general, however, environmental problems are not solved effectively in practice [19-23], and do not figure prominently in the minds of the nationals' elites, especially in raw materials producing countries like Russian Federation etc. [24-28]. The problem of re-thinking of the role of scientific-and-technological advance concerns speeding up the machinery development [29-37]. Joining the efforts of different countries in this area is very slow, and concerted actions of different states are still insufficient.

Studying the causes of environment degradation and measures for its protection and improvement, social philosophy contributes to the expansion of human freedom through the formation of a more humane attitude toward nature and toward other people. Philosophical thought is involved in the development of eco-humanism in consciousness and behavior of a human being and various social strata. The advancement of eco-humanism in society greatly influences the development of new approaches to understanding the importance of environmental problems solution.

The relation of human being to nature is always mediated by relationships between members of society and its structures, especially the political system and economic relations. Therefore, clarifying the nature of interaction between society and nature, philosophical thought should provide its guidance and ideological influence both on individuals, and society that is to adopt environmental views, values, norms, attitudes to spiritual culture in general. Any social system needs to be in harmony with the natural environment, and adapt its instrument of labor, production, consumption, and ways of using nature. It also should be environment friendly in its way of life as well as its population, including population size.

Philosophers determine ecological philosophy as a set of social and philosophical researches of nature and society coordination. In today's knowledge it is philosophy that can combine different approaches to environmental problems, all the diversity of its faces and bases. Philosophy provides this generalization, solves the traditional problem of methodical search, philosophical understanding of ecological conditions in order to form general methodological principles of research and problem solving.

Nature acts as a postulate and a condition of society development. Society is an isolated part of nature; Man is the pinnacle of evolution. The problem "Society – Nature" is the problem of relationship of a biologically social creature, human being, with the world around him.

Ecosystem is a unity that includes all the organisms of the area and interacts with the physical environment in such a way that the flow of energy creates a clearly defined trophic structure, species diversity and circulation of substances. In the functional characteristics of living organisms, man is a consumer of the first and second order, as he uses vegetable and animal products for food. However, the main feature of the person and its relations to the prosperity of the Earth's ecosystem at this stage is an anthropogenic industrial activity that keeps on changing natural ecosystems while developing. Modern ecology is inconceivable without a biosphere perspective in philosophical studies.

3. Results and discussion

Currently the threat to the environment has become global in nature, it has created a pre-crisis situation, that may escalate into a crisis under certain conditions and become dangerous to people and the world community as a whole. Therefore, the situation should be changed radically. Philosophy can play an important role in the solution of this task.

Many thinkers see the reason of ecological disasters in Man himself, in his ability to create and use technologies destroying life in nature; in his unlimited desire for consumerism and enrichment at the expense of nature.

Though Man is not the king of nature; he depends on natural resources, the state of the biosphere. Nature resources are not unlimited, and many of them are close to exhaustion. To change the attitude to nature, save natural resources, recycle waste are the tasks that have come out to the forefront today. Preservation of the biosphere is a necessary condition for the survival of the mankind. Ecologically sustainable development of the mankind is such development that meets people's needs currently, but it doesn't threaten a possibility of future generations to satisfy their requirements.

Modern ecological situation arose quite spontaneously and regardless of human consciousness. It can be improved by organizing an environmental management system well thought out, developing the structure of social relations. Besides people are to understand and take into account the objective laws of the biosphere existence. Unfortunately, consciousness of many people is often clouded with the outdated ideas about the possibility of nature passively endure any human actions.

The global nature of today's environmental challenges requires global and systematic approach to overcome them when a set of particular solutions has one common ground. It should be a universal philosophical principle rather than some particular statement claiming for global status.

Philosophy focuses on changing the human environment, exploring the means of its protection and improvement, helping to expand the scope of human freedom. It creates more moral and humane attitude towards people and the world around them. Philosophical thought is involved in forming moral relationship to nature in human mind and behavior, as well as different strata of society. Society becoming kinder and more compassionate contributes to developing relationships between Nature and human Society, and, hence, solving environmental problems.

Many philosophers believe the source of the environmental disasters to be Man himself, his ability to create and use technologies that can be destructive for nature. Philosophers note his irrepressible urge to enrichment at the expense of nature and philistine attitude towards it from the standpoint of the owner, 'the king of nature'. This standard of thinking and behavior needs changing.

Ensuring human security is becoming more important than ongoing technological progress. We should move from increasing to modifying production processes with special attention to environmental requirements. It is important to make the transition to a more "natural", less parasitic lifestyle. Moreover, economy should take into account the need for revival and protection of nature.

Thus, ecology confronts philosophers with new global challenges: they are to find some spiritual meaning in the problems of interaction of Man, Society and Nature, and identify competitive ways of their solution.

The main environmental task of philosophy is to understand the model of future society as an environment information culture, able to be in a harmonious relationship with the natural environment. This is, undoubtedly, the main strategic task for the future, the challenge for philosophical thought. Another fundamental task of philosophy is to refocus social consciousness from a parasitical to a sensible relationship to nature thus balancing the system "Society – Nature".

As the main reason for current environmental problems results from man, it is he who needs changing, and mostly his spirituality. To solve challenging environmental problems facing the world community it is necessary to study, define and use spiritual reserve of Man and Society.

Ultimately, man himself needs to be protected. Nowadays one of the most dangerous threats is intensifying negative changes in the genetic basis of human being that happen under the influence of adverse environmental conditions as well as the use of various harmful substances. When developing solutions to these challenges philosophers can and must take an active part for the good of the individual and of all humanity.

Being both a biological and social creature, human being is simultaneously connected with two mutually intersecting ecological systems – natural and social ecology. Therefore, philosophy should form ecological consciousness in society as a peculiar form of social consciousness, reflecting the mutual influence of natural and social systems in terms of their current imbalance. According to its objectives and areas ecological consciousness should be focused on working out a large-scale comprehensive plan aimed to prevent a bio-social ecological disaster.

Due to developed ecological consciousness the future society will be able to admit the inevitability of the influence of scientific and technological progress and innovative technologies on the environment as well as to use them for developing resource-saving and waste-free technologies which do not harm nature.

Equally important global environmental issue that requires rational philosophical understanding is the uncontrolled population growth. The birth rate and population growth are not purely biological processes; they are not subjects to exclusively biological laws. Determinants of the demographic transformation are social and cultural mechanisms and regulations. Population growth in countries with low level of life and culture is much higher than in economically and culturally developed countries. This is because in richer countries the costs of education and training of children are extremely high. The costs of child care and training in developing countries with low social and cultural standard of living is much lower. Equally significant is the fact that consciousness and traditions denying birth quotas do not allow society to take actions on keeping in check rapid population growth in these countries. Therefore, control of population and growth rate is becoming a worldwide problem without any philosophical understanding of birth quotas.

It is the rationality of Man modern societies rely upon while solving the demographic problem. Though we have to admit that abstract rationality is a rather weak mechanism that can be successful only on the basis of a thorough and exhaustive analysis of social and cultural order of a country, region, historical and national traditions, it can help to exercise rational management of population growth rates and population structure. At these points, there is the intersection of the interests of philosophy and psychology, ethnology, sociology, demography and cultural studies that is likely to lead to formation of independent demographic consciousness.

While studying global environmental problems philosophers should also focus on wars as a special kind of armed conflict resolution. At the moment, the major nations have nuclear bombs, missiles that can reach any point on the Earth. The existence of various types of bacteriological and chemical weapons, as well as constantly increasing power of conventional weapons make it possible to assert that the third world war – nuclear missile war – may become the last one. The strong increase of radioactivity will quickly lead to the death of all living things, the survivors of the nuclear strike. Most major cities, scientific and industrial centers will be eliminated. The entire biosphere will be contaminated by radiation. Thus, the life on our planet will be forever impossible. To prevent a nuclear war it is necessary to reduce or even better to stop local and regional armed conflicts. This is the idea that gives the ground for new political thinking. That is why the task of philosophy is to comprehend the essence of these changes and their direction, as well as to indicate the way leading to the social and historical progress without wars.

The philosophical view of current ecological situation may be very beneficial for the correct formulation of environmental problems and the development of the best ecological strategy. The current environmental problems have resulted in deeper and multidimensional understanding of the world, humanity, and universal values. It requires reconstructing society and production, as well as developing environmental education as a new morality and forming ecological consciousness as a set of opinions, knowledge and beliefs, reflecting the system "Society – Nature" and aiming at treating it reasonably [38-39].

At present, philosophy can help to solve environmental problems in different aspects, because it forces to form new social consciousness, focused on overcoming the ecological contradictions caused by the former cultural traditions. It contributes to negotiating limitations of private scientific statements, dissociation of opinions, and unilateralism of spiritual and practical interests of Man in his relations with Nature.

The time of carelessness passed, the era of environmental ethics has come [40-41]. Human beings should feel the necessity to change the role of conqueror of nature to the role of its ordinary member. It especially concerns development geotechnical activities and mineral resource extraction [42-44]. Designing and producing powerful mining machines, constructors and engineers must analyze ecological consequences of their implementation [45-47]. Fighting harmful trends will strengthen environmental ethics, we must stop thinking only about how much we can extract from the ground.

Every action will be considered correct if it does not violate the order and beauty of the biotic community, and naught if the opposite happens.

5. Conclusion

The period of development has come when the human safety becomes even more important than further technological progress. The further expansion of production is not a priority any more. It's obligatory to change it in the way to meet current environmental requirements. The mankind should transfer to a more natural and less consumptive lifestyle and practice the eco-technical type of economy that takes into account the need for restoration and conservation of the natural environment.

The development of ecologically oriented thinking is the most important condition for the society to become able to harmonize the system "Society – Nature" and stabilize the ecological situation in the world, in regions and individual countries. Since the main cause of environmental problems comes from human being himself therefore it is he and his spirituality that needs changing. Thus, it is necessary to study, define and use spiritual reserve of Man and Society to solve challenging environmental problems facing the world community.

References

[1] Huesemann M H and Huesemann J A 2011 Technofix: Why Technology Won't Save Us or the Environment (Washington: New Society Publishers) 388 p
[2] Hopwood N 2014 Haeckel's Embryos: Images, Evolution, and Fraud (Chicago: University of Chicago Press) 330 p
[3] Raudsepp-Hearne C, Peterson G D, Tengo M and Bennett E M et al. 2010 Untangling the Environmentalist's Paradox: Why is Human Well-Being Increasing as Ecosystem Services Degrade, BioScience 60 (8) pp 576–589
[4] Hobson J M 2004 The Eastern origins of Western civilization (Cambridge: Cambridge University Press) 366 p
[5] Keulartz J 1999 The Struggle for Nature: A Critique of Environmental Philosophy (Routledge: Routledge Pub.) 288 p
[6] Mannison D, McRobbie M and Routley R 1980 Environmental Philosophy (Canberra: Australian National University) 404 p
[7] Devall W and Sessions G 1985 Deep Ecology: Living As if Nature Mattered (Salt Lake City: Gibbs M. Smith, Inc.) 390 p
[8] Oelschlaeger M 1993 The Idea of Wilderness: From Prehistory to the Age of Ecology (New Haven: Yale University Press) 658 p
[9] Drengson I 1995 The Deep Ecology Movement (Berkeley: North Atlantic Books) 322 p
[10] Derr P G and McNamara E 2003 Case Studies in Environmental Ethics (Hampshire: Bowman & Littlefield Publishers) 490 p
[11] Baird J C and Nelson M 1998 The Great New Wilderness Debate (Atlanta: University of Georgia Press) 352 p
[12] Foltz B V and Frodeman R 2004 Rethinking Nature (Indiana University Press, Bloomington North Morton Street) 601 p
[13] Vogel S 1999 Environmental Philosophy After the End of Nature Environmental Ethics 24 (1) pp 23-39.
[14] Weston C 1999 An Invitation to Environmental Philosophy (New York: Oxford University Press) 472 p
[15] Zimmerman M J, Callicott M and Sessions G et al. 2010 Environmental Philosophy: From Animal Rights to Radical Ecology (New Jersey: Prentice-Hall, Inc.) 632 p
[16] Lapo A V 2001 Vladimir I. Vernadsky (1863–1945), founder of the biosphere concept (Moscow: NOW Pub.) 288 p
[17] Schweitzer A 1931 On the Edge of the Primeval Forest (New York: Macmillan) 622 p

[18] Moiseev N. N 1986 Man, nature and the future of civilization: "nuclear winter" and the problem of a "permissible threshold" (Moscow: Novosti Press Agency) 211 p

[19] Lesin Y V Luk'yanova S and Tyulenev M 2010 Mass transfer of dispersed particles in water filtration in macro-grained media J. Journal of Mining Science 46 (1) pp 78-81

[20] Tyulenev M A, Gvozdkova T N and Zhironkin S A et al. 2016 Justification of Open Pit Mining Technology for Flat Coal Strata Processing in Relation to the Stratigraphic Positioning Rate Geotechnical and Geological Engineering 34 (6) doi:10.1007/s10706-016-0098-3

[21] Lesin Y V, Luk'yanova S Y and Tyulenev M A 2015 Formation of the composition and properties of dumps on the open-pit mines of Kuzbass IOP Conference Series: Materials Science and Engineering 91 (1) 012093

[22] Tyulenev M, Zhironkin S and Litvin O 2015 The low-cost technology of quarry water purifying using the artificial filters of overburden rock Pollution Research 34 (4) pp 825-830

[23] Tyulenev M, Zhironkin S, Kolotov K and Garina E 2016 Background of innovative platform for substitution of quarry water purifying technology Pollution Research 35 (2) pp 221-226

[24] Tyulenev M A, Zhironkin S A, Garina E A 2016 The method of coal losses reducing at mining by shovels International Journal of Mining and Mineral Engineering 7 (4) DOI: 10.1504/IJMME.2016.10000781

[25] Zhironkin S A, Khoreshok A A, Tyulenev M A, Barysheva G A, Hellmer M C 2016 Economic and technological role of Kuzbass industry in the implementation of national energy strategy of Russian federation IOP Conference Series: Materials Science and Engineering 142 (1) 012127

[26] Zhironkin S A 2002 Prospects and new possibilities investment attracting to Kuzbass coal mining industry Ugol' 6 pp 31-36

[27] Zhironkin S A 2002 About measures of vexel circulation development and vexelability definition of fuel-and-power complex' enterprises Ugol' 4 pp 47-48

[28] Tyulenev M A, Lesin Yu, Vik S and Zhironkin S 2016 Methodological Bases of Advanced Geo-ecological Problems Resolving in Neo-industrial Clusters Proceedings of the 8th Russian-Chinese Symposium "Coal in the 21st Century" pp 333-336

[29] Khoreshok A A, Mametyev L E, Borisov A Y, Vorobyev A V 2016 Influence of the Rigid Connection between Discs in the Tetrahedral Prisms on Equivalent Stresses When Cutting Work Faces IOP Conference Series: Materials Science and Engineering 127 012039

[30] Khoreshok A A, Buyankin P V, Vorobiev A V, Dronov A A 2016 Simulation of Stress-Strain State of Shovel Rotary Support Kingpin IOP Conference Series: Materials Science and Engineering 127 012014

[31] Khoreshok A A, Mametyev L E, Borisov A Yu and Vorobyev A V 2015 The distribution of stresses and strains in the mating elements disk tools working bodies of roadheaders. IOP Conference Series: Materials Science and Engineering 91 (1) 012084

[32] Kovalev V A, Gerike B L, Khoreshok A A and Gerike P B 2014 Preventive maintenance of mining equipment based on identification of its actual technical state Symposium of the Taishan academic forum – Project on mine disaster prevention and control pp 184-189

[33] Aksenov V V, Khoreshok A A and Beglyakov V Y 2013 Justification of creation of an external propulsor for multipurpose shield-type heading machine - GEO-WALKER Applied Mechanics and Materials 379 pp 20-23

[34] Lekontsev Yu M., Sazhin P V, Temiryaeva O A, Khoreshok A A and Ushakov S Yu 2013 Two-side sealer operation Journal of Mining Science 49(5) pp 757-762

[35] Khoreshok A A 2002 On side cutting bit when operating at sheerer drums Ugol' 7 pp 10-11

[36] Tyulenev M A, Khoreshok A A, Garina E A, Danilov S and Zhironkin S 2016 Adaptive technology of using backhoes for full coal extraction Proceedings of the 8th Russian-Chinese Symposium "Coal in the 21st Century: Mining, Processing, Safety" pp 111-115

[37] Khoreshok A, Tyulenev M and Vöth S 2016 Conditions for Minimum Dynamic Loading of Multi-brake Hoists Proceedings of the 8th Russian-Chinese Symposium "Coal in the 21st Century" pp 239-245
[38] Fakhrislamova E I., Dolgikh T V, Pevneva I V and Formulevich I V 2015 Trends and Prospects of Higher Education Development in Russia by the Example of Training of Specialists in the Field of Economics Indian Journal of Science and Technology 8 (S10) pp 1-9
[39] Khoreshok A A, Zhironkin S A and Tyulenev M A et al. 2016 Innovative technics of managing engineers' global competencies IOP Conference Series: Materials Science and Engineering 142 (1) 012122
[40] Pojman L P and Pojman P 2000 Environmental Ethics (Orlando: Thomson-Wadsworth) 478 p
[41] Sherer D and Attig T 1983 Ethics and the Environment (New Jersey: Prentice-Hall, Inc.) 554 p
[42] Aksenov V V, Efremenkov A B and Beglyakov V Y 2013 The influence of relative distance between ledges on the stress-strain state of the rock at a face Applied Mechanics and Materials 379 pp 16-19
[43] Golik V I, Rasorenov Y I and Efremenkov A B 2014 Recycling of metal ore mill tailings Applied Mechanics and Materials 682 pp 363-368
[44] Golik V I and Efremenkov A B 2016 Physicochemical Processes of Metal Lixiviation in the Disintegrator IOP Conference Series: Materials Science and Engineering 125 (1) 012038
[45] Efremenkov A B 2011 Forming the subterranean space by means of a new tool (geohod) Proceedings of the 6th International Forum on Strategic Technology IFOST 6021037
[46] Efremenkov A B and Timofeev V Y 2012 Determination of necessary forces for geohod movement Proceedings - 2012 7th International Forum on Strategic Technology IFOST 6357729
[47] Efremenkov A B and Aksenov V V et al. 2012 Force parameters of geohod transmission with hydraulic drive in various movement phases Proceedings - 2012 7th International Forum on Strategic Technology IFOST 6357716

IOP Conf. Series: Earth and Environmental Science **50** (2017) 012023 doi:10.1088/1755-1315/50/1/012023

International Cooperation in Environmental Management and Rational Use of Natural Resources

E A Fedulova[1, a], I V Korchagina[1], S V Vik[2, b], O I Kalinina[2, c], V L Martyanov[2, d]

[1] Kemerovo state university, Krasnaya st., 6, Kemerovo, Russian Federation, 650043
[2] T.F. Gorbachev Kuzbass State Technical University
Vesennyaya st., 28, Kemerovo, Russian Federation, 650000

E-mail: [a]fedulovaea@mail.ru, [b]viksvetlanavs@ya.ru, [c]koi2008@list.ru, [d]martvic2005@yandex.ru

Abstract. The progress in technologies is developing towards the unlimited growth of production and consumption, wasteful use of natural resources and biosphere. These problems require adequate response such as international cooperation and integration of the efforts of authorities, scientists, representatives of educational system. Such cooperation is important to ensure the transition to the sustainable, ecologically-oriented practices of natural resources rational use. This is impossible without establishing a new environmental management system based upon formation of ecological competence of all scientific and technological progress participants among which the higher school scholars must play a leading role.

1. Introduction

The globalization processes of political, economic, technological, human life actively taking place today are caused by such problems, challenges and risks as uneven economic and social development of the regions of the world, demographic problems, international terrorism and the threat of large-scale military conflict. In the second half of the XX century the most important - environmental problems of mankind - become global, and therefore the question of saving civilization, the survival of the world as a result of human-induced destruction of biological balance became really important.

2. Material and Method

The first United Nations Conference devoted to the Environment (Stockholm, 1972) drew the attention of the Heads of States and Governments, the international community to the impossibility of further economic growth without taking into account the environmental and social and economic consequences [1]. The International Commission for Environment and Development in the report "Our Common Future" (1987) proposed the concept of sustainable development as a strategic solution to this problem. In its basis - the balancing of the natural biosphere opportunities and social and economic development for human needs without damaging the living conditions of future generations [2]. Conservation and sustainable reproduction of life on the Earth is understood as a specific function of human biosphere.

The Second United Nations Conference devoted to the Environment and Development (Rio de Janeiro, 1992) declared in the outcome document "Agenda 21" the program of the implementation of sustainable development, which was confirmed at the Sustainable Development World Summit (Johannesburg, 2002) [3]. In its materials it is emphasized that the most important mechanism for ensuring the sustainable development of nature and society is education. The United Nations declared the decade 2005-2014 years as the Decade of Education for Sustainable Development, which aims to promote public understanding of the importance of education for sustainable development [4].

The European Economic Commission adopted a "Strategy for Education for Sustainable Development" (Vilnius, 2005). Its main objective is to contribute to sustainable development and the inclusion of education in sustainable development of nature and society, in the national systems of formal and informal education [5]. What is fundamentally important, the strategy provides the conceptual, information and organizational expansion of environmental education as an educational base for sustainable development of nature and society.

In Russia the idea of sustainable development are reflected in the "Concept of the Russian Federation transition to sustainable development" (1996) [6] and "The Russian Federation National Security Strategy" (2009) [7]. The role of education in the solution of sustainable development problem is reflected in the concept of modernization of Russian education for the period up to 2010 and earlier - in the "Principles of the Russian system of universal, comprehensive and continuous environmental education and training" (2002).

It should be noted that in Europe the extensive work on the introduction into the educational process of sustainable development courses in higher education in all countries of "Bologna area" is carried out. For example, in 50% of universities in Scandinavia and in 96% of higher schools in the UK and Sweden the lectures on Sustainable Development are delivered [8-10]. Almost at all leading universities in the US and Canada educational programs in the field of sustainable development in a variety of forms (lectures, research) are implemented.

However, today in Russia the most educational problems listed in the documents above have not been solved yet. The existing system of environmental education does not give the desired result for a number of reasons. They are:

- *Regulatory and strategic* - the lack of comprehensive government regulations, organizational and methodical documents in the field of education for sustainable development [11-13]. In the standards of higher professional education (2010), approved by the Ministry of Education of the Russian Federation and in the content of the educational programs the aspects of sustainable development are not included;

- *Social and economic* – the lack of effectiveness of environmental education in environmental issues (especially on air and water pollution [14-19], using mining equipment [20-28), despite all the evidence of its crucial role. As a result the ecological aspect of education is considered as "rudimentary", the number of environmentally oriented and creative public associations is being reduced;

- *Methodological* - a priority in the environmental educational process is given to a professional function of environmental education (the student is given the required level of knowledge, skills, competences). As a result, the formation of ecological outlook (moral and environmental imperatives in the human mind, when the latter gets an environmentally-centric orientation) suffers.

The higher school plays the most important role in the realization of the idea of nature and society sustainable development, in the becoming of this kind of personality. It forms the basis of environmentally oriented professional world outlook, establishes value-semantic sphere of the person, and helps to develop in practice environmental standards and requirements. One of the most important problems of environmental education in the higher education institution, which is rather relevant today, is creating a culture of acceptance of environmentally sound management decisions, personal involvement in solving environmental problems in order to improve environmental quality.

As a result, it is necessary to develop the environmentally oriented professional, capable and ready to solve the problems of sustainable development. At the same time environmental competence as developing personal characteristic can and should become an indicator of the formation of such characteristics of environmentally oriented person as an environmentally-centric consciousness (and its central component – ecologically oriented world outlook) and environmental culture.

Conceptual bases of realization of environmental education for sustainable development at the university level are not fully developed. Despite the wide range of the terms "education for sustainable development", "sustainable development of the education system", the available studies on the higher school level focus on individual, narrow span issues of environmental education for sustainable development.

Thus, at the present time the problem solution of the development of the concept of environmental education for sustainable development in general and its implementation in higher education in particular is at the preliminary stage of the analysis, the description of specific situations, their classification, highlighting the relationships and regularities.

At the same time there is a shortage of professionals able to carry out professional activities taking into account the environmental viability of decisions, i.e. professionals with environmentally-oriented consciousness, capable of implementing the concept of sustainable development in their professional activities.

3. Results and Discussion

We believe that the system of environmental education for sustainable development at universities can be implemented in case of the theoretical development of its conceptual basis and practical actualization in the sphere of innovation activity of environmentally-oriented higher education institution.

The essential features of environmental education for sustainable development include the following:

- *Categorical* - a process based on the integration of natural sciences, social and human, environmental and cultural knowledge. Environmentally-oriented values should be formed on their basis as an integral part of universal human values, ecologically-oriented world outlook. The principles of the process should become eco-centrism, ecological culture, active approach to solving the problems of sustainable development at local and regional levels, as well as responsible behavior in the environment not violating the interests of the future generations;

- *Targeting* – grounding the objective of including in all formal and informal education the ideas and principles of sustainable development to provide students with knowledge, specialized skills, competencies, values and beliefs in the field of environment and sustainable development. This knowledge, values and competencies will increase their ecological awareness and self-confidence, and give them an opportunity to lead a healthy and productive life in harmony with nature, to follow environmental protection in their behavior;

- *Structural and content* - include the following components:

a) natural science block with the fundamentals of ecology, environmental management, biodiversity conservation, and technological principles of the interaction between man, nature and society in terms of escalating ecological crisis; cultural and ethical aspects of education and formation eco-centric world outlook;

b) moral and legal block, considering the human rights for a dignified life and the responsibility for the wellbeing of the people living today and the future generations; gender equality as a means of solving demographic problems and the elimination of illiteracy and poverty; corporate responsibility;

c) social and economic block, comprising the formation of consumer culture, the development of rural and urban settlements; the formation of culture of peace and non-violence in resolving territorial, ethnic and religious problems.

The result of the implementation of the essential features of environmental education in modern higher education should become the formation of the ecological competence of the individual [29]. It must be considered in the following forms:

- as a link in the system of the most important education of environmentally-oriented person (environmentally-oriented world outlook, environmental awareness, environmental culture). This link appears in the ecology-oriented activities and carries out ideological, methodological, environmental, value, predictive, social, cultural, professional functions;

- as one of the most important kinds of professional competence, actually having inter-professional, and ideally - over-professional nature. Also, the environmental competence of the person is an inherent component of the high level of professionalism, which allows to judge matters of professional activity sphere taking into account the environmental aspect, as well as the quality of the individual, making it possible to carry out professional activities with regard to its environmental viability;

- as a form of personal integration, reflecting the ability of acquiring environmental knowledge and skills, methods of their use in various kinds of practical activity, readiness for implementation of environmental activities. It also includes the experience in preserving the surrounding natural world and solving environmental problems. In the system of an ecology-centric person his competence is the result of acquiring other key competencies - general culture; values and cognitive; competence of personal self-improvement.

The conditions of formation of ecological competence of a person include the following:

- *Macro-social* - the demand for specialists who are ready to the environmentally-oriented professional activity and making environmentally sound management decisions [30];

- *Social and psychological* - the development in specific professional and educational environment a single content knowledge, skills, professionally important individual, personal and subjective qualities of ecologically-oriented specialist as a basis of pedagogical influence on the student;

- *Didactic* - expansion of teachers' methodical techniques of educational influence on students; the development and use of technologies for creating situations of personal importance of ecologically-oriented values for sustainable development of nature and society, and regulation ways aimed at the nature activity in cooperation with it, and the other person as a part of it. Equally important is the continuity of ecological and educational process and the appropriate distribution of its content in the course of higher school training [30-31]; the organization of students practical implementation of ecologically-oriented activities and its projecting in cooperation with teachers. Also the organization of diverse opportunities for the acquisition and improvement of students and teachers personal experience of practical ecologically-oriented activities is in demand. It is strongly recommended for studying geotechnical science and mineral resource extraction [32-36]. Obtaining the knowledge of designing mining machines, constructing of mining objects, the future engineers must learn how to estimate ecological consequences of their implementation [37-39].

The principles of environmental education for sustainable development are:

1) *the principle of comprehensiveness and continuity*, which consists in the fact that based on the fundamental ecological knowledge, ethical standards set forth at school environmental education at higher school is subdivided into the following types of training:

- Classical university (which gives special professional education with the basis of scientific research);

- Education in the humanities and technical higher schools for non-ecological specialties;

- Education at higher educational institutions in the field of environmental (natural-resource) profile and associated with it special interdisciplinary education (Environmental law, Ecopolitology, Ecopsychology, Environmental linguistics, Industrial ecology, Health and safety and environmental protection, etc.).

The continuity of environmental education by courses as a part of educational programs in these types of higher education institutions is characterized by a multiplicity of targets, objectives and content of environmental knowledge, skills and competences;

2. *the principle of differentiated approach* to students, which consists in the fact that on the basis of a certain level of self-education and self-improvement motives and incentives should be developed, the ways to enhance the process of self-education and self-improvement of ecological values should be outlined;

3) *The principle of theory and practice correlation*, according to which the efficiency of training the ecologically-oriented person depends on his involvement in the relevant work and is

determined by the content, views, forms, and the orientation of the latter. In other words, the more perfect the system of labor and productive activity of students, in which the correlation between theory and practice is realized, the higher the quality of their training, the more successful their adaptation to modern conditions of the professional activity.

4. Conclusion

Environmental education for sustainable development plays a fundamental role in the formation of ecological competence of the individual, which is a part of the integration of psychological and pedagogical and akmeological knowledge. We have considered it in the system of ecologically-oriented person: as a result of acquiring by the individual in the process of formation the key competencies (general culture, values and meaning, competent personal self-improvement), as his integral feature performing ideological, methodological, environmental, value, prognostic, social, cultural, professional functions, and a system-link of his education (environmentally-oriented world outlook, ecological consciousness, ecological culture).

References

[1] Atkinson G, Dietz S and Neumayer E 2009 Handbook of Sustainable Development (London: Edward Elgar Publishing) 308 p
[2] Bakari M K 2013 Globalization and Sustainable Development: False Twins? New Global Studies 7(3) pp 23-56
[3] Rogers P, Jalal K F, and Boyd J A 2007 An Introduction to Sustainable Development (Routledge: Routledge Pub.) 380 p
[4] Edwards A R and McKibben B 2010 Thriving Beyond Sustainability: Pathways to a Resilient Society (Washington: New Society Publishers) 688 p
[5] Wallace B 2005 Becoming part of the solution: the engineer's guide to sustainable development (Washington: American Council of Engineering Companies) 592 p
[6] Zhironkin S A 2002 Prospects and new possibilities investment attracting to Kuzbass coal mining industry Ugol' 6 pp 31-36
[7] Khoreshok A A, Zhironkin S A and Tyulenev M A et al. 2016 Innovative technics of managing engineers' global competencies IOP Conference Series: Materials Science and Engineering 142 (1) 012122
[8] Van der Straaten J and Van den Bergh J C 1994 Towards Sustainable Development: Concepts, Methods, and Policy (Reykjavik: Island Press) 466 p
[9] Farah P D and Rossi P 2015 Energy: Policy, Legal and Social-Economic Issues Under the Dimensions of Sustainability and Security. World Scientific Reference on Globalisation in Eurasia and the Pacific (Rim: Consul Pub.) 450 p
[10] Huesemann M H and Huesemann J A 2011 Technofix: Why Technology Won't Save Us or the Environment (Washington: New Society Publishers) 388 p
[11] Zhironkin S A 2001 Governmental factoring development of TEK Kuzbass Ugol' 6 p 62.
[12] Zhironkin S A 2001 Factoring and leasing development at coal mining industry of Kuzbass as an important element of its financial part Ugol' 4 pp 29-30
[13] Zhironkin S A 2002 About measures of vexel circulation development and vexelability definition of fuel-and-power complex' enterprises Ugol' 4 pp 47-48
[14] Lesin Y V, Luk'yanova S and Tyulenev M 2010 Mass transfer of dispersed particles in water filtration in macro-grained media J. Journal of Mining Science 46 (1) pp 78-81
[15] Tyulenev M A and Lesin Y V 2014 Justification complex purification technology open-pit mines wastewater Symposium of the Taishan academic forum – Project on mine disaster prevention and control pp 441-444
[16] Tyulenev M, Zhironkin S and Litvin O 2015 The low-cost technology of quarry water purifying using the artificial filters of overburden rock Pollution Research 34 (4) pp 825-830

[17] Lesin Y V, Luk'yanova S Y and Tyulenev M A 2015 Formation of the composition and properties of dumps on the open-pit mines of Kuzbass IOP Conference Series: Materials Science and Engineering 91 (1) 012093
[18] Tyulenev M, Zhironkin S, Kolotov K and Garina E 2016 Background of innovative platform for substitution of quarry water purifying technology Pollution Research 35 (2) pp 221-226
[19] Tyulenev M A, Lesin Yu, Vik S and Zhironkin S 2016 Methodological Bases of Advanced Geo-ecological Problems Resolving in Neo-industrial Clusters Proceedings of the 8th Russian-Chinese Symposium "Coal in the 21st Century" pp 333-336
[20] Khoreshok A A, Mametyev L E, Borisov A Y, Vorobyev A V 2016 Influence of the Rigid Connection between Discs in the Tetrahedral Prisms on Equivalent Stresses When Cutting Work Faces IOP Conference Series: Materials Science and Engineering 127 012039
[21] Khoreshok A A, Buyankin P V, Vorobiev A V, Dronov A A 2016 Simulation of Stress-Strain State of Shovel Rotary Support Kingpin IOP Conference Series: Materials Science and Engineering 127 012014
[22] Zhironkin S A, Khoreshok A A, Tyulenev M A, Barysheva G A, Hellmer M C 2016 Economic and technological role of Kuzbass industry in the implementation of national energy strategy of Russian federation IOP Conference Series: Materials Science and Engineering 142 (1) 012127
[23] Kovalev V A, Gerike B L, Khoreshok A A and Gerike P B 2014 Preventive maintenance of mining equipment based on identification of its actual technical state Symposium of the Taishan academic forum – Project on mine disaster prevention and control pp 184-189
[24] Aksenov V V, Khoreshok A A and Beglyakov V Y 2013 Justification of creation of an external propulsor for multipurpose shield-type heading machine - GEO-WALKER Applied Mechanics and Materials 379 pp 20-23
[25] Lekontsev Yu M., Sazhin P V, Temiryaeva O A, Khoreshok A A and Ushakov S Yu 2013 Two-side sealer operation Journal of Mining Science 49(5) pp 757-762
[26] Khoreshok A A 2002 On side cutting bit when operating at sheerer drums Ugol' 7 pp 10-11
[27] Tyulenev M A, Khoreshok A A, Garina E A, Danilov S and Zhironkin S 2016 Adaptive technology of using backhoes for full coal extraction Proceedings of the 8th Russian-Chinese Symposium "Coal in the 21st Century: Mining, Processing, Safety" pp 111-115
[28] Khoreshok A, Tyulenev M and Vöth S 2016 Conditions for Minimum Dynamic Loading of Multi-brake Hoists Proceedings of the 8th Russian-Chinese Symposium "Coal in the 21st Century" pp 239-245
[29] Fakhrislamova E I, Dolgikh T V, Pevneva I V and Formulevich I V 2015 Trends and Prospects of Higher Education Development in Russia by the Example of Training of Specialists in the Field of Economics Indian Journal of Science and Technology 8 (S10) pp 1-9
[30] Sianipar C P, Dowaki K, Yudoko G and Adhiutama A 2013 Seven Pillars of Survivability: Appropriate Technology with a Human Face European Journal of Sustainable Development 2(4) pp 1-18
[31] Raudsepp-Hearne C, Peterson G D, Tengo M, Bennett E M et al. 2010 Untangling the Environmentalist's Paradox: Why is Human Well-Being Increasing as Ecosystem Services Degrade, BioScience 60 (8) pp 576–589
[32] Efremenkov A B 2011 Forming the subterranean space by means of a new tool (geohod) Proceedings of the 6th International Forum on Strategic Technology IFOST 6021037
[33] Efremenkov A B and Timofeev V Y 2012 Determination of necessary forces for geohod movement Proceedings - 2012 7th International Forum on Strategic Technology IFOST 6357729
[34] Efremenkov A B and Aksenov V V et al. 2012 Force parameters of geohod transmission with hydraulic drive in various movement phases Proceedings - 2012 7th International Forum on Strategic Technology IFOST 6357716

[35] Tyulenev M A, Zhironkin S A, Garina E A 2016 The method of coal losses reducing at mining by shovels International Journal of Mining and Mineral Engineering 7 (4) DOI: 10.1504/IJMME.2016.10000781

[36] Tyulenev M A, Gvozdkova T N and Zhironkin S A et al. 2016 Justification of Open Pit Mining Technology for Flat Coal Strata Processing in Relation to the Stratigraphic Positioning Rate Geotechnical and Geological Engineering 34 (6) doi:10.1007/s10706-016-0098-3

[37] Aksenov V V, Efremenkov A B and Beglyakov V Y 2013 The influence of relative distance between ledges on the stress-strain state of the rock at a face Applied Mechanics and Materials 379 pp 16-19

[38] Golik V I, Rasorenov Y I and Efremenkov A B 2014 Recycling of metal ore mill tailings Applied Mechanics and Materials 682 pp 363-368

[39] Golik V I and Efremenkov A B 2016 Physicochemical Processes of Metal Lixiviation in the Disintegrator IOP Conference Series: Materials Science and Engineering 125 (1) 012038

IOP Conf. Series: Earth and Environmental Science **50** (2017) 012024 doi:10.1088/1755-1315/50/1/012024

From Mining to Post-Mining: The Sustainable Development Strategy of the German Hard Coal Mining Industry

J Kretschmann[1, a], A B Efremenkov[2, b], A A Khoreshok[2, c]

[1] TH Georg Agricola University, Herner Straße 45, Bochum, Germany
[2] Yurga Technological Institute (branch) of National Research Tomsk Polytechnic University, Leningradskaya 26, Yurga, Russian Federation, 652055

E-mail: [a]Juergen.Kretschmann@thga.de, [b]ytitpu@tpu.ru, [c]haa.omit@kuzstu.ru

Abstract. By the end of the 1950s, the German coal mining industry produced 150 million tons of hard coal per year in 170 collieries with 600,000 employees. At that time, 70% of the primary energy demand of the Federal Republic of Germany was covered by domestic coal. Since the advance of oil, later of natural gas, in the world energy market and with the growth of world coal trade, domestic coal stood under a long-term restructuring pressure. This decision required a new strategy for the coal mining industry. Now German coal mining will be strictly finalized and will be prepared for the post-mining era. Within a sustainability strategy the long-term impacts of mining activities before and after the mine closures concerning the environmental, economic and social dimensions will be analyzed systematically and forward-looking.

1. Introduction

Principally, mining raw materials is a finite business that reaches its limits once the deposits mined are depleted or the resources can no longer be mined in an economically profitable way. However, the impact of winning raw materials by mining is potentially an infinite one – at least if you measure it at a human time scale because mining always means to interfere with geological and ecological structures that cannot be reversed to their original condition. If we look at the entire process of mining end-to-end, we have to consider the period after active mine works have finished as well.

Therefore, the mining cycle can be divided into three stages:

1. The exploration phase, in which the deposit is investigated to determine its technical and economic profitability. The period of these undertakings is relatively short and can lead to the launch of mining operations.

2. What follows is the actual mining period which usually lasts for a long time and ends latest when all deposits are fully depleted. The operational lifespans of German hard coal mines for example reached up to 120 years. This point of time, however, may be brought forward if the economic conditions (production cost or market price) deteriorate. Nevertheless, mining may be resumed even after a longer period of downtime if those conditions become favorable once more (examples are silver mining in Germany or rare earths mining in the USA).

3. The longest stage, however, is the post-mining stage as the human interference into geology and nature is usually intensive and irreversible. Mining can lead to consequences which have a permanently adverse impact on people and the environment; therefore, they have to be managed as best as we can. In Germany, what has to be done at the post-mining stage is called "Ewigkeitslasten" or "Ewigkeitsaufgaben" – which translates as "perpetual obligations" or "perpetual tasks". The literal translation for "Ewigkeit" is "eternity", expressing that we will have to cope with post-mining endlessly [1].

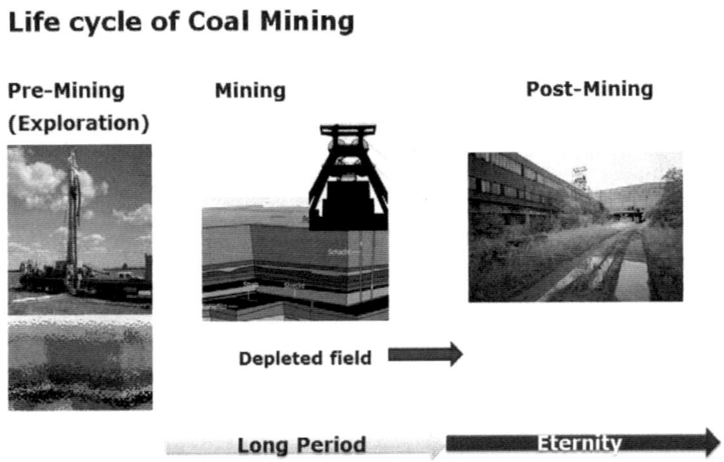

Figure 1 – The mining cycle

In the past, mining companies were mostly busy with the first two stages as those were the ones in which they could act profitably on the market. Moreover, most mining nations had not created a legal framework that would oblige companies to take care of the impact of active mining at the post-mining stage.

As a result, the medium and long term impacts the environment and the people living in the mining regions have to endure were often neglected – and even the economic impact was neglected. Mining damage that occurred years later such as the permanently necessary drainage of mine water and the restoration of land and water bodies are what the entire society has to cover the costs if the funding of such activities is not properly defined. In the end, such impacts can lead to a loss of acceptance that mining often faces in politics and among the population.

2. Material and Method

Thus the aim has to be to strive for mining that is as sustainable as possible. In other words, all three stages of the entire process have to be integrated, and the requirements of post-mining stage have to be included in the strategic planning and the operations. Especially as mining cannot be done without impact, there is the need to minimize the negative consequences as much as possible.

The Brundlandt Report, published by the United Nations in 1987, defines the term sustainable development as follows: "the development that meets the needs of the present without compromising the ability of future generations to meet their own needs." [2]

The origins of the idea of sustainability or rather, its German expression of *Nachhaltigkeit*, to provide for later, lie in the eighteenth century. The German pioneer of forestry, Hans Carl von Carlowitz, described in his "Sylvicultura oeconomica", which was the first standard work of this scientific discipline, the "continuous and sustainable use" of wood as essential for economic, social and ecological development. [3] Wood was the major energy source for mining and metallurgy during this period. Because of the intensive use of wood the industry risked depleting this resource, creating an energy crisis that would threaten not only production but the fast growing population as well that also needed wood as fuel for heating. By establishing a sustainable forest economy, Carlowitz claimed, the cultivation and use of wood could be balanced, so that an enduring supply could be ensured as a foundation of general well-being.

Sustainability is sometimes shortened to ecological issues, especially in political debates. But it is by no means limited to ecology. Of equal importance are social and economic aspects. Sustainability is about three basic aims: the share of wealth for as many people as possible (social sustainability), a durable positive economic development (economic sustainability), and the preservation of nature (environmental sustainability). [4]

With particular regard to post-mining, the mining authorities – in close collaboration with the mining companies – have developed a risk management system that facilitates recognition of all risks and definition of suitable mitigation measures.

The risk fields of abandoned mine sites can be divided into the following categories:

- Close-to-surface extractions [5-8];
- Surface openings, shafts [9-12];
- Subsidence, uplifts;
- Discontinuous faults;
- Mine gas emission at the ground surface;
- Surface water, pumping of pit water [13-14];
- Refuse dumps [15-17];
- Operation areas.

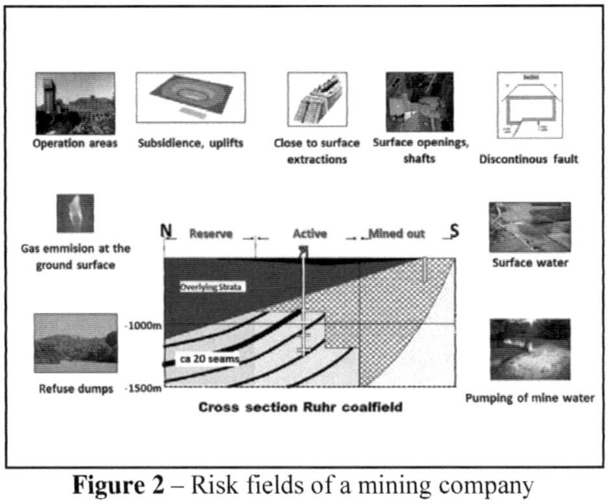

Figure 2 – Risk fields of a mining company

In the risk context, sustainability requires a holistic view of the mining cycle, including the opportunities that arise from abandoned mining sites. Effective post-mining provides numerous opportunities for avoiding, mitigating or even utilizing the risks by reinventing brownfields to create new jobs. What principally matters is the successful control and management of post-mining risks and the effective use of the opportunities it provides. [1]

3. Results and Discussion

In Germany, industrial underground mining of hard coal began in the early 19th century. In the roughly 200 years since there were times when several hundred mining companies of different sizes existed, mostly located in the Ruhr area and in the Saar region where Germany's largest hard coal deposits can be found.

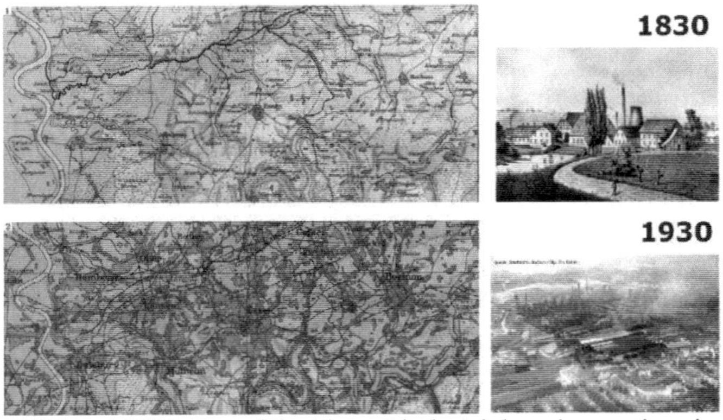

Figure 3 – The Ruhr 1830 and 1930, one of the biggest industrial agglomerations in Europe was built on coal

For more than hundred years, hard coal from domestic production was the basis for industrial success in Germany, even after World War II, when Germany started its "economic miracle", becoming one of the leading economies in the world. But since the 1960s hard-coal mining in Germany has been in the state of continuous decline. Because of relatively high production costs hard-coal mining in Germany seemed not to be capable of competing in the world market any more.

Since the end of mining has been definitely decided at the beginning of the 21st century, RAG is turning from an active hard-coal producer to a post-mining group. Its business activities today can be defined according to the three dimensions of sustainability. The structural change that has been going on in the last decades already aimed at achieving a balance of social and economic sustainability and included certain ecological aspects, too. In today's era of post-mining we need to master the organizational and technical challenges of mine closure in such a manner that any harm to people or damage to the environment can be avoided or offset: we need to strive for environmental and social sustainability. The question of how to master the perpetual tasks is inevitably linked with the question of the running cost, in particular that of permanent mine water drainage. In 2007, the German government passed a bill on funding the termination of hard-coal mining. Based on this act, the "old" RAG was split into three parts: a newly set up foundation, the RAG Stiftung (1), the subsidized coal mining unit plus coal trade, land management, site development, and a few other coal related service companies, still named ("new") RAG (3), and profitable business units mainly the subsidiary Evonik Industries, one of the world´s leading specialty chemicals companies, beside other interests (3). The foundation has to ensure that the proceedings from the profitable business units will be used to provide sustainable funding for the eternity tasks so that German tax payers will not have to pay for them. In addition, the RAG foundation has the responsibility for promotion and support of education, science and culture in the mining regions. This corporate structure mainly aims at economic, ecological and social sustainability. [18]

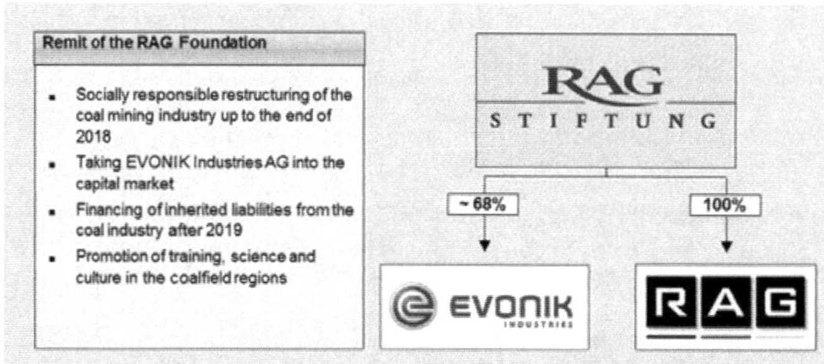

Figure 4 – Company structure of the RAG Stiftung / RAG

The general public only perceives the "tip of the iceberg" when it comes to the impact of underground mining, as the challenges of the development of the entire post-mining underground remain hidden.

Figure 5 – Challenges in Post-Mining: The Iceberg Model

Regarding the pit water drainage, probably the biggest challenge among the perpetual tasks, the RAG has presented a concept in 2015 that is currently being reviewed by the state authorities. This concept intends to optimize the pumping out of the pit water in the coal fields. As the Chief Executive of the RAG, Bernd Tönjes, has explained: "Our plans on how to optimize the underground water drainage intend to reduce the number of pit water sites successively and to discharge waterbodies from the pit water. Respective measures consider a sustainable protection of drinking water supply and, in addition, also another aim of protection concerned with how to avoid higher emissions of methane, heavy subsidence caused by ground heaving and hazards caused by caving to the surface." [19]

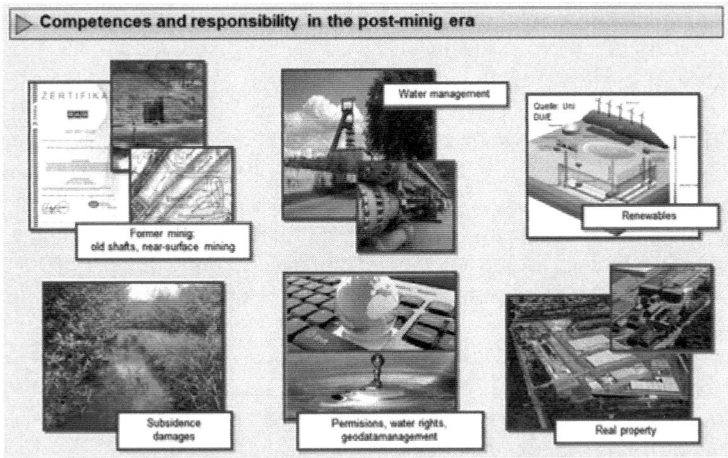

Figure 6 – Burdens and perpetual obligations

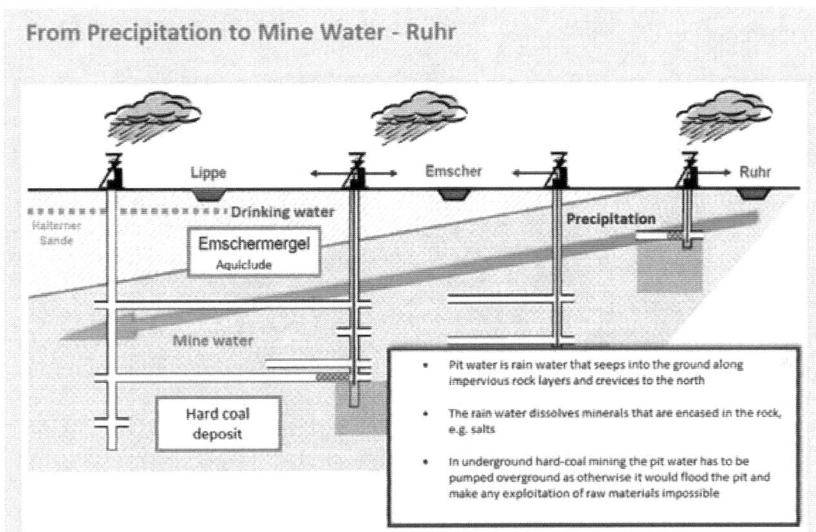

Figure 7 – Pit water handling at the Ruhr

Mine water is rain water that seeps into the ground along impervious rock layers and crevices. The rain water dissolves minerals that are encased in the rock, e.g. salts. In underground hard-coal mining the mine water has to be pumped over the ground as otherwise it would flood the pit and make any exploitation of raw materials impossible. The mine water concept intends to reduce this number to eight water handling systems by connecting adjacent pit works. On the one hand, this project would lead to a reduction in cost, and on the other hand, it would help to discharge some water bodies that still receive strongly mineralized mine water today. Moreover it is planned to raise the pit water level in order to reduce the effort for the pumps. For example, in the Ruhr area the pumping level is to be reduced from 800 m to 600 m whereas a distance of minimum 150 m will be maintained to the drinking water levels above. So far, the drainage is conventionally done in the mine workings which need to be kept open for this purpose; in the future, this system is to be replaced by a well drainage where immersion pumps will be used. These pumps can be installed in the shaft, lowered from the surface, and can be operated and serviced without the need to guarantee continuous access to the mine workings. [20]

There are a number of applications already in use, research projects, prototypes, ideas and visions included in this:

- Photovoltaic plants on mining areas. Apart from their height, mine dumps have another advantage: there are many free areas and hardly any shading. Therefore, they are ideal locations for photovoltaic systems; like-wise, the large roofs of factory buildings can be used for these, too. (Figure 8.1)
- Heat from mine water or exhaust heat. Every year, approximately 90 m cubic meters of pit water are pumped in the Ruhr area. The temperature of this pit water is 35-40°C (95-105°F) and can be used for supplying heating to buildings by means of heat exchangers or for accelerating the bio mass production when generating energy. (Figure 8.2)
- Wind wheels on refuse dumps. The mine dumps in the Ruhr area are often 80-100m above ground surface; thus, they often feature high wind speeds which allow for an economically reasonable use of wind wheels, and three wind wheels have already been erected. (Figure 8.3)
- Energy-producing utilization of methane that is released from coal beds
- Production of biomass on former mining areas
- Pump-storage power plants either on refuse dumps or underground in existing mine structures
- Geothermal energy. [1]

Figure 8 – Creating Renewable Energy, an Opportunity of Post-Mining. 1 Photovoltaic plant on a mud pond; 2 Heat from mine water; 3 Windmill on a dump

Post-mining excellence requires a high motivation as well as a high level of abilities und skills. Without key elements like ideas and visions, research and development, integration of surface and underground, and risk management post-mining will be nothing but inefficient short-term thinking of muddling through, or lip service where no one practices what they preach.

In order to competently implement post-mining management, a sufficient number of experts and executives have to be trained in this field. For example, TH Georg Agricola University offers a unique master program in geoengineering and post-mining to train specialists who will then be able to deal with the challenges provided by legacy and eternity tasks. [21] This focal point in teaching and research also helps to preserve the know-how and intellectual heritage of underground hard-coal mining in Germany: a knowledge transfer which is also urgently required because, due to the closure of coal mine operations in Germany, many experts in post-mining have either retired or are nearing retirement and the company RAG will not undertake the training of any new specialists. As part of its geoengineering activities the TH Georg Agricola University has opened a research-focused competence center for post-mining where new methods of geoengineering will be developed and tested for a sustainable management of the mining impact. Both the master program and the research center are supported by the RAG foundation which also provides funding for an endowment professorship for geo-engineering and post-mining.

On a global level the perpetual tasks of governments, universities and companies are to interconnect knowledge and experiences from different countries and regions in order to jointly collaborate on finding solutions for post-mining problems that are country-specific, international and global. Excellent post-mining experts and staff members are needed in many countries. That´s why collaborations should aim at long-term continuous improvements step-by-step based on the "plan-do-check-act" circle from Quality Management (Figure 9).

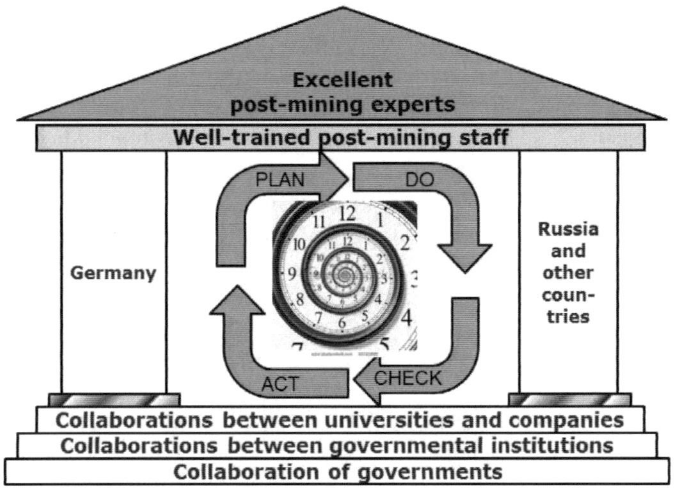

Figure 9 – International Collaboration Strategy

4. Conclusion

Being established in 1968 RAG will stop producing hard coal after a 50 years long period in 2018. To avoid frictions during the transition period from the mining to the post-mining era RAG has realized a sustainable development strategy that includes numerous measures. The experiences RAG has made and the knowledge the company has gained can be used in many mining regions all around the world that have to face similar transition processes in the future. To develop and transfer the knowledge and the experiences in post-mining the TH Georg Agricola University has established a specific master program and a research center.

References

[1] Kretschmann J and Hegemann M 2012 New chances from old shafts. Risk management in abandoned mine sites in Germany. In Proceedings of the Annual Meeting of the Society for Mining, Metallurgy & Exploration (Seattle, Washington: Red Hook, NY: Curran Associates) pp 153-158

[2] United Nations. 1987. Report of the World Commission on Environment and Development: Our Common Future. URL: http://www.un-documents.net/wced-ocf.htm.

[3] Carlowitz H 2000 Sylvicultura oeconomica. Anweisung zur wilden Baum-Zucht (Freiberg: TU Bergakademie Freiberg and Akademische Buchhandlung)

[4] Kretschmann J 2014 Stakeholder orientated sustainable land management: The Ruhr area as a role model for urban areas Vestnik 02 pp 127-132

[5] Tyulenev M A, Zhironkin S A, Garina E A 2016 The method of coal losses reducing at mining by shovels International Journal of Mining and Mineral Engineering 7 (4) DOI: 10.1504/IJMME.2016.10000781

[6] Tyulenev M A, Gvozdkova T N and Zhironkin S A et al. 2016 Justification of Open Pit Mining Technology for Flat Coal Strata Processing in Relation to the Stratigraphic Positioning Rate Geotechnical and Geological Engineering 34 (6) doi:10.1007/s10706-016-0098-3

[7] Tyulenev M A, Lesin Yu, Vik S and Zhironkin S 2016 Methodological Bases of Advanced Geo-ecological Problems Resolving in Neo-industrial Clusters Proceedings of the 8th Russian-Chinese Symposium "Coal in the 21st Century" pp 333-336

[8] Tyulenev M A, Khoreshok A A, Garina E A, Danilov S and Zhironkin S 2016 Adaptive technology of using backhoes for full coal extraction Proceedings of the 8th Russian-Chinese Symposium "Coal in the 21st Century: Mining, Processing, Safety" pp 111-115

[9] Kovalev V A, Gerike B L, Khoreshok A A and Gerike P B 2014 Preventive maintenance of mining equipment based on identification of its actual technical state Symposium of the Taishan academic forum – Project on mine disaster prevention and control pp 184-189

[10] Khoreshok A A, Mametyev L E, Borisov A Yu and Vorobyev A V 2015 The distribution of stresses and strains in the mating elements disk tools working bodies of roadheaders. IOP Conference Series: Materials Science and Engineering 91 (1) 012084

[11] Khoreshok A, Tyulenev M and Vöth S 2016 Conditions for Minimum Dynamic Loading of Multi-brake Hoists Proceedings of the 8th Russian-Chinese Symposium "Coal in the 21st Century" pp 239-245

[12] Lesin Y V, Luk'yanova S and Tyulenev M 2010 Mass transfer of dispersed particles in water filtration in macro-grained media J. Journal of Mining Science 46 (1) pp 78-81

[13] Tyulenev M A and Lesin Y V 2014 Justification complex purification technology open-pit mines wastewater Symposium of the Taishan academic forum – Project on mine disaster prevention and control pp 441-444

[14] Lesin Y V, Luk'yanova S Y and Tyulenev MA 2015 Formation of the composition and properties of dumps on the open-pit mines of Kuzbass IOP Conference Series: Materials Science and Engineering 91 (1) 012093

[15] Tyulenev M, Zhironkin S and Litvin O 2015 The low-cost technology of quarry water purifying using the artificial filters of overburden rock Pollution Research 34 (4) pp 825-830

[16] Tyulenev M, Zhironkin S, Kolotov K and Garina E 2016 Background of innovative platform for substitution of quarry water purifying technology Pollution Research 35 (2) pp 221-226

[17] RAG-Stiftung 2016 About us URL: http://www.rag-stiftung.de/en/about-us/

[18] Tönjes B 2015 Nachbergbau – Aufgaben und Verantwortung (Post-mining – Tasks and responsibility), speech given at the post-mining conference (TFH Georg Agricola Bochum)

[19] Fischer P 2015 Grubenwasserkonzept der RAG Aktiengesellschaft (Pit water concept of RAG), speech given at the post-mining conference (TFH Georg Agricola Bochum)

[20] Georg Agricola T H 2016 Master program: Geotechnical engineering and post-mining. URL: https://www.thga.de/en/faculties-of-the-tfh/faculty-i-geotechnical-engineering-mining-and-technical-business-management/master-programs/geotechnical-engineering-and-post-mining/

IOP Conf. Series: Earth and Environmental Science 50 (2017) 012025 doi:10.1088/1755-1315/50/1/012025

The Concept of Ecologically Oriented Progress and Natural Resource Preservation

M A Gasanov[1, a], **K A Kolotov**[1, b], **K A Demidenko**[2, c], **E A Podgornaya**[2, d], **O V Kadnikova**[2, e]

[1] National Research Tomsk Polytechnic University
Lenina av. 30, Tomsk, Russian Federation, 634034
[2] Kemerovo Institute (branch) «Plekhanov Russian University of Economics»
Kuznetskiy avenue 39, Kemerovo, Russia, 650992

E-mail: [a]hursud1@yandex.ru, [b]594950@bk.ru, [c]ksenan@mail.ru, [d]katje.siberia@gmail.com, [e]ksanita@list.ru

Abstract. The most important issue of scientific and technological progress is considering the environment challenges of industrial development. It means that the progress must be ecologically oriented and environmentally friendly. The most adequate concept for the approach to the issue of "man - society – nature" relations is the ontology of the noosphere - the idea of a common space for human beings and nature. It presents an ideal example of an optimistic attitude towards the coordination between accelerating the scientific and technological development and natural resource saving. However, to maintain the balance between human needs and environmental processes determined by this concept, it is essential to include the lean production training into technological development of society.

Introduction

A low level of the environmental education development in modern universities can be explained as follows. By historical standards, environmental education, especially in its modern form, is yet very young. In fact, it is still in its infancy; many universities do not have complete and comprehensive courses on the environment considering its social component. Ecology as a scientific field and educational subject still exhibits more problems than solutions. Social background plays an important role too, revealing critical underestimation of the environmental crisis with the modern society being occupied with the crisis another of another origin.

As a result, environmental education and promotion of a moderate approach to the use of nature for satisfying people's needs has neither consistent concept, nor explicit method. The noosphere concept is aimed at filling the gap in teachers' and students' outlook on environment. The concept can determine the essence of the conveyed knowledge of the environmental management.

Material and method

Nowadays the planet's wildlife is involved in human activities and the very life support of human society. This state of affairs determines another priority of the modern environmental education which should be paid particular attention to. The wildlife is being destroyed gradually resulting in the loss of biological diversity, the irrevocable losses of animal and plant species, etc. This destruction is not so obvious and it seems to be not of people's concern. It can be demonstrated with a simple example. The consequences of industrial wastewater pollution are rather evident, so this topic can be considered as early as at elementary school level. However, it is more difficult to observe how living organisms provide the purity of natural water, the atmospheric composition, recycling of household and industrial waste, and restoration of disturbed biosphere communities.

Implementing the idea that the diversity of life forms is the only condition of the effectiveness of the said processes is a complex but vital task in modern environmental education, requiring a high level of worldview as well as art of teaching.

The primitive form of what we call "environmental education" was implemented in the ancient Western and, in particular, Eastern civilizations. The ecological component was a part of the worldview of philosophy founders: Confucius [1-2], Lao-Zi [3-4], Democritus [5], Aristotle [6]. In modern history general environmental education accompanies the genesis and the development of ecology as a science.

Environmental science originates from biology as a knowledge about the connection between living organisms and the environment. The idea of the necessity of a specific research of relations between species and their environment was developed along with understanding that living organisms have been evolving and that the environment is critical for this process. Such an understanding is known to be completed and presented in Charles Darwin's "struggle for existence" [7].

A closer connection of ecology as a science and environmental education was formed coincidentally with the introduction of the word "ecology" into the scientific community in 1866 by Ernst Haeckel. Not only living organisms and their communities, but also the biosphere itself became the object of environmental knowledge [8]. Finally, a critical progress shift in the development of environmental education occurred concurrently with the foundation of social ecology in the 1970s. This science considers environmental education to be its priority.

Modern environmental education has a long-standing tradition of forming the knowledge about the relationship between people and nature. Initially it was based on natural science, which was introduced in the school curriculums of Western Europe and Russia in the late XVIII century. Subsequently environmental education in our country has evolved significantly.

In the 1920-30s, mass implementation of the training forms and methods activating students' cognitive and practical activity in studying and protecting environment was started. In particular they were implemented at natural science lessons and during extracurricular activities. In the 1930s nature-oriented knowledge was included into educational program in biology and geography. Many of these programs remained unchanged until the late 1970s. In the 1950-60s environmental education and training of young people expanded rapidly around the world. All the former USSR republics adopted Nature Protection Law and; environment protection themes were enhanced in the learning process. In 1970-80s education in the field of environmental management and conservation was emphasized. An innovation in environmental education was the development of educational programs in nature protection elective courses.

Environmental education received a new development impetus at the beginning of the 1990s. The formation of the system of environmental education was based on a number of methodological principles formulated in the 1980s [9]. At present stage the following set of statements concerning the ecology of human society is accepted:

- humankind is connected with the nature by its origin, existence and future;
- human history is a part of the history of nature and the integrity of the environment is the foundation of life in the biosphere;
- ecologically healthy environment along with social conditions are the basis of people's physical and mental health;
- labor is the basis of the interaction of people and society with the nature, a factor of the transformation of nature;

- transformation of environment is inevitable but limited, so the objectives of reasonable human activity must take full account of the manifestations of nature laws etc. [10-11]

Results and discussion

The four interrelated components can be identified as the core of the modern environmental education:

1. Cognitive - basic ideas about the principles of the interaction between nature and society, global environmental problems and their solutions etc.
2. Axiological - the system of values concerning diversified social and personal significance of nature.
3. Regulatory - the basis of ethical and legal standard of nature management, the rules of conduct in the environment.
4. Activity-related - the types of students' activities aimed at forming cognitive and practical skills of ecological character.

Existing principles of environmental education include the principle of the unity of knowledge-adjustment-action; the principle of continuity; the principle of interconnection of global, national and local approach to the analysis of environmental problems and the ways of solving them; the principle of inter-disciplinarity etc., which formed the basis of environmental education along with the ones abundantly applied in didactics.

At present, the position of a new humanistic approach to ecological culture has been strengthening. Along with it, it is essential to decide which perceptions of bio-social system "man-society-nature" should be first and foremost applied to environmental education and training, and which principles of former consumer culture demand a critical review.

Global environmental crisis requires a new world outlook to be integrated in the younger generation education and upbringing. The concept of "nature and society" system should be transformed into the "nature-society" system. The scientific and technological revolution and the rising degree of mechanization [12-17] reveals human's complete dependence on the animated and inanimate nature resources. This especially influence on the environment during mineral resources extraction [18-23]. This dramatic situation is to be the focus of teachers' attention.

Admittedly, the methodological foundation, required for research and development in students' environmental education, as well as for writing teaching materials, has yet to be developed. Even in developed countries the system of secondary and higher education does not offer well-established principles and criteria for conveying environmental knowledge. There is a certain gap between the relatively high level of school environmental education and the level of development of ecological pedagogical education.

In recent years, humanitarian science, natural science and technological culture have been facing an increasing opposition within each other. On account of this opposition environmental education contains rather full natural science content but classical one has been falling behind. In particular, social ecology and ecology philosophy are largely deprived of professional attention.

The noosphere concept (the theory of humankind possibility to balance their own needs, technological opportunities and natural resources) helps to eliminate the lack of humanitarian knowledge in environmental education. The primary role in noosphere studying, however, belongs to human knowledge that resulted not only in conquering nature, but also in creating conditions for harmonious coexistence of people and environment. These very components should be the determinants of the environmental education development, form its new principles.

F. Bacon payed attention to the special role of scientific knowledge in developing relations between society and nature [24]. He believed that knowing the laws of nature would enable people to meet their basic needs and interests. In this respect, F. Bacon considered that increasing human power was impossible without subordination to these natural laws. While cognizing them, science must help human beings move towards their goals using the shortest way, avoiding obstacles, and preventing mistakes, many of which may threaten the very existence of the human race.

These views were fleshed out and developed in the noosphere concept, which was widespread and scientifically recognized in the 20th century. The noosphere concept was introduced into science in 1927 by a French philosopher, mathematician and anthropologist Edouard Le Roy (1870-1954), who proposed defining the shell of the Earth including human society and its industry, languages and other attributes of sentient activity as noosphere [25]. However, it is a Russian scientist and philosopher V.I. Vernadsky, who is rightfully consedered the main founder/creator of the noosphere concept, having developed the idea of noosphere as "biospherized society".

V.I. Vernadsky was one of the first to realize that humanity had become a powerful geological and, perhaps, cosmic force, capable of transforming nature in a large scale [26]. He noted that people encompassed the whole biosphere with their lives and cultures, and was trying to deepen and expand the sphere of influence. The biosphere, from his perspective, was gradually transformed into the noosphere – the sphere of mind. V. I. Vernadsky considered the noosphere to be the highest stage of the biosphere development, where cognitive human activities became the determining factor. He suggested that the biosphere-noosphere transition was related to developing science, deepening scientific insight in natural processes, and organizing a rational human activity thereupon. The study of the noosphere is intrinsically optimistic, implying that humanity will find a way to restore and preserve the ecological balance of the planet, will develop and put into practice the strategy of crisis-resistant development of nature and society. Thus, the idea that people can manage the environmental development of the whole planet is rather optimistic and corresponds with the contemporary achievements in biochemistry, genetic engineering, mining and metallurgy. And it is precisely due to optimism, that the noosphere concept is so valuable for modern environmental education.

A prominent French paleontologist and philosopher Pierre Teilhard de Chardin (1881-1955) was developing the noospheric genesis problem simultaneously with V.I. Vernadsky [27]. He believed that the noosphere is originally shaped like a thin layer of thinking substance ("mental sheath or envelope of thought") which is detached from all of the Earth's shells.

However, as human thought penetrates deeper into the essence of the processes occurring in all the Earth spheres, the noosphere transforms into its higher state - the "Earth Spirit". Like V.I. Vernadsky, Pierre Teilhard de Chardin emphasized the role of science in accomplishing this task. Science, in his opinion, should be not only an effective tool of the world learning, but also a means of unity, synthesis of ideas.

Throughout the 20th century, many thinkers and representatives of various scientific specialties turned to the problem of the noosphere, trying to identify its origin and ways of subsequent development. In general, the followers of the noospheric future of humanity are characterized by an optimistic view on the perspective of developing the relationship between society and nature. However, there are only so many environmentalists ready to share such optimism nowadays. They consider that humankind has yet to prove, whether it is capable of playing the role of the organizer of a new world, free from the devastating crises, ecological disasters, wars and violence [28]. The problem of water resources technogenic pollution stands out with the special sharpness disturbing the harmony of the humans and the nature [29-32].

It is important for the development of environmental education that the idea of the noosphere does not contradict the concept of sustainable development, which is gradually achieving leading positions in explaining the prospects of modern civilization. Since the second half of the 1970's many prominent environmentalists, sociologists, economists, politicians, etc. concerned about the current situation, combined their efforts in order to design a new approach to building relationships between people and their environment. It is of high importance for studying geotechnical activities and mineral resource extraction [33-37]. Designing and producing powerful mining machines, constructors and engineers must ensure minimizing negative ecological consequences of their implementation [36-40].

As defined by the World Commission on Environment and Development, sustainable development is such development of Nations that meets the needs of modern humanity without compromising the well-being of future generations and their ability to meet their own basic needs. This implies that some parameters, such as key physical constants (composition of air, water, soil, mechanical properties of the Earth's surface, gravity, etc.), gene pool, parts of major ecosystems in their original forms, human health are supposed to comprise a constant value over time. In this respect, environmental protection is becoming the task of utmost importance. Its ultimate goal is, on the one hand, to ensure safety of the environmental qualities which are supposed to remain unchanged, and, on the other hand, to ensure a continuous growth of useful plants, animals and other essential resources through balanced cycles of withdrawal and renewal.

The concept of sustainable development gained strong support not only with the specialists in the field of social ecology and human ecology, but also Governments and leaders of most countries in the world, which was reflected in the decisions of the United Nations Conference on Environment and Development held in 1992 in Rio de Janeiro. The need to move the entire global community to take a balanced and integrated approach to environment and development questions was declared at the Conference. In addition, the decision to form Green Cross International environmental organization was adopted there. The main purpose of the new organization was ecological education as a basis for sustainable development and changing system of values, as well as elimination of the consequences of the cold war for the environment.

Thus, the concept of the noosphere corresponding with the mainstream of sustainable development may contribute significantly to the formation of a coherent scientific structure of environmental education. Such structure should include a number of elements:

1. Environmental education of modern university students as a component of their organically built-in general culture.
2. Formation of humanistic University environment as a condition of development of ecological culture and practical preparedness of graduates to environmental education in its modern and progressive understanding.
3. Teacher ecologist training to improve their professionalism.

A scheme of consistent environmental education based on this structure should be built in higher schools. At the initial stage self-development and personality development of the students occurs, along with in-taking of the cultural and environmental priorities and values on impressional and emotional levels. In accordance with this dominant education in the ecological program is organized. At the second stage students gain necessary educational and methodical knowledge and develop professional competence of future specialists. As a result, a number of subjects of basic environmental education are formed: general ecology, social ecology, applied ecology, etc. At the third stage, the conditions for professional identification of students are created, based on environmental imperative. At the fourth stage students know paradigms of the system man-society-nature, integrate in research and academic activities in this field. In addition, situations stimulating students' creative self-development are created at this stage.

Conclusion

Understanding the specifics of sustainable development begins with accepting that a human, who has an inalienable right to a healthy and meaningful life is in the center of such development; environmental protection should be realistically the most important part of sustainable development; society moves to a co-evolutionary method of interaction with nature. This fully reflects the key point of the noosphere concept - focusing collective intelligence on ensuring the security of humankind in every respect and realization of human striving to personal and social well-being. To achieve these objectives it is necessary to intensify full potential of the noosphere concept, to make radical humanistic reorientation of the whole system of modern education values, to fully disclose the significance of nature for people's existence.

Acknowledgement

The research was made in National Research Tomsk Polytechnic University with financial support of the Ministry of Education and Science of the Russian Federation in the framework of scientific research "The evaluation and improvement of the social, economic and emotional prosperity of senior citizens" contract № 14.Z50.31.0029.

References

[1] Hobson J M 2004 The Eastern origins of Western civilization (Cambridge: Cambridge University Press) 366 p.
[2] Bonevac D and Phillips S 2009 Introduction to world philosophy (New York: Oxford University Press) 296 p.
[3] Henricks R G 1992 Lao Tzu: Te-Tao Ching – A New Translation Based on the Recently Discovered Ma-wang-tui Texts (Classics of Ancient China) (New York: Ballantine Books) 488 p.
[4] Roberts M 2004 Dao De Jing: The Book of the Way (Berkeley: University of California Press) 322 p.
[5] Guthrie W K 1979 A History of Greek Philosophy - The Presocratic tradition from Parmenides to Democritus (Cambridge: Cambridge University Press) 362 p.
[6] Richo K 2008 Ancilla to the Pre-Socratic Philosophers (London: Forgotten Books) 316 p.
[7] Desmond A and Moore J 2009 Darwin's sacred cause: race, slavery and the quest for human origins (London: Allen Lane) 290 p.
[8] Hopwood N 2014 Haeckel's Embryos: Images, Evolution, and Fraud (Chicago: University of Chicago Press) 508 p.
[9] Zhironkin S A, Khoreshok A A, Tyulenev M A, Barysheva G A, Hellmer M C 2016 Economic and technological role of Kuzbass industry in the implementation of national energy strategy of Russian federation IOP Conference Series: Materials Science and Engineering 142 (1) 012127.
[10] Tyulenev M, Zhironkin S, Kolotov K and Garina E 2016 Background of innovative platform for substitution of quarry water purifying technology Pollution Research 35 (2) pp 221-226.
[11] Tyulenev M A, Lesin Yu, Vik S and Zhironkin S 2016 Methodological Bases of Advanced Geoecological Problems Resolving in Neo-industrial Clusters Proceedings of the 8th Russian-Chinese Symposium “Coal in the 21st Century” pp 333-336.
[12] Khoreshok A A, Mametyev L E, Borisov A Y, Vorobyev A V 2016 Influence of the Rigid Connection between Discs in the Tetrahedral Prisms on Equivalent Stresses When Cutting Work Faces IOP Conference Series: Materials Science and Engineering 127 012039.
[13] Khoreshok A A, Buyankin P V, Vorobiev A V, Dronov A A 2016 Simulation of Stress-Strain State of Shovel Rotary Support Kingpin IOP Conference Series: Materials Science and Engineering 127 012014.
[14] Khoreshok A A, Mametyev L E, Borisov A Yu and Vorobyev A V 2015 The distribution of stresses and strains in the mating elements disk tools working bodies of roadheaders. IOP Conference Series: Materials Science and Engineering 91 (1) 012084.
[15] Kovalev V A, Gerike B L, Khoreshok A A and Gerike P B 2014 Preventive maintenance of mining equipment based on identification of its actual technical state Symposium of the Taishan academic forum – Project on mine disaster prevention and control pp 184-189.
[16] Aksenov V V, Khoreshok A A and Beglyakov V Y 2013 Justification of creation of an external propulsor for multipurpose shield-type heading machine - GEO-WALKER Applied Mechanics and Materials 379 pp 20-23.
[17] Lekontsev Yu M., Sazhin P V, Temiryaeva O A, Khoreshok A A and Ushakov S Yu 2013 Two-side sealer operation Journal of Mining Science 49(5) pp 757-762.
[18] Zhironkin S A 2001 Factoring and leasing development at coal mining industry of Kuzbass as an important element of its financial part Ugol’ 4 pp 29-30.
[19] Zhironkin S A 2001 Governmental factoring development of TEK Kuzbass Ugol' 6 p 62.
[20] Zhironkin S A 2002 About measures of vexel circulation development and vexelability definition of fuel-and-power complex' enterprises Ugol' 4 pp 47-48.

[21] Zhironkin S A 2002 Prospects and new possibilities investment attracting to Kuzbass coal mining industry Ugol' 6 pp 31-36 .
[22] Tyulenev M A, Khoreshok A A, Garina E A, Danilov S and Zhironkin S 2016 Adaptive technology of using backhoes for full coal extraction Proceedings of the 8th Russian-Chinese Symposium "Coal in the 21st Century: Mining, Processing, Safety" pp 111-115.
[23] Khoreshok A, Tyulenev M and Vöth S 2016 Conditions for Minimum Dynamic Loading of Multi-brake Hoists Proceedings of the 8th Russian-Chinese Symposium "Coal in the 21st Century" pp 239-245.
[24] Bacon F 2014 The Essays and Counsels, Civil and Moral of Francis Bacon: all 3 volumes in a single file (Moscow: B&R Express) 302 p.
[25] Le Roy 1958 E Essai d'une philosophie première: l'exigence idéaliste et l'exigence morale (Paris: Sirocco Pub.) 366 p.
[26] Lapo A V 2001 Vladimir I. Vernadsky (1863–1945), founder of the biosphere concept (Moscow: Science) 278 p.
[27] Aczel A 2007 The Jesuit and the Skull: Teilhard de Chardin, Evolution and the Search for Peking Man (London: Riverhead Hardcover) 408 p.
[28] Khoreshok A A, Zhironkin S A and Tyulenev M A et al. 2016 Innovative technics of managing engineers' global competencies IOP Conference Series: Materials Science and Engineering 142 (1) 012122.
[29] Lesin Y V Luk'yanova S and Tyulenev M 2010 Mass transfer of dispersed particles in water filtration in macro-grained media J. Journal of Mining Science 46 (1) pp 78-81.
[30] Tyulenev M A and Lesin Y V 2014 Justification complex purification technology open-pit mines wastewater Symposium of the Taishan academic forum – Project on mine disaster prevention and control pp 441-444.
[31] Tyulenev M, Zhironkin S and Litvin O 2015 The low-cost technology of quarry water purifying using the artificial filters of overburden rock Pollution Research 34 (4) pp 825-830.
[32] Lesin Y V, Luk'yanova S Y and Tyulenev MA 2015 Formation of the composition and properties of dumps on the open-pit mines of Kuzbass IOP Conference Series: Materials Science and Engineering 91 (1) 012093.
[33] Tyulenev M A, Zhironkin S A, Garina E A 2016 The method of coal losses reducing at mining by shovels International Journal of Mining and Mineral Engineering 7 (4) DOI: 10.1504/IJMME.2016.10000781.
[34] Tyulenev M A, Gvozdkova T N and Zhironkin S A et al. 2016 Justification of Open Pit Mining Technology for Flat Coal Strata Processing in Relation to the Stratigraphic Positioning Rate Geotechnical and Geological Engineering 34 (6) doi:10.1007/s10706-016-0098-3.
[35] [Efremenkov A B 2011 Forming the subterranean space by means of a new tool (geohod) Proceedings of the 6th International Forum on Strategic Technology IFOST 6021037.
[36] Efremenkov A B and Timofeev V Y 2012 Determination of necessary forces for geohod movement Proceedings - 2012 7th International Forum on Strategic Technology IFOST 6357729
[37] Efremenkov A B and Aksenov V V et al. 2012 Force parameters of geohod transmission with hydraulic drive in various movement phases Proceedings - 2012 7th International Forum on Strategic Technology IFOST 6357716.
[38] Aksenov V V, Efremenkov A B and Beglyakov V Y 2013 The influence of relative distance between ledges on the stress-strain state of the rock at a face Applied Mechanics and Materials 379 pp 16-19.
[39] Golik V I, Rasorenov Y I and Efremenkov A B 2014 Recycling of metal ore mill tailings Applied Mechanics and Materials 682 pp 363-368.
[40] Golik V I and Efremenkov A B 2016 Physicochemical Processes of Metal Lixiviation in the Disintegrator IOP Conference Series: Materials Science and Engineering 125 (1) 012038.

Water Resources Management In Support Of Raw Region Based On Decoupling Effect

Yu V Lesin[1], E A Fedulova[2], A O Akulov[2], S V Vik[3], R L Korchagin[2]

[1] Yurga Technological Institute (branch) of Tomsk Polytechnic University
Leningradskaya 26, Yurga, Russian Federation, 652000
[2] Kemerovo state university
Krasnaya st. 6, Kemerovo, Russian Federation, 650043
[3] Kuzbass State Technical University named after T.F. Gorbachev
ul. Vesennyaya 28, Kemerovo, Russian Federation, 650043

E-mail: akuanatolij@yandex.ru

Abstract. It determines the presence of the decoupling effect in the Russians raw regions materials by using water. Developed models that explain the relationship between the gross regional product and water intake. It proved no effect on the growth of water consumption in most regions of the reference commodity. Recommendations for the decoupling effect development in support of Russians raw regions.

1. Introduction

Rational water resources use and sufficiently complete waste water treatment – one of the key areas of sustainable development. The traditional type of economic development has led to the depletion of natural resources, environmental pollution, lack of clean water (Smith, 2003). It is important to the effective use water resources in the framework of the "green" sustainable development (Azad et al., 2015). The criterion of sustainable development is to use a minimum of resources to get the result unit (Bleischwitz, 2010). Only in this case it is possible to combine the growth and environmental requirements. Approach to water management must be differentiated according to the type of region. This takes into account parameters such as the current level of economic development, the load on the environment, the conditions of natural resources, strategic development. There are different approaches to the classification of regions (Stepanov, 2013; Zubarevich, 2012). In this study, we are interested in supporting raw materials regions. They are characterized by a significant proportion of the mining sector, the relatively high level of economic development, have a significant impact on the national economy of Russia as a whole.

Successful economic development of the raw regions was accompanied by increasing load on the environment. Water resources are scarce in most regions of the reference commodity. Extraction and processing of raw materials had a negative impact on water basin. The big problem is the contamination of drinking water (Dolnicar et al., 2012; Molinos-Senante et al., 2011). Therefore, you need to increase the efficiency of use of natural resources and the capacity of the biosphere. At present, the impact of economic growth in the regions supporting commodity on the environment is studied to a limited extent. Virtually no models to explain the dependence of environmental damage, in particular, water consumption and pollution, the economic dynamics. It is difficult to form a sound environmental policy.

Sustainable development support raw regions achieved when decoupling effect. Decoupling means mismatch trends of various indicators, among which the previously assumed a close connection, for example, the growth and decline of developed and developing economies (Benjamin, 1992). In the modern environmental economics decoupling effect means a reduction in the environmental burden while maintaining or increasing the rate of economic growth. This effect is based on the more efficient use of natural resources, in particular water (Cahill, 1997). Achieving decoupling effect allows you to create a "green" economy, where growth in production does not lead to negative environmental consequences. Decoupling – the most attractive way of development for the country and individual regions. However, its existence is not currently shown. In fact, there are no empirical studies on the relationship between economic growth and environmental damage in the regions of Russia. In particular, we studied the effect of weak economic growth in the regions supporting commodities for consumption and water pollution. Thus, it is necessary to determine the relationship between economic growth and water consumption. This will allow predicting the damage to water resources and regulating water use in support of primary regions of Russia.

Thus, the purpose of the article – to identify the presence or absence of the effect of decoupling the use of water resources in support of primary regions of Russia, to develop a model that explains the impact of the economic growth in these regions on water consumption.

2. Materials and Methods

Polygon study are supporting commodity regions of the Russian Federation – Kemerovo region, Nenets Autonomous District, Komi Republic, Republic of Sakha, Sakhalin region, Tyumen region, the Khanty-Mansi Autonomous District, Yamalo-Nenets Autonomous District (of 8 subjects of the federation). These federated entities are classified as raw regions support in the official government records of the Russian Federation. The study period includes the years 2005-2014. The period of 10 years seems to be sufficient to produce statistically reliable estimates.

To conduct the study were used official statistical publications of the Federal Service of State Statistics (Rosstat of the Russian Federation), which presents the main indicators of economic development of all regions of the country. In particular, to determine the explanatory (independent) variable data on the production of gross regional product was used in support of primary regions (Table 1).

Table 1 – Production of gross regional product in support of primary regions of Russia, the index of physical volume, 2005-2014, per cent

	2005	2006	2007	2008	2009	2010	2011	2012	2013	2014	Growth rate
Kemerovo region	106.8	107.2	106.8	102.1	92.9	102.7	102.3	95.8	96.0	102.1	114.4
Nenets Autonomous District	108.8	114.6	118.7	86.7	122.6	95.9	88.7	94.9	99.4	103.6	130.8
Komi Republic	104.0	108.6	99.6	103.3	98.5	102.6	105.7	101.8	96.7	95.7	116.9
Republic of Sakha	104.9	103.4	104.5	107.2	97.6	101.6	107.1	103.2	100.9	103.2	138.7
Sakhalin region	108.8	112.2	126.3	95.7	110.9	108.0	104.3	97.5	101.4	100.7	172.6
Tyumen region	110.8	106.0	103.1	103.8	94.3	106.0	103.1	99.8	102.4	98.1	129.9
Khanty-Mansi Autonomous District	112.4	106.7	103.2	102.9	95.2	103.0	100.9	98.1	100.0	98.7	122.0
Yamalo-Nenets Autonomous District	101.2	104.6	103.5	105.5	89.8	110.1	102.7	102.2	103.9	105.7	131.7

Given the need for elimination of the effects of inflation and the inter-regional comparisons, the gross regional product is estimated on the basis of data on the index of the physical volume (in constant prices). The index of physical volume of the gross regional product is an integral feature of the economic development of the region as a whole. Since the reference commodity region's largest share in the production of gross value added of the industrial production, the index of physical volume of the gross regional product reflects the state of the relevant sectors of the economy.

As explanatory variables – characteristics of the water consumption in the region can be used indicators such as: water extraction; the use of fresh water; wastewater, transit, mine, mine waters; discharge of polluted wastewater, transit, mine, mine waters. The study used the indicator "water abstraction" because it best reflects the needs of the region's economy in water resources. Data on water intake in support of primary regions of Russia for the years 2005-2014 are presented in Table 2.

Table 2 – Water intake in support of primary regions of Russia, sampling volume, 2005-2014, million cubic meters

	2005	2006	2007	2008	2009	2010	2011	2012	2013	2014	Growth rate
Kemerovo region	2057	2098	2071	2102	2083	1751	2009	1992	1731	1726	0.84
Nenets Autonomous District	11	12	11	18	28	30	30	29	33	9	0.82
Komi Republic	568	568	543	542	519	506	495	486	456	463	0.82
Republic of Sakha	120	130	140	166	151	164	159	201	161	161	1.34
Sakhalin region	295	306	300	263	256	263	223	249	194	162	0.55
Tyumen region	1822	1929	1878	2095	1749	1825	2026	1980	2040	3022	1.66
Khanty-Mansi Autonomous District	1199	1340	1351	1560	1184	1185	1439	1388	1468	2411	2.01
Yamalo-Nenets Autonomous District	166	168	135	131	122	235	220	223	225	218	1.31

As a basic research methods used in the calculation of the Pearson correlation coefficient of paired at the level of significance $\alpha = 0.05$, build equations Simple Linear Regression to estimate the reliability parameters and criteria Student Fisher score determination, analysis of variance, analysis of the index. Along with these generally accepted statistical methods for grouping regions in terms of manifestations of decoupling it is proposed to use the metric D_{ef} (1):

$$D_{ef} = GRP/WR \qquad (1)$$

where GRP – the growth rate of the physical volume of the gross regional product over the period, per cent; WR – the growth rate of consumption of water resources (water intake) for the same period, per cent.

The economic meaning of the metric is to demonstrate the correlation between economic growth and changes in water consumption. If it takes a value less than 1.00, the water consumption is growing at a faster pace compared to the gross regional product (or decreases more slowly than the reduced gross regional product). If the metric is equal to 1.00, the gross regional product and water consumption vary to the same extent. The value of the metric over 1.00 indicates a rapid increase in the gross regional product compared to water consumption and indirect evidence of the manifestation of the effect of decoupling.

Research procedure included the following steps.

1. Calculation of paired correlation coefficients Pearson to estimate the significance level $\alpha = 0.05$

between the indices of "index of physical volume of gross regional product" and "withdrawals" assessment of the nature connection, the wording of the preliminary finding of the existence and nature of the statistical relationship between these indicators.

2. Calculation of the linear regression equations of the pair, where the independent variable used indicator "Index of physical volume of the gross regional product," explained variable – "water intake", the selection of the type of constraint equations, evaluation of reliability of the regression equation for the Fisher test, Student's t, explained variance (coefficient of determination).

3. Drawing conclusions about the nature of the relationship of indicators, identifying the presence or absence of a decoupling effect.

4. Develop a set of recommendations for rationalizing the use of water resources on the basis of the best available technologies.

3. Results and Discussion

Data on the correlation coefficients between the indicators' index of physical volume of gross regional product and withdrawals are presented in Table 3.

Table 3 – Correlation coefficients of indicators "index of physical volume of gross regional product" and "the volume of water intake" of support raw Russian regions

	The value of the correlation coefficient	The number degrees of freedom	The critical level of the correlation coefficient	The character of the statistical relationship
Kemerovo region	0.2265	8	0.6319	Absent
Nenets Autonomous District	-0.4045	8	0.6319	Absent
Komi Republic	0.6623	8	0.6319	Positive relationship
Republic of Sakha	-0.0529	8	0.6319	Absent
Sakhalin region	0.5829	8	0.6319	Absent
Tyumen region	-0.3168	8	0.6319	Absent
Khanty-Mansi Autonomous District	-0.2908	8	0.6319	Absent
Yamalo-Nenets Autonomous District	0.5329	8	0.6319	Absent

The data in Table 3 show that a statistically significant positive correlation between economic growth and water intake observed in only one supporting the commodities of the Russian Federation of the eight (Komi Republic) regions. In other regions there is no statistically significant association, which indirectly indicates the existence of decoupling effect in the implementation of water. Economic growth has no rigid connection with the consumption of water. Thus, in seven of the eight regions, correlation coefficients are below a statistically significant level. However, their absolute values differ greatly from region to region. So, in the Sakhalin Region and the Yamalo-Nenets Autonomous District, the relationship between the index of gross regional product and water intake was positive and only a little not reached the level of statistical significance. In the Kemerovo region and the Republic of Sakha correlation coefficients are close to zero, indicating that the independent variables. In the Nenets, Khanty-Mansi Autonomous District, Tyumen region is also correlation coefficients below a critical level, but has a negative sign.

An analysis of correlation coefficients shows that certain manifestations of decoupling effects take place in seven of the eight regions. Increase water intake with an increase in the gross regional product is observed in only one region. At the same time in two other regions of the correlation between the index of physical volume of the gross regional product and water intake remains positive, but does not reach a statistically significant level. In five other regions there is no positive correlation, indicating a need for a thorough study the effect of decoupling on the basis of regression models. On the basis of data on production growth rate of the gross regional product and the water intake for the entire range studied regions (Figure 1) regression equation (2) was obtained.

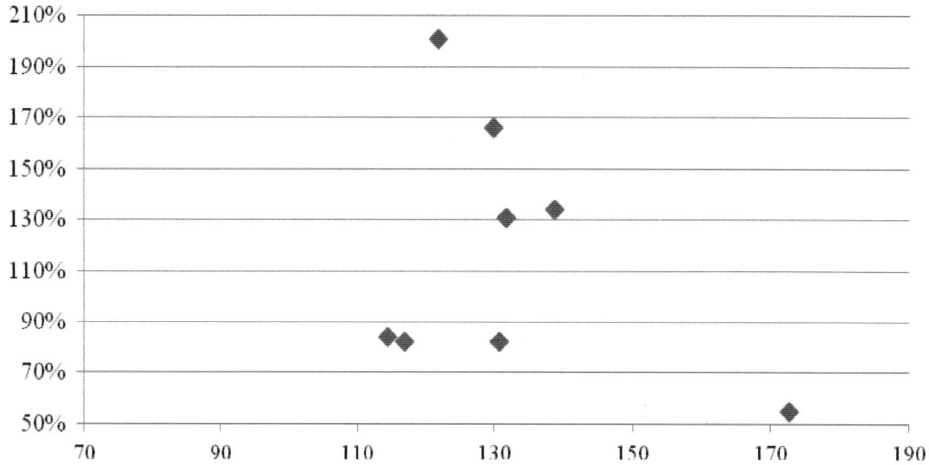

Figure 1 – Diagram of the scattering medium growth rate of gross regional product and water intake in support of primary regions of Russia, the average values for the period 2005-2014

As can be seen from Figure 1, the overall growth rate of the gross regional product is characterized by positive dynamics in all regions and a much smaller variation. The rate of growth of water intake during the period was a positive (in four regions out of eight), or negative.

The resulting regression equation is (2):

$$Y = -0.94X + 2.41, \quad (2)$$

where X – index of physical volume of gross regional product, per cent; Y – the volume of water abstraction, millions cubic meters.

At the same time, evaluation of the significance of the regression equation indicating its lack explanatory capacity (Table 4).

Table 4 – Dispersion analysis of the regression equation (2)

Source of variation	Sum of squares	The number degrees of freedom	Dispersion 1 degree of freedom	F-statistics
Explained model dispersion	0.21	1	0.21	0.81
The residual dispersion	1.53	6	0.26	1
The total dispersion	1.73	8-1		

As shown in Table 4, the equation (2) explains the small portion of the dispersion that does not allow its use for prediction. T-statistics (t-test) of 0.9 at a table value 2.447, F-statistics (Fisher's exact test) was 0.81 at the table 5.99. These values indicate a lack of statistical significance of the coefficients of the equation and forms of communication. Also it restricts the analytical and predictive capabilities of the equation (2) the high value of approximation errors (34.75%). Consequently, the assessment of water intake depending on the dynamics of the gross regional product on the whole range of regions demonstrates a lack of statistical relationship. At the level of the reference commodity regions of Russia as a whole there was a mismatch trend of growth and water consumption.

Consequently, in the period under decoupling trend, economic growth does not automatically lead to an increase in the need to use water. At the same time dynamics of these variables is almost independent of each other. Dynamics of consumption of water could "decouple" from the economic dynamics. But to reduce the need for water resources at the same time economic growth is not yet possible.

Analysis of individual regions to determine the features of display decoupling effect. The results of calculation of the metric (2) for the studied regions are shown in Figure 2.

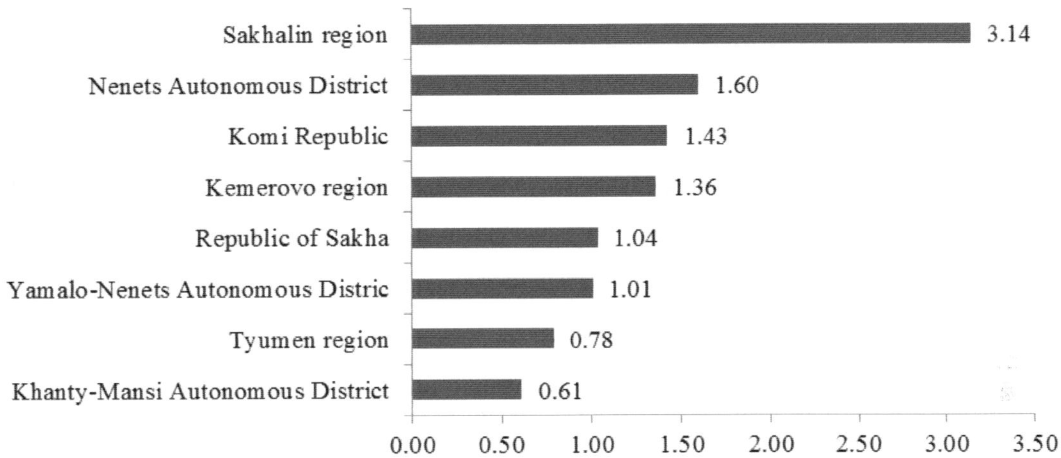

Figure 2 – The values of D_{ef} metrics to support raw regions of Russia, 2005-2014

From the data in Figure 2 shows that the increase in water consumption outpaced the increase in gross regional product in only two regions – the Khanty-Mansiysk Autonomous District and the Tyumen region. The rest of the increase in demand for water lagged behind the growth of the gross regional product (from 1.01 to 1.60 times). Results of the analysis of correlation coefficients (Table 3) and Def metrics give different results in terms of the ratio of the rate of economic growth and changes in water consumption in the regions. This is due to the use in the first case of differentiated data, in the second – the period of integrated data.

Regression equations of the form (2) for each of the regions in question are shown in Table 5. The table shows that for all the studied regions, with the exception of the Republic of Komi, lacking both positive and negative relationship between economic growth and water intake.

As for the 10 observation periods critical Fisher criterion is 5.32, and the t-test – 2.31, seven of the eight regression equations presented in Table 5 are not statistically significant. This is also indicated low values of the coefficient of determination and the big error of approximation. Consequently, in the majority of regions of Russia of raw support economic growth did not lead to a corresponding increase in water consumption. At the same time formed decoupling effect has also been observed as a negative relationship between the change in the volume of gross regional product and water abstraction has been identified in the study.

For the Republic of Komi is characterized by a positive impact of economic growth on the consumption of water, one percent increase in the gross regional product growth causes the water needs of 6.6 million cubic meters. At the same time revealed the dependence explains only about 44% of the changes in water consumption. In all likelihood, you can expect a weakening of the statistical relationship and separation of the two trend lines, as is the case in other regions.

Table 5 – Regression equations that determine the dependence of water abstraction from the dynamics of the physical volume of gross regional product in support of primary regions of Russia

	Type the equation	*F*-test	*T*-test	Error approximation percent	The coefficient of determination, percent
Kemerovo region	$Y = 7.20X + 1231.71$	0.43	0.66	6.91	5.13
Nenets Autonomous District	$Y = -0.32X + 53.78$	1.25	1.56	48.32	16.36
Komi Republic	$Y = 6.59X - 155.48$	6.25	2.50	5.04	43.86
Republic of Sakha	$Y = -0.41X + 197.70$	0.02	0.15	10.8	0.28
Sakhalin region	$Y = 3.06X - 74.80$	4.12	2.03	13.7	33.98
Tyumen region	$Y = -25.11X + 4616.42$	0.89	0.94	8.54	10.04
Khanty-Mansi Autonomous District	$Y = -21.62X + 3659.76$	0.74	0.86	12.6	8.45
Yamalo-Nenets Autonomous District	$Y = 4.55X - 283.66$	3.17	1.78	17.98	28.4

4. Practical Recommendations

Study the relationship between water consumption and economic growth in the support of primary regions of Russia showed that in seven of the eight regions observed mismatch relevant trends. Economic growth ceased definitely lead to an increase in water consumption. A statistically significant association between these indicators, described by the equation of linear regression, preserved only in one region (Komi Republic). In other regions, there is no statistically significant relationship. In general, the range of the reference commodity regions proved no effect of economic growth to improve the flow of water. Therefore, in modeling the economic development of the region of the reference raw economic growth and use of water resources can be considered independent. Formation of system of indicators of development of the region may take account of the environmental load limitations are less (Fedulova et al., 2016).

These evaluation studies indicate that the growth of the gross regional product of the consumption of water does not change significantly. This can be viewed as a positive trend, which determines the beginning of the formation decoupling effect in support of primary regions of Russia. The further development of these regions and improving its sustainability requires further development of the decoupling effect. It is a negative relationship between economic growth and water consumption. In this case, a statistical study of the impact of economic growth on the use of water does not show lack of communication, and a significant negative relationship.

Formation pronounced decoupling effect requires a change in environmental policy. It is based must be put to the transition to the best available technology. With respect to existing businesses is necessary to carry out a gradual greening activities, modernization of technology and production, the introduction of modern technology for emissions and discharges clean. During the construction of the new commodity producers should ensure minimum environmental damage, to demand to achieve world-class environmental performance using best available techniques.

When the greening of existing enterprises must be based on planning, targets to reduce emissions and discharges into the environment by applying the best available technologies. Currently, commodity businesses do not have a clear objective to reduce water consumption and wastewater discharge. Existing limits clearly exceed the capacity of the environment and require adjustments. In determining the target water consumption values should focus on maximum environmental capacity of bearing raw regions. Strategic regional plans should take account of these constraints. In front of them should be given the task in the medium term to keep within the set limits. To do this within the framework of the investment programs should be provided for activities to transition to the best available technology to reduce environmental damage, with the assessment of the financial and economic impact and environmental performance. After the approval of the program at the level of the Commission, consisting of representatives of federal and regional authorities, businesses could qualify for tax breaks. This will give positive incentives to an active environmental policy. During the construction of new businesses should be required to use the best available technologies that should be a mandatory licensing requirement. Financial, technical and organizational capacity of the transition to the best available technology is limited. Therefore, the best available technology must be ranked in ecological efficiency. In particular, to reduce the discharge of waste water is most expedient to use biogidrobotanic cleaning method, active and passive treatment of wastewater discharge reduction technologies in the water.

To reduce the impact on the environment is necessary to change the terms of payment for environmental management, to expand the range of economic incentives. Currently, environmental fees and fines imposed by Rosprirodnadzor, are not tangible for violators. In this regard, it is necessary a significant increase in environmental sanctions, so that they have a real impact on natural resources. Today Rosprirodnadzor fines are negligible compared to the revenue and profit. Therefore, it is advisable to reconsider the approach to their establishment. It is proposed to levy fines in the amount of a certain percentage of the revenue of the coal enterprises, which make tangible punishment. Otherwise, nature will continue "as usual", with no systematic work to reduce environmental damage and without the use of best available technologies.

It is necessary to introduce a differentiated fee for a negative impact on the environment, taking into account regional specificities. The current level of differentiation is clearly insufficient; it requires corrections with the views of the authorities of the Federation, the expert community and the general public. For example, the coefficient taking into account the state of the water bodies of the Kemerovo region is only 1.16 – roughly the national average, and it is at an extremely high degree of water pollution. It seems that to establish the same rate for all federal district unreasonably regions. However, a radical increase in fees for use of natural resources alone cannot change the situation. To change it must be prepared to do commodity companies and enterprises. Undoubtedly, a reasonable regulation of fees for use of natural resources - a very complex issue, coupled with the collision of the plurality of interests. But gradually, gradually increase payments for the use of natural resources with the prospect of 5-10 years will create for nature predictable working conditions, will give an opportunity to plan environmental activities, long-term investments.

5. Conclusions

There is a need to change the environmental management system and environmental protection on the basis of the transfer of powers to the level of the Federation. According to Constitution of the Russian Federation, issues of ownership, use and disposal of natural resources and environmental management issues of environmental protection are in joint conducting the Russian Federation and its subjects. But in fact, the regions have the right to dispose of only subsoil, containing widespread mineral deposits. With regard to the subsoil with deposits of coal, oil and metals, the subjects of the Federation have not only the right of disposal of the joint, but even the right to use matching. Federal authorities, providing rights for subsoil use, come mainly from the fiscal interests.

It is therefore necessary for a redistribution of the functions of the federal center to the regions in terms of mineral resources and subsequent supervision of the mining companies. It is expedient to provide as a condition of admission to competitions in the field of subsurface distribution agreement with the Commission from the representatives of the authorities of the Federation. Such a commission could carry out the selection of nature and form requirements for their technical and environmental policies more carefully. It is also proposed to assign to sub-federal authorities, the right to apply to the authority which granted the license, with the obligatory consideration of the appeal on the merits at the optimum time. In this case it is possible, without waiting for the results of periodic monitoring; in case of complex environmental situation immediately appeal to the federal authority that issued the license, for the consideration of the merits. This practice exists in principle, but it is necessary to issue legally.

Supporting raw regions should be eligible to participate in the establishment of technical standards for emissions and emission limit values. They should vary from region to region and be based on the standards that exist for the best available techniques. Features of these regions with the very specific structure of the economy and at the same time severe environmental conditions require the differentiation of legal regulation of nature management and environmental protection. Environmental standards in such regions of the Federation should be more stringent nature – the more expensive, and the control system should increasingly focus on the regional level, than in the whole country. Implementation of these measures will reduce the adverse environmental impact of supporting raw regions form a background to the development of decoupling effect.

References

[1] Azad M, Ancev T & Hernández-Sancho F. 2015, «Efficient Water Use for Sustainable Irrigation Industry», Water Resources Management, vol. 29, no. 5, pp. 1683–1696.
[2] Benjamin D. 1992, «Household composition, labour markets and labour demand: testing for separation in agricultural farm-household models», Econometrica, vol. 60, no 2, pp. 287–322.
[3] Bleischwitz R. 2010, «International economics of resource productivity – Relevance, measurement, empirical trends, innovation, resource policies», International Economics and Economic Policy, vol. 7, pp. 227–244.
[4] Cahill S. 1997, «Calculating the rate of decoupling for crops under CAP/oilseeds reform», Journal of Agricultural Economics, vol. 48, no 3, pp. 349–378.
[5] Dolnicar S, Hurlimann A & Grunn B 2012, «Water conservation behavior in Australia», Journal of Environmental Management, London, vol. 105, pp. 44–52.
[6] Fedulova E, Medvedev A, Kosinskiy P, Kononova S. & Pobedash P. 2016. «Modeling of the agribusiness enterprise activity on the basis of the balanced scorecard», Foods and Raw Materials, vol. 4, no. 1, pp. 154–162.
[7] Molinos-Senante M, Hernández-Sancho F & Sala-Garrido R. 2011, «Cost-benefit analysis of water-reuse projects for environmental purposes: A case study for Spanish wastewater treatment plants». Journal of Environmental Management, London, vol. 92, no. 12, pp. 3091–3097.
[8] Smith A 2003 «Transforming technological regimes for sustainable development: a role for appropriate technology niches?», Science and Public Policy, vol. 30, no. 2, pp. 127–135.
[9] Stepanov A 2013 «Classification of Russian regions by the degree of their orientation on external and internal markets», Regional Research of Russia, vol. 3, no. 4, pp. 324–327.
[10] Zubarevich N 2012 «Cities as the Centers for the Modernizatuin of the Economy and Human Capital», Sociological Research, no. 4, pp. 3–27.

Environmental Management: the Ideology of Natural Resource Rational Use

V M Zolotukhin, V A Gogolin[b], M Yu Yazevich[c], M I Baumgarten[d], A V Dyagileva[e]

T.F. Gorbachev Kuzbass State Technical University
Vesennyaya st. 28, Kemerovo, Russian Federation, 650000

E-mail: [a]zvm64@mail.ru, [b]gva.pm@kuzstu.ru, [c]Maria762003@list.ru, [d]bmi45@mail.ru, [e]dyagileva1952@mail.ru

Abstract. The article presents an analysis of the ontological and methodological principles of environmental management. These principles form the united ideology of natural resource rational use as the environment preservation basis. Consideration of environmental issues from the environmental management point of view is stipulated by the concern of the scientific community about the existence of mankind and the sphere of its inhabiting. The need to overcome the stereotypes existing in mass consciousness about safe and environmentally friendly consumption is stressed. The process of forming environmental management policy should contribute to the stabilization (balancing) of the consumers' expectations and collective decision-making based on a public ecological consensus.

Introduction

The relationship between a human being and nature has always been controversial, but in the 20th century, human-induced impact on the natural environment has become particularly significant, and changes in this area today, as is known, are irreversible [1]. The absence of a clear and coherent strategic policy that can dramatically affect the activity of modern humanities and society is one of the reasons underlying the crisis in the field of environmental and social relations. In recent decades, social reality has undergone significant changes that were taking place under intense influence of diverse information processes, political and economic innovations, and also due to the rapid growth of scientific knowledge and technological progress [2]. It is worth mentioning that given transformations affected not only social practice, but also contributed to a serious ecological imbalance, stimulated worsening of social and natural problems on the global and local levels.

Material and Method

Under current conditions the need to build and enhance a culture of self-preservation of both human being and his environment in the framework of his philosophical paradigms is very urgent. One of the most important factors is technological, including compliance and the establishment of new regulatory standards of human activity in the industrial production. According to O. Spengler, every culture has its own quite an individual way of seeing and understanding the world-as-nature, or every culture has its own, peculiar nature. But even in higher degree every culture has its exceptionally own type [3]. In the culture itself we observe the presence of anxiety feeling, waiting for something, uncertainty (when we are still in the state of environmental comfort, but the risk of transition to unfavorable situation is already appearing).

The practice of double standards in the field of environmental policy creates certain standards of behavior from the part of the ruling elites at the state and international levels. On the one hand it is the degree of interest of society as a whole, its classes, social groups, territorial communities and their mentality. On the other hand it is the political and social attitude of authorities, contributing to environmental conservation. The degree of development of the environment by human being depends on how focused and structured the public opinion is [4]. This refers not only to the implementation of mandatory standards and symbols, but also to those standing in opposition to the official ideology, and drew the natural instincts of man, in all their aggressiveness, which bears the destructive nature for human being and his natural environment.

It must be recognized that organizational issues based on a specific regulatory framework also deserve attention. The reasons for the formation of institutional mechanisms should become different models of management practices. With the help of simulation, including mathematical, there is a possibility of comparison of basic indicators of "green economy" - the investments in green growth, i.e. the transition to green economy in the countries importing and exporting hydrocarbons. These countries, the "champions" of green economy, are Korea (80.5% of all financial measures), China (37.8%) and France (21.2%) [4]. One of the drivers of green growth in these countries is the preservation of national energy security [5-8].

In terms of social, cultural and legal aspects of definition of the environmental policy the issues of creating and maintaining the appropriate for human living environment are at the topside. In this environment, a person must be guaranteed a certain environmental quality standards. This raises the question of the formation of ecological thinking, as the learned facts by themselves mean nothing. The developed thinking without necessary minimum of theoretical knowledge does not provide the required performance of daily activities too. To determine the level of ecological thinking development it is necessary to assess objectively the level of training of the population according to the established criteria of environmental education. In terms of social and cultural needs satisfaction its types are identified. For example, primitive thinking (De Broglie), as the most ancient and simple reflection of the world in thinking is described [9-11]. Currently the peculiarities of thinking of representatives of different knowledge branches (natural sciences, mathematical, biological, humanitarian, etc.) are being considered.

Currently, the situation tends to get worse and threatens the existence of human society and the natural environment on a planetary scale. This is evidenced by the conclusions of various international symposiums, the APEC summit (Vladivostok), the UN Conference in Rio de Janeiro and Japan, publications and reports of scientists from different countries. These events highlight the overall trends - increased attention to the principles of sustainable development in environmental, economic and social spheres, as well as improved maintenance practices of the balance between nature and the social world [12-13]. It is clear that without active measures in environmental policy it is impossible to maintain the delicate balance between natural processes and the realization of modern humanity needs. Environmental problems of the nearest future will be caused by water pollution [14-18], speeding up the mining machinery development [19-27], settling the neo-industrial clusters [28]. So the special attention must be paid to sustainable development of methodology [29-32] and environmental ethics [33-34]. Hence the necessity to find new solutions, the development of agreements about measures of environmental policy based on both common ideological and axiological principles and uniform legal and technical standards appears.

Results and Discussion

It should be stressed that the measures in the sphere of nature preservation policy cannot be limited by initiatives or actions from the part of individual state governments. A special need is to create a unified ecological space with global perspectives. Activities of this kind should be directed to the development of environmentally-oriented, economically viable and socially acceptable solutions relevant to the representatives of the various communities. But the peculiarity of this situation is that pollution and natural resources are geographically localized and situated in specific geographical areas, and have not only natural, but also state and administrative borders. Thus, the management of social and environmental processes must take place within the framework of large-scale and efficient environmental policy drawn up both at the global level and by international, national, regional and municipal authorities. Greening of state procurement, primarily in the public sector, should be aimed at promoting the use of voluntary environmental certifications for the formation of a unified ecological space.

The scope of integrated environmental policy should include issues related to the improvement of the ecological situation on the planet. In its structure it is important to determine the main direction of the positive development of different levels of diverse and contradictory social and natural processes, based on the principles of environmental world outlook. Formation of such strategy may take place on the international environmental platforms, based on the independent international organizations, in environmental programs of political parties and civil society organizations. This will establish a set of legal, economic social and political normative multi-purpose documents in the environmental regulatory system. At the same time, it must be emphasized that conflict between the legislative and executive branches is dangerous for the stability of the legal system. Overcoming of this conflict is possible in the process of finding compromises among all the subjects of law-making and law-enforcing activity in the conditions of social agreement. Formation of the environmental policy should contribute to the stabilization (balancing) the behavioral expectations and collective decision-making based on public consensus.

One of the urgent tasks of modern management today is to create an information space using ideological practices. Ideology has traditionally had a strong orientation and mobilization potential in the mass consciousness. In the context of globalization, the growth of ecological contradictions, social tension, the problem of social identity and self-identification is particularly acute, and the humanity needs definite objectives, absolute values and mobilizing ideas. So the phenomenon of a new ideological project can provide power structures and civil society institutions philosophical platform for regulation of social and environmental crises and stabilization of current conflicts in general.

Ideology has traditionally been an effective means to cause large-scale action of political and social practices, to produce the values and meanings that are relevant for specific social and historical circumstances. For example, on the basis of the state report "On the state and environmental protection of the Russian Federation in 2013", prepared by the Ministry of Natural Resources and Environment of RF: "The observation of air pollution in Russia was conducted in 252 cities, 694 posts, of which regular observations by Roshydromet were performed in 225 cities on 631 posts, additional sporadic observations were carried out in 4 settlements [35]. According to Roshydromet, in 123 towns (57% urban) the level of air pollution is estimated as high and very high and only in 22% of towns as low. In cities with high and very high levels of air pollution 54.2 million people live accounting for 52% of Russia's urban population. In 204 cities (81% of cities where observations were conducted) the average annual concentration of one or more pollutants exceeds MPC. 66.6 million people live in these cities" [36]. In this situation, when the solution of environmental problems is of vital importance, the development and implementation of global ideological project will, on the one hand, consolidate the efforts of various social groups at international and local levels; on the other, change the ethical-axiological component of social action, outlining environmental imperative as the basis of social outlook.

Today in modern social theory postmodern worldview oriented on pluralism of values, ideas, cultures and ways of life is of special interest. This tradition is opposed to the classical paradigm of modernism, which dates back to antiquity and follows the description of the world based on the metaphysical rationalism and monistic model of the universe. Richard Rorty, criticizing the rationalist bias of modernism, believes that under the Age of Enlightenment the formation of "essentialist" type of world view with the characteristic features of logocentrism, fundamentalism and universalism that led to a special relationship of man and nature was completed [37]. Earlier we noted that in these circumstances classical for the Western European practices scheme of relations is establishing, where a special type of person - rational, autonomous individual with political rights and claims for management is appearing. Nature in this system of relations was seen as imperfect, in need of change and reorganization in accordance with the requirements of the human mind. In classical axiological system nature was seen as a value of the secondary order, and had significance only in the context of human activities, as the only resource for the implementation of human interests. As a result social and natural relations have been reduced to a narrow utilitarian level, focused on the creation of economic advantages and material benefits.

Postmodern tradition seriously criticizes and deconstructs the anthropocentric model of being. It rejects man's position as a privileged being, a starting point of world processes. From this perspective, the interpretation of the chain relationship between man-society-nature must be based on the principles of self-worth of the natural world, the multiplicity of ways of knowing the environment and dialogical communication with Nature (Richard Rorty [37]. According to J. Habermas, one of the ways to form a correct understanding is a debate [38]. The decision of the majority can only be taken in such a way that its content is considered to be rationally motivated (although not immune from error) outcome of discussions, which is conventionally completed, because it is necessary to accept final decision. This adoption of a majority-based environmental policy is considered to be provisionally approved by the minority. J. Habermas said that nobody demands of the minority that it renounces its will, that it announces its opinion wrong; the minority is not even required to reject its purpose. But it is required from the minority to refuse practical implementation of its beliefs as long as it manages to present its arguments better and to collect the necessary number of proponents [39-40]. That is the correct understanding and application of environmental policy and environmental legislation which ensures the achievement of socially significant goals. It especially concerns development geotechnical activities and mineral resource extraction [41-43]. Designing and producing powerful mining machines, constructors and engineers must analyze ecological consequences of their implementation [44-46].

This requires reorientation of society for new social ideals and values, in which, as the priorities, it is offered to adopt the principles of unity cultural and natural environment, expressing the features of modern human being. In recent decades pluralism, multiculturalism, environmentalism having been considered as an alternative basis for the establishment of a new system of values with dominating environmental imperative as opposite to industrial civilization of modern. This will give in theoretical aspect the special importance to nature and appreciate the diversity of its forms, as well as determine as a value that any natural phenomenon (animate and inanimate natural objects) has an absolute significance. This kind of idea about the value of life diversity may be the basis for a new type of relations "human being - world", providing an opportunity to further survival of the civilization.

Conclusion

Thus, the "solidarity" should be the most important principle of modern society, which is based on the ideals and values of consensus, tolerance, pluralism and freedom. One of the conditions for the implementation of these guidelines may be rethinking of theoretical and ideological bases, inherited from the modern society, which will determine the right of existence for many voices in the "conversation" of humanity with nature. At the same time environmental imperative can become the principle of universal connection, unifying link between "human being and human being", "human being and society" and "human being and nature". Future social world, in this way, can be represented in the form of "mutual stories" of individuals united by "the desire of solidarity." This kind of communication will be the basis for adaptive behavior of individuals, allow adapting to modern realities and non-standard situations by adopting new values and norms of existence, developed under environmental ideology.

References

[1] Rogers P, Jalal K F, and Boyd J A 2007 An Introduction to Sustainable Development (Routledge: Routledge Pub.) 380 p.

[2] Bakari M K 2013 Globalization and Sustainable Development: False Twins? New Global Studies 7(3) pp 23-56.
[3] Hardy S 2001) Oswald Spengler et Gabrielle Roy: quelques pistes de lecture Cahiers franco-canadiens de l'Ouest 13 (2) pp 143–56.
[4] Kahle L R and Gurel-Atay E 2014 Communicating Sustainability for the Green Economy (New York: M.E. Sharpe) 478 p.
[5] Zhironkin S A 2001 Governmental factoring development of TEK Kuzbass Ugol' 6 p 62.
[6] Zhironkin S A 2001 Factoring and leasing development at coal mining industry of Kuzbass as an important element of its financial part Ugol' 4 pp 29-30.
[7] Zhironkin S A 2002 About measures of vexel circulation development and vexelability definition of fuel-and-power complex' enterprises Ugol' 4 pp 47-48.
[8] Zhironkin S A 2002 Prospects and new possibilities investment attracting to Kuzbass coal mining industry Ugol' 6 pp 31-36.
[9] Warnant L 1987 Dictionnaire de la prononciation française dans sa norme actuelle (Gembloux: Duculot) 466 p.
[10] Pierret J M 1994 Phonétique historique du français et notions de phonétique générale (Louvain-la-Neuve: Peeters) 780 p.
[11] Leroy F 2003 A Century of Nobel Prize Recipients: Chemistry, Physics, and Medicine (Birmingham: CRC Press) 562 p.
[12] APEC roadmap on FTAAP a historic decision 2014 Xi Xinhua November 11 pp 55-58.
[13] Elite Talk: A talk with APEC chief Alan Bollard on China, FTAAP, New Silk Road 2014 People's Daily Online November 10 pp 91-99.
[14] Lesin Y V Luk'yanova S and Tyulenev M 2010 Mass transfer of dispersed particles in water filtration in macro-grained media J. Journal of Mining Science 46 (1) pp 78-81.
[15] Tyulenev M A and Lesin Y V 2014 Justification complex purification technology open-pit mines wastewater Symposium of the Taishan academic forum – Project on mine disaster prevention and control pp 441-444.
[16] Tyulenev M, Zhironkin S and Litvin O 2015 The low-cost technology of quarry water purifying using the artificial filters of overburden rock Pollution Research 34 (4) pp 825-830.
[17] Lesin Y V, Luk'yanova S Y and Tyulenev M A 2015 Formation of the composition and properties of dumps on the open-pit mines of Kuzbass IOP Conference Series: Materials Science and Engineering 91 (1) 012093.
[18] Tyulenev M, Zhironkin S, Kolotov K and Garina E 2016 Background of innovative platform for substitution of quarry water purifying technology Pollution Research 35 (2) pp 221-226.
[19] Tyulenev M A, Gvozdkova T N and Zhironkin S A et al. 2016 Justification of Open Pit Mining Technology for Flat Coal Strata Processing in Relation to the Stratigraphic Positioning Rate Geotechnical and Geological Engineering 34 (6) doi:10.1007/s10706-016-0098-3.
[20] Khoreshok A A, Buyankin P V, Vorobiev A V, Dronov A A 2016 Simulation of Stress-Strain State of Shovel Rotary Support Kingpin IOP Conference Series: Materials Science and Engineering 127 012014.
[21] Khoreshok A A, Mametyev L E, Borisov A Yu and Vorobyev A V 2015 The distribution of stresses and strains in the mating elements disk tools working bodies of roadheaders. IOP Conference Series: Materials Science and Engineering 91 (1) 012084.
[22] Tyulenev M A, Zhironkin S A, Garina E A 2016 The method of coal losses reducing at mining by shovels International Journal of Mining and Mineral Engineering 7 (4) DOI: 10.1504/IJMME.2016.10000781.
[23] Aksenov V V, Khoreshok A A and Beglyakov V Y 2013 Justification of creation of an external propulsor for multipurpose shield-type heading machine - GEO-WALKER Applied Mechanics and Materials 379 pp 20-23.
[24] Lekontsev Yu M., Sazhin P V, Temiryaeva O A, Khoreshok A A and Ushakov S Yu 2013 Two-side sealer operation Journal of Mining Science 49(5) pp 757-762.

[25] Khoreshok A A 2002 On side cutting bit when operating at sheerer drums Ugol' 7 pp 10-11.
[26] Tyulenev M A, Khoreshok A A, Garina E A, Danilov S and Zhironkin S 2016 Adaptive technology of using backhoes for full coal extraction Proceedings of the 8th Russian-Chinese Symposium "Coal in the 21st Century: Mining, Processing, Safety" pp 111-115.
[27] Khoreshok A, Tyulenev M and Vöth S 2016 Conditions for Minimum Dynamic Loading of Multi-brake Hoists Proceedings of the 8th Russian-Chinese Symposium "Coal in the 21st Century" pp 239-245.
[28] Tyulenev M A, Lesin Yu, Vik S and Zhironkin S 2016 Methodological Bases of Advanced Geo-ecological Problems Resolving in Neo-industrial Clusters Proceedings of the 8th Russian-Chinese Symposium "Coal in the 21st Century" pp 333-336.
[29] Foltz B V and Frodeman R 2004 Rethinking Nature (Indiana University Press, Bloomington North Morton Street) 601 p.
[30] Vogel S 1999 Environmental Philosophy After the End of Nature Environmental Ethics 24 (1) pp 23-39. .
[31] Weston C 1999 An Invitation to Environmental Philosophy (New York: Oxford University Press) 472 p.
[32] Zimmerman M J, Callicott M and Sessions G et al. 2010 Environmental Philosophy: From Animal Rights to Radical Ecology (New Jersey: Prentice-Hall, Inc.) 632 p.
[33] Pojman L P and Pojman 2000 P Environmental Ethics (Orlando: Thomson-Wadsworth) 478 p.
[34] Sherer D and Attig T 1983 Ethics and the Environment (New Jersey: Prentice-Hall, Inc.) 554 p.
[35] Ministry of Natural Resources and Environment of the Russian Federation URL: http://www.mnr.gov.ru/english/.
[36] Federal service for hydrometeorology and environmental monitoring (Roshydromet) 2014 Assessment report on climate change and its consequences in Russian Federation (Moscow: Roshydromet) 28 p.
[37] Rorty R 1967 The Linguistic Turn, Essays in Philosophical Method (University of Chicago press) 360 p.
[38] Habermas J 2002 Religion and Rationality: Essays on Reason, God, and Modernity (Mass.: MIT Press) 276 p.
[39] Raymond D 1986 Nascent Environmental Protection in the Second Empire // Germ. Stud. 9 pp 262-267.
[40] Giddens A 1990 The Consequences of the Modernity (Cambridge: Camb. Univ. Press) 302 p.
[41] Aksenov V V, Efremenkov A B and Beglyakov V Y 2013 The influence of relative distance between ledges on the stress-strain state of the rock at a face Applied Mechanics and Materials 379 pp 16-19.
[42] Golik V I, Rasorenov Y I and Efremenkov A B 2014 Recycling of metal ore mill tailings Applied Mechanics and Materials 682 pp 363-368.
[43] Golik V I and Efremenkov A B 2016 Physicochemical Processes of Metal Lixiviation in the Disintegrator IOP Conference Series: Materials Science and Engineering 125 (1) 012038
[44] Efremenkov A B 2011 Forming the subterranean space by means of a new tool (geohod) Proceedings of the 6th International Forum on Strategic Technology IFOST 6021037.
[45] Efremenkov A B and Timofeev V Y 2012 Determination of necessary forces for geohod movement Proceedings - 2012 7th International Forum on Strategic Technology IFOST 6357729.
[46] Efremenkov A B and Aksenov V V et al. 2012 Force parameters of geohod transmission with hydraulic drive in various movement phases Proceedings - 2012 7th International Forum on Strategic Technology IFOST 6357716.

Guaranteeng Of Ecological Safety Of Wastewater Tretment By Phytoremediation

E E Nefedieva, Yu N Kartushina, V F Zheltobryukhov, G A Sevriukova, E V Baybakova

Volgograd State Technical University
Pr. V.I. Lenina 28, Volgograd, Russian Federation, 400005

E-mail: sevrykova2012@yandex.ru

Abstract. This article is devoted to the experimental identification of the copper ions (II) uptake by the remediation plant. Wheat and barley were used as plants for phytoremediation. They were cultivated in the aqueous culture on the Hellriegel nutritive medium prepared without phosphorus. The copper was dissolved in the medium in the concentration 0 mg/l (control variant), 0.5 mg/l, 1.0 mg/l, 2.5 mg/l, 5 mg/l, 10 mg/l, 20 mg/l. The concentrations of the copper were identified in solutions by the ion meter in 2 weeks of cultivation. The moving arrangement with a plant as a biological bed is elaborated for phytofiltration. That is a technology for decontamination and final treatment of the wastewater from pollutants using plants. Particularly it is possible to remove ions of heavy metals. The device will be inexpensive, simple and easy for exploitation at the enterprise. It is offered to complete standard rectangular settler by the mentioned device.

Introduction.

The growth of industry influences on the contaminant ingress to the superficial water which is the main depository of pollutants.

Industrial and residential wastewater contains wide variety of contaminants. Mineral resource and concentrating industry, non-ferrous metallurgy, chemical and petrochemical enterprises, mechanical engineering, machine-tool industry, electrical and electronic engineering, and heat-power industry are the sources of wastewater, which contains the most toxic substances such as ions of heavy metals. All of heavy metals have biological activity. Dangerous related to the accumulation of heavy metals is conditioned by their ability to step in the food chains and to change the geological cycling, to concentrate primarily in microorganisms and plants. Then heavy metals are able to contaminate people. In conjunction requirement strengthening to methods of water treatment take place in the whole world.

Modern methods of water purification such as reagent and charge-transfer methods, pressure and electrical dialysis, reversed osmose are effective, but they have significant limitations. For example, reagent method needs the investment for reagents. Besides a lot of residuals are formed in the process of purification. Charge-transfer method demands the preliminary removal of suspended substances and cations of calcium. Also that method needs the division of waste water into flows with different kinds of ions. High-toxic and highly-concentrated unutilizes eluates appears. Pressure dialysis and electrodialysis have high purification efficiency, but they are hardscrabble methods.

Phytoremediation is a relatively new and actual method of purification of waste and natural waters and soils as well as decontamination of air pollutants. It is use of green plants to absorb or decontamination of organic or inorganic toxicants in the environment [[5]; [10]]. It is a promising technology for the revegetation of polluted territories.

Researches [[19]; [20]; [21]; [14]; [16]; [15]; [3]; [22]; [30]; [7]; [9]] are devote to the quantitative characteristics of uptake of heavy metals, sulfates, nitrates, and low-molecular substances by plants from industrial and residential wastewater and model solutions. Experimental and field researches demonstrate that plants can be used for purification of water. They can uptake and transform heavy metals, inorganic anions, and pesticides [[1]; [2]; [8]; [4]; [24]; [23]; [25]; [26]; [27]; [12]; [11]; [18]; [6]; [34]; [32]; [17]; [31]; [13]].

The following kinds of phytoremediation are divided due to the peculiarities of remediation processes and localization of contaminants, such as phytoextraction, phytostabilization, phytofiltration, and phytovolatization.

In this case we tell about phytofiltration because the method of absorption of heavy metals from water by roots of plants was elaborated. The process includes the growing of plants in hydroponics and replant to wastewater which contains ions of metals. Plants adsorb and concentrated metals in their tisssues. Besides, the extrusion of organic and inorganic substances, production of oxygen and changes of pH due to the transport of protons by roots or generation of hydrocarbonate anions are the reasons of the sedimentation of metals on the surface of plant or in the growth solution [[29]]. This technology is more inexpensive than technologies based on technical approach because the main source of energy is sun light which is transformed into chemical energy by plants. Those benefits are the evidences of the necessity of the elaboration and the search of new possibilities of use of phytoremediation.

Phytoremediation is applied to treat wastewater in the world [[33]; [28]].

The purpose of present work was experimental assessment of the ability of phytoremediants to absorb the ions of copper (Cu^{2+}) and their tolerance to the contamination for design of the special equipment based on standard rectangular settlers completed by a moving arrangement.

Materials and methods.

Wheat and barley were the model plants in this research because they have the following adventures, they have high absorption capacity as well as easy cultivation, desiccation and utilization. Their ontogenesis is well described.

The experiments were divided for two parts.

Germinating seedlings of wheat and barley were planted as a water culture in six glasses with growth medium which contained different concentrations of copper cations (table 1). It was prepared using the feeding solution of Hellriegel (table 2) without phosphorus because phosphates form insoluble compounds with copper. So copper would be unavailable for plants in the full Hellriegel medium.

Table 1 – Concentrations of Cu^{2+} in the test solutions

Wheat		Barley	
Variant	Concentration of Cu^{2+}, mg/l	Variant	Concentration of Cu^{2+}, mg/l
1	10.0	1	20.0
2	5.0	2	10.0
3	2.5	3	5.0
4	1.0	4	2.5
5	0.5	5	1.0
6	Control	6	Control

Table 2 – Composition of the growth medium (in 1 liter of distilled water)

Substance	Content in Hellriegel solution, g	Content in Hellriegel solution without phosphorus (control), g	Content in used medium
$Ca(NO_3)_2$	0.492	0.492	0.492
KH_2PO_4	0.136	0.0	0.0
$MgSO_4$	0.060	0.060	0.060
KCl	0.075	0.075	0.075
$FeCl_3$	0.025	0.025	0.025
$CuSO_4$	0.0		0.5-20.0

The seedlings were put on the plates with perforations, and their roots were immersed into the growth medium. The volume of glasses was 200 ml, the amount of seedlings was 10 (fig 1).

Figure 1 – Plants before the experiment

In the second part of the experiment plants adsorbed Cu^{2+} from the solution during 2 weeks. After the planting the concentration of Cu^{2+} was estimated by Ecotest 2000 as well as the length of shoots and mass of roots were assessed.

The activity of adsorption was calculated as

$$A = K\ 1000 / M\ V, \text{ if} \quad (1)$$

K is the mass of adsorbed Cu^{2+}, M is the mass of roots, V – is the volume of glass (200 ml)

We used the electrode system with 2 ion-selective measurement electrodes and electrical converter. The ***electromotive force*** of electrode system depends on the activity of ions in the solution.

Results and discussion.

Maximal growth of plants was in control (fig 2, 3). Despite the cation of copper is a necessary feeding microelement for plants, their growth was considerably inhibited in variants with the concentration of Cu^{2+} more than 0,5-1,0 mg/l. The lethal concentrations of Cu^{2+} for wheat and barley were 10 mg/l and 20 mg/l correspondingly.

Table 2 – Concentration of Cu^{2+} in growth medium

Concentration of Cu^{2+} before planting, mg/l	Concentration of Cu^{2+} after planting, mg/l	Concentration of Cu^{2+} before planting, mg/l	Concentration of Cu^{2+} after planting, mg/l
Wheat		Barley	
Control	-	Control	-
0.5	0.3	1.0	0.0
1.0	0.5	2.5	1.0
2.5	2.2	5.0	4.0
5.0	4.8	10.0	9.0
10.0	10.0	20.0	20.0

The diagram was built using the data obtained in the tests with wheat and barley (fig. 3 a, b). The length of shoots and the mass of roots as well as the adsorption activity depended on the concentration of Cu^{2+}. The optimal concentration of Cu^{2+} was 1.0 mg/l and 2.5 mg/l for wheat and barley correspondingly. Cu^{2+} was absorbed by barley totally if the concentration of Cu^{2+} was 1 mg/l.

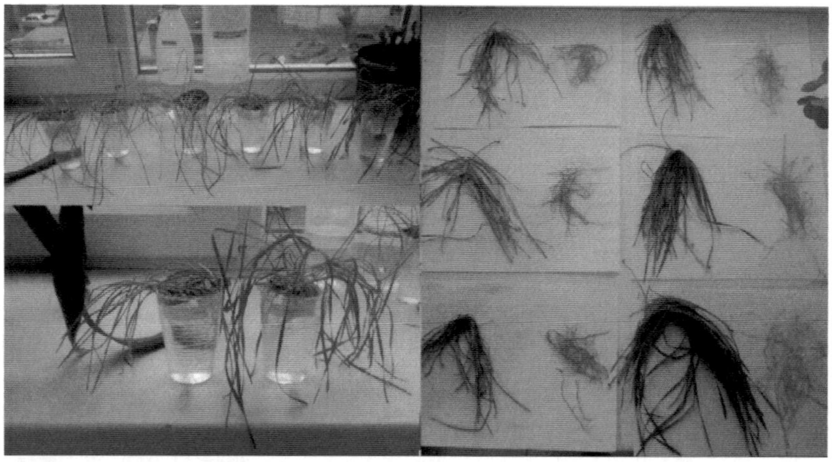

a

b

Figure 2 – Plants after the experiment: a – wheat, b – barley

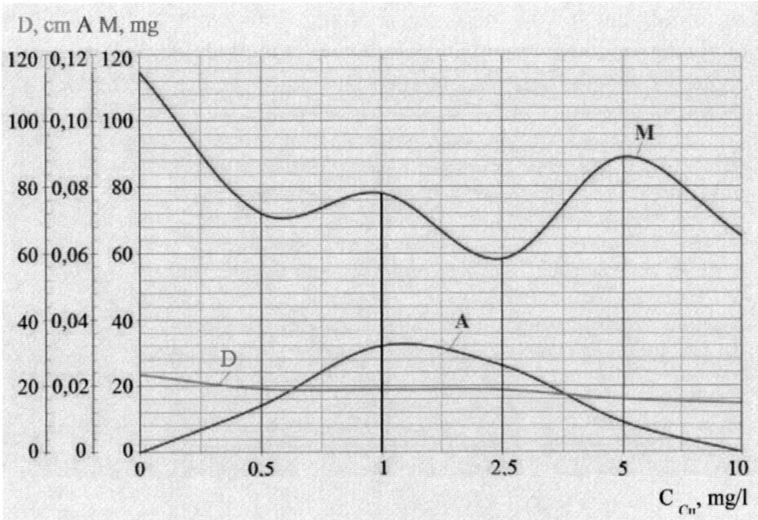

a

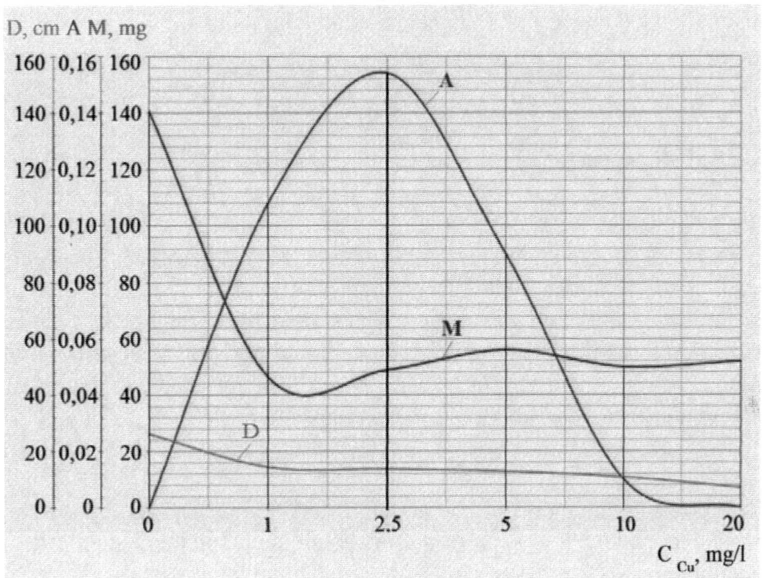

b

Figure 3 – Dependence of the morphological features and adsorption capacity on the concentration of Cu^{2+}: a – wheat, b – barley; D is the length of shoots, M is the dried mass of roots, and A is adsorption activity.

Basing on the described results we can conclude that barley is better plant for phytoremediation then wheat. It was more tolerant for the contamination. Use of barley allows getting the maximum permissible concentration (1.0) at the end of the experiment when the concentration of Cu^{2+} in growth medium (model wastewater) was 2.5 mg/l.

The rectangular settler is one of the possible devices which can be adapted to phytoremediation. It has good capacity and volume, but also it needs law operating costs. We propose to provide the standard device by moving arrangement. This arrangement is inexpensive, simple and easy to use. It can be applied on the enterprises.

Moving arrangement for the treatment of wastewater by phytofiltration is shown on fig. 4. It consists of bearing member (8) for biological bed (6). It contains the frame with overlay (2), side walls

(7), bottom (5), and turning hatch (4). The hatch is attached by hinge straps (3) to the side wall (7). The overlay is held on the suspended machinery (1). The bearing member is made of micromesh.

The size and shape of the moving arrangement depends on the parameters of the settler and biological bed.

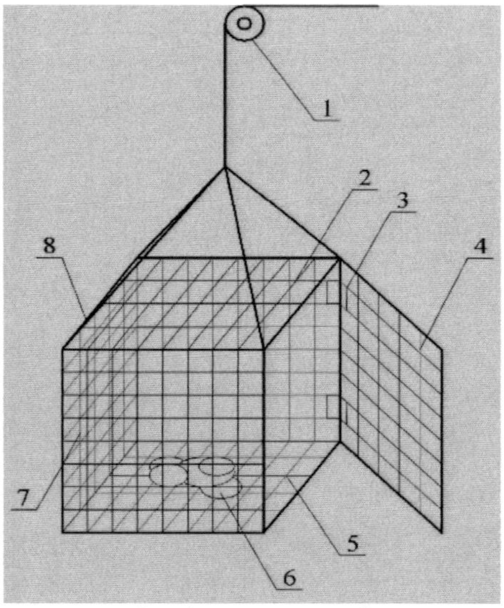

Fig. 4 – Moving arrangement for the treatment of wastewater by phytofiltration

It is the way the arrangement works. Plants for phytofiltration are put to the bearing member. Then the hatch is being closed. The arrangement is being immersed to the settler. The depth of submersion depends on the peculiarities of plants, particularly, their requirements for light. Also the depth depends on the characteristics of the settler and wastewater. Plants distributes regularly in the volume of settler, water contacts with plants, so the plants adsorb contaminations. After the process of treatment the moving arrangement is being drawn up. Plants contain heavy metals in their tissues, so the best way is to dry or burn up them and remove contaminants from the ash.

Conclusion.

The experiment demonstrated the effective extraction of the copper ions (II) by the remediation plants such as wheat and barley. Barley was more effective and tolerant than wheat. The optimal concentration of copper was 2.5 mg/l, in this variant we can get the maximum permissible concentration (1.0) at the end of the experiment.

The moving arrangement with plants as a biological bed based on standard rectangular settler was elaborated for phytofiltration. That is a technology for decontamination and final treatment of the wastewater from pollutants using plants. Particularly it is possible to remove ions of heavy metals. The device will be inexpensive, simple and easy for exploitation at the enterprice.

References

[1] Aquatic phytoremediation of CCA and copper contaminated water /C. Keith, H. Borazjani, S.V. Diehl, et al. //37th Annual Mississippi Water Resources Conference [Электронный ресурс] – Режим доступа: http://www.wrri.msstate.edu/pdf/keith07.pdf.

[2] Assessment of the efficiency of duckweed (Lemna gibba) in wastewater treatment [Text] /W.A. El-Kheir, G. Ismail, F. A. El-Nour et al. //International journal of agriculture & biology. 2007. V. 9. №. 5. P. 681-687.

[3] Capacity of the macrophytes Lemna minor and Eichhornia crassipes to remove nickel /P. Bres, D. Crespo, P. Rizz et al. //Agricultural Research Journal. – Режим доступа: http://ria.inta.gov.ar/english/wp-content/uploads/2012/04/Bh-10067-Bres-inglespress.pdf (27.10.2014).

[4] Cedergreen, N. Nitrogen uptake by the floating macrophyte Lemna minor [Text] /N. Cedergreen, T.V. Madsen //New Phytologist. 2002. № 155. P. 285–292.

[5] Cunningham S., Berti W. Remediation of contaminated soils with green plants: an overview In Vitro Cell. Dev. Biol., 29, 207-212 (1993).

[6] Diritgen, N. Effects of zinc and copper on growth and metal accumulation in duckweed, Lemna minor [Text] /N. Diritgen, Y. Inel Bulletin Environmental Contamination and Toxicology. 1994. № 53. P. 442-449.

[7] Dixit, A. Process and plants for wastewater remediation: a review [Text] /A. Dixit, S. Dixit, C.S. Goswami //Scientific Reviews & Chemical Communications. 2011. № 1(1). P. 71-77.

[8] Effects of copper and cadmium on heavy metal polluted waterbody restoration by duckweed (Lemna minor) [Text] /W. Hou, X. Chen, G. Song et al. //Plant physiology and biochemistry. 2007. № 45. P. 62-69.

[9] Effects of nickel on the chloroplasts of the duckweeds Spirodela polyrhiza and Lemna minor and their possible use in biomonitoring and phytoremediation [Text] / K.-J. Appenrotha, K. Krecha, Á.Keresztesb et al. // Chemosphere. – 2010. – V. 78. – P. 216–223.

[10] Garbisu G.C., Alkorta I. Garbisu G.C., Alkorta I. Phytoextraction: A cost-effective plant-based technology for the removal of metals from the environment //Bioresources Technol., 77, 229-236 (2001).

[11] Garg, P. The duckweed Wolffia globosa as an indicator of heavy metal pollution: sensitivity to Cr and Cd [Text] /P. Garg, P. Chandra //Environmental monitoring and assessment. 1994. № 29. P. 89-95.

[12] Gürtekin, E. The role of duckweed (Lemna minor L.) in secondary clarifier tanks /E. Gürtekin, N. Şekerdağ // [Электронный ресурс] – Режим доступа: http://oaji.net/articles/2014/1084-1418117778.pdf.

[13] Heavy metal levels and esterase variations between metal-exposed and unexposed duckweed Lemna minor: field and laboratory studies [Text] /S. Mukherjeea, S. Mukherjeea, P. Bhattacharyya et al. //Environment International. 2004. №. 30. P. 811-814.

[14] Jafari, N. Effect of pH and metal concentration on phytoaccumulation of zinc by three duckweeds species [Text] / N. Jafari, M. Akhavan // American-Eurasian Journal of Agricultural & Environmental Sciences. 2011. V. 10 (1). P. 34-41.

[15] Kaur, L. Role of pH in the accumulation of lead and nickel by common duckweed (Lemna minor) [Text] /L. Kaur, K. Gadgil, S. Sharma //International Journal of Bioassays. 2012. № 1. P. 191-195.

[16] Khellaf, N. Growth response of the duckweed Lemna gibba L. to copper and nickel phytoaccumulation [Text] /N. Khellaf, M. Zerdaoui //Ecotoxicology. 2010. V. 19. P. 1363–1368.

[17] Khellaf, N. Growth response of the duckweed Lemna minor to heavy metal pollution [Text] /N. Khellaf, M. Zerdaoui //Iranian journal of environmental health science & engineering. 2009. V. 6. №. 3. P. 161-166.

[18] Miranda, M.G. Uptake of lead by Lemna gibba L.: Influence on specific growth rate and basic biochemical changes [Text] /M.G. Miranda, K. Llangovan //Bulletin Environmental Contamination and Toxicology. 1996. № 56. P. 1000-1007.

[19] Mkandawire, M. Are Lemna spp. effective phytoremediation agents? [Text] /M. Mkandawire, E.G. Dudel //Bioremediation, Biodiversity and Biovailability. 2007. V. 1(1). P. 56-71.

[20] Mojiri, A. Phytoremediation of heavy metals from municipal wastewater by Typha domingensis [Text] /A. Mojiri //African Journal of Microbiology Research. 2012. V. 6(3). P. 643-647.

[21] Nayyef, M. A. Efficiency of duckweed (Lemna minor L.) in phytotreatment of wastewater pollutants from Basrah oil refinery [Text] /M.A. Nayyef, A.A. Sabbar //Journal of Applied Phytotechnology in Environmental Sanitation. 2012. V. 1. № 4. P. 163-172.

[22] Ozengin, N. Performance of duckweed (Lemna minor L.) on different types of wastewater treatment [Text] /N. Ozengin, A. Elmaci //Journal of Environmental Biology. 2007. V. 28. P. 307-314.

[23] Patel, D.K. Phytoremediation potential of duckweed (Lemna minor L.: a tiny aquatic plant) in the removal of pollutants from domestic wastewater with special reference to nutrients [Text] /D.K. Patel, V.K. Kanungo //The bioscan. 2010. № 5(3). P. 355-358.

[24] Physiological responses of Lemna trisulca L. (duckweed) to cadmium and copper bioaccumulation [Text] / M.N.V. Prasad, P. Malec, A. Waloszek et al. //Plant Science. 2001. № 161. P. 881-889.

[25] Phytotoxicity of atrazine herbicide to fresh water macrophyte duckweed (Lemna perpusilla Torr.) in Thailand [Text] /O-A. Phewnil, N. Tungkananurak, S. Panichsakpatana et al. //Environment and Natural Resources Journal. 2012. V. 10. № 1. P. 16-27.

[26] Potential of macrophyte for removing arsenic from aqueous solution [Text] /F.P. Guimaraes, R. Aguiar, J.A. Oliveira et al. //Planta Daninha, Viçosa-MG. 2012. V. 30. №. 4. P. 683-696.

[27] Removal of chromium from wastewater using macrophyte Lemna minor as biosorbent /L.K. dos S. Lima, S.J. Kleinübing, E.A. da Silva et al. //[Электронный ресурс]. – Режим доступа: http://www.nt.ntnu.no/users/skoge/prost/proceedings/pres2011-and-icheap10/PRES11/23Lima.pdf.

[28] Saifullah, Sadia Bibi, Ejaz Ahmed Waraich, Communications in Soil Science and Plant Analysis, 45, 3150–3160 (2013).

[29] Salt D., Smith R., Raskin I. Phytoremediation. Annu Plant Physiol Plant Mol Biol., 49, 643-668 (1998).

[30] Singh, D. Phytoremediation of lead from wastewater using aquatic plants [Text] /D. Singh, A. Tiwari, R. Gupta //Journal of Agricultural Technology. 2012. V. 8(1). P. 1-11.

[31] Steveninck, R.F.M. Van. Heavy-metal (Zn, Cd) tolerance in selected clones of duckweed (Lemna minor) [Text] /R.F.M. Van Steveninck, M.E. Van Steveninck, D.R. Fernando //Plant and Soil. 1992. V. 146. P. 271-280.

[32] Tolerance to heavy metals in the duckweed, Lemna minor [Text] /N. Khellaf, M. Zerdaoui, O. Faure et al. [Электронный ресурс] – Режим доступа: http://centrostudinatura.it/public2/documenti/290-60266.pdf.

[33] Valipour A., Hamnabard N., Woo K.-S., Ahn Yo.-H. Performance of high-rate constructed phytoremediation process with attached growth for domestic wastewater treatment: Effect of high TDS and Cu. Journal of Environmental Management, 145, 1–8 (2014).

[34] Wolf, A.J. Effects of copper on the growth of Lemna minor (duckweed) [Text] /A.J. Wolf, B.S. Rynearson, J.M. Sperry //Beloit College Journal of Plant Ecophysiology. 2004. V. 3. P. 9-12.

IOP Conf. Series: Earth and Environmental Science **50** (2017) 012029 doi:10.1088/1755-1315/50/1/012029

Visualization Of Electromagnetic Environment Near GSM Antennae

M Yu Zvezdina, Yu A Shokova, N I Kundryukova, V D Kutukova and A V Pozdnyakova

Radioelectronic Department, Don State Technical University
1 Gagarin Square, Rostov-on-Don, Russian Federation, 344000

E-mail: zvezdina_m@mail.ru

Abstract. Web-enabled mobile phone enhancement has determined a sharp increase of mobile base station quantity, which has led to a surge of electromagnetic noise in metropolitan areas. Electromagnetic noise permanently influences adult and child population; the impossibility to remove this impact has promoted the electromagnetic monitoring. The most accessible form of socially oriented monitoring is a visualization of energy flux density for GSM base station antennae. Forecast calculation routines take antenna construction and location into an account; they also allow giving the recommendations on electromagnetic environment improvement basing on physical interpretation the obtained results.

1. Introduction

In modern mobile phones – smartphones - there are web-enabled applications available [1] along with voice transmission. The increase of smartphone users (353 million users in 2015 [2]) has led to a surge of GSM base station quantity. In Russia alone in 2013 the number of base stations has totaled to 82 211, with packing density equal to 11 km^{-2} and 90 km^{-2} in metropolitan areas. As a result, in densely populated areas a complete cover of nonionizing radiation from base station antenna is generated. Electromagnetic fields excited by 100 watt transmitters give about 60% of a total exposure in radio frequencies; they have a permanent around the clock impact on population health, including children, elderly and unhealthy persons. Moreover, changing the rooftop location of base station antennae from high-rise apartment buildings to stand-alone supermarkets (on the level of 2-3 floor of apartment buildings) aggravates electromagnetic environment of residential area even more.

Electromagnetic field radiated by an antenna can be controlled in various ways, like, by coating antenna support-structures with composites described in [3]. The field structure changes are shown in [4]. But it is impossible to exclude all the antenna support-structure diffraction. Hence, regulatory documents [8-10], wherein the energy flux density (EFD) for frequencies ranging from 300 MHz to 300 GHz was limited, were issued basing on theoretical and experimental results, like the ones given in [5-7]. Maximum permissible level of power flux density (PFD), which causes no irreversible changes on human body, is 10 μW/cm^2 for GSM base station antenna. It's important to stress that this level is of sanitary and not ecological nature, because various environment components are more sensitive to electromagnetic field than a human being; they degrade or peter out really fast [11].

Maximum permissible level of radiation introduction implies regular influence factor monitoring of GSM stations locations, forecasting the environmental assessment, i.e. electromagnetic field monitoring and its sub-type, which is socially-oriented monitoring [12]. This kind of monitoring encircles large areas and promotes timely and situational provision of information to the public concerning electromagnetic pollution level; it limits sanitary protection zones (SPZ) measured at height of 2 m above ground or above top floor of building if antenna is located at the rooftop. Sanitary protection zone limits correspond to maximum permissible level of 10 $\mu W/cm^2$.

Population comprehensibility and capability of continuous adjustment to new electromagnetic environment problems are core requirements to socially-oriented monitoring results. The best graphic representation of results is their visualization with software graphical functionalities as it is shown in [13].

The paper objective is to estimate electromagnetic environment near GSM base station antennae with the visualization of the results obtained with forecast calculation method.

2. Characteristics of electromagnetic environment visualization

Proximity of electromagnetic environment visualization in GSM antenna locations is defined by both input data accuracy and graphical package functionality. The first factor is influenced greatly by a method for obtaining energy flux density distribution data. The second factor defines the capability to give population-friendly interpretation of the results. In-line adjustment of visualization results is also linked to the method for obtaining electromagnetic environment data.

2.1. Methods for obtaining electromagnetic environment data

To obtain information about electromagnetic environment near GSM antenna, as it is shown in [13], two methods are used: instrumental procedure and forecast calculation technique. The instrumental procedure has two major deficiencies in meeting the visualization requirements, i.e. low responsiveness to radiating objects location changes within the area of investigations and impossibility of measuring at every desired point. Forecast calculation technique is based on application of strict electrodynamic models; it allows obtaining electrodynamic field structure data for every point in antenna near-field zone and real time updating the measurement results.

EFD calculation technique for isolated points for various antenna design and antenna deployment depends on a specific antenna construction, its location and bandwidth. The technique is described in [10]. Its peculiarity is an application of "preventive principle", which is recommended by the World Health Organization [14]. This principle allows for preventive actions to reduce possible harmful effect of electromagnetic radiation on human beings even if the effect itself is not proven, but is assumed with contemporary level of scientific knowledge. As a result, sanitary protection zone limits would exceed the real ones.

2.2. Software products

There are specific requirements imposed on software for visual representation of electromagnetic environment near GSM antennae under socially-oriented monitoring.

- It should use forecast calculation technique, while the mathematical models should consider electrodynamic effects, i.e. antenna design and deployment.
- The routine output should present EFD near antenna location on surface plots in horizontal and vertical sections. One should set the section position relative to antenna phase center.
- Sanitary protection zone limits (10 $\mu W/cm^2$ level) must be visible.

Well-known software commercial products like SanZone [15] and PR2 [16], which are used mostly to calculate certificate of hygiene for radiating objects, do not meet all of these requirements of socially-oriented monitoring. Hence, the necessity of specialized program development arises [17]. The descriptions of MathCad and MATLAB-based EFD distribution visualization programs in horizontal and vertical sections are given in other works of the authors, like [18-20].

3. Results and Discussion

3.1. Electromagnetic environment estimation by forecast calculation

Electromagnetic environment has been estimated near GSM antennae with the routine developed by the authors. Specific antenna types have been selected according to mobile antenna market research given in [19]. That paper has showed that old standards like IMT-2000 have been used along with the new ones, i.e. 3G and 4G. The following frequencies are given for mobile radio of IMT-2000 standard under the recommendations of International Telecommunication Union ITU-R M.1036-3: 806-960 MHz, 1720-1885 MHz and 2500-2690 MHz. Antenna electrical characteristics for the most common mobile communication system in Russia are given in table 1.

Table 1. Electrical characteristics of considered base station antennae [19]

Antenna model	Frequency range, MHz	Gain, dB	Aperture angle, degree		Input power rating, W
			H-plane	E-plane	
TETRA antennae					
RAV-2UL-90	400-430	8	37	90	400
RAV-4UL-90	400-430	11	18	90	400
RAO-4U-120	400-470	8	36	120	400
RAX-2UL-70	400-430	9	36	70	400
RAX-4UL-70	400-430	12	18	70	400
RAO-2U-60	380-440	10	36	60	500
GSM (3G) antennae					
RAO-11GL-60	860-970	11	30	60	50
RAO3-10GH-60	1710-1880	13	18	60	50
RAO-14GL-70	860-970	14	15	70	50
LTE (4G) antennae					
RAX-14Yota-70	2400-2700	14.2	8	65	50

Upper sanitary protection zone limits calculated for maximum radiation level according to the "preventive principle" are shown on figure 1.

Figure 1 analysis shows that the worst case antenna could be pinpointed for each communication standard. It is trunking antenna RAX-4UL-70, 4G antenna RAX-14Yota-70 and 3G antenna RAO-14GL-70.

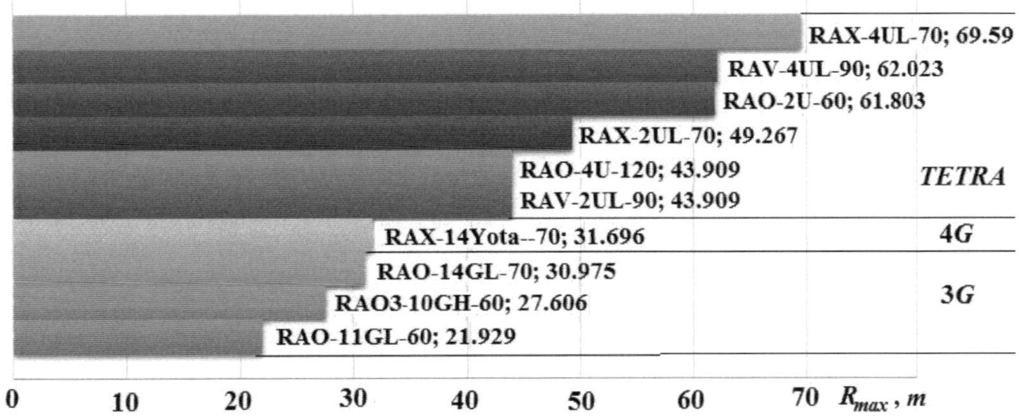

Figure 1. SPZ limits according to the "preventive principle"

EFD distribution in vertical section calculated for 3G antenna RAO-14GL-70 deployed 5 m above the rooftop, if antenna is located in the roof edge center of 30 m tall building, is shown on figure 2. Roof corner point coordinates were defined by $A(0,0,30)$, $B(-7,4,30)$, $C(-13.5,-4,30)$, $D(-2.5,-8.5,30)$. All measures are given in meters.

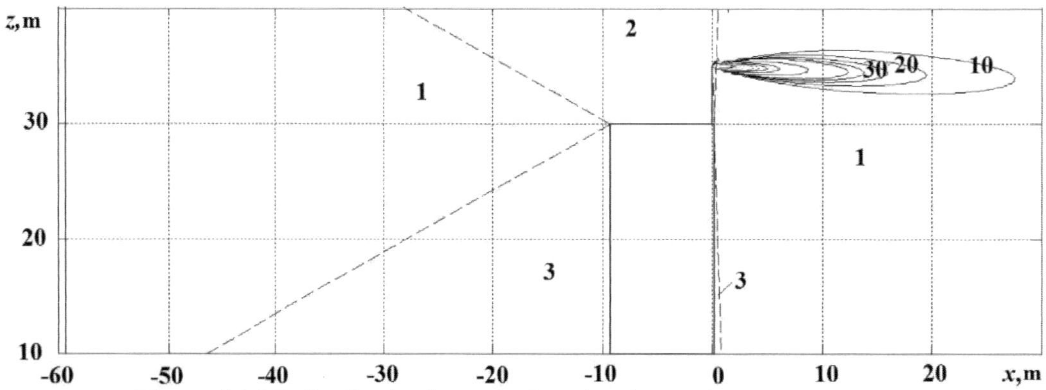

Figure 2. EFD distribution in vertical section for RAO-14GL-70 antenna

As a comparison, figure 3 and figure 4 show omnidirectional antenna field distribution in vertical section containing antenna phase center, and in horizontal section at 2 m above the rooftop. Power and location of omnidirectional antenna are equivalent to the case shown on figure 2. At figures 2 and 3 number 1 stands for the one-beam zone (incident beam), 2 – for the two-beam zone (incident beam and the one reflected by the roof), 3 – for shadow zone.

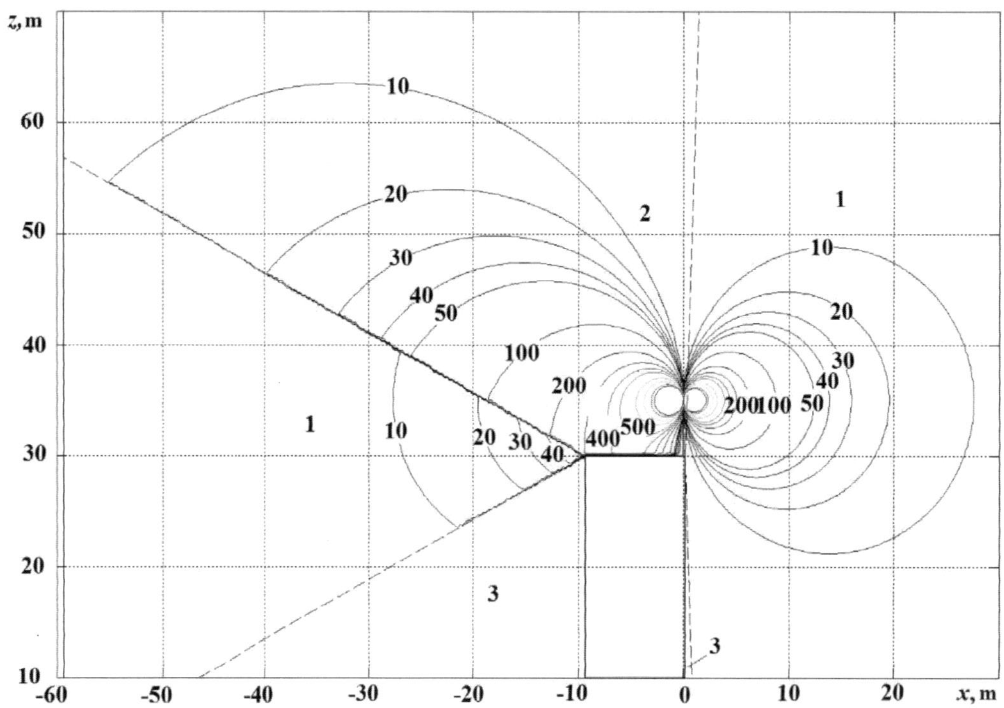

Figure 3. Vertical section of EFD distribution for omnidirectional antenna

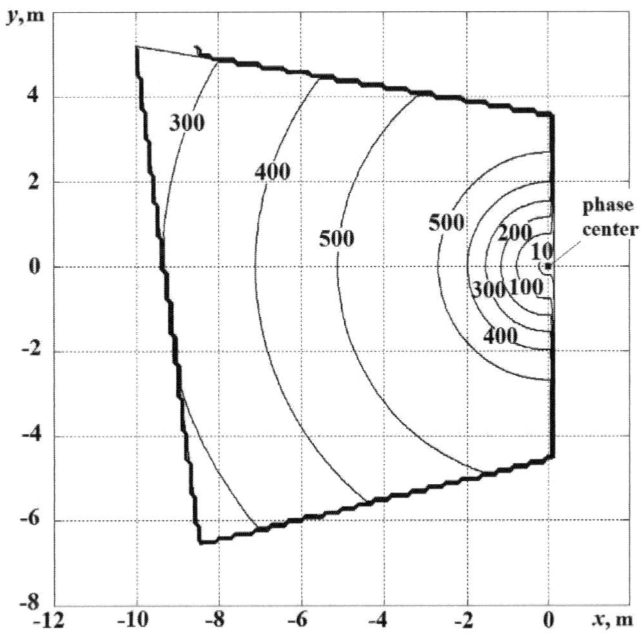

Figure 4. Horizontal section of EFD distribution for omnidirectional antenna

The analysis of the obtained EFD distributions shows that omnidirectional antennae have EFD level of 500 μW/cm^2 in horizontal section at 2 m above the rooftop; it exceeds maximum permissible level by several orders. Moreover, a zone with radiation level greater than maximum permissible level appears in antenna back half-space area sized about 60 m horizontally and 30 m in vertically. SPZ horizontal limits in antenna front half-space are only 30 m. Directional GSM antennae with back-lobe level of less than -20 dB do not exceed the radiation level of 0.021 μW/cm^2 on the rooftop. Even if GSM antennae are deployed with standard configuration – three in a circle spaced at intervals of 120 degrees – EFD level in horizontal section at 2 m above the rooftop would be less than maximum permissible level.

3.2. Comparison of electromagnetic environment estimation results obtained with forecast calculation technique and with instrumental procedure

Let's compare the results given in the section 3.1 and ecological environment data obtained with instrumental procedure and taken form [21]. In the last paper the results have been measured within a 300 m radius of base station antenna at levels of 0.5 m and 2 m above bearing surface. It should be noted that the level of 0.5 m above the bearing surface characterizes electromagnetic environment ecologically. Data on antenna position upon a roof, height of a building, and antenna directional properties are not given in [21]. The EFD measurements from [21] for rooftop antenna are shown in table 2.

Table 2. Electromagnetic environment data obtained with instrumental procedure [21]

Measurement point locations	Mean EFD value, μW/cm^2	Maximum permissible EFD level, μW/cm^2
Roof	2-10	500
Accommodation under base station antenna	<2	10
Adjacent grounds intended for building	0.2-0.7	5
In first and second line buildings	0.5-1	30

Let's compare the results from table 2, figure 2 and figure 4, which are obtained with different methods. As is easy to see, that for omnidirectional antenna the values of highest EFD level by forecast calculation and the one by instrumental procedure coincide and are equal to 500 $\mu W/cm^2$. It proves the correctness of the software routine, which has been developed by the authors. The routine itself is given in [18-20].

The discrepancy within the results can be easily explained. The research area of 300 m radius, which has been considered in [21], holds for ecological estimation of maximum permissible level of 1 $\mu W/cm^2$ that has been suggested to use in the paper. If the value of maximum permissible level is taken from the standards [10] of sanitary and hygienic estimation and is considered to be 10 $\mu W/cm^2$, the research area can be decreased by about an order; it was shown on figures 2-4. If we the visualization results of EFD distribution for rooftop antenna obtained with a strict electrodynamic model (figure 2) are considered, it is easy to notice that the accommodations under the antenna are in the shadow zone. Hence, the EFD value will be minimal in this direction. GSM antenna field affects only the open balconies of the upper floors, if the balconies are large enough. This reasoning is supported with instrumental procedure results. It is impossible to evaluate the instrumental procedure results for adjacent grounds and in first and second line buildings, because in the paper [21] there are no data on antenna height above the ground, distance to the first and second line buildings and their height.

Thus, forecast calculation, which was implemented in the EFD distribution visualization routines developed by authors for horizontal and vertical sections, may be said to give consistent results in an accessible form; their physical interpretation promotes recommendations on electromagnetic environment improvement. Hence, the routines can be used in socially-oriented electromagnetic monitoring.

4. Conclusion

The increase of GSM base stations quantity causes a serious decline of electromagnetic environment, especially in metropolitan areas. The impossibility to change this makes electromagnetic environment monitoring necessary. The most accessible form of giving out results of socially oriented monitoring is their visualization with specially developed software routines.

The developed EFD visualization routines for GSM base station antennae are based on forecast calculation method with an account of electrodynamic effects caused by antenna construction and location. The calculations are also based on Russian standards for land-based mobile communication location and operation.

The analysis of EFD distributions has shown that omnidirectional GSM antennae have EFD of 500 $\mu W/cm^2$ in horizontal section at 2 m above the ground level, which exceeds maximum permissible level by several orders. A zone with radiation level greater than maximum permissible level appears in antenna back half-space area sized about 60 m horizontally and 30 m in vertically. SPZ horizontal limits in antenna front half-space are only 30 m. Directional GSM antennae with back-lobe level of less than -20 dB do not exceed the radiation level of 0.021 $\mu W/cm^2$ on the rooftop. Even if GSM antennae are deployed with standard configuration – three in a circle spaced at intervals of 120 degrees – EFD level in horizontal section at 2 m above the rooftop would be less than maximum permissible level.

The results obtained with the developed software routines and their comparison to instrumental measurements have shown that the forecast calculation, which had been implemented in the developed routines for EFD visualization in horizontal and vertical sections, gives consistent results in an accessible form. Their physical interpretation promotes recommendations on electromagnetic environment improvement. Hence, the routines can be used in socially-oriented electromagnetic monitoring.

References

[1] Ali S, Khusro S 2016 Indian J. of Sci. and Techn. 9 (19) doi: 10.17485/ijst/2016/v9i19/53088

[2] Bechet P, Miclaus S 2013 Meas. Sci. and Techn. J. 24 015003 doi: 10.1088/0957-0233/24/1/015003
[3] Senyk I V, Barsukov V Z, Savchenko B M, Shevchenko K I, Plavan V P, Shpak Yu V, Krukova O A 2016 IOP Conf. Series: Materials Science and Engineering 111 012026 doi: 10.1088/1757-899X/111/1/012026
[4] Gabriel'yan D D, Zvezdina M Y, Kostenko P I 2001 J. Commun. Techn. and Electron. 46 (8) 875-78
[5] Martinez-Burdalo M, Anguiano M, Villar R 2005 Phys. Med. Biol. 50 (17) 4125 doi: 10.1088/0031-9155/50/17/015
[6] Raja D S 2015 Indian J. Sci. Techn. 8 (32) doi: 10.17485/ijst/2015/v8i32/87451
[7] Hand J W 2008 Phys. Med. Biol. 53 R243-86
[8] ICNIRP (International Commission on Non-Ionising Radiation Protection) 1998 Guidelines for limiting exposure to time-varing electric, magnetic and electromagnetic fields (up 300 GHz) Health Phys.74 494-522
[9] IEEE (Institute of Electrical and Electronic Engineers) 2005 C95.1-2005 IEEE Standart for Safety Levels with respect to Human Exposure to Radio Frequency Electromagnetic Fields, 3 kHz to 300 GHz (New York IEEE) Web-site: https://standards.ieee.org/findstds/standard/C95.1-2005.html (accessed September 4, 2016)
[10] Ministry of Health of the Russian Federation 2003 SanPin 2.1.8/2.2.4.1190-03 Hygienic requirements on land-based mobile communication location and operation (Moscow)
[11] Grigoriev Yu 2011 Environmentalist 32 (2) 193-200 doi: 10.1007/s10669-011-9371-4
[12] Maslov M Yu, Spodobaev M Yu, Spodobaev Yu M 2014 Telecommunications (10) 39-42
[13] Ministry of Health of the Russian Federation 2002 MG 4.3-1167-02 Methodological guidelines for energy flux density estimation for electromagnetic field for location of radio facilities with operating band of 300 MHz – 300 GHz (Moscow)
[14] WHO Library Cataloguing-in-Publication Data 2004 Establishing a dialogue on risk from electromagnetic fields. http://www.who.int/peh-emf/publications/en/russian_risk.pdf
[15] SanZone: Web-site: http://www.ing-tv.ru/index/sanzone/0-53 (accessed September 4, 2016)
[16] RPS2: Radio Planning System 2. Web-site: http://www.rps2.ru/ (accessed September 4, 2016)
[17] Murelli A, Milotti E 2014 Europ. J. Phys. 35 (1) 015014 doi: 10.1088/0143-0807/35/1/015014
[18] Zvezdina M Y, Shokova Yu A, Shokov A V 2015 Proc. Int. Conf. on Application of Information and Communication (Rostov-on-Don) 399-403 doi: 10.1109/ICAICT.2015.7338588
[19] Zvezdina M Yu, Shokova Yu A, Shokov A V, Dymchenko A A, Parhomenko P A 2015 Int. J. Theor. & Appl. Sci. 03 (23) 10-17 doi: 10.15863/TAS.2015.03.23.3
[20] Zvezdina M Yu, Shokova Yu A, Shokov A V, Andronov M A, Tyul'pin V I 2015 Int. J. Theor. & Appl. Sci. 08 (28) 56-64 doi: 10.15863/TAS.2015.08.28.8
[21] Grigoriev O A, Merkulov A V 2012 Bulletin of medical internet conferences 2 (6) 458-60

IOP Conf. Series: Earth and Environmental Science **50** (2017) 012030 doi:10.1088/1755-1315/50/1/012030

Environmental-Socio-Economic Monitoring as a Tool of Region's Environmental-Economic System Management

T V Galanina, M I Baumgarten, V G Mikhailov, T G Koroleva and G S Mikhailov

T. F. Gorbachev Kuzbass State Technical University
street Vesennyaya28, Kemerovo, Russian Federation, 650000

E-mail: galanina64@mail.ru

Abstract. The paper deals with the region's environmental-economic system management through a tool such as the environmental-socio-economic monitoring. The purpose of research – is analysis and development of theoretical assumptions of environmental-socio-economic monitoring system for the effective management of geographically distributed environmental-economic system. The main elements of environmental-socio-economic monitoring are identified, taking into account the characteristics of the studied area. The main result of the research is the development of multi-functional integrated monitoring system for the evaluation of the indicators "gross domestic product" and "gross national product", taking into account the influence of environmental factors. The results of the study conducted may be recommended to the regional and federal governments to support the effective, environment-friendly management decision-making consistent with the overall development concept.

Introduction

Sustainable development of society presupposes the harmonization of relations in the system "nature - society - economy". The overall design of sustainable development concept is the alternative to the nature-destructive forms of economic activities.

The relevance of the sustainable development strategy at the regional level, taking into account the market changes, proves the feasibility of the construction of environmental-economic system management (EES) model.

The concept of the EES is used by many researchers, along with the allied concepts of "bio-economic system", "natural-economic system", "ecological and economic system". The main essence of the EES concept is the integration of the economy and the nature that is the interrelated and interdependent functioning of social production and the flow of natural processes. There are two main interpretations of environmental-economic system concept - the global and regional-territorial, or local.

The current state of environmental-socio-economic development of any region as a system of the interrelated factors have a direct impact on the adjacent areas of public life [1,2]. In particular, the unstable state of the environmental-socio-economic system of Russia is characterized by three main interrelated negative trends:

• shrinkage of the process of reproduction of means of production and of life benefits (economic downturn);
• violation of the process of reproduction of natural resources (both organic and inorganic) and the living environment conditions (specific degradation of ecosystems);
• shrinkage of the process of population reproduction (depopulation) [3].

Significant changes affecting the natural environment in the last century, in the course of human life require fundamental changes in the comprehensive assessment of the region. Due to the negative environmental situation, both in individual regions and the country as a whole, an important aspect is the adequate assessment of the state of the environment.

To develop the regional EES management model, in our opinion, the considerable amount of research is required, the basis of which is the environmental-socio-economic monitoring (ESEM). Special attention should be paid to the environmental component due to the fact that the errors in this area in the future, as practice shows and as is noted by some scientists [4, 6], have a negative impact on social and economic indicators of the state of territories.

Material and methods

The object of research of the environmental-socio-economic monitoring as a part of the organizational-economic mechanism of the region's environmental-economic system management. The subject of research - social, ecological and economic ties and processes that determine the existing monitoring system. The study is based on an analysis of domestic and foreign literature on the problems of the organization of the environmental-socio-economic monitoring. Particular attention is paid to the study of the methodology for the calculation of macroeconomic indicators, taking into account environmental constraints. The elements of the system analysis and the results obtained by the specialists in the field of environmental-economic system management are also used in the paper.

Results and Discussion

Currently, in most regions there are differentiated monitoring system, designed for the monitoring of particular field of activity: biomedical, economic, environmental, seismic, social, socio-environmental and others. Each of these systems has the individual methodological framework, criteria for selection of indicators and targets, the software and the certain segment of consumers of the received information.

The efficiency of the implementation of the regional development strategy depends largely on the establishment of the regional system of environmental-socio-economic monitoring. The performance and practical importance of monitoring depend on the reliability and operability of the evaluation of the processes in socio-economic and natural areas.

The term "monitoring" appeared in the XX century in foreign science and was used to determine the system of targeted repeated observations of one or more elements of the environment in space and time. Broadly speaking, monitoring (from the Latin "monitor" -warning) is the activity of monitoring (tracking) of certain objects or events [7]. Despite the priority of foreign scientists in the development of monitoring methodology the term was first introduced in the economic life by Russian economist A. Nikonov, who argued that "... monitoring is a set of methods for tracking, analysis, evaluation and forecasting of socio-economic processes, related to the reforms, as well as the collection and processing of information and preparation of recommendations for the development of reform ... " [8]. Further refinements of the "monitoring" concept for social, environmental and economic performance are given in [9-12].

The subject of regional monitoring is the complex set of socio-economic processes and phenomena in the country's regions. Here are the following areas of the complex regional monitoring: environmental; economic; biomedical; social and political; social; science and technology.

On the basis of these areas the main tasks of the regional socio-economic monitoring can be formulated:

• management of observation, acquisition of reliable and objective information about the course of the socio-economic processes on the territory;

• assessment and systems analysis of the received information, the identification of factors that influence the character of the social and economic processes;

• identification of the factors that cause economic and social threats now and in the future;

• ensuring government, enterprises, institutions and organizations irrespective of their subordination and forms of ownership, citizens with the information obtained in the exercise of ESEM;

• development of forecasts of socio-economic situation;

• preparation of recommendations aimed at overcoming the negative trends and supporting the positive ones, bringing them to the knowledge of regional management bodies.

Due to the fact that economic, social and environmental processes are the basis of life in the region, it is advisable to identify the concept of "regional monitoring" and "ESEM" (hereinafter - monitoring). The main purpose of such monitoring is to monitor and regulate the processes of regional development.

The subject of monitoring in this case is the set of economic, environmental and social processes in the region in their interdependence and interrelation. The environmental processes affect the health of population, recreational capacity of the territory of the region and the potential for the self-regeneration of environment. In addition, the environmental state of the region significantly affects the employment, migration and demographic situation. Economic processes determine the standard of living of the population, including its political activity, work and migration. Social processes are characterized by social and political activity of the population, its ecological literacy, the general level of education and other aspects. The above processes are interrelated, since their subject is a person actively implementing certain policies and defining a vector of development. It is therefore necessary to consider the region in terms of the constant interaction of these processes: environmental - natural environment, economic - industrial and economic activity and social - the society and its activities.

Establishing monitoring operation of the first stage generates the following tasks:

• acquisition of information on the indicators and indices of the processes;

• verification of the reliability and the objectivity;

• systematic analysis of the information received and its evaluation;

• definition of correlations;

• construction of the mathematical model of the integrated monitoring system;

• identification of representative factors;

• corrsponding activities.

The greatest attention should be paid to the creation of information databases of monitoring (IDM), which is challenging, widely discussed task in the community of ecologists and economists. [13] ESEM can contain a different set of indicators and indices depending on the task, the model used, the regulatory framework, economic specifics of the region. In addition, it is advisable to correlate it with other databases, such as the state statistics database. The monitoring is also used as a regulator of the above processes in the region. It can serve as part of regional development management system, carrying out the functions of regular monitoring and short-term forecasting.

On the basis of the study the following basic principles of IDM of socio-economic transformation in the region can be outlined:

1. IDM of socio-economic transformation in the region should be organically integrated in the system of information support, control and regulation of social and economic development of the region, contain the information required by the federal bodies of state administration, reflecting the specificity of the region and especially its development.

2. The mandatory presence of full of information about the purposes of the system of targets in each specific area of reforming, factors, indicators and indices of reformed object development facilities at IDM of socio-economic transformation of the region.

3. The basis of the formation of the system of targets and the indices that describe them are laws, regulations, government programs, as well as instructional and teaching materials in the directions of reforming.

4. The aggregated indicators characterizing the progress and results of socio-economic transformation in the region should be clear, simple to interpret, available for analysis and use in the practice of management.

5. To create IDM of social and economic reforms, in addition to the state statistics system it is required to use the specialized information systems (banking, stock exchange, and employment systems).

ESEM must contain not only certain elements, but also unique to it traits:

• high performance of ESEM requires a long time period;

• ESEM as the system includes a plurality of environmental, social, and economic elements;

• information acquisition process is implemented on the basis of the developed technique with specific indicators;

• the data obtained as a result of information acquisition are summarized, analyzed and reflect the dynamics of the current processes.

As a result, we can formulate the ESEM priorities:

• improving the regulatory framework governing the interaction of federal executive bodies, exercising state environmental, social and economic monitoring, including the formation of the fund of information resources;

• improving the system of indices, establishment of the methodology for environmental, social and economic monitoring of the Russian Federation;

• ensuring the reliability and comparability of data of environmental, social and economic monitoring of certain sectors of the economy and regions of the country;

• ensuring transparency of environmental, social and economic information.

Schematically, the elements of environmental-socio-economic development monitoring are presented in Figure 1 [1, 2].

Despite the fact that it is impossible to take into account the impact of all environmental factors on the level and dynamics of all aggregate indices of socio-economic development, it is extremely important to consider those calculable. In the future, it is advisable to evaluate the performance of the national wealth and Gross Domestic Product (GDP), calculated taking into account and excluding the impact of environmental factors. The level of GDP, not taking into account the impact of socio-economic activities on the environment, generates an overestimation of economic development. At present, on the basis of accounting of natural capital consumption costs the methodology of calculation of environmentally adjusted GDP is being developed in order to identify the challenges of sustainable development.

The above shortcomings of the existing monitoring system in Russia suggest the absence of complete and reliable information, which requires its structuring into the information flows necessary for the efficient functioning of the regional economy.

The authors believe that without the use of ESEM the regional management will be ineffective with the "deficit" of optimal conditions for the improvement of the socio-environment state of the region. In addition, the analysis of the state and the dynamics of social and environmental factors in the process of social and environmental decision-making is necessary to determine the trend of development of the system "man-society-nature" in the region.

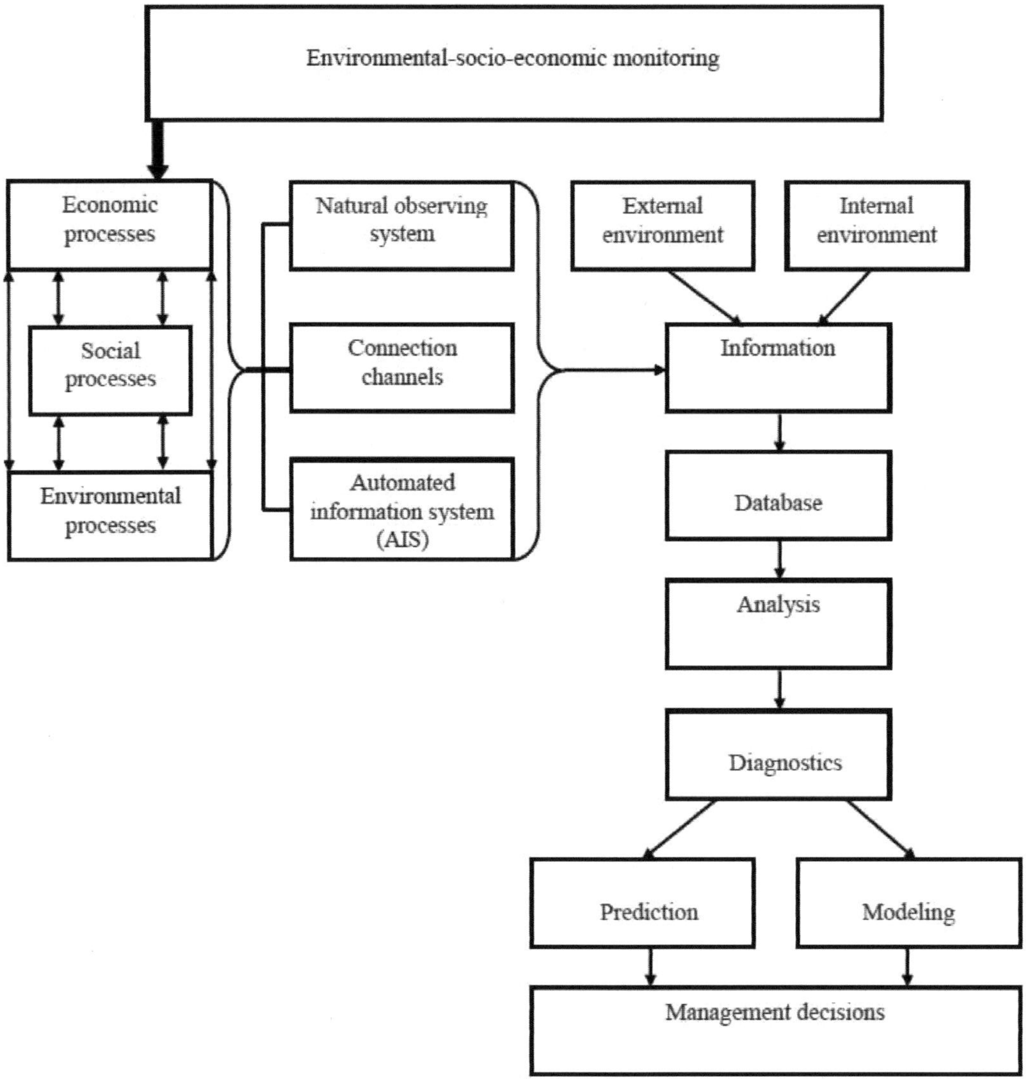

Figure 1 – The basic ESEM elements

The efficiency of the integrated ESEM system forming process is hampered by a number of restrictions, the most significant of which are as follows:

• lack of access to the information for a wide number of users;
• the need to integrate the monitoring entities;
• lack of comprehensive system of indicators.
• the problem of data reliability.

Especially important is the absence of developed comprehensive system of indicators, resulting in the fact that such indicators as gross national product (GNP) and the value of individual resources or pollution flows do not provide sufficient information about the sustainability of development. In such circumstances it is necessary to use all kinds of information for the development of indicators that characterize the natural or noosphere resistance [14, 15].

The degree of "participation" of environmental and economic factors in the environmental-socio-economic model may be different. In some cases - in the "pure" economic model (for example, along with the release of the main product) the release of "by-products" - waste as environmental pollutants is taken into account and; in others - the relationship between environmental factors are modeled and the results of the calculations are used in forecasting and planning of production tasks. Of similar importance is the model of the natural ecosystem, comprising the living organic matter (biomass) balance equation. Some researchers are engaged in the construction of complexes and models of environmental-economic systems for environmental planning and management.

A number of countries put in practice the advanced models of input-output balance, including not only the manufacturing industry but also industry destroying hazardous waste for the prediction of the impact of the economical structure on the environment. The approach to the environmental-economic balance coming from the laws of thermodynamics is of global importance nowadays: the amount of material taken from nature for the production of goods is compared to the total number of human waste (materials balance principle). In general, the development of environmental-socio-economic (regional) monitoring provides the opportunity to build an optimal EES model in the region.

Conclusions

The conducted study of environmental-socio-economic monitoring as a tool for the region's environmental-economic system management allowed the following conclusions:

• the determinants and trends characterizing the unstable state of socio-economic and environmental-economic systems are outlined;
• the main tasks of the regional socio-economic monitoring are formulated;
• the principles of information monitoring framework are developed;
• the system-defined elements of environmental-socio-economic monitoring are determined;
• the proposals for the evaluation of macroeconomic indicators, taking into account the environmental characteristics obtained from an integrated monitoring system are made.

References

[1] Magomadova H A 2012 Problems of socio-environmental and economic efficiency of interaction between society and nature Engineering Journal of Don Vol 19 Issue 2 pp 239–248 URL: http://www.ivdon.ru/en/magazine/archive/n1y2012/666
[2] Losevskaya E A and Rossinskaya M V 2012 Monitoring as instrument of regulation sustainable development of the region Engineering Journal of Don Vol 20 Issue 2 pp 555–560 URL: http://www.ivdon.ru/en/magazine/archive/n2y2012/838
[3] Pavlov B S and Sentyurina L B 2006 Demographic heal of population as a basis of social and economic development of the region Journal of Economic Theory Issue 04 pp 151
[4] Konchakova N V, Ushakova N S and Aikina T Yu 2016 Ecological risk assessment of Tomsk region groundwater used for drinking purposes IOP Conf. Series: Earth and Environmental Science 33 012022
[5] Shinkevich A I, Lubnina A A, Koryakov A G, Mikhailov VG and Vodolazhskaya EL 2016 Economic aspects of risk management of stakeholders activities International Review of Management and Marketing Vol 6 Issue 2 pp 328–332
[6] Galanina T V and Ovsyannikova S V 2012 Ecological conditions in Kuzbas coal-mining branch: problems and decision ways Mining informational and analytical bulletin No 3 pp 187–194
[7] Golofastova N N, Mikhailov V G and Galanina T V 2016 Environmental safety management of a coal mining enterprise The 8th Russian-Chinese Symposium. Coal in the 21st Century: Mining, Processing and Safety pp 372–376
[8] Shishkin A I 2004 Essence, objectives and principles of monitoring of socio-economic processes in the regional Economy of the North-West: problems and prospects Vol 19 Issue 04 pp 16
[9] Antosenkov E and Petrov O 1998 Monitoring the social and labor sphere Economist Issue 4 pp 34–45

[10] Chesnokova A V, Radina O I and Serdyuk R I 2014 Consumer loyalty as a factor of establishing the competitive advantages in a company under the market conditions Asian Social Science Vol 10 Issue 23 pp 255–260
[11] Rychihina E N 2008 Monitoring role in the formation of long-term plan of socio-economic development of the municipality Regional economy and management: electronic sientific journal Issue 13 pp 23–33 URL: http://eee-region.ru/article/1303/
[12] Chen F, Ishwaran N and Brito Pezzuti J C 2016 Deforestation monitoring in the Amazon River estuary by multi-temporal Envisat ScanSAR data IOP Conf. Series: Earth and Environmental Science 34 012003
[13] Fuentes-Yaco C, Caverhill C, Maass H, Porter C, and White G N III 2016 Monitoring the North Atlantic using ocean colour data IOP Conf. Series: Earth and Environmental Science 34 012008
[14] Mikhailov V G, Koryakov A G and Mikhailov G S 2015 Ecological risk management in coal mining and processing Journal of Mining Science Vol 51 No 5
[15] Efremenko V M, Baumgarten M I and Galanina T V 2012 Power ecological audit at the fuel and energy complex enterprises Mining informational and analytical bulletin No 12 pp 195–198

The Study Of Soil And Agrochemical Features Of Zonal Soils Of Coal Mining Enterprises In Kemerovo Region

M A Yakovchenko, A A Kosolapova, V A Ermolaev

Kemerovo State Agricultural Institute
Stroiteley Av. 50-616, Kemerovo, Russian Federation, 650003

E-mail: mara.2002@mail.ru

Abstract. The paper represents the results of the study of soil and agrochemical features of zonal soils: the grain-size composition, the content of humus, phosphorus and potassium, and heavy metals, the reaction of soil solution of the territory of the open-pit coal mine No12 of Kemerovo region in the areas of the working enterprise. The species composition of the lignose and herbaceous vegetation of the undisturbed territories has been studied. It has been revealed that the fertile soil layer of the studied areas of the open-pit coal mine is characterized as fertile but can't be removed and stored because the surface of the whole area under study is forest-covered very much, rumpled, there are gullies and a lot of wind-fallen trees.

Introduction

Huge coal reserves are concentrated in Kemerovo region, which are estimated as 524.4 billion tons. The total area of Kuznetsk Basin accounts for 27 thousand km^2. Nowadays the coal extraction in the region is conducted by more than 50 mines and 30 open-pit coal mines which extracted more than 310 million tons in 2015. Coal extraction is followed by a significant ecological damage. About 5 800 ha of forests, meadows and fields are destroyed in Kuzbass for a year, and then borrow cuts, rock dumps, technological roads, depositing tanks, etc. appear instead of them.

Therefore, restoration of economic and ecological value of the lands disturbed by mining should be undoubtedly considered as an important question for the territory of Kemerovo region.

The dumps and open pits which are often situated nearby and within the settlements reduce the green area around cities and towns, pollute the environment and herewith worsen living conditions of people. The territories situated in the immediate proximity to the open pits gradually become unsuitable for life [1].

Among negative consequences of the open pits there is withdrawal of huge land areas from agricultural production and their disturbance, change of hydrogeological conditions of the region, development of erosion processes, as well as mixing of rock along with removal of infertile and even toxic rock to the surface [2].

The process of restoration of the disturbed lands is called reclamation, that is, the measures directed to productivity restoration of the disturbed lands, as well as to improvement of the environmental conditions.

The main tasks of the biological reclamation are formation of the soil layer, soil aggregation, humus and nutritional substances accumulation [3,4].

Since the beginning of the biological reclamation of the disturbed lands there begin complicated processes which in total influence on trend and intensity of soil formation with an active participation of plants and microorganisms [5,6].

The research objective

In this context the question of pollution and loss of soil fertility becomes urgent for the territories of the so-called environmentally unfriendly regions which Kemerovo region can be reckoned among. The research objective is to study the agrochemical features of the undisturbed soils nearby the rock dumps of the open-pit coal mines and to determine their quality.

The research object and methodology

The open-pit coal mine Co Ltd «Mine No12» is a working coal mining enterprise which is situated in the town of Kiselevsk in Kemerovo region. In the north the territory of the enterprise borders with the closed mine «Krasnokamenskaya», in the west – with Co Ltd «Uchastok Koksoviy», in the south – with the field of the closed mine «Cherkasovskaya». The region has a developed infrastructure: approach roads, loading dead ends, electric power substations and so on.

The full capacity of the mine accounts for 1 million tons of coal a year. Coal reserves account for 76 million tons of coal. Nowadays coal extraction is done both by underground method at the depth of 250-350 m and by open-cut method – there is a production area of coal extraction by open-cut method.

The enterprise includes: the areas of underground works, the areas of open-pit mining, the crushing-and-sorting complex (CSC), the central and buffer coal stores, technological roads, the boiler plant, the machine shop, the garage, ventilation shafts, degassing units.

A production activity of the subdivisions of the enterprise is performed on the land areas leased from the municipal government. Co Ltd «Mine No12» is one of the enterprises forming a company town of Kiselevsk. The number of people working at mine No 12 accounts for 1755 persons.

The distance from the border of the allotment to the border of the nearest residential areas of Kiselevsk accounts for:

- in the northern direction – the region «Afonino», in the distance of 90 m;
- in the northeastern direction – the region «Boynya», in the distance of 100 m;
- in the eastern direction – the region «Zelenaya Kazanka» and the settlement Shakhta No12, in the distance of 100 m and 20 m correspondingly;
- in the southern direction – the region «Surtaikha», in the distance of 290 m.

The climate of the region is acutely continental. The lowest temperatures are in December and January. The absolute minimum at this time reaches -43.9°C, and in some winters -50-54°C.

The snow cover is stable from the beginning of November till the end of April, and on the northern slopes – till the middle of May. The thickness of the snow cover on the northern slopes of ravines reaches 2 m. The hottest month is July, its average monthly temperature reaches +19.4°C, maximum is +36.7°C.

The average annual precipitation total accounts for 400-700 mm. Prevailing winds in the region are southern and south-western. The winds of these directions have a maximum speed of 17-24 m/s, the average annual speed – 4.9-5.2 m/s.

According to the soil and geographical zoning of Kemerovo region, the lands of the open-pit coal mine №12 are situated in the south-west of the Kuznetsk Alatau, belong to group E – Kuznetsk-Alatau high soil district with four belts of vertical soil zonality.

The zonal soil cover of the soil and geographical region, including the area of ecological investigation on fund materials and soil map of Kemerovo region, is represented by zonal types and subtypes of soil for the given geographical region:

- sod-medium-podzol;
- mountain-sod-podzol;
- light-grey forest;
- grey forest.

Soil as any natural substance has some external features - morphology. The most important morphological feature of the soil is an external structure of its genetic horizons forming soil section [7,8].

On the area under study nearby the working enterprise of the open-pit coal mine No 12 the soil sections were done: plot No 1 (picture 1, 2), plot No 2, plot No 3. The samples for agrochemical analysis were selected in July 2016 according to State Standard 17.4.4.02-84 «Nature protection. Soils. Methods of samples selection and preparation for chemical, bacteriological, helminthological analysis».

Results and discussion

Plot № 1. The square is 47 ha. The relief: a slope of the northern aspect (5-7°). The surface is rumpled, with gullies along the dells. There are tussocks, and anthills. The soil-forming and bed rock: heavy brown noncalcareous covering loams and clays. The soil: deep-sod, deep-podzolic, heavy loam.

Figure 1 – A general view of the landscape(plot 1) deep-podzolic heavy loam

Figure 2 – The soil – deep-sod

In the soil profile the following soil horizons are revealed:
$A_1 - A_2 - A_2B - B - BC - C$.

Table 1 – Soil horizons of plot No 1

A_1 - 0-18	Dry, grey, heavy loam regarding granule-size composition, fine clot structure, loose, plenty of plant roots, noticeable transition in density.
A_2 18-25	Dry, whitish-grey, heavy loam, clot-laminated structure with whitish overburden, compacted, rare plant roots, noticeable transition in density.
A_2B-25-40	Fresh, light-brown with whitish shade, heavy loam, rough-nuciform with noticeable siliceous overburden, compact, rare roots, gradual transition.
B - 40-61	Damped, light-brown, heavy loam, prism-like structure, very compact, gradual transition in colour.
BC – 61-90	Damped, light-brown, heavy loam, cloddy, compact, gradual transition.
C 90 - 154	Damped, light-brown, heavy loam, unstructured, compact, carbonate horizon is not opened.

According to the soil classification and diagnosis (1977) the names of soils with respect to the granule-size composition are given regarding top humus soil horizon.

The solid phase of soils and soil-forming rocks consists of the particles of different size which are called mechanical elements. With respect to correlation of physical clay (particles <0.01mm) and physical sand (particles > 0.01mm) a granule-size (mechanical) soil composition is determined.

The species composition of plant communities in the studied territory of plot №1 of the field of mine No12 is represented by 37 species of 21 families. The territory is forest-covered, with silver birch prevailing (*Betulapendula*), with rare representatives of *Píceaábies,* bird cherry *(Padusavium),* cranberry high (*Viburnumopulus, Acernegundo*), red raspberry (*Rubusidaeus*). A significant part of grass vegetation is represented by perennials [8, 9].

Plot No 2. The square is 6.5 ha. The soil profile No 2. The relief: a slope of the southern aspect (4-6°). The surface is flat, hollows are found. The soil-forming and bed rock: stone field. The soil: light-grey podzolized undeveloped-skeleton.

In the soil profile the following soil horizons are revealed:

A_l – B – BC – C_c

Table 2 – Soil horizons of plot No 2

A_l - 0-14	Dry, grey, medium-loam, fine-nuciform structure, loose, plenty of plant roots, noticeable transition in colour and inclusion of skeleton rock.
B 14-32	Dry, grey, nonhomogenously coloured, medium-loam, pulverescent-cloddy-friable structure, loose, solid inclusions in the shape of aleurites and others, gradual transition.
BC-32-63	Dry, brown-grey with brown inclusions, cloddy-friable, loose-compact, solid inclusions in the shape of small and large aleurites, noticeable transition in density.
C_c63 cm.	Dry, a mixture of solid inclusions in the shape of small and large aleurites with sand and clay.

The vegetation of plot No 2 is represented by 30 species of 16 families. The whole territory of the field is forest-covered with silver birch (*Betulapendula)* forming dense stand of trees, along the edge of the field there are vast bushes of pea shrub (*Caragana pygmaea)* [8,9].

Plot №3. The square is 4.5 ha. The relief: a slope of the northern and north-eastern aspect (4-6°). The surface is cut, by gullies and hollows. The soil-forming and bed rock: heavy brown noncalcareous covering loams and clays. The soil: deep-sod fine-podzolic heavy loam.

In the soil profile the following soil horizons are revealed:

A_l – A_lA_2 – B – BC – C

Table 3 – Soil horizons of plot No 3

A_l - 0-26	Dry, grey, heavy loam, cloddy-nuciform structure, loose, plenty of tree and grass roots, noticeable transition in structure.
A_lA_2 -26-37	Dry, grey, heavy loam, cloddy-laminated structure, loose-dense, plant roots, noticeable transition in density.
B – 37-51	Damped, brown-grey, heavy loam, coarse-cloddy-prismlike, siliceous overburden (SiO_2), dense, gradual transition.
BC – 51-80	Damped, brown, heavy loam, cloddy-prismlike structure, dense, gradual transition.
C – 80 -130	Damped, brown, heavy loam, unstructured, dense, with rusting in the top part of the horizon, with gleization in the bottom part of the horizon.

The vegetation of plot No 3 is represented by 30 species of 16 families. The whole territory of the field is forest-covered with silver birch (*Betulapendula)* forming dense stand of trees, along the edge of the field there are vast bushes of pea shrub (*Caragana pygmaea)* [8,9].

According to the classification and analysis data the granule-size composition of medium-sod fine-podzolic soils of plots No 1, No 2, No 3 of the field of mine No 12 by the method of N.A. Kachinskiy is presented in table 4.

Table 4 – Granule-size composition of the plots by the method of N.A. Kachinskiy

Plot	Soil name	Content of fractions, %	
		Fine sand	dust
No1	Coarse-pulverescent silty heavy loam	15.3	39.4
No2	Light loam	14.3	30.3
No3	Coarse-pulverescent silty heavy loam	15.3	39.4

The availability of basic elements of nutrition in the soil was determined according to State Standard 26204-91 «Determination of labile phosphorus, potassium and nitrogen compounds by the method of Chirikov in the modification of Central Institute of Agrochemical Service of Agriculture». The investigation results are presented in table 5.

The reaction of soil solution in the top horizon 0-14 cm was determined according to State Standard 26483-85 «Soils. Preparation of salt extraction and determination of its pH by the method of Central Institute of Agrochemical Service of Agriculture». The investigation results are presented in table 5.

Table 5 – Agrochemical parameters of the plots

Plot	Humus, %	pH	Content of nutrition elements		
			Nitrogen, %	Phosphorus, %	Potassium, mg/kg
No1	6.2	5.8	0.48	0.13	50.2
No2	3.1	5.2	0.31	0.13	50.0
No3	5.0	5.6	0.12	0.13	48.2

Conclusions. The analysis of the investigation data shows that humus amount in the top soil horizon (A) 3.1-6.2% falls sharply as the depth increases. In the transitional horizon BC only humus traces were found.

Total phosphorus content 0.13% in the soil is low, it falls sharply as the depth increases. Total potassium content is medium through the whole profile, and nitrogen content in all the plots varies from 0.12 to 0.48%, which is undoubtedly connected with a natural-climatic location of the plots.

According to the study results of the plots of the open-pit coal mine of the territory of Co Ltd Mine No12 of the town of Kiselevsk in Kemerovo region the conclusion is as follows:

- the surface of soil cover of plots No1 and No3 is represented with deep-sod deep-podzolic and fine-podzolic heavy loam soils, and that of plot No2 – with light-grey undeveloped-skeleton light-loam soils;
- the reaction of soil solution is from weak-acid to strong-acid in the bottom horizons.

The content of labile forms of heavy metals in the soil samples doesn't exceed APC (mg/kg) their content in the soil taking into account Clarke [10].

The fertile soil layer of the studied areas of the open-pit coal mine is characterized as fertile but can't be removed and stored because the surface of the whole area under study is forest-covered very much, rumpled, there are gullies and a lot of wind-fallen trees.

References

[1] Yakovchenko M. The Study of Soil Protection in the Sistem of the Cultivated Lands of Kemerovo Region / IOP Conf. Series: Materials Science and Engineering 91(2015)012078 doi:10.1088/1757-899x/91/1/012078.

[2] Růžek L. Chemical and biological characteristics of reclaimed soils in the Most region (Czech Republic) / Plant Soil Environ., 49, 2003 (8):346-351.

[3] Micanova O. Utilization of Microbial Inoculation and Compost for Revitalization of Soils / Soil and Water Res., 4,2009 (3): 126-130.

[4] Zhonqiu Zhao. Soils development in opencast coal mine spoils reclaimed fo 1-13 years the West-Northern Loess Plateau of China / European Journal of Soil Biology 55 (2013) 40-46.

[5] Blaylock M J. Enhanced accumulation of Pb in Indian mustard by soil-applied chelating agents / M.J. Blaylock, D.E. Salt, S. Dushenkov et al. // Environ. Sci. Technol. 1997. V.31. 3. P.860-865.

[6] Remes J. Biological activity of anthropogenic soils after spoil-bank forest reclamation / Journal of Forest Science, 53, 2007 (7): 299-307.

[7] Pietrzykowski M. Soil and plant communities development and ecological effectiveness of reclamation on a sand mine cast / Journal of Forest Science, 54, 2008 (12): 554-565.

[8] Růžek L. Slit seeded grass-legume mixture improves coal mine reclamation / Plant Soil Environ., 58,2012 (2): 68-75.

[9] Ibanez J J. Future of soil science / J.J. Ibanez // The future of soil science / Ed. A.E. Hartemink. Wageningen: IUSS,2006. P.60-62.

[10] Pietrzykowski M. Linking heavi metal bioavailability (Cd, Cu, Zn and Pb) in Scots pine needles to soil properties in reclaimed mine areas / Science of the Total Environment 470-471 (2014) 501-510.

Main Pipelines Corrosion Monitoring Device

Bazhenov Anatoliy, Bondareva Galina, Grivennaya Natalia, Malygin Sergey, Goryainov Mikhail

Technological Institute of Service (branch) Don State Technical University
Prospect. Kulakov d.41/1, Stavropol, Russian Federation, 355035

E-mail: katrinastenton@yandex.ru

Abstract. The aim of the article is to substantiate the technical solution for the problem of monitoring corrosion changes in oil and gas pipelines with use (using) of an electromagnetic NDT method. Pipeline wall thinning under operating conditions can lead to perforations and leakage of the product to be transported outside the pipeline. In most cases there is danger for human life and environment. Monitoring of corrosion changes in pipeline inner wall under operating conditions is complicated because pipelines are mainly made of structural steels with conductive and magnetic properties that complicate test signal passage through the entire thickness of the object under study. The technical solution of this problem lies in monitoring of the internal corrosion changes in pipes under operating conditions in order to increase safety of pipelines by automated prediction of achieving the threshold pre-crash values due to corrosion.

1. Introduction

Nowadays oil and gas industry gives more and more attention to the safety of production, transportation and storage of oil and gas worldwide. It's important both for the human beings and their environment. At the same time, despite the development of methods and ways of ensuring industrial safety in oil/gas industry and success achieved, oil and gas pipeline accidents and failures are still a problem, and the associated environmental problems still occur [1].

So, research in protection against corrosion [2] [3] and creating devices for the pipeline corrosion control are quite urgent [4] [5] [6] [7] [8] because of a significant number of accidents due to pipeline corrosion. The possibility to determine the critical corrosion level and changes in a pipe on the basis of instrumental and analytical forecast [9] allows not only to provide well-timed replacement of pipeline sections but also to substantiate the use of specialized electrical, chemical or mixed corrosion prevention methods.

The main methods of gathering information about the corrosion changes in the linear part of oil / gas main pipelines are smart pigging and electrometric measurements with subsequent instrumental survey of the condition of metal and pipe insulation at the inspection points. Pipeline pigging is the most informative method for obtaining data on the extent of damage in the pipeline inner walls but only 40% of gas pipelines are prepared to pass the corrosion detection pig. Besides, the main pipeline is out of service while pigging. Electrometric measurements are aimed at determining the main indicator of corrosion – the corrosion rate, expressed as weight loss of metal per unit area per unit time Kм,г/(m^2·year) or a rate of corrosion penetration per time unit KΓ,mm/year [10] [11]. Automated remote corrosion monitoring on the basis of the electrometric measurements is able to only monitor external corrosion processes and therefore it is not accurate enough to perform predictive analytics [12] [13].

The main problem of monitoring corrosion changes in pipelines is the difficulty of measuring the thickness of ferromagnetic conducting pipes directly when under service. Primary data collection devices must ensure continuous monitoring of a wall thickness corrosive wear in automatic mode without using the pipeline pigs and digging. As such smart transmitter the main pipeline corrosion monitoring device [14] can be used.

2.Method and structural scheme

The essence of technical solution is that the drive (exciting) and measuring inductance coils covering the pipeline are placed at the most deteriorated section of the main pipeline. So, the pipe acts as a magnetic conductor of a transformer eddy-current transducer (eddy-current probe), and EMF induced in the measuring coil will depend on the pipe parameters and its thickness in particular. Periodic EMF measurements, accumulation of measured values in digital memory and comparison with design /target/ values make it possible to determine the residual life of a pipe wall and to predict the moment of dramatic loss of wall thickness up to failure.

The technical implementation of the device involves using two drive inductance coils (exciting coils) connected in series in Helmholtz arrangement and measuring coil placed between them; frequency-tunable test-signal generator, pipe temperature sensor and digital computer. The EMF induced in a measuring coil is pre-amplified by a broadband amplifier and is filtered by a tunable bandpass filter to increase noise immunity and protect from random interference.

Figure 1. gives the main pipelines corrosion monitoring device block diagram.

The main pipelines corrosion monitoring device (Figure 1) includes a digital computer consisting of central processor 1, random access memory 2, read only memory 3, analog-to-digital converter (ADC) 4 and input/output port 5; tunable harmonic oscillator 6, broad-band amplifier 7, narrow-band tunable filter 8, first 11 and second 13 driving coils, measuring coil 12 and digital temperature sensor 9.

After placing the measuring and drive coils on a pipeline (winding can be performed together with pipe insulation) the monitoring device is set up, it involves production of harmonic oscillations in a given frequency range (e.g., 20 to 2000 Hz). The choice of frequency range is determined by the necessity of penetration of the magnetic field into the depth of the pipe wall. Frequency tuning of a test signal is performed by a central processor. Voltage, proportional to the EMF of mutual induction is amplified by a broadband amplifier and fed to a narrow-band filter which is frequency tuned along with the generator. Next, the current or peak stress voltage value is converted into digital code and stored in RAM of a digital computer. As specific conductivity of structural pipe steel of depends on its temperature, together with the measurement of output voltage, measurement of the pipe temperature by the digital sensor takes place. The measured temperature is stored in RAM. The operating mode of the device is periodic, short-time or continuous. Data exchange with external devices is provided upon request of external devices or without the request via the programmed time intervals. The device supports interface Ethernet, Modbus protocols, as well as data transfer using a cellular GPRS.

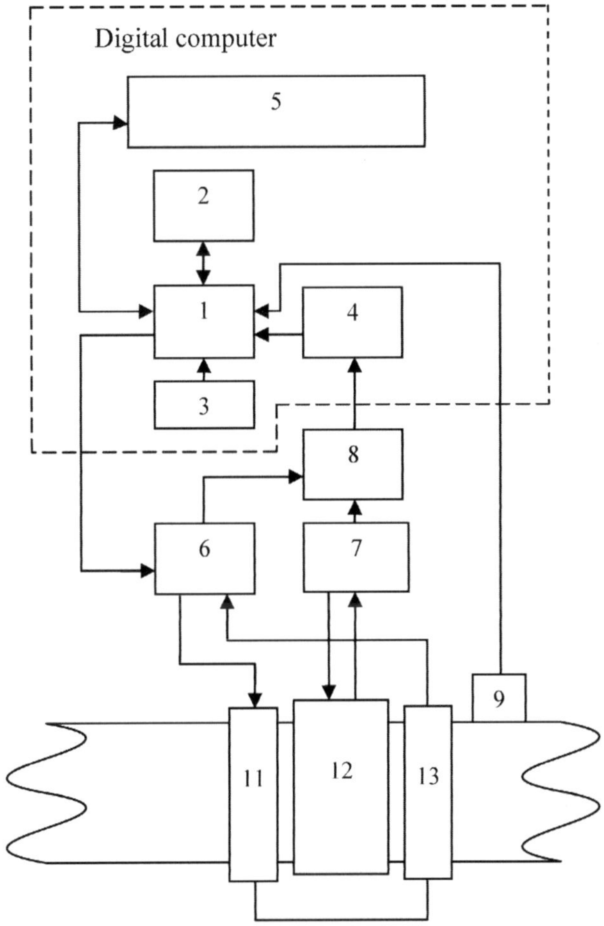

Figure 1 – The main pipelines corrosion monitoring device block diagram (explanation in the text).

3. Numerical modeling

Next, the calculation of EMF using the formula (1) is performed for the same frequency range for which the measured data were obtained,

$$\dot{E} = -j\pi\mu_0\omega w_u \frac{\dot{I}w_b}{l_b} R_u^2 \left(1-\eta+\eta\mu_a\dot{\mu}_{aff}\right), \tag{1}$$

where $\dot{I}$ – the excitation (drive) current; l_b – length of the exciting (drive) coil; w_u, w_b – the number of turns in the measuring coil and the exciting coil; ω – angular (circular) frequency of harmonic excitation signal; μ_a – relative magnetic permeability of pipe steel; $\dot{\mu}_{aff}$ – effective magnetic permeability, which determines the degree of attenuation of the magnetic flux due to eddy currents; $\eta = \left(R_1 / R_u\right)^2$ – the fill factor defined by the ratio of the cross sectional area of the test object and tube of magnetic flux, linked with the measuring coil; R_1 – the external radius of the pipe; R_u – the radius of the measuring coil.

Figure 2 shows approximate graphs of the calculated by the formula (1) and measured EMF induced in the measuring winding.

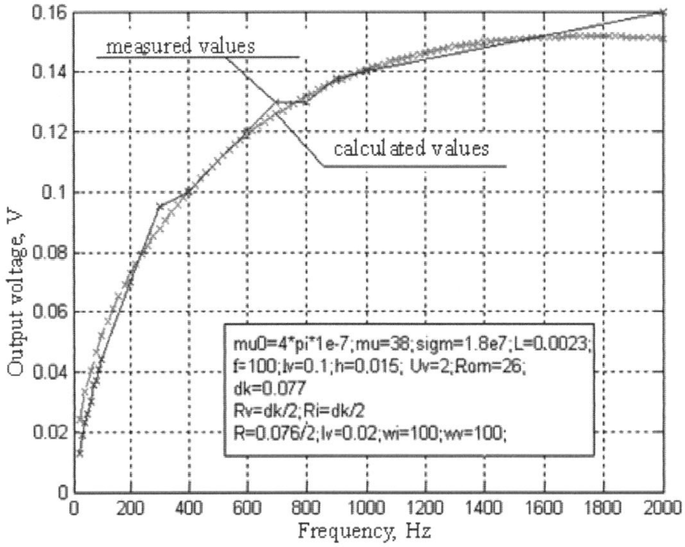

Figure 2 – The dependence of the EMF induced in the measuring coil upon the frequency of the harmonic excitation signal.

To monitor corrosive changes in the pipe wall it's necessary to periodically measure the voltage induced in the measuring winding in the same frequency range as that used in setting up. After that the calculation by the formula (2) is carried out, and with that the value of specific conductivity σ is corrected taking into account the pipe temperature at the time of measurement. One must try to achieve coincidence of the measured and expected graphs by changing the pipe wall thickness of a tube in calculating equations. According to several cycles of measuring the thickness of the pipe we construct a regression model that makes it possible to predict the moment of the wall thickness loss to pre-failure condition. The results of the automatic control of the pipeline internal corrosion changes can be transferred to the operator or radio modem via the input/output port.

The eddy currents density is maximum at the surface of the object in circuit, with a diameter close to the diameter of the exciting coil and decreases to zero on the axis of the eddy-current probe and at $r \to \infty$. Eddy current density also decreases with depth of the test object. For approximate estimation of the penetration depth of the electromagnetic field of surface eddy-current probe one can use the formula for the depth of penetration δ (м) of a plane wave:

$$\delta = \sqrt{\frac{2}{\omega \mu_a \sigma}}, \quad (2)$$

where ω – angular (circular) frequency of the excitation current of eddy current probe; μ_a – absolute magnetic permeability, H/m; σ – specific electrical conductivity of the test object material, Cm/m.

Figures 3, 4 present the results of calculating the penetration depth of the electromagnetic wave into a conductive medium, calculated according to the formula (2).

The value δ corresponds to the attenuation of the magnetic field by a factor of e compared to the value of magnetic field intensity on the surface of the object.

As follows from the content of the Figures 3,4 at the operating frequencies of the monitoring device (80–200 Hz) penetration depth is sufficient for the control of corrosion processes in pipes with up to 2 cm wall thickness.

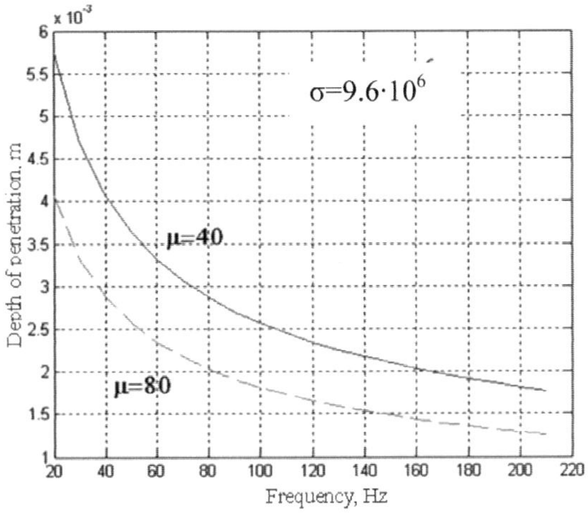

Figure 3 – The penetration depth of an electromagnetic wave depending on the relative magnetic permeability.

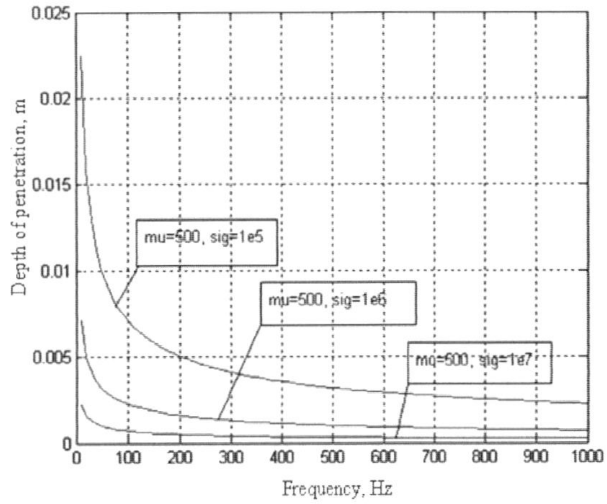

Figure 4 – The penetration depth of electromagnetic waves depending on the specific conductivity.

4.Conclusions.

The pilot studies of the main pipelines corrosion monitoring device have been carried out; the studies confirmed the possibility of detecting local and extensive pipe defects, including the inside of a pipe, in a wide range of experimental conditions.

References

[1] Evgeny A Shvarts, Alexander M Pakhalov and Alexey Yu Knizhnikov 2016 Assessment of environmental responsibility of oil and gas companies in Russia: the rating method J. of Cleaner Production. V 127 pp 143-151

[2] C Sean Brossia 2014 11–The use of probes for detecting corrosion in underground pipelines In Underground Pipeline Corrosion ed Mark E Orazem (Woodhead Publishing) pp 286-303

[3] Branko N Popov and Swaminatha P Kumaraguru 2012 25 – Cathodic Protection of Pipelines In Handbook of Environmental Degradation of Materials ed Myer Kutz (William Andrew Publishing/Oxford) pp 771-798

[4] Carvalho A A, Rebello J M A, Souza M P V, Sagrilo L V S, Soares S D 2008 Reliability of non-destructive test techniques in the inspection of pipelines used in the oil industry International Journal of Pressure Vessels and Piping. V 85 pp 745-751

[5] Glisic B 2014 14 – Sensing solutions for assessing and monitoring pipeline systems Sensor Technologies for Civil Infrastructures (Woodhead Publishing Series in Electronic and Optical Materials). V 56 pp 422-460

[6] Hopkins P 2014 3 – Assessing the significance of corrosion in onshore oil and gas pipelines In Underground Pipeline Corrosion, ed Mark E. Orazem (Woodhead Publishing) pp 62-84

[7] Shantanu Datta and Shibayan Sarkar 2016 A review on different pipeline fault detection methods, J of Loss Prevention in the Process Industries V 41 pp 97-106

[8] Jiles D C 1990 Review of magnetic methods for nondestructive evaluation (Part 2) NDT International. V 23 I 2 pp 83-92

[9] Lam Hong Lee, Rajprasad Rajkumar, Lai Hung Lo, Chin Heng Wan and Dino Isa 2013 Oil and gas pipeline failure prediction system using long range ultrasonic transducers and Euclidean-

Support Vector Machines classification approach Expert Systems with Applications. V 40 I 6 pp 1925-34

[10] James G Speight, Chapter 2014 6 – Corrosion Monitoring and Control In Oil and Gas Corrosion Prevention (Gulf Professional Publishing/Boston) pp 109-149

[11] Carlos Eduardo de Almeida Souza Torres, Cíntia Gonçalves Fonseca Costa, Amanda Petronilha Pereira, Maria das Mercês Reis de Castro and Vanessa de Freitas Cunha Lins 2016 Corrosion failure analysis in a biodiesel plant using electrical resistance probes Engineering Failure Analysis. V 66 pp 365-372

[12] Yunze Xu, Huang Yi, Xiaona Wang, Xuanqin Lin 2016 Experimental study on pipeline internal corrosion based on a new kind of electrical resistance sensor Sensors and Actuators B: Chemical. V 224 pp 37-47

[13] Scherbinin V E; Shur M L 1987 Experience gained in the use of the magnetographic method of NDT of butt welds in oil and gas pipelines Proceedings of the 4th European Conference on Non-Destructive Testing (London/United Kingdom). V 2 pp 1110–16

[14] Pat. 2012125083 Russian Federation MPK F17D 5/00. Method to monitor internal corrosive changes of manifold pipeline and device for its realisation / Bazhenov A V, Malygin S V, Fedorenko V V, Kurilov A I. Claimer and owner ZAO KIEhP "Ehnergomera" – № 2012125083/06 claimed 15.06.12; published 20.12.2013 Bull. № 35.

IOP Conf. Series: Earth and Environmental Science **50** (2017) 012033 doi:10.1088/1755-1315/50/1/012033

Dynamics Analysis of Wind Energy Production Development

V I Berg[a], A G Zakirzakov[b], E F Gordievskaya

Tyumen Industrial University
Volodarsky str. 38, Tyumen, Russian Federation 625000

E-mail: [a]Allbert-@mail.ru, [b]Petrov@mail.ru

Abstract. The paper presents the analysis of the introduction experience and dynamics development of the world wind energy production. Calculated the amount of wind energy sources investments and the production capacity growth dynamics of the wind turbines. The studies have shown that the introduction dynamics of new wind energy sources is higher than any other energy source.

Introduction

In 2015, the government of the Russian Federation approved the concept of energy-efficient low-hydrocarbon way of industrial economy sectors development for the period till 2020. During this period the reduction of the greenhouse gas emissions is being planned up to the level not less than 25% of the emissions' level in 1990. Taking into consideration the current economic and technological conditions of the development in Russia, namely, the record hydrocarbon production and the established embargo on the high-tech equipment purchase in foreign companies, the realization of the greenhouse gases level reduction concept is extremely difficult [1].

One of the ways of the emissions reduction and at the same time the solution of the economy diversification is the development and use of alternative (renewable) energy sources on the territory of the Russian Federation with the possibility to use the electricity for the internal needs and for sale to the neighbouring countries, including the Scandinavian Peninsula [2].

The countries of the Scandinavian Peninsula are major importers of electricity from the Russian Federation. During the period from 2006 to 2012 Russia has exported in total more than 70 billion kW·h of electrical energy to Finland, however since 2012 there has been a decrease in electricity demand from Russia, due to the constant cost increasing of electricity in Russia and the comparative cheapness of electricity on the Nordic market thanks to the generating capacity increasing the cost of alternative energy sources use [3]. The growth of prices in Russia is stipulated by the liberal reforms in the energy sector and by the formation of free prices.

The analysis of the electricity prices growth in Russia has shown that the average annual price increase is almost 9%; this fact indicates that there is a necessity to improve energy efficiency. If we take into account the requirements to the greenhouse gas emissions reduction, there will be further increase of electricity prices.

The wind energy industry is a promising direction to develop. According to the Russian Association of Wind Energy Industry (RAWI) the Russian Federation has the world's largest wind energy capacity, constituting more than 40 billion kWh of electricity per year. So the Northern regions such as the Kola Peninsula, located near the Scandinavian countries, the Gulf of Ob, the coastal territory of the Far Eastern Federal district are the wind zones with the highest wind capacity coefficient [4, 5]. The wind speed at the height of modern wind equipment, calculated according to the annual average values, is 11-13 m/s. It exceeds the economic payback threshold of wind energy twice. Also it is very important to take into account the low rating of environmental damage from wind energy, so the total environmental damage from the wind energy is significantly lower compared to "traditional" ways of energy generating. In Europe the external negative socio-environmental impact of 1 kWh of the produced electricity is estimated at 0.15 cents for wind energy, 1.1 cents for gas power plants and 2.5 cents for coal.

The aim of the article is the analysis of the introduction experience and the dynamics development of the world wind energy production.

Materials and methods

The analysis of the empirical material the period 2004 – 2015 and news articles were used as our primary empirical data to study: renewables - global status report and renewable energy capacity statistics.

Results and Discussion

From 2004 to 2015, the total amount of investments in the wind energy sources amounted to more than $ 2.6 trillion, and the dynamics of investments is constantly increasing, with the exception of 2012 and 2013. In the table 1 and figure 1 there are shown the annual investments and the topics of their growth.

Table 1 - The amount of wind energy sources investments.

Year	The volume of investments, bln. $	The increase as compared to the previous year
2004	62	-
2005	88	42
2006	128	46
2007	175	36
2008	206	18
2009	207	0.8
2010	274	32
2011	318	16
2012	297	-7
2013	272	-8
2014	316	16
2015	329	4

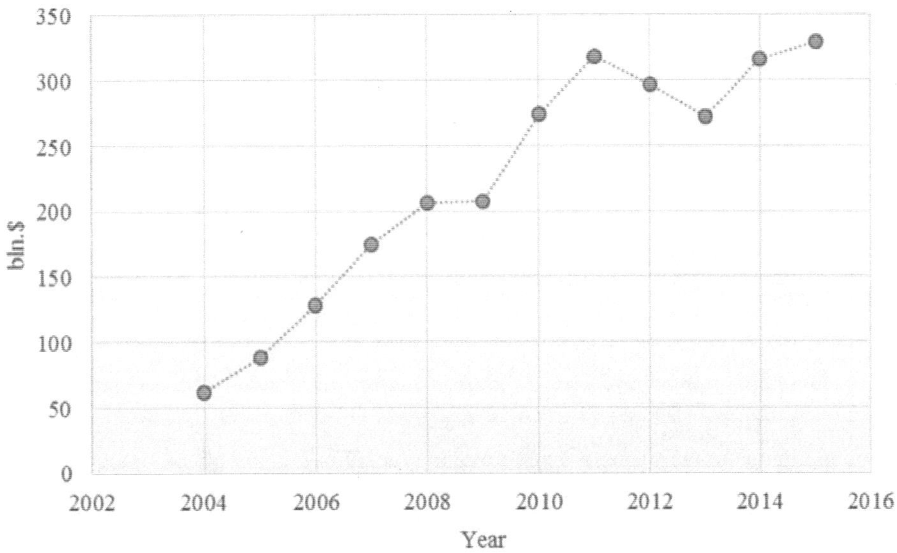

Figure 1 - The amount of wind energy sources investments.

The graph shows that investments have positive dynamics and they change depending on the political courses of the countries, developing wind energy and world energy prices. In contrast to the investment's dynamics, the growth of wind turbines production capacity has only a positive trend. In 2014 there was introduced a record number of production capacity equal to 51 GW, which is 44% more than in 2013 [6]. Currently, the global production capacity of wind energy industry can be estimated as 370 GW. Figure 2 presents the production capacity growth dynamics.

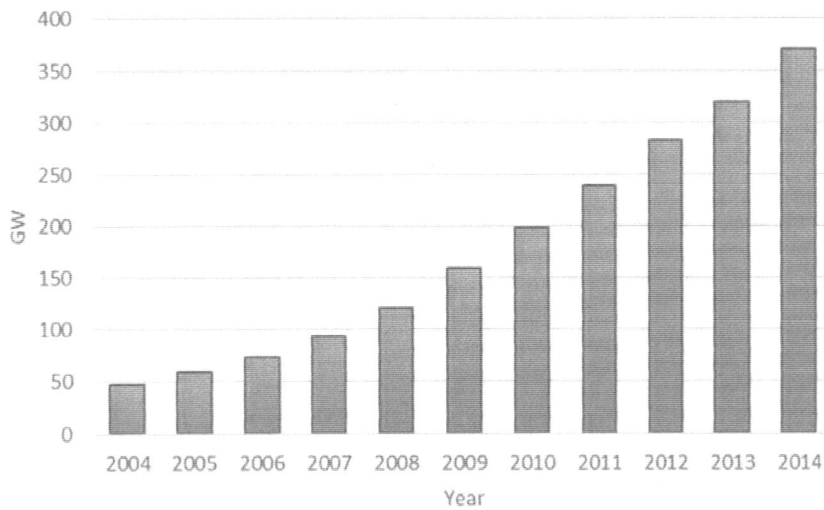

Figure 2 - The production capacity growth dynamics of the wind turbines

85% of all wind energy industry production capacity is given to the top ten leading countries, but the growth analysis has shown that about 80 countries put into operation on-site wind energy sources at the end of 2014 [7, 8]. So, Asian countries remain the largest producers of wind energy accounting for 50% of all productivity in the world, the second place is given to the EU with 23% in 2014, the third place ranks North America with 13%. Figure 3 presents the dynamics of production capacity growth of ten leading countries for 2014 year.

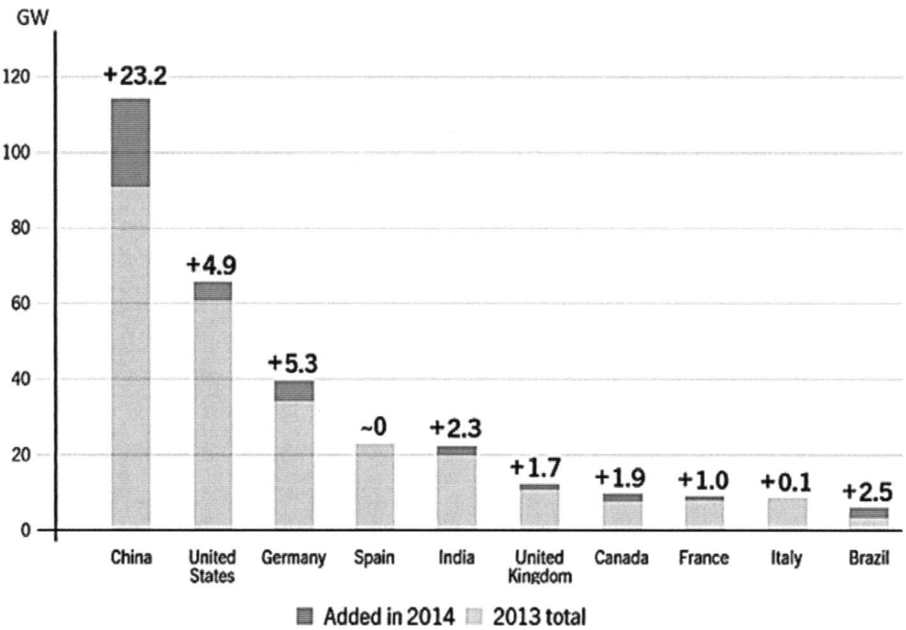

Figure 3 - The production capacity growth dynamics of ten leading countries in wind energy industry

At the end of 2014 Asian countries ranked the first place for production capacity amount of wind power plants. The EU countries despite the increase by 4% took the second place, but in general, the wind energy generation took about 44% of the market among all the rest alternative sources of energy. Figure 4 show the distribution of production capacity growth in the total amount of energy.

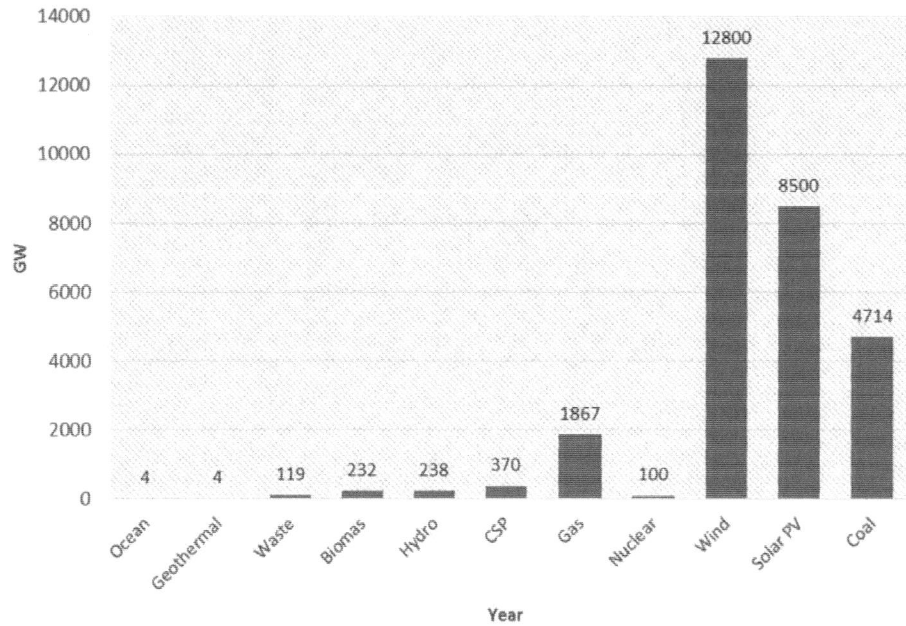

Figure 4 - The distribution of production capacity growth in the EU in the total energy sources amount.

The largest increase is given to Germany and the UK. These countries provided an increase of 59% of the production capacity of the entire Eurozone. This increase is provoked by the policy aimed at the development of "green energy" and relatively calm political situation. Thanks to the coordinated actions and large investments wind energy in 2014 took 7.5% of the total energy balance of the Eurozone, and according to 2015 year data it is more than 10%. The countries such as Denmark, Ireland, Portugal, Spain provided more than 20% of their own energy consumption by means of wind energy. Denmark, Sweden, Spain, Ireland and Germany became the leading countries in the growth of generation per inhabitant [9,10]

In 2015 the European Union introduced 29 GW of new production capacity of wind energy. It is interesting to mention that in 2015 there was not built any station for generating energy from peat and fuel oil, but total energy generation from coal reduced by 8 GW, gas by 4 GW, fuel oil by 3 GW, nuclear power by 1.8 GW.

In 2014, China introduced additional 23.2 GW of production capacity of wind power plants, 20.7 GW of which were connected directly into the national energy system, which is a record for the growth among all the countries. Currently in China, there are about 115 GW of energy generated by wind energy plants, or it is 3% of the total production. So, the region of Inner Mongolia provides 21% of energy by means of wind turbines, Gansu provides in such a way 10.3%, and Xinjiang - 8.1%.

The main problems, affecting the development of wind energy in China, are the difficulties with the lack of the infrastructure preparedness and delayed building, also, here we can include the necessity to integrate a large amount of energy in the national energy network.

Starting with 2011 the capital expenditures for the construction of wind power plants have had a downward trend, which is ensured by the development of competition and technological innovations introduction that have led to the capacity increase. It is estimated that wind turbines located on the land are economically competitive with coal and gas power plants without taking into account the compensation costs and state support. The most economical from the operation point of view are high-power wind turbines located on the sea shelf.

Conclusion

As it can be observed from the analysis, the introduction dynamics of new wind energy sources is higher than any other energy source. Governmental support of European and Asian countries in the field of renewable energy, high rates of technological progress, as well as the instability of the prices for hydrocarbons, set wind energy up in the rank of the future traditional energy in the medium term. Globally, in 2014 the production capacity generated by the wind stations around the world is enough to satisfy at least 3% of the total electricity consumption in the world that tells about the growing influence of "green energy".

References

[1] European Automobile Manufacturers Association, “Natural Gas Vehicles Increased in Europe,” press release (Brussels: 30 April 2015).

[2] Isabel Lane, “Navy’s NAVAIR Complete Supersonic ATJ Flight with Gevo’s Fuel,” Biofuels Digest, 29 December 2014.

[3] Yuning Zhanga, Ningning Tanga, Yuguang Niub, Xiaoze Dua 2016 Wind energy rejection in China: Current status, reasons and perspectives J. Renewable and Sustainable Energy Reviews 66 322–344.

[4] China National Renewable Energy Centre. Renewable energy data manual 2015 (in Chinese). 1st ed. Beijing: China National Renewable Energy Centre; 2015.

[5] Akdağ S A, Güler Ö 2015 A novel energy pattern factor method for wind speed distribution parameter estimation J. Energy Conversion and Management 106 1124–1133.

[6] Heydarian-Forushani E, Golshan M E H, Moghaddam M PShafie-khahc M, Catalão J P S 2015 Robust scheduling of variable wind generation by coordination of bulk energy storages and demand response J. Energy Conversion and Management 106 941–950.

[7] Nykampa S, Molderinka A, Hurinka J L, Smita G J M 2012 Statistics for PV, wind and biomass generators and their impact on distribution grid planning J. Energy 45 (1) 924–932.
[8] Kazeta M Y, Mouangue R, Kuitche ANdjaka, J M 2016 Wind Energy Resource Assessment in Ngaoundere Locality J. Energy Procedia 93 74–81.
[9] Sedefian L 1980 On the vertical extrapolation of mean wind power density J. Applied Meteorology 19 488–493.
[10] Bataineh K M, Dalalah D 2013 Assessment of wind energy potential for selected areas in Jordan J. Ren Energy 59 75–81.

IOP Conf. Series: Earth and Environmental Science **50** (2017) 012034 doi:10.1088/1755-1315/50/1/012034

The Analysis of the Development Dynamics and Structural Balance of Solar Energy in the World

A E Brand[a], S M Chekardovskiy and K A Akulov

Tyumen Industrial University
Volodarsky str. 38, Tyumen, Russian Federation 625000,

E-mail: [a]Allbert-@mail.ru

Abstract. The paper presents data the analysis of the development dynamics and structural balance of solar energy in the world. In the article presents information about total installed production capacity of solar energy, the world solar energy production capacity distribution and the European Union energy market structure in 2000 and 2015 years.

Introduction

Currently, the countries potentially or de facto politically dependent on the world prices for hydrocarbon resources and politics of the countries exporting hydrocarbons, are actively developing and introducing renewable energy sources such as solar, wind, water, biomass, with the aim of improving their energy security, the ecological situation and introducing the so-called "Green economy"[1].

According to the 2015 year data more than 100 countries are introducing renewable energy sources in their national energy systems. The most popular among these sources are solar and wind energy. Thanks to the technological progress the following important factors have been obtained:

1. In 2016, there has been created solar battery, converting solar energy with an efficiency of 34.5%, this fact was considered impossible previously and thought to be reached not earlier than in 2050.
2. In 2016, there has been achieved a surplus of electricity generation from renewable energy sources in Germany, thanks to it, gas-fired power plants were temporarily disabled.

According to [2] solar energy is more promising and high-tech than wind or water energy. The beginning of the solar energy birth is considered to be 1973 - the year of the first energy and oil crisis. During the crisis all the Arab countries, the members of the OAPEC, Egypt and Syria ceased to supply oil to the countries such as the UK, Canada, Netherlands, the USA and Japan because they supported Israel during the hostilities with Syria and Egypt [3-5].

The oil crisis of 1973 was the first and the biggest energy crisis in the history, when the hydrocarbon resources recovery rates were used to create political pressure on the world community. Later in 1975, 1979 and 1985 the EEC adopted a program to stimulate the development of non-nuclear energy; the central position in the program was given to solar energy. Currently, the budget of the Russian Federation directly depends on the world hydrocarbons prices. The revenue diversification and share expanding of Russia on the European energy market, particularly solar energy, will contribute to improving of the investment climate and a more dynamic development of the internal energy market [6, 7].

The aim of the article is the analysis of the development dynamics and structural balance of solar energy in the world.

Materials and methods

The analysis of the empirical material the period 2004 – 2015 and news articles were used as our primary empirical data to study: renewables - global status report and renewable energy capacity statistics

Results and Discussion

The using of the world experience and available technological advances is an integral aspect of the country development[8]. A significant solar energy aspect is the market dynamics and its total production capacity. There were about 179 GW of solar energy in 2015 according to the European Association of solar energy. Figure 1 presents the development and increasing production capacity dynamics.

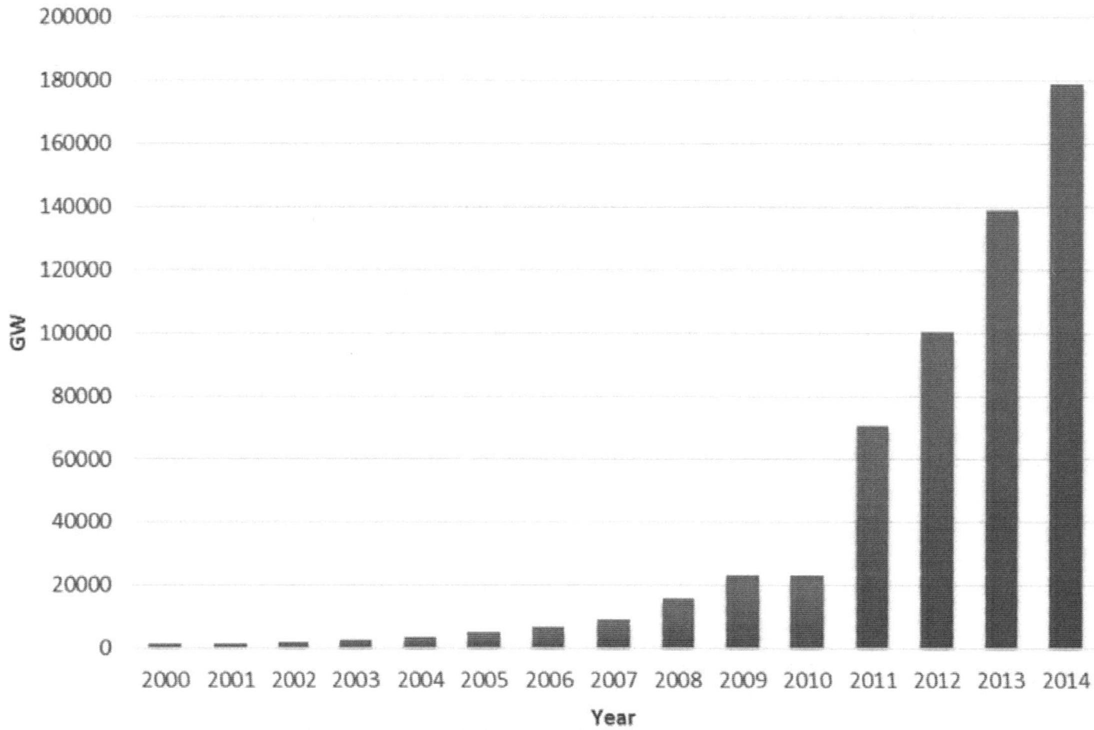

Figure 1 - Total installed production capacity of solar energy

As the graph shows, the production capacity growth dynamics is positive that indicates priority and dynamic development. China, Japan, Germany, Italy, the USA are the leading countries in solar energy production. Figure 2 shows the percentage distribution of solar energy production capacity all around the world.

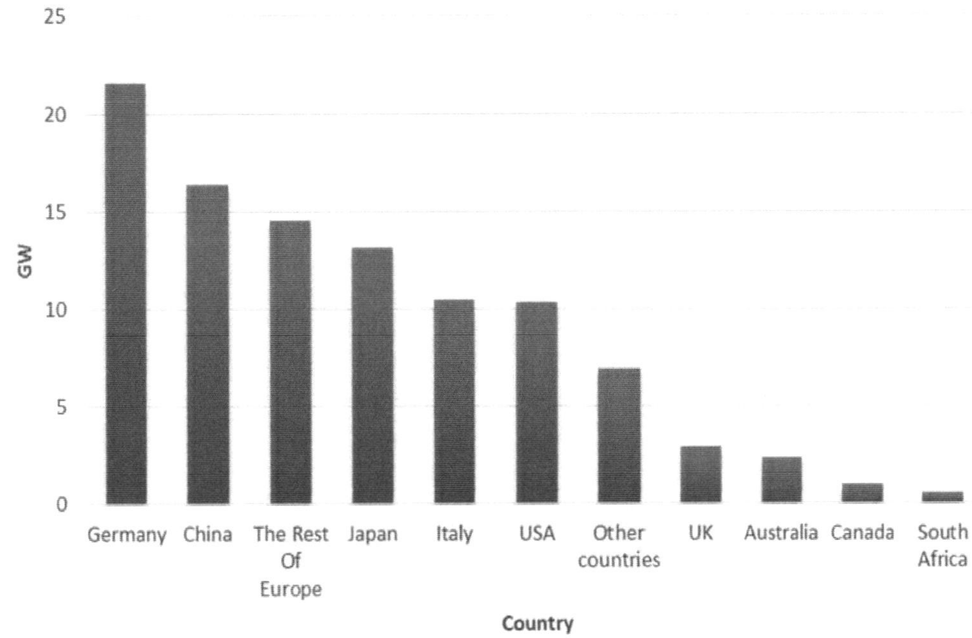

Figure 2 - The world solar energy production capacity distribution

Prior to 2011-2012, a large part of the production capacity growth was provided by such countries as Germany and Italy, however, starting with the 2012 year the leadership in the capacity growth shifted to China, the USA and Japan[9]. There were built more than 18 GW of solar power plants in China; more than 12 GW in the USA and 14 GW in Japan thanks to the government support and rapid economy diversification strategy development and state energy security. Figure 3 shows the growth dynamics in 2014 relative to 2013.

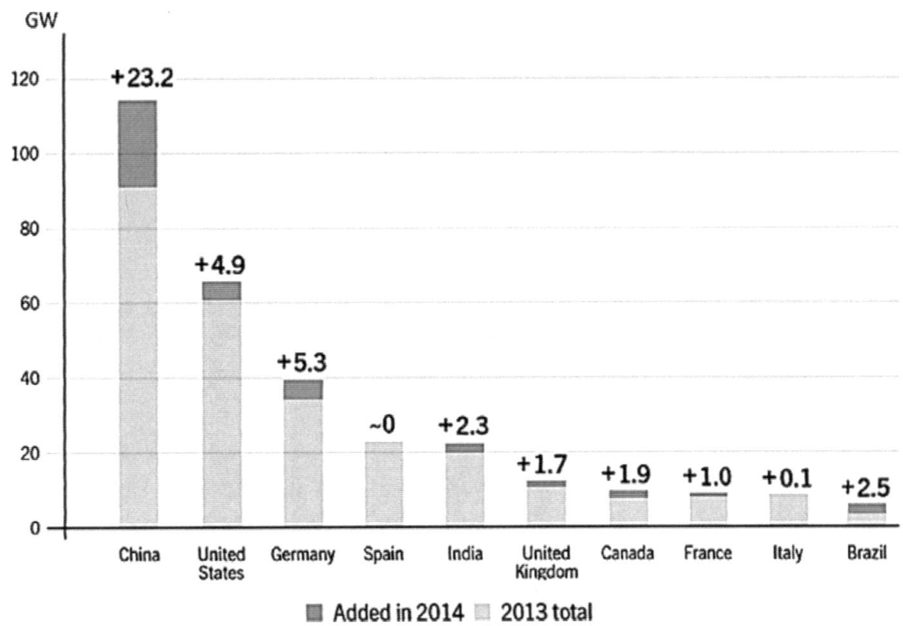

Figure 3 – Growth dynamics

As it can be seen, the increase above 5% is observed in China, Japan and the USA. The introduced state program of the alternative energy support in India and the fixing of quotas for the installations that produce not less than 6 GW of solar power a year starting with the 2016 year, will directly affect the growth leaders' rating. According to the experts [10] India will be able to enter the top 5 leaders on the solar energy production capacity in 2017.

The decision of the necessity to diversify the economy and active support at the state level contributed to the rapid development of solar energy in China. So at the end of 2014, the total production capacity of solar power plants was 28 GW, and the growth amounted to 10.6 GW, which accounts for 25% of global growth[11]. Also, China has occupied a niche of global manufacturer of solar modules and panels that are the key components of solar power plants. In 2014, China produced 33 GW of solar batteries, more than 70% of which were exported to the other countries. It is expected that in 2017 more than 70 GW of electricity, generated by solar power plants will be exported. The growth dynamics of solar energy in China is presented in figure 4.

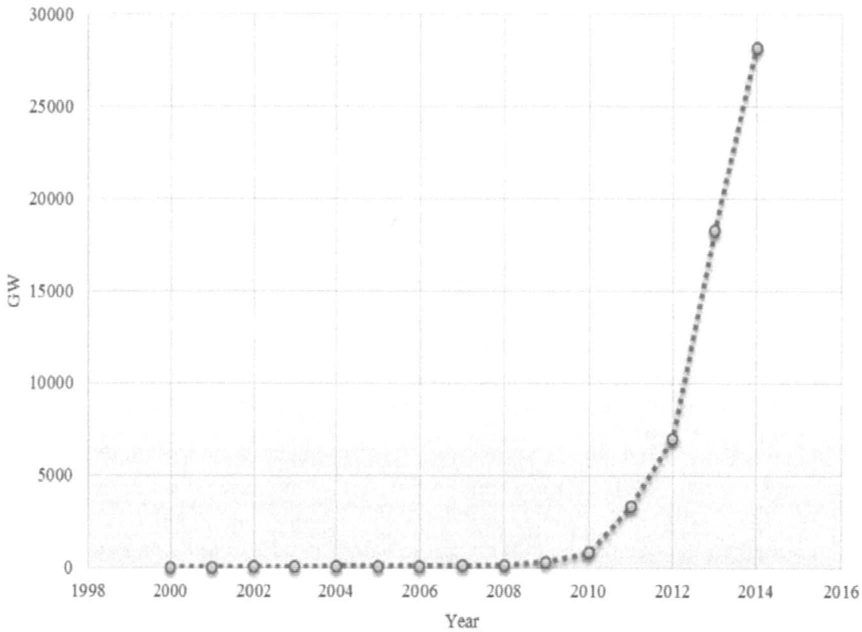

Figure 4 - Solar generation introduction dynamics in China

The rise of China, Japan and the United States have affected the distribution of leadership and the displacement of Germany, which provided the increase in 2014, less than 2%. This indicator can be explained by the fact that solar energy has been developing in this country for a long time and there is a competition and the race of technological advances in the market, therefore a quantitative indicator of the introduction goes into a quantitative indicator of the solar panels' efficiency and the indicator of the operation reduction. The unit cost reduction for solar panel up to 10 kW of power has a downward trend and, according to the 2012 reached the mark of 2 Euros per watt of energy.

The decrease in growth rate is also observed in Italy, which in 2011 put into operation more than 9 GW of solar power plants; in 2012 it was 3.6 GW, and in 2013 only 1.5 GW. This decrease indicates the high competition in the market of renewable energy sources. The analysis of changes between 2000 and 2015 has shown that the European Union market considers the development of wind energy priority. Wind energy has provided the increase more than 13.2% for the last 15 years compared to solar energy, the growth of which is equal to 10.4 %. The data, concerning the European Union energy market modification, are presented in figure 5.

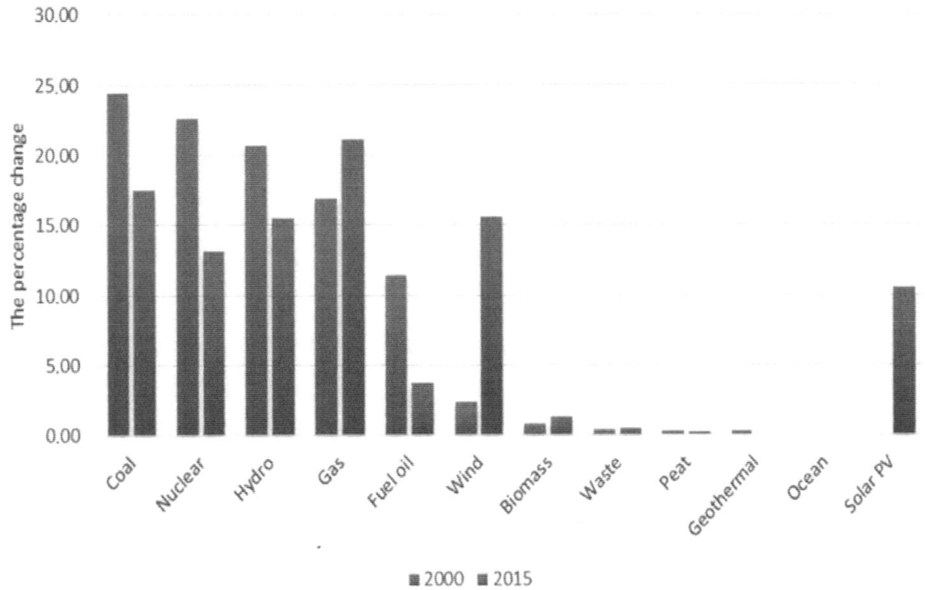

Figure 5 - The European Union energy market structure in 2000 and 2015 years.

An interesting fact is the entry into the race, concerning the introduction of solar panels, such country as the Republic of South Africa, where the main development factor is the introduction of the grid parity of solar energy relatively to the other forms of energy. Thus in South Africa the cost of a kW of electricity of solar power plants is equal to the rate with traditional power plants with the additional function of fixing the tariff for the purchase.

During the analysis a special attention should be given to the legislative aspects of the United States, supporting the development of green energy by means of traditional companies of energy sector. So according to the U.S. law, the generating companies of fuel and energy complex should assume the obligation for purchase of facilities, generated from renewable energy or buy "Green certificates". At the end of 2014 the production capacity of solar power plants in the USA amounted to more than 18 GW of energy and in 2017 it is expected the introduction of another 40 GW. Due to the dynamic development, large amount of introduced capacities and technological development the cost of capacity decreased almost 4 times, from 8$ to 2.2 $, in the period from 2005 to 2015.

Conclusion

The world experience analysis of solar energy introduction to the national energy system has shown that the dynamic development of solar energy is the integral attribute of the modern state development. The top ten are the countries which are the drivers of progress and development of the world economy. The world experience and the use of legislative acts as the basis for the development of renewable energy sources will contribute to the promotion of Russian energy resources, green economy, export diversification and role enhancement of the Russian Federation in the world.

References

[1] Ivancic A, Mugnier D, Stryi-Hipp G, Weiss W. 2014 Solar Heating and Cooling Technology Roadmap Renewable heating and cooling European technology platform.

[2] European Commission. Renewable Energy: Progressing towards the 2020 target COM(2011) 31 final; 2011.

[3] Martinopoulos G, Tsalikis G 2014 Active solar heating system for energy efficient buildings in Greece: A technical economic and. environnmental evaluation J. Energy and Buildings 68 130–137.

[4] Argiriou A, Klitsikas N, Balaras C A, Asimakapoulos D N 1997 Active solar space heating of residential buildings in northern Hellas - a case study J. Energy and Buildings 26 215–221.
[5] Badescu V, Staicovici M D 2006 Renewable energy for passive house heating: Model of the active solar heating system J. Energy and Building 38 129-147.
[6] Rojas D P 2013 Atrium building design: key aspects to improve their thermal performance on the Mediterranean climate of Santiago de Chile J. Low- Carbon Technology 1–4.
[7] Mottard J M, Fissore A 2007 Thermal simulation of an attached sunspace and its experimental validation J. Solar Energy 81 305–315.
[8] Mihalakakou G 2002 On the use of sunspace for space heating/cooling in Europe J. of Renewable Energy 26 415–429.
[9] Urban F, Geall Sam, Wangc Yu 2016 Solar PV and solar water heaters in China: Different pathways to low carbon energy J. Renewable and Sustainable Energy Reviews 64 531–542
[10] Lewis J I 2013 Green innovation in China: China's wind power industry and the global transition to a low carbon economy Columbia University Press, New York.
[11] Giampietro M, Sorman A H 2012 Are energy statistics useful for making energy scenarios? J. Energy 37 (1) 5–17.

A Method of Effective Quarry Water Purifying Using Artificial Filtering Arrays

M Tyulenev[1,a], E Garina[2,b], A Khoreshok[1,2,c], O Litvin[2,d], Y Litvin[2,e], E Maliukhina[3,f]

[1]Yurga Technological Institute (branch) of National Research Tomsk Polytechnic University, Leningradskaya 26, Yurga, Russian Federation, 652055
[2]T.F. Gorbachev Kuzbass State Technical University, Vesennyaya st. 28, Kemerovo, Russian Federation, 650000
[3]National Mineral Resources University (University of Mines), Faculty of Mining, Department of Surveying, 21 Line 2, St. Petersburg, Russian Federation, 199106

E-mail: [a]tma.geolog@kuzstu.ru, [b]garina_e@mail.ru, [c]haa.omit@kuzstu.ru, [d]litvinoi@kuzstu.ru, [e]ormpi@kuzstu.ru, [f]elenamaliukhina@spmi.ru

Abstract. The development of open pit mining in the large coal basins of Russia and other countries increases their negative impact on the environment. Along with the damage of land and air pollution by dust and combustion gases of blasting, coal pits have a significant negative impact on water resources. Polluted quarry water worsens the ecological situation on a much larger area than covered by air pollution and land damage. This significantly worsens the conditions of people living in cities and towns located near the coal pits, and complicates the subsequent restoration of the environment, irreversibly destroying the nature. Therefore, the research of quarry wastewater purifying is becoming an important mater for scholars of technical colleges and universities in the regions with developing open-pit mining. This paper describes the method of determining the basic parameters of the artificial filtering arrays formed on coal pits of Kuzbass (Western Siberia, Russia), and gives recommendations on its application.

Introduction

Industrial complexes concentrated in the most developed part of the biggest Russian coal basin – Kuzbass (Kemerovo Region, Western Siberia) – have the decisive influence on the ecological condition of environmental objects [1]. The decrease of production from the mid-1990s in the coal-mining areas of Kemerovo region has not led to the improvement of its environment [2-3]. The negative effect made by coal-mining enterprises on the natural environment has complex origination. We observe an intensive pollution as a result of the influence of coal deposits open-pit mining on the atmosphere, water resources and landscape complexes. Complex processes of environmental anthropogenic changes caused by open pits' operations have brought to light the problem of surface water pollution near large mining segments [4-5]. Currently, the volume of coal production in Kemerovo region amount to more than 210 million tons per year [6-7]. According to the report of supervisory environmental authorities in 2015, coal-mining enterprises of Kemerovo region disposed 1.239 billion m3 of wastewater, including inadequately treated – 228.8 million m3 [8-9].

The main source of pollutants' transportation into the surface waters near the territories of open-pit mines is quarry water, resulting from massive extraction of groundwater of developed coal seams for draining purpose [10-11].

The issue of wastewater treatment in open pit mining of mineral deposits causes great ecological problems [12-15]. Due to the constant tightening of the requirements to the quality of discharged wastewater their treatment in settling basins and purifying ponds does not provide the parameters of maximum permissible concentration (MPC). Moreover these treatment facilities occupy large areas of land. Also it should be noted that usually there are several distanced weir wastes, which location is changed during development of mining. For example, JSC "KuzbassRazrezUgol" (Kuzbass, Western Siberia, Russia) has 32 places of water use, taking into consideration that the structure of this coal company consists of 7 large open pits [16-18].

As an alternative coal enterprises may use expensive wastewater purifying equipment, which include different devices like hydroclassifiers, electrocoagulators, modules of water clarifying and sludge thickening, modules of sludge dewatering, finalizers, aerators and so on [19-24]. Complex of such equipment for purifying highly voluminous quarry wastewater may totally cost up to several million dollars. So because of these large expenses at present time even prosperous Kuzbass coal open pits conduct wastewater treatment in artificial filter arrays made of waste coal [9]. But the quality of purifying this way is quite low and does not match MPC [6].

Taking into account the upward trend in open pit mining of coal in Kuzbass, and as a consequence inevitable growth of the volume of discharged water by the coal enterprises, further study of improvement the environmental safety of mining operations is needed.

Materials and methods

The main pollutants of waste waters sent from the open-pit territory to the water reservoirs are suspended solids and oil. However, the qualitative composition of mining waste water is specific in every open-pit mine of Kuzbass and depends on the formation conditions, climatic factors, and the way of coal seams mining within the boundaries of a particular open-pit area. Reducing the level of anthropogenic impact on water bodies adjoining the territories of Kuzbass open-pit mines is one of the urgent problems of water use ecological safety.

Currently the most common method of wastewater treatment on the Kuzbass open-pits is a mechanical purifying by clarification in settling ponds and filtration through arrays of overburden rocks [9].

Except specially piled rock arrays, already existing open-pit dumps can be also used as the filters.

For using waste dumps or any other similar arrays of rock in water purification schemes it is strictly necessary:

- to choose one or another particular dump (filtering array) and identify the ways of supplying and discharging water;
- to test the ability of the dump to pass all the clarified water at the given water disposal in the period of maximum flow;
- to test the ability of the selected dump to clarify water to the maximum permissible concentration rates.

It is desirable to use both active and inactive truck and railway external dumps, located near the water disposal. The content of the slacking rocks in the array must not exceed 30%. When choosing a dump the following issues should be observed: topography of the bedding surface in order to determine the most appropriate places of clarified water supply, the direction and length of the filter, the area of water seepage from the dump [25-31]. Water may be supplied to the dump by gravity or by pumping. To avoid infiltration into the bed of the dump rock the bases should be low permeable [32-36].

The check of the selected dump's (filtering array's) ability to purify water from suspended solids to the required extent is made by the following formula:

$$C_{out} = C_{in} \exp(-\eta L) \le C_{mpc}, \quad (1)$$

where C_{out} – the concentration of suspended solids in the water outlet from the filter array, mg/l; C_{out} – the concentration of suspended solids in contaminated water, mg/l; L – filter length, m; C_{mpc} – the maximum permissible concentration of suspended solids , set for the water disposal, mg/l .

So when $C_{out} < C_{mpc}$ it can be assumed that this dump meets water purification requirements.

The technical solution to the problem of wastewater treatment from suspended solids in the filters of coarse rocks can be searched in two ways. The first of them is to use the discrete rocks arrays existing at enterprises as filters, after a preliminary examination. Such arrays include overburden rock arrays, refuse heaps and various technological dumps.

The second way is the purification of water in specially constructed filtering arrays of rocks and semi-rocks, which are, as a rule, the mining waste.

The filters' design depends on the terrain and the properties of the upper layers of the ground of the underlying surface, and the parameters are determined by the volume of water supply, its pollution density and filtration characteristics of the filtering material.

The main filter elements are the unit for supplying polluted water, filtering array, filter body, a device for collecting and draining the purified water. If there are any natural or artificial hollows (ravines, logs, river beds of dried-up rivers, trenches, ditches, abandoned mines and others) near the water disposal and low permeable rocks that lie at their sides and bottom, the latter can be the body of the filter, fig. 1. In this case, water may be supplied to the filter array by gravity on a surface or pipe and by pumping. The filter array is piled from rocks that meet the relevant requirements using trucks and bulldozers.

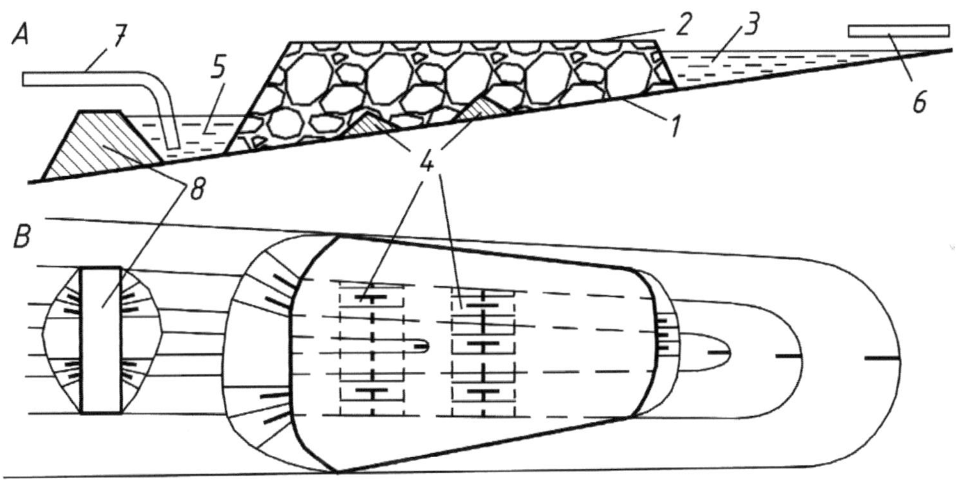

Figure 1 - Construction of the filtering array with waterproof stopping: A – longitudinal section; B – plan view; 1 – the bottom of the filter; 2 – filtering array; 3 – receiver for contaminated water; 4 – waterproof stopping; 5 – lodgement of purified water; 6 – conduit for feeding for contaminated water; 7 – conduit for purified water removing; 8 – water retention levee.

A pond appeared as the result of water infiltration becomes a device for supplying and routing water in the filter. The retaining dam for collection and turning out the purified water must be constructed of the rocks with low hydraulic permeability lower filter array. The disposal of pure water from water reservoir is produced by gravity in chutes or pipes in order to prevent the erosion of soil and new water pollution. When water discharge by gravity is impossible or there is a necessity of its subsequent use for enterprise needs pumping station should be installed.

If rocks of the sides and bottom of the hollows are characterized by high hydraulic permeability, to prevent infiltration of water into the ground and possible pollution of aquifers bottom filter forming is necessary. The bottom is arranged by dumping the low permeable rocks, followed by their layout and thickening by bulldozers. To prevent losses the water for cleaning must be supplied by chutes or pipeline.

When surface is horizontal and rock foundation has low filtration rate, for receiving polluted water and direction of seepage the filter body in the filter array is used (Fig. 2).

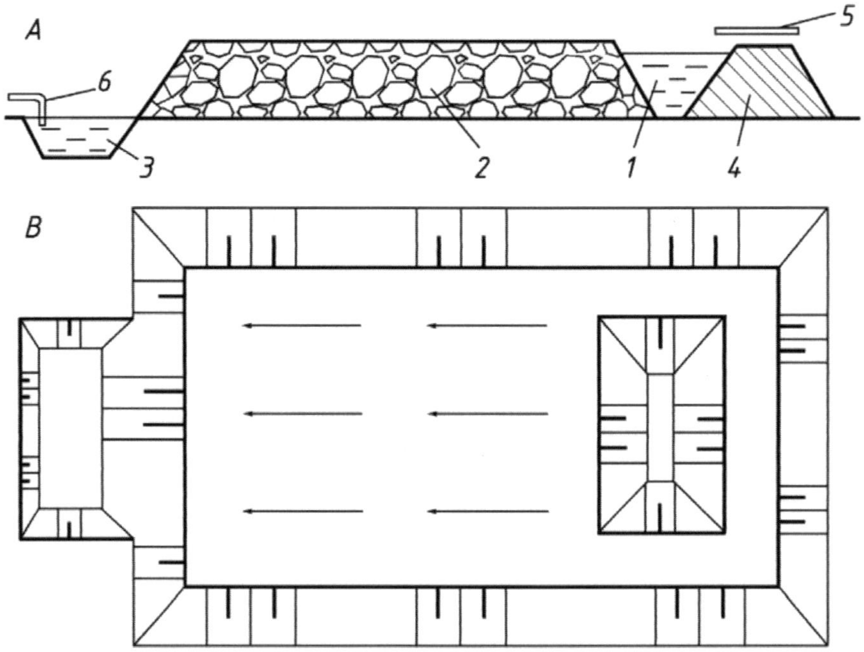

Figure 2 - Artificial filtering array on the flat surface: A – longitudinal section; B – plan view; 1 – receiver for contaminated water; 2 – filtering array; 3 – lodgment of purified water; 4 – water retention levee; 5 – conduit for feeding for contaminated water; 6 – conduit for purified water removing.

The filter body is a water-holding dam, piled of rocks with low hydraulic permeability. For supplying purified water to the filter array at one of its ends with the help of the dam a container is formed. Water reservoir with purified water is a pit created by the excavators and bulldozers at the base of the filter at the filter array open end. The movement of water in the filter is carried out by the pressure created by a column of water in the receptacle. The purified water is pumped from the water reservoir, and supplied to water bodies spillage or to meet the needs of the enterprise.

If the base rocks have low filtration coefficient, to prevent the infiltration and loss of water the filter bottom is made.

In this case water reservoir is formed by piling water retaining dam. Dam is formed of the same rocks that are used for the construction of filter body and bottom. The movement of water in the filter is carried out by the pressure generated by the difference in water levels in the receiving tank and water reservoir.

Results and discussion

Calculation of water filtration in filter rock array is performed on the basis of two conditions:

- first, the length of the filtration should provide water purification up to the required condition;
- secondly, the width and depth of the filtering flow should match the size of the filter array, i.e. array must provide a pass for the whole water inflow at this water disposal.

The minimum length of the filtration, providing water purification up to required degree, is found out of the equation (1) when $C_{out} = C_{mpc}$:

$$L = \frac{1}{\eta} \ln \frac{C_{out}}{C_{mpc}}, \quad (2)$$

where C_{mpc} – maximum permissible concentration of suspended solids in the clarified water.

Water filtration in the array is considered to be a steady uneven movement of water in a wide rectangular channel, which makes it possible to solve the problem as a two-dimensional one.

The equation that determines the change in the depth of filtration flow h in rock array with a linear filtration has the following form:

$$\frac{dh}{dx} = i - \frac{q}{kh}, \quad (3)$$

where x – distance from the water entry points to the array; i – inclination of water pressure (the underlying surface of the array); q – specific water flow rate per unit width; k – filtration rate for linear filtration.

Within the quadratic dependency filtration equation (3) is written as:

$$\frac{dh}{dx} = i - \left(\frac{q}{k_q h}\right)^2, \quad (4)$$

where k_q – filtration rate for quadratic filtration.

Equations (3) and (4) should be solved with the following limiting conditions: at the site of the water entering the filter array $x = 0$, and the seepage depth $h = h\varphi$. Filtrating flow should be at a distance of L, that is under $x = L$, $h = 0$.

When filtering water three situations are possible: the first, when underlying surface is horizontal, i.e. water pressure gradient i is equal to zero; the second, when water pressure gradient has a positive value ($i > 0$), i.e., elevation point of water entry to an array is higher than elevation point of the water exit; the third, with negative water pressure gradient ($i < 0$) .

The solution of equation (3) with the imitating conditions described above is given by the following formulae, depending on the incline of water pressure plate:

a) with $i = 0$:

$$L = \frac{h_1^0}{2} z_1^2, \quad (5)$$

where $h_1^0 = q/k$ – normal flow depth resulting steady water movement; $z_1 = h_\varphi / h_1^0$ – relative flow depth;

b) $i > 0$

$$L = -\frac{h_1^+}{i}\left[z_1^+ + \ln\left|z_1^+ - 1\right|\right] \quad (6)$$

where $h_1^+ = q/k$; $Z_1^+ = h_\varphi / h_1^+$;

c) $i < 0$

$$L = \frac{h_1^-}{i}\left[Z_1^- - \ln\left|Z_1^- - 1\right|\right], \quad (7)$$

where $h_1^- = -q/k_i$; $z_1^- = h_\varphi / h_1^-$.

The solution of equation (4) is expressed by the following formulae under the same limitating conditions:

a) with i = 0

$$L = \frac{h_2^0}{3} Z_2^3 , \quad (8)$$

where: $h_2^0 = q / k_к$; $Z_2^3 = h_φ / h_2^0$;

b) with i > 0

$$L = \frac{h_2^+}{i}\left(\frac{1}{2}\ln\frac{Z_2^+ + 1}{Z_2^+ - 1} - Z_2^+\right), \quad (9)$$

where $h_2^+ = q / k_k\sqrt{i}\,; z_2^+ = h_φ / h_2^+$;

c) with i < 0

$$L = -\frac{h_2^-}{i}\left(Z_2^- - arctg\, Z_2^-\right), \quad (10)$$

where $h_2^- = q / k_k\sqrt{-i}\,; z_2^- = h_φ / h_2^-$.

Thus, for a given length of filter *L* and for a linear filtration the depth and width of the filtrating flow, and also the height and width of the filtering array are determined by formulae (5-7). For quadratic filtration these parameters should be calculated by formulae (8-10). They provide passage of all incoming water without overflowing on the array surface.

Conclusions

Development of effective quarry water purifying technology requires cooperation of technical universities, research organizations and mining enterprises for saving the environment in coal basins. Geoecological effect of this technological platform is connected with the growth of wellbeing of communities located near open pits. It can be achieved with decrease of harmful contaminants' content in quarry water disposed to the rivers to MPC during filtering through artificial overburden arrays. Technology of forming and using filter arrays allows substituting expensive water treatment equipment what was proved by conducted research. Despite low capital expenditures of filtering arrays creation (equally to the costs of peripheral dozer dumping), further research within technological platform is required for improving the methods of selective arrays' filling due to raise the quality of filtering. In general high perspectives of the new technologies of quarry water treatment implementation at surface mining are expected to expand the limits of using mining machines [37-39].

Acknowledgment

The research was made in T.F. Gorbachev Kuzbass State Technical University under State Assignment N10.782.2014 K "Development of high-performance process for complex processing of low-grade coals and waste coal with producing a low ash coal-and-oil concentrate, composite fuels, rare earth and trace elements".

References

[1] Barysheva G A and Novoselova E G 2014 Methodology of Application of the Structural Shift Mechanism for Regulation of the National Economic Management System J. Applied Mechanics and Materials 682 pp 550-554.
[2] Zhironkin S A, Khoreshok A A, Tyulenev M A, Barysheva G A, Hellmer M C 2016 Economic and technological role of Kuzbass industry in the implementation of national energy strategy

of Russian federation IOP Conference Series: Materials Science and Engineering 142 (1) 012127.

[3] Tyulenev M A, Gvozdkova T N and Zhironkin S A et al. 2016 Justification of Open Pit Mining Technology for Flat Coal Strata Processing in Relation to the Stratigraphic Positioning Rate Geotechnical and Geological Engineering 34 (6) doi:10.1007/s10706-016-0098-3.

[4] Tyulenev M, Zhironkin S and Litvin O 2015 The low-cost technology of quarry water purifying using the artificial filters of overburden rock Pollution Research 34 (4) pp 825-830

[5] Tyulenev M A, Zhironkin S A, Garina E A 2016 The method of coal losses reducing at mining by shovels International Journal of Mining and Mineral Engineering 7 (4) DOI: 10.1504/IJMME.2016.10000781.

[6] Tyulenev M A, Lesin Yu, Vik S and Zhironkin S 2016 Methodological Bases of Advanced Geo-ecological Problems Resolving in Neo-industrial Clusters Proceedings of the 8th Russian-Chinese Symposium "Coal in the 21st Century" pp 333-336.

[7] Khoreshok A A, Zhironkin S A and Tyulenev M A et al. 2016 Innovative technics of managing engineers' global competencies IOP Conference Series: Materials Science and Engineering 142 (1) 012122.

[8] Zhironkin S A, Khoreshok A A, Tyulenev M A et al. 2016 Economic and Technological Role of Kuzbass Industry in the Implementation of National Energy Strategy of Russian Federation IOP Conference Series: Materials Science and Engineering 142 (1) 012127.

[9] Lesin Y V, Luk'yanova S Y and Tyulenev MA 2015 Formation of the composition and properties of dumps on the open-pit mines of Kuzbass IOP Conference Series: Materials Science and Engineering 91 (1) 012093.

[10] Antony Z et al. 2005 Effects of changing surface heat flux on atmospheric boundary-layer flow over flat terrain J. Boundary-Layer Meteorology 116 pp.331–361.

[11] Akhmetshina A S 2013 Assessment of thermal structure of boundary layer atmosphere of Western Siberia J. Bioclimland (biota,climate, landscapes), 1b pp 5-8.

[12] World Energy Council 2013 Coal: Strategic insight (London: World Energy Council) 211 p.

[13] Tyulenev M A, Khoreshok A A, Garina E A, Danilov S and Zhironkin S 2016 Adaptive technology of using backhoes for full coal extraction Proceedings of the 8th Russian-Chinese Symposium "Coal in the 21st Century: Mining, Processing, Safety" pp 111-115.

[14] Zhironkin SA 2002 Prospects and new possibilities investment attracting to Kuzbass coal mining industry Ugol' 6 pp 31-36.

[15] Gasanov M, Gasanov E and Egorova M 2016 The Technologies of the Network Prosperity in Russia Procedia - Social and Behavioral Sciences 166 pp 103-106.

[16] Zhironkin S A 2001 Factoring and leasing development at coal mining industry of Kuzbass as an important element of its financial part Ugol' 4 pp 29-30.

[17] Zhironkin SA 2001 Governmental factoring development of TEK Kuzbass Ugol' 6 p 62.

[18] Zhironkin SA 2002 About measures of vexel circulation development and vexelability definition of fuel-and-power complex' enterprises Ugol' 4 pp 47-48.

[19] Kovalev VA, Gerike BL, Khoreshok AA. and Gerike PB 2014 Preventive maintenance of mining equipment based on identification of its actual technical state Symposium of the Taishan academic forum – Project on mine disaster prevention and control pp 184-189

[20] Ryzhkov YA, Gogolin VA, and Karpenko NV 1992 Modelling the structure of solid masses of lump and granular materials (plane problem). Journal of Mining Science 28(1) pp 6-12

[21] Prokopenko S A 2014 Multiple service life extension of mining and road machines' cutters J. Applied Mechanics and Materials 682 pp 319-323.

[22] Ryzhkov YA, Lesin YV, Gogolin VA, Karpenko NV 1996 Modeling the structure of fragmented and granular material: Three-dimensional problem. Journal of Mining Science 32(3) pp 188-191.

[23] Khoreshok A A, Buyankin P V, Vorobiev A V, Dronov A A 2016 Simulation of Stress-Strain State of Shovel Rotary Support Kingpin IOP Conference Series: Materials Science and Engineering 127 012014.
[24] Khoreshok A, Tyulenev M and Vöth S 2016 Conditions for Minimum Dynamic Loading of Multi-brake Hoists Proceedings of the 8th Russian-Chinese Symposium "Coal in the 21st Century" pp 239-245.
[25] Lei L, Jian L, Yutao W, Nvjie W and Renqing W 2011 Cost-benefit Analysis and Payments For Watershed-scale Wetland Rehabilitation: A Case Study in Shandong Province China Int. J. Environ. Res. 5 (3) pp 787-796.
[26] Areepitak T and Ren J 2011 Model simulations of particle aggregation effect on colloid exchange between streams and streambeds Environ. Sci. Technol. 45 (13) pp 5614-5621.
[27] Arnon S L, Marx P, Searcy K E and Packman A I 2010 Effects of overlying velocity, particle size, and biofilm growth on stream-subsurface exchange of particles Hydrol. Processes 24 pp 108-114.
[28] Esakkimuthu T, Sivakumar D and Akila S 2014 Application of nanoparticles in wastewater treatment Pollution Research 33 (3) pp 567-571.
[29] Gupta S K, Ramesh K S and Shaik Sameer 2015 Decentralised wastewater treatment, a sustainable approach for use in developing country environment Pollution Research 34 (1) pp 111-120.
[30] Jones J I, Murphy J F, Collins A L, Sear D A, Naden P S and Armitage P D 2011 The impact of fine sediment on macro-invertebrates River Res. Appl. 9 pp 67-73.
[31] Karwan D L and Saiers J E 2012 Hyporheic exchange and streambed filtration of suspended particles Water Resour. Res. 48 pp 15-19.
[32] Sharma M M, Yortsos Y C 1987 Transport of particulate suspensions in porous media: Model formulation AIChE J. 33 pp 1636–1643.
[33] Xu S, Gao B, Saiers J E 2006 Straining of colloidal particles in saturated porous media Water Resour. Res. 42 pp 12-16.
[34] Karwan D L, Gravelle J A and Hubbart J A 2007. Effects of timber harvest on suspended sediment loads in Mica Creek, Idaho For. Sci. 53(2) pp 181–188.
[35] Zamani A and Maini B 2009 Flow of dispersed particles through porous media - deep bed filtration Journal of petroleum science & engineering 69(1-2) pp71-88.
[36] Zheng Xi-lai, Shan Bei-bei and Chen Lei et al. 2014 Attachment–detachment dynamics of suspended particle in porous media: Experiment and modeling Journal of Hydrology 511 pp 199-204.
[37] Nedjah N, Mizi A, Daas D, Laskri N and Baccouche M 2016 Urban wastewater phosphorus removal: mixed treatment Pollution Research 35 (2) pp 229-234.
[38] Efremenkov A B 2011 Forming the subterranean space by means of a new tool (geohod) Proceedings of the 6th International Forum on Strategic Technology IFOST 6021037.
[39] Efremenkov A B and Timofeev V Y 2012 Determination of necessary forces for geohod movement Proceedings - 2012 7th International Forum on Strategic Technology IFOST 6357729.
[40] Efremenkov A B and Aksenov V V et al. 2012 Force parameters of geohod transmission with hydraulic drive in various movement phases Proceedings - 2012 7th International Forum on Strategic Technology IFOST 6357716.

IOP Conf. Series: Earth and Environmental Science **50** (2017) 012036 doi:10.1088/1755-1315/50/1/012036

Heuristic Model Of The Composite Quality Index Of Environmental Assessment

A N Khabarov, A A Knyaginin, D V Bondarenko, I P Shepet, L N Korolkova

Technological Institute of Service (branch) DSTU,
Kulakov 41/1, Stavropol, Russian Federation 355000

E-mail: habrw@yndex.ru

Abstract. The goal of the paper is to present the heuristic model of the composite environmental quality index based on the integrated application of the elements of utility theory, multidimensional scaling, expert evaluation and decision-making. The composite index is synthesized in linear-quadratic form, it provides higher adequacy of the results of the assessment preferences of experts and decision-makers.

1. Introduction

The peculiarity of the assessment the ecological status of the territorial development programs is the need to analyze large amounts of heterogeneous information, conflicting goals of different actors involved in the industrial activities, and the complexity of evaluating alternative projects. The solution of problems like these is impossible without using modern computer-based decision support systems that contain databases, mathematical models, interactive and graphical user interfaces, analytics, data mining tools, etc.

Decision support systems provide the possibility to generate or choose the desired alternative from a variety of options when making decisions, give timely and comprehensive information. Such systems are a class of integrated intelligent systems that combine rigorous mathematical methods and solution search models with heuristic search models, logical-linguistic models and knowledge-based expert techniques.

2. Methods used to assess the quality of ecological status

So far, the quantitative and expert methods can be used to assess the environmental quality. Quantitative methods include differential, integrated and mixed methods. Differential method of quality assessment lies in a separate comparison of individual indices (indicators) of the environmental quality with the reference "benchmark" indicators. For this purpose we determine relative indicators q using the formula:

$$q = \frac{P_i}{P_{i\delta}}, \tag{1}$$

where P_i – individual indicator of the considered state; $P_{i\delta}$ – individual "benchmark" indicator.

However, the application of the differential method does not reduce the amount of information analyzed by the decision maker (DM) and, therefore, it is ineffective in assessing the ecological status quality.

The integrated method of quality assessment involves the use of composite, generalized indices. While the individual quality indicators of the considered state are expressed in scores and are assigned some weighting coefficients K, a composite quality index is determined by the formula:

$$Q = \sum_{i=1}^{n} P_i K_i, \quad (2)$$

where P_i – individual indicator of the considered state; n – number of individual indicators under consideration.

In the integrated method of quality assessment quality is expressed as ratio of calculated composite index to predetermined reference index.

The mixed method of quality assessment is applied in cases when composite quality index isn't sufficient enough to take into account all the essential characteristics of the ecological status. Then the obtained composite indices and distinguished individual indices are considered using the differential method of quality assessment.

Expert evaluation methods used to assess the quality of ecological status are based on the experts' knowledge, experience and intuition.

In recent years, in assessing the quality of the ecological status, methods based on the use of mathematical apparatus of the choice and decision-making theory are used. Integrated mathematical means of solving practical problems of environmental quality assessment in interactive mode on the computer using the multi-criteria decision making methods [1, 2] are being developed now.

The proposed heuristic model of the composite quality indices is based on the convolution of the individual indicators after the exclusion of non-competitive solutions, through unconditional preference criterion (the Pareto principle). To represent the system of individual quality indicators in the form of a convolution of a multi-purpose index components, we use empirical reasoning that leads to the functional relationships.

The system of quality indices for the heuristic model is formed according to nomenclature of indices groups, used for the environmental quality assessment, on the basis of study of documents and expert polls. In this case, the heuristic model reflects the system as a research object.

At the first step, the individual quality indicators are normalized. To equalize metric scales of heterogeneous quality indicators, methods of rationing and centering, or the procedure of the conditional utility function construction has been applied [3, 4]. The procedure of balancing individual indicators allows not only to formally come to the dimensionless parameters but also to take into account the decision-makers (DM) preferences. Besides, there can also be determined the values of the utility of non- numeric qualitative indicators; it makes possible to use such indicators in the environmental assessment procedures.

Next, the problem of determining composite indices by groups of indices is being solved. For that purpose we apply the methods of multidimensional scaling [5] and the "ideal point" model (Figure 1).

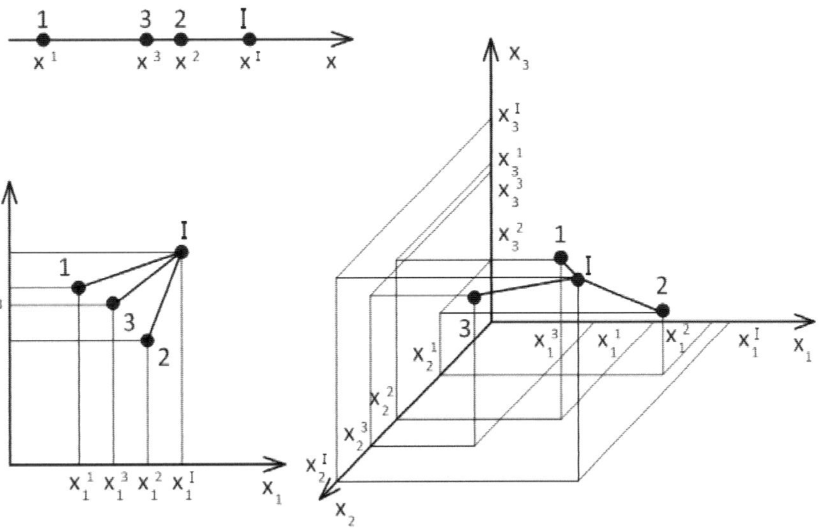

Figure 1 - Geometrical interpretation of the "ideal point" model

To quantitatively determine the proximity of the objects to the "ideal point" in a coordinate space we use metrics.

The following metrics are used [3].

The Helder metric:

$$u = 1 - \left(\sum_{i=1}^{n} \lambda_i \left| x_i^* - x_i \right|^n \right)^{1/n}, \quad n \geq 1, \tag{3}$$

where x_i^* – value of the "ideal state"; x_i – value of indicators of the compared states; λ_i – coefficient of relative importance (weight) of the i-th indicator; n – number of the considered individual indicators.

Euclidian metric:

$$u = 1 - \left(\sum_{i=1}^{n} \lambda_i \left| x_i^* - x_i \right|^n \right)^{1/2}, \tag{4}$$

Using these metrics is useful for groups that includes quality indicators of unequal importance, however, considerable difficulty is the assignment of corresponding non-negative weight to each of the indicators; the weight represents a coefficient of relative importance of the indicator from the point of view of the objectives of the evaluation.

Hamming metric:

$$u = 1 - \sum_{i=1}^{n} \lambda_i \left| x_i^* - x_i \right|, \tag{5}$$

This metric is used mainly to differentiate objects that are measured in scales of names and order.

Introduction of weighted metrics into the space of individual indicators sets the task of determining the relative importance of the indicators; the task can be solved by one of the methods of expert evaluation.

The comparative analysis of these methods in characteristics of the labor intensity, the required qualification of experts and reliability of the results has shown that the pairwise comparison and pairwise substitution methods are the most preferable for determining the coefficients of relative importance.

We use the method of pairwise comparisons, its essence is as follows.

There are P indicators in the group. Experts compare all pairs of indicators in the group to each other, for example, in the form of statements "more – less", "better – worse" [6]. Using the symbol of preference >, we can write $\langle x_i > x_j \rangle$ - x_i is more preferable than x_j.

The δ - function is assigned to each pair $\langle x_i, x_J \rangle$:

$$\delta_{ij} = \begin{cases} \mathbf{1, if}\ x_i > x_j, i, j = \overline{1, p}, i \neq j \\ \mathbf{0, otherwise} \\ \mathbf{0{,}5, if}\ x_i \infty x_j \end{cases} \tag{6}$$

where ∞ means equivalency.

It is required to determine the ranking of objects by each expert, to determine mathematical expectation (l_{ij}) of a random variable δ_{ij}. On the basis of $L = |l_{ij}|$ mathematical expectations of the pairs of indicators to determine the relative importance coefficients of the indicators and get a normalized ranked list of indicators in the group according to the results of expert poll.

The analysis of existing polling methods showed that to improve the reliability of the results automation of the process is needed, through the use of computing technologies during all stages of the process because the traditional techniques base on the questionnaires followed by the computer processing, and don't take into account some important psychological characteristics of the experts.

It is known that there are two factors that significantly affect the expert assessments during the pairwise comparison of indicators. The order of presenting alternatives in a pair obviously affects the assessment of their relative importance. Influence like this is called a spatial effect.

It is considered that spatial effects are balanced for the given alternative if in one half of the pairs that include this alternative it goes first, and in the other it goes second. The effects associated with the ordering of pairs of alternatives in the list of ranked pairs are called temporary. It is believed that the temporary effects for the given alternative are balanced if pairs that include this alternative, are arranged in the list evenly.

It should be noted that the widely used form of questionnaires for the survey of experts like the "tournament style" table, maximize the negative impact of the spatial and temporary effects on the reliability of the polling results.

To compensate for the spatial and temporary effects, an algorithm of presenting the pairs of the alternatives to the experts [7] was developed and computer-implemented. Its scheme is shown on Figure 2. During the procedure the expert poll is executed using of a multi-user or personal computer system in several rounds according to a technique similar to Delphi method [8].

The following procedures were automated: the generation of the expert poll, processing of polling data, management of a multi-round expert survey, preparing and mailing the results of the next round to the experts' terminals, messaging, access by the experts to additional and reference background information on the subject of the poll.

In order to improve the reliability of the expert poll results the experts' answers are automatically checked on internal consistency. With this purpose the each set of three alternatives (triplets) $\langle x_i, x_J, x_k \rangle$ /irrespective of the order/ is checked for correspondence of the obtained expert estimates to the rules of logical decision:

$$\begin{gathered} \{(x_i > x_J) \cap (x_J > x_k)\} \Rightarrow x_i > x_k; \\ \{(x_i > x_J) \cap (x_J \infty x_k)\} \Rightarrow x_i > x_k; \\ \{(x_i \infty x_J) \cap (x_J \infty x_k)\} \Rightarrow x_i \infty x_k, \end{gathered} \tag{7}$$

where $>$ – greater preference of alternatives; ∞ – equivalence of alternatives.

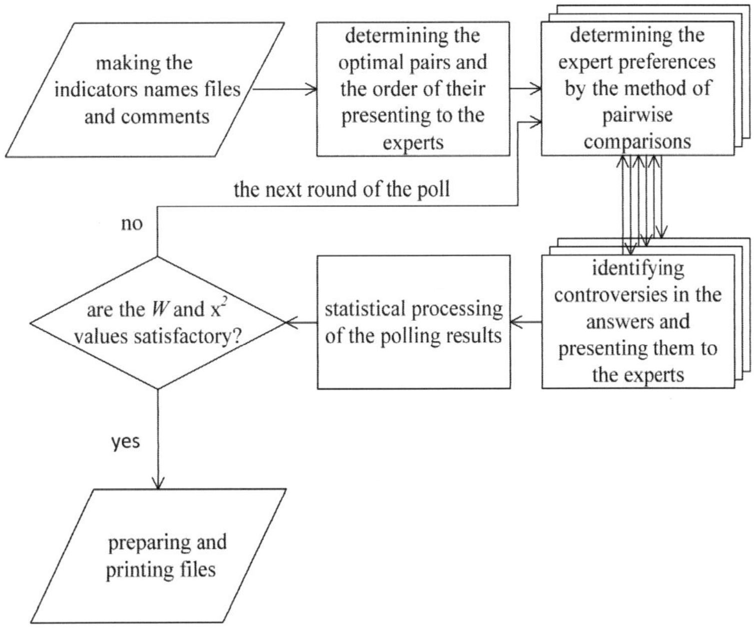

Figure 2 - Automated polling algorithm

Preferences for the triplets in which contradictions are revealed, are presented for joint correlation.

Using automated interactive polling procedure increases the accuracy and efficiency of determining relative importance coefficients in the groups.

The polling results are processed by the computer in the following way.

If while estimating the preference $\langle x_i, x_J \rangle$, m_i experts expressed in favor of the preference $x_i > x_J$, m_j experts were against, and m_k experts were for the equivalence of indicators, then the mathematical expectation of a random variable:

$$l_{ij} = M[\sigma] = \frac{1m_i + 0{,}5m_k + 0m_j}{m_i + m_k + m_j} \tag{8}$$

The total number of respondents

$$m = m_i + m_k + m_j \tag{9}$$

From (8) and (9) we obtain

$$l_{ij} = \frac{m_i + 0{,}5m_k}{m}, i \neq j, i = j = \overline{1,p} \tag{10}$$

It is obvious that $l_{ji} = 1 - l_{ij}$.

The set of quantities l_{ij} forms a matix *L*, of dimension $p \times p$.

The relative importance coefficient vector of t order is determined by the formula:

$$\lambda^t = \frac{1}{K^t} LK^{t-1}, t = \overline{1,p} \tag{11}$$

where $L = \left|\left| l_{ij} \right|\right|, K^0 = (1,1,\ldots,1); K^t = (K_1^t, K_2^t, \ldots, K_p^t)$. Then $\lambda^t = \sum_{i=1}^{p}\sum_{j=1}^{p} l_{ij}K_j^{t-1}$.

We confine ourselves to the definition of the first-order relative importance coefficient (t = 1), which gives a sufficient approximation accuracy of the results of pairwise comparisons and simplifies all the calculations. These coefficients represent the relative sums of the elements of the matrix L rows.

Indeed, when t = 1, we obtain from (11):

$$\lambda_i^1 = \frac{\sum_{j=1}^{p} l_{ij}}{\sum_{i=1}^{p}\sum_{j=1}^{p} l_{ij}} \qquad (12)$$

where $\sum_{i=1}^{p}\sum_{j=1}^{p} l_{ij} = \frac{p \cdot p}{2}$.

Note that using the matrix $X = \|\omega_{ij}\|$ made of the K-th expert answers as matrix L, we obtain the relative importance coefficients offered for the indicators by the given expert.

The experts opinion consistency is evaluated with the concordance coefficient W; to test the significance of the polling results, the Pearson criterion χ^2 is determined.

The determined after experts' survey relative importance coefficients of the individual environmental quality indicators are used to calculate metrics values by the (3), (4), (5) formulas. Using metrics makes it possible to come from a set of individual quality indicators to composite group indices.

3. Synthesis of heuristic composite environmental quality index in a linear-quadratic form

Unlike other earlier papers where the additive form of objective function was used, we suggest synthesis of heuristic composite environmental quality index in a linear-quadratic form:

$$K = \sum_{i=1}^{m}(\alpha_i \gamma_i u_i(x_i) + \frac{1}{2}\alpha_{ii} u_i^2(x_i)) + \sum_{i,j=1}^{m} \alpha_{ij} u_i(x_i) u_j(x_j), \qquad (13)$$

where γ_i – relative importance coefficients for the groups of indicators; $\alpha_i, \alpha_{ii,} \alpha_{ij}$ – scaling coefficients; u_i, u_j – metric functions of the the groups of indices.

While synthesizing the composite quality index K it is necessary to take into account that the decision about correspondence of ecological status to a certain quality level is usually taken by a group of people (expert group) [9] but the final decision belongs to the decision-maker. With this consideration in mind, the procedure of synthesizing must be based on the decision-maker preferences [10] with due consideration to the opinion of other experts about the importance of indicators of one or another group.

Considering the experts' opinions can be reached by introducing the relative importance coefficients γ of the groups of indices in the first-order terms (13). The problem of determining the relative importance coefficients of the groups of indicators is solved by the method of pairwise comparisons.

The pairwise comparison of group indices differs from that of individual indicators because in the latter case the expert has to deal not with tangible, physically perceivable indicators but with generalized composites, like hazardous chemical content in the atmosphere due to storage, transportation and consumption of products; possibility of emission into the atmosphere of pollutants (into water, air, earth, closed or ventilated room); harmful radiation level (high-frequency radiation, light radiation, etc.) while manufacturing, storage, transportation, consumption of products; hazardous noise- and vibration level, etc. In situations like these rating with "better – worse – equal"-scale becomes difficult.

To make the polling procedure more precise, the continuous preference scale is used, on which unit 1 corresponds to unconditionally greater preference of the first group, ½ corresponds to equivalency,

IOP Conf. Series: Earth and Environmental Science **50** (2017) 012036 doi:10.1088/1755-1315/50/1/012036

and 0 is unconditional preference of the second group. Using the continuous scale allows to more precisely describe expert preferences.

Synthesis of heuristic composite environmental quality index in a linear-quadratic form makes possible to take into account the fact that with the improvement of the individual quality indicators, quality cost increases to a much greater extent than at a lower index value. Taking into consideration this circumstance provides higher adequacy of the results of the experts and DM preference evaluation.

To determine coefficients a_i, a_{ii}, a_{ij} we use technique based on obtaining additional information from the decision-maker; this leads to the following system of equations and inequations:

$$\begin{aligned}
&\sum_{i=1}^{m}(\alpha_i+\frac{1}{2}\alpha_{ii})+\sum_{i=1}^{m}\sum_{j=i+1}^{m}\alpha_{ii}=1,\\
&\Delta K_S=0,\\
&m_S<\Delta K_S<M_S,\\
&\alpha_i+(\delta^k,\alpha^i)>0,
\end{aligned} \tag{14}$$

where ΔK_S – the environmental quality assessment; m_S, M_S – the result of more accurate estimates after obtaining additional information from the decision-maker; δ^k – m-dimensional vectors with components equal to zero and one.

To solve the system (14), the $\sum_{s=p+1}^{p}\Delta K_S$ is maximized, it leads to maximum "resolving power" of K-function.

This statement leads to a linear programming problem:

$$\sum_{s=p'+1}^{p}\Delta K_s(\alpha_1,...,\alpha_m,\alpha_{11},...,\alpha_{mm})\to\max \tag{15}$$

To solve the problem of linear programming the simplex method is used.

Lack of solution to the linear programming problem means inconsistency of the DM's preferences and then it will require the DM to be resurveyed.

Then, in order to verify the decisions stability we organize the iteration procedure of correction the function found, based on the analysis by the DM of the quantitative quality scores, as well as the difference between their scores.

As zero iteration of the definition $K_{(\bullet)}$ the linear programming problem (14), (15) is considered. The result is $K^{(0)}(u)$ function with *coefficients* $\alpha_1^0,...,\alpha_m^0,\alpha_{11}^0,...,\alpha_{mm}^0$.

On the basis of the analysis of the values $\Delta K_s=\left|K_s^{(0)}(\overline{u_s})-K_s^{(0)}(u_s)\right|, s=\overline{p'+1,p}$, the decision-maker must report sets of S $S<$ and $S>$ $(S<\cap S>\in\overline{p'+1,p})$ ranking numbers S of those pairs of systems, for which he/she isn't satisfied with the difference of estimates, or these pairs have insufficiently expressed, from his point of view, distinctions in preferences of alternatives $\overline{u_s}, u_s$ (set of $S<$), or these differences are too large (set of $S>$).

For each $s\in S<\bigcup S>$ the procedure of correcting the estimate (15) is applied, where in the capacity of m_s^0, M_s^0 one should choose $\Delta K_s^{(0)}+\varepsilon, M_s$ for $s\in S<$ and $m_s,\Delta K_s^{(0)}-\varepsilon$ for $s\in S>$ respectively.

After that we solve the linear programming problem (14) once again with constraints (15), wherein the last group of constraints is replaced with inequalities:

$$\begin{aligned}
&\Delta K_s^{(0)}-\varepsilon\le\Delta K_s\le\Delta K_s^{(0)}+\varepsilon, s\in S<\bigcup S>;\\
&m_s\le\Delta K_s\le M_s, s\in S<\bigcup S>.
\end{aligned}$$

The described iterative procedure continues until the decision-maker is satisfied with the obtained differences between the *K* values for the alternatives $\overline{u_s}$ and u_s $(s = \overline{p' + 1, p})$.

To improve the quality of preference description with a function the decision-maker is required to indicate as much as possible pairs of alternatives. While correcting the estimates of the alternatives the decision-maker can report new pairs $\overline{u_s}, u_s$ $(s = \overline{p + 1, p''})$, which at this stage may be compared by preference. Then the next iteration is carried out with additional inequalities with $s = \overline{p + 1, p''}$ in constraints (14).

4. Conclusion

Synthesized in such a way objective function appears as the composite environmental quality index; it describes in a qualitative manner the preferences of the experts and decision-makers and establishes ranking of consequences which are described by the initial set of individual indicators.

References

[1] Jin Si, Ljiljana Marjanovic-Halburd, Fuzhan Nasiri, Sarah Bell, Assessment of building-integrated green technologies: A review and case study on applications of Multi-Criteria Decision Making (MCDM) method, Sustainable Cities and Society, Available online 16 June 2016, ISSN 2210-6707.

[2] Xiaoguang Ouyang, Fen Guo, Dan Shan, Huanyun Yu, Jian Wang, Development of the integrated fuzzy analytical hierarchy process with multidimensional scaling in selection of natural wastewater treatment alternatives, Ecological Engineering, Volume 74, January 2015, Pages 438-447, ISSN 0925-8574.

[3] Leoneti A B. Utility Function for modeling Group Multicriteria Decision Making problems as games, Operations Research Perspectives, Volume 3, 2016, Pages 21-26, ISSN 2214-7160.

[4] Wilson K J, Quigley J. Allocation of tasks for reliability growth using multi-attribute utility, European Journal of Operational Research, Volume 255, Issue 1, 16 November 2016, Pages 259-271, ISSN 0377-2217.

[5] Jung and K, Takane Y. Multidimensional Scaling I, In International Encyclopedia of the Social & Behavioral Sciences (Second Edition), edited by James D. Wright,, Elsevier, Oxford, 2015, Pages 34-39, ISBN 9780080970875.

[6] Rezaei J. Best-worst multi-criteria decision-making method: Some properties and a linear model, Omega, Volume 64, October 2016, Pages 126-130, ISSN 0305-0483.

[7] Podinovski V V, Decision making under uncertainty with unknown utility function and rank-ordered probabilities, European Journal of Operational Research, Volume 239, Issue 2, 1 December 2014, Pages 537-541, ISSN 0377-2217.

[8] Seddiki M, Anouche K. Amar Bennadji, Prince Boateng, A multi-criteria group decision-making method for the thermal renovation of masonry buildings: The case of Algeria, Energy and Buildings, Volume 129, 1 October 2016, Pages 471-483, ISSN 0378-7788.

[9] Gupta P, Mehlawat M K, Grover N. Intuitionistic fuzzy multi-attribute group decision-making with an application to plant location selection based on a new extended VIKOR method, Information Sciences, Volumes 370–371, 20 November 2016, Pages 184-203, ISSN 0020-0255.

[10] Reichert P, Schuwirth N, Langhans S, Constructing, evaluating and visualizing value and utility functions for decision support, Environmental Modelling & Software, Volume 46, August 2013, Pages 283-291, ISSN 1364-8152.

IOP Conf. Series: Earth and Environmental Science **50** (2017) 012037 doi:10.1088/1755-1315/50/1/012037

Investment Appeal of the Recreational Potential of the Region

T V Galanina, V G Mikhailov, N N Golofastova, T G Koroleva

T. F. Gorbachev Kuzbass State Technical University
Vesennyaya st. 28, Kemerovo, Russian Federation, 650000

E-mail: galanina64@mail.ru

Abstract. The article deals with the issue of environmental and economic assessment of natural and recreational potential of Kuzbass. The purpose of the research is to study the trends in the development of the regional recreational potential, as well as to develop the methodological bases for creation of resort and recreational areas to improve the health of the population and increase the investment appeal of Kemerovo region. The constraints to improve the investment, environmental and economic performance of regional resorts are identified. The main result of the research is the development of the project to improve the stability of the socio-economic system of the region, on the basis of the mitigation of environmental and economic risks and the systematic development of recreational infrastructure. The practical significance of the study consists of the recommendations for regional and federal authorities concerning the development of socio-economic plans and forecasts, including the provision of a rationale for the status of priority development territories.

Introduction

Historically Kuzbass is positioned as the industrially developed region of the country, the potential of which is represented by fuel and energy industries, including the extraction and processing of minerals, mainly coal, electricity generation, metallurgy, chemical industry and other basic industries. However, the investment appeal of the region declined in recent years, owing to the negative socio-economic and environmental factors. The region is experiencing high anthropogenic impacts, which generates considerable environmental and economic risks and reduces profitability due to the irrational use of nature [1, 2, 3]. The solution to this problem may be related to several areas of activity:

- development of a comprehensive program for the development of natural resources,
- development of ecological and economic risk management strategies,
- search for new environmentally friendly activities.

In this situation the search for reserves, which include recreational potential, which is not less important than the investment appeal, seems urgent [4]. The most important recreational resource of Kemerovo region is the ski resort Sheregesh, located in Tashtagol area and well-known not only in Russia but also abroad. In addition, the Kuzbass has significant reserves of medicinal mud, peat, clay, mineral water springs, forests and medicinal herbs. The development and widespread use of such resources in the course of study of the environmental conditions of the area allows using these balneal factors to create the resort and recreation areas for the purpose of the improvement of the health of population and the increase of the investment appeal of the region. [5]

Material and methods

The object of the study is the recreational infrastructure of Kemerovo region. The subject of the research is the factors, conditions and constraints that affect the efficiency of the regional recreational system. The study is based on an analysis of domestic and foreign literature, including primary data, on the issues of investment and ecological-economic substantiation of recreational areas. The particular attention is paid to the differentiation of positive effects of the target realization of the recreational potential of Kemerovo region. The elements of the system analysis and the results obtained by specialists in the management of protected areas and recreational systems are used in the work.

Results and Discussion

Recreational resources are diversified components of the environment: climate, terrain, surface and underground waters, parks, forest parks. In turn, the natural recreational resources are set of physical, biological and energy-informational elements and forces of nature, which are used in the process of restoration and development of physical and spiritual powers of man, his ability to work and health. Most of natural resources have recreational and tourist potential, but the extent of its usage varies and depends on the demand for the recreation and the specialty of the region. The rational use of recreational resources determines the security of natural features, which allow using them for the purpose of recreation, tourism and treatment [6, 7].

The specially protected natural areas designed to preserve the typical and unique natural landscapes, flora and fauna diversity, to protect natural and cultural heritage, play a huge role in the resource recreational potential. They are fully or partially withdrawn from economic use, have special protection, and on the adjacent land plots and water areas the protected zones or districts with regulated regime of economic activity can be established [8, 9]. Many specially protected natural areas are the objects of national heritage, for example, in Kemerovo region – they are the state nature reserve "Kuznetsky Alatau"; Shor National Park; natural monument "Lime Island"; museum-reserve "Tomsk Pisanitsa"; Kuzbass Botanical Garden (Department of Ecology of Plant Resources of Human Ecology Institute of the Russian Academy of Sciences); state natural reserves; mass recreation, sport and tourism areas; green areas of cities and settlements; territories with natural medical factors (Borisov, Tersinka); protected zones of history and culture monuments.

The total area of specially protected natural areas (SPNA) of Kemerovo region is more than 15% of its entire territory, which is one of the highest rates in Siberia. The SPNA system allows saving not only the habitat of animals and plants, but also of human, helping to maintain the ecological balance of the biosphere as a whole, ensuring the safety of biological diversity and the restoration of natural systems. There are 13 wildlife reserves of regional importance with a total area of over 500 thousand hectares in the region according to statistical studies as of 31.12.2015.

The medicinal plants are of great importance in the structure of plant resources used by humans [10]. More than a third part of the extensive list of drugs currently used in medical practice for prevention and treatment are the drugs, the composition of which includes the plant-based ingredients. Their role is especially great in the treatment of cardiovascular diseases, nerve, kidney and liver diseases and some other common diseases (Minaeva, Zhanaeva 1983, Krylov and others, 1989; Nekratova and others, 1989; Kazarinova and others, 1991).

The floristic composition of Kemerovo region vegetation is very diverse: not less than a half of 3380 species described in the twelve volumes of "Flora of Western Siberia" by P.N. Krylov and L.P. Sergievkaya grows on the territory of Kuzbass; and about 450 species of wild and cultivated shrubs and herbaceous plants are used in scientific and folk medicine. The medicinal plants, the exploitation reserve of which is not less than 500 kg of air-dried raw materials have the practical importance. Table 1 shows the differentiation of medicinal plants depending on their exploitable reserve volume.

Table 1 - Grouping of medicinal plants according to their exploitable reserve volume

Group	exploitable reserve volume, tons
First (1)	more than 1000
Second (2)	500 - 1000
Third (3)	100 - 500
Forth (4)	50 - 100
Fifths (5)	10 - 50
Sixth (6)	1 - 10
Seventh (7)	less than 1

The analysis shows that the raw material of the species within 1-4 groups is suitable for industrial use to meet the needs of Russia. In turn, the fifth, sixth and seventh groups can be used to provide Western Siberia, the Kemerovo region and the regional network of pharmacies respectively with medicinal plants.

The most important recreational resource of the region are the natural mud, which include natural formations variable in their genesis, including water, minerals and organic substances, which are finely divided homogeneous plastic mass with certain thermal and other physical and chemical properties. In different countries these "mud-like" substances are historically used in heated state for the medical procedures and have different local names. For the purpose of uniformity and the designation of medicinal mud the Committee of the International Society of medical hydrology proposed in 1938 the common name "peloid" (from the Greek “pelos”), which includes different kinds of mud: peat, sapropel, lake, mixed or other. In the early 60-ies a new genetic classification of therapeutic muds was developed (Ivanov, V.V. Malakhov A.M., 1963), according to which they are divided in the 3 separate groups according to their genesis (peat, silt, pseudovolcanic), types and subtypes, having different chemical composition and physical properties.

In the Kemerovo region the siliceous sapropel of lake Berchikul are recommended for the balneal development. In the region there are also peat mud deposits. On the territory of Kuzbass the explored peat deposits are located in 6 areas with total reserves of peat within 57 million tons. The experience in the use of peat in balneal practice shows that the peat treatment is as effective as other natural mud. This type of mud treatment has its advantages: peat has higher thermal properties, adsorption capacity of absorbing pathogenic microorganisms; the presence of specific microflora contributes to the accumulation of biologically active substances. Therefore, the study of the microbiological composition of peat used both in native and in perforated form is prospective.

Within Kemerovo region three deposits of mineral waters are explored, which are operated at the present time - Tersinskoe in the 1957-68 with reserves of 172 m^3 / day; Borisovskoye in 1977, with reserves of 42 m^3 / day; Berezovoyarskoe in 1998-2000 with reserves of 138 m^3 / day. In addition, in 2000, Barzasskoye mineral water show with a maximum capacity of the core intersection of 70 m^3 / day was revealed. In 2004 in Barzas area the first phase of work was completed, but since 2005 due to difficult economic situation the financing of works was suspended [11, 12, 13].

The considered recreational resources determine the broad resort capabilities of Kemerovo region, in particular, today Sheregesh and Tanai have the status of the resort areas in the region. Sheregesh (Mountain Zelyonaya (“Green”)), located in the south of Kemerovo Region - is one of the most furnished Western Siberia resorts. Another resort is Tanai - young ski resort in the northwest part of the region.

The main disadvantage of these resorts is their remoteness, which requires the establishment of more eco-friendly resort and recreational areas to meet the basic conditions required for such areas (environmental, transport infrastructure, beautiful landscapes, favorable climate features, the presence of stocks of natural therapeutic factors).

Under the unfavorable environmental conditions, the sufficiently high level of morbidity, mortality and low life expectancy, the creation of the area answering the purpose of resort and consisting of a set of ecological, balneal and climatic characteristics is relevant and timely in Kemerovo region

The territory of Salair ridge, located in the center of the Kemerovo region, including good access roads, housing and utilities infrastructure to meet the demand for food and human resources, as well as the low industrialization, meets the considered requirements. The biodiversity studies were carried out, and the most significant factors were considered to study the possibility of using the territory Salair Ridge as a resort area.

The aesthetic factors that determine the attractiveness of the rest in the resort and recreational areas are of great importance, including the diversification of recreational activities from the point of view of the possibility of organizing mass sports and swimming, fishing and other activities.

The social and public importance of creating the resort area near Salair ridge is that the part of the population of Kuzbass will repair the health within the region limits, minimizing the transportation costs and using its own balneal factors (mineral water, clay, mud), which will increase the recreational opportunities of Kemerovo region.

In the study area there are some sanatory facilities, which are the possible basis for the further development of resorts, including sanatorium "Forest Lake", sanatorium "Kuzbass Pearl", ski complex "Gold Mountain", sanatorium "Magistral", sanatorium "Salair Silver".

The practical value of the existing infrastructure is the ability to build a network of sanatoriums and resorts in the Salair ridge area with a view to health care, restorative treatment and rehabilitation of the population of Kemerovo region and the surrounding regions. Another important socio-economic aspect of this process is the employment of the part of population of Guryevsk and Belovo areas, which is important taking into account the low employment in the sectors of economy [14].

Rich recreational resources of Kemerovo region require the economic evacuation for their efficient use, as shown in table 2.

Table 2 - Differentiation of positive effects from the achievement of the recreational potential of Kemerovo region

Evident effects	Implicit effects
The weakening or elimination of the negative effect of single-industry towns	Increase of investment and innovation appeal of the region
Additional demand for goods and services, including those produced in the region	Improvement of regional infrastructure (roads, transport facilities, life support systems).
The increase in tax deductions from the tourist services	Improvement of the image of the region, including at the expense of environmental "green" goodwill
Minimizing the cost of tourism, including those environmental spent from the personal savings of citizens	
Foreign tourist service substitution	

Table 2 shows that the tourism effect in terms of additional demand for goods and services is of great importance. This tourism feature generates a new form of consumer demand supported by "external" tourists for various goods and services, including, most importantly, the local production.

In modern conditions of the complex geopolitical and economic situation, the most important factor is savings of cost, planned to be spent in the process of consumption of recreational services. In particular, the significant difference in transport costs (up to 10 times or more) determines the priority of regional recreational services. At the same time there is the process of the substitution of foreign recreational services, which is also motivated by not only economic, but also geopolitical component. Thereat, the complex international situation may cause a multiplier effect, when the refusal of foreign recreational services will significantly increase domestic market segment, regional tourism and other components of recreational resources [15].

If we consider the implicit effects of the development of the recreational potential of Kemerovo region, the most important factor is the increase of the investment appeal of the region and avoiding "one-industry investment dependence" when the determining factor of investing is associated only with the development of coal mining and other related sectors of economy [16, 17].

Assessment of the adequacy of regulatory and procedural framework and strategic documents for the development of tourism and recreation activities in Kemerovo region has shown that the regulatory support of the industry meets the modern requirements. The main problem is the lack of regional vision and tourism development strategy, the implementation of which is necessary for the efficient development of tourism and recreation cluster of Kemerovo region.

Conclusions.

The study of investment appeal of the region's recreational potential has allowed the following conclusions:

1. the necessity of maintaining and developing the specially protected natural areas as the basis of the recreational potential of the Kemerovo Region is proved;
2. the factors and conditions that characterize the efficient functioning and development of the recreational system of Kemerovo Region are determined;
3. the analysis of recreational resources of Kemerovo Region is made;
4. the ecological and economic assessment of the Salair ridge territory with a view to the possibility of creating a resort zone is made;
5. the economic assessment of existing recreational resources with the differentiation of revealed positive effects is performed.

References.

[1] Galanina T V and Baumgarten M I 2013 Ecological consequences of technogenic action at carrying out of mountain operations Mining informational and analytical bulletin (scientific and technical journal) Issue 7 pp 288–292.
[2] Mikhailov V G, Koryakov A G and Mikhailov G S 2015 Ecological risk management in coal mining and processing Journal of Mining Science Vol 51 No 5 pp 930–936.
[3] Golofastova N N, Mikhailov V G and Galanina T V 2016 Environmental safety management of a coal mining enterprise The 8th Russian-Chinese Symposium. Coal in the 21st Century: Mining, Processing and Safety pp 372–376.
[4] Bulgakova O A 2009 Methodological approaches to evaluation of investment attractiveness of a recreational territory Problems of modern economics Vol 30 Issue 2 pp 437–441
[5] Vukovich G G and Belyaea TA 2014 The mechanism of investment appeal development of the industrial enterprises in the Krasnodar territory Theory and practice of social development Issue 11 pp 104–111.
[6] Zaitseva N A and Mekush G E 2015 Formation of favorable economic zones for tourism and recreation as a promising direction for tourism development in regions (based on the example of Kuzbass) Russian regions: looking into the future Issue 3 pp 90–93
[7] Tolstykh E V and Litvak A I 2011 Evaluation of the impact of tourism on the socio-economic development of the Kemerovo Region Actual problems of humanities and natural sciences Issue 5 pp 68–73.
[8] Mekush G E and Kalugin A V 2010 Indirect estimation of citizens willingness to pay for past environmental damage elimination: case study of Kemerovo city Journal of Siberian Federal University. Humanities & Social Sciences Vol 3 Issue 6 pp 891–902.
[9] Ermoshkin V V and Galanina T V 2006 Recultivation of disturbed lands and solution of environmental problems in Kemerovo region Mining Journal Issue 11 pp 88–89.
[10] Kovyazin V, Belyaev V, Pasko O and Romanchikov A 2014 Taxation indices of forest stand as the basis for cadastral valuation of forestlands Conf. Series: Earth and Environmental Science 21 012026.

[11] Adilov V B, Eliseev V A and Puzanov A V 2001 Siliceous mineral waters of the south of Western Siberia, and their therapeutic use Russian regions: looking into the future Issue 2 pp 41–43.
[12] Konchakova N V, Ushakova N S and Aikina T Yu 2016 Ecological risk assessment of Tomsk region groundwater used for drinking purposes IOP Conf. Series: Earth and Environmental Science 33 012022.
[13] Glyzina T S, Matugina E G, Bagamaev B M, Tokhov Yu M, Kolbysheva Yu V, Gorchakov E V, Sotnikova T V and Shilova A S 2016 Environmental monitoring of natural waters in Krasnodar and Stavropol Territories IOP Conf. Series: Earth and Environmental Science 33 012021.
[14] Santikayasa I P 2016 Development of an Integrated Agricultural Planning Model Considering Climate Change Conf. Series: Earth and Environmental Science 31 012042.
[15] Shinkevich A I, Lubnina A A, Koryakov A G, Mikhailov V G and Vodolazhskaya E L 2016 Economic aspects of risk management of stakeholders activities International Review of Management and Marketing Vol 6 Issue 2 pp 328–332.
[16] Mamzina T F 2011 Future development of small and medium enterprises in the city depending on one company (industry production) Journal "Economics: Yesterday, Today and Tomorrow" Issue 2 pp 31–52.
[17] Pavlov P V 2013 Special economic zones as a mechanism for efficient development of international investment and innovative activities World Politics Issue 1 pp 51–144.

Environmental-Economic Assessment Of Generation, Flow And Efficiency Of Use Of Production And Consumption Waste

V G Mikhailov, N N Golofastova, T V Galanina, T G Koroleva, Ya S Mikhailova

T. F. Gorbachev Kuzbass State Technical University
Vesennyaya st. 28, Kemerovo, Russian Federation, 650000

E-mail: mvg.eohp@kuzstu.ru

Abstract. The article deals with the issues of environmental and economic analysis of industrial and economic activities of an enterprise to assess the generation, flow and efficiency of production and consumption waste. The purpose of research is the analysis and the development of theoretical propositions for the functioning of the system of environmental and economic indicators for the effective management of production and consumption waste in the enterprise. The analysis of the existing systems of environmental and economic indicators taking into consideration the industry characteristics and the types of negative impacts is carried out. The main result of the study is the development of the system of environmental and economic indicators of production and consumption waste, adapted to the modern requirements. The results of the study can be recommended to support the effective management decision-making concerning waste management and the establishment of appropriate infrastructure.

Introduction

Management of waste treatment is a complex process, which includes legal, technical, economic and environmental aspects. The efficient management of waste treatment, ensuring the achievement of the objectives should be in accordance with the management theory when management is understood as a system of various forms and methods of influence of the subject of management on the object of management to achieve the objectives [1]. To improve the efficiency of waste treatment management treatment of management of waste is necessary to build a system of environmental and economic indicators characterizing the production process [2].

Material and methods

The object of study is the current system of accounting, flow and use of industrial and consumer waste. The subject of research is the environmental and economic efficiency of generation, flow and use of industrial and consumer waste. The study is based on the analysis of domestic and foreign literature on the problems of environmental-economic analysis of enterprise. Particular attention is paid to the methodology for calculating the environmental and economic indicators related to the dynamics of waste flow and usage. The statistical processing of environmental data of an industrial enterprise on the basis of the reporting form № 2-TP (wastes) – Report on generation, use, neutralization, transportation and disposal of industrial and consumer waste – has been carried out to test the proposed methodology. The elements of the system analysis and the results obtained by specialists in the field of environmental-economic analysis of the enterprise are also used in the paper.

Results and Discussion

Currently, there are different approaches to the assessment and interpretation of environmental and economic performance of an enterprise, including the study of waste flow, the efficiency of its use and the analysis of production waste volume. A number of authors determine the environmental performance indicators describing the negative impact on the environment:

• profits and revenue per 1 ton of production waste;

• profit and revenue value attributable to 1 ruble of charges for the negative impact on the environment;

• specific profit and sales revenue per unit of environmentally hazardous materials used [3].

Other experts have developed a set of criteria of environmental and economic sustainability of the enterprise, including the following elements:

• criterion of ecological safety of an enterprise;

• criteria of eco-efficiency of company as a comprehensive characterization of nature use efficiency – resource- and energy consumption, production waste volume, efficient use of consumed "external" and "internal" resources of an enterprise;

• criterion of constant improvement of indicators of development of environmental and economic system of an enterprise [4].

The environmental and economic indicators are of great importance for the comprehensive assessment of the environmental safety of the enterprise [5]. If we consider the current activities of the company, the developed schemes of production environmentalization, including the environmental-economic assessment of human impact on the environment, can be used to make the environmentally-oriented economic decision [6].

Environmental and economic indicators of production and consumption waste should be maximally adapted to the particular industry, which has its own characteristics. Due to the fact that the coal industry is a major source of waste generation, the environmental performance indicators of mining companies are divided into 3 groups: environmental, social and economic. Environmental indicators reflect the level of use of various natural resources and the environment state before and after the environmental protection measures, including:

• the level of achievement of the science-based standards or the established limits of use of natural resources and the environment state;

• the areas of damaged and reclaimed lands, land consumption of production;

• the amount of water consumption, the amount of recycled water, re-use of water in sequence, water consumption of products;

• the ratios of mineral resource recovering, the complexity of used mineral raw materials and waste-free production, mineral consumption of products;

• the total and specific (per unit of output) volumes of generation education, capture, utilization and entry of harmful substances into the environment;

• the productivity of agricultural lands, water bodies and forest lands located within the area of mining operations [7].

Other features are the environmental and economic indicators, designed for refineries. In this case, the main (see table) criteria of differentiation of these indicators are the following:

• the cohesion of economic and environmental aspects in one indicator at a time;

• the ability to determine the value of the indicator not only in volume but also in terms of value.

Table 1 - Classification of environmental and economic performance of manufacturing processes (as exemplified in refinery)

Criterion	Indicator types	Indicator examples
Purpose of indicators	Core operation indicators	Consumption of resources in core operations
	Auxiliary operation indicators	Consumption of resources in auxiliary operations
Application range	General	Energy intensity, material consumption
	Particular	Octane and cetane number (refining)
Evaluation method	Qualitative	The contents of harmful impurities in the waste water
	Quantitative	The volume of wastewater discharged
Aggregation	Special	Environmental capacity, energy consumption
	Integral	Index of environmental friendliness of production
Innovativeness of manufacturing processes	Indicators, characterizing the product output	Oil refining ratio
	Indicators, characterizing waste generated by technology	The volume of waste generation per unit of output, measured in terms of value

The table shows that the manufacturing process the oil refining industry is its differentiation into core and auxiliary operations. On this basis, all environmental and economic indicators can be attributed, respectively, to core and auxiliary operations. These indicators which are significant for other industries will determine the proportion of useful product and the waste volume of core and auxiliary operations, as well as to relate them to each other [1].

The efficiency of ecological and economic performance calculation depends significantly on the used tools. In particular, the flexibility and the power of the methods of fuzzy set theory can be considered as a promising and effective tool for the evaluation of environmental and economic security of industrial enterprises on the basis of the calculation of the system of ecological and economic indicators [8, 9, 10, 11, 12].

For most enterprises, the actual problem is the estimation of flow and efficiency of production and consumption waste. To this end, the system of basic ecological and economic indicators of production and consumption waste, presented in Figure 1, has developed on the basis of official statistical reporting forms.

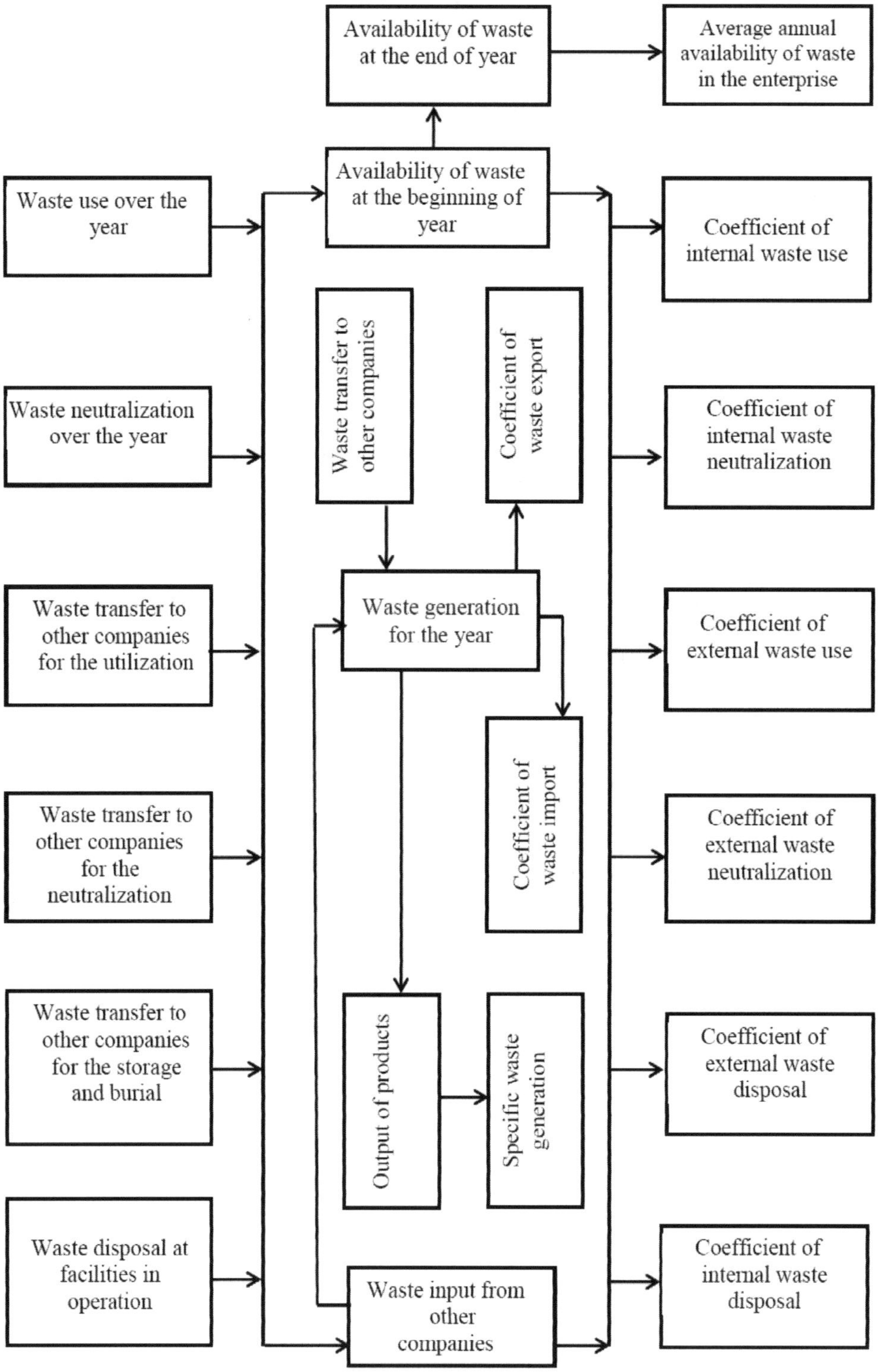

Figure 1 - Relationship of the basic environmental and economic waste indicators.

The formulas for the environmental and economic indicators of generation, flow and efficiency of use of production and consumption waste have been developed on the basis of the proposed scheme.

The average annual availability of waste in the enterprise (AAAWE) is determined from the formula:

$$AAAWE = \frac{AWBY + AWEY}{2}, \tag{1}$$

where AWBY – availability of waste at the beginning of year, tons; AWEY – availability of waste at the end of year, tons. This indicator "averages" the waste flow stream in the company during the year.

The specific waste generation (SWG) is computed using the following formula:

$$SWG = \frac{WGY}{OP}, \tag{2}$$

where WGY – waste generation for the year, tons; OP – output of products, thousands of tons. The presented indicator shows the waste content of the manufacturing processes of the enterprise

The coefficient of internal waste use (CIWU):

$$CIWU = \frac{WUY}{AWBY + WGY + WIOC}, \tag{3}$$

where WUY – waste use for the year, tons; WIOC – waste input from other companies, tons. This coefficient shows the level of autonomous use of waste at the enterprise

The coefficient of internal waste neutralization (CIWN):

$$CIWN = \frac{WN}{AWBY + WGY + WIOC}, \tag{4}$$

where WN – waste neutralization for the year, tons. The relative measurement of the coefficient reflects the internal waste neutralization capabilities of the enterprise

The coefficient of waste import (CWI):

$$CWI = \frac{WIOC}{WGY}. \tag{5}$$

This coefficient indicates the ratio of the "external" waste mass inflowing from outside companies and the own waste generated during the reporting year

The coefficient of waste export (CWE):

$$CWE = \frac{SWTOC}{WGY}, \tag{6}$$

where SWTOC – the summary waste transfer to other companies, tons. The presented indicator shows the ratio of the mass of waste directed to other companies, and the amount of own waste generated during the reporting year. The coefficients of the import and export of waste can be used in the analysis of the total waste mass balance.

The coefficient of external waste use (CEWU):

$$CEWU = \frac{WTOCU}{AWBY + WGY + WIOC}, \tag{7}$$

where WTOCU – waste transfer to other companies for the utilization, tons. This indicator represents the ratio of the amount of waste sent to third-party companies for the utilization, and own waste, including those obtained from external sources.

The coefficient of external waste neutralization (CEWN):

$$CEWN = \frac{WTOCN}{AWBY + WGY + WIOC}, \tag{8}$$

where WTOCN – waste transfer to other companies for the neutralization, tons. The coefficient shows the ratio of waste sent for the neutralization to outside companies, and own wastes, including those obtained from external sources.

The coefficient of internal waste disposal (CIWD):

$$CIWD = \frac{WDFO}{AWBY + WGY + WIOC}, \tag{9}$$

where WDFO – waste disposal at facilities in operation, tones. The coefficient under consideration characterizes the level of technical capacity of waste disposal coming from internal and external sources at own facilities in operation (sites, polygons).

The coefficient of external waste disposal (CEWD):

$$\mathrm{CEWD} = \frac{\mathrm{WTOCS} + \mathrm{WTOCB}}{\mathrm{AWBY} + \mathrm{WGY} + \mathrm{WIOC}}, \qquad (10)$$

where WTOCS – waste transfer to other companies for the storage, tones; WTOCB – waste transfer to other companies for the burial, tones. This coefficient shows the ratio of waste sent to third parties for the storage and burial and own waste, including those obtained from external sources.

The model calculation of environmental and economic indicators for the specific enterprise - JSC "Open-pit mine Inskoy" is shown to test the methods proposed in Table 2

Table 2 - The model calculation of basic environmental and economic indicators of production and consumption waste of JSC "Open-pit mine Inskoy" in 2014.

Indicator name	Unit of measurement	Value
Availability of waste at the beginning of year	tones	10.351
Availability of waste at the end of year	tones	444.037
Average annual availability of waste in the enterprise	т	227.194
Output of products	thousand of tones	2190
Specific waste generation	tones / thousand of tones	0.322
Waste use for the year	tones	0.025
Waste generation for the year	tones	706.206
Waste input from other companies	tones	0.000
Coefficient of internal waste use		$3.49 \cdot 10^{-5}$
Waste neutralization for the year	tones	0.000
Coefficient of internal waste neutralization		0.000
Coefficient of waste import		0.000
Summary waste transfer to other companies	tones	272.495
The coefficient of waste export		0.386
Waste transfer to other companies for the utilization	tones	0.000
Coefficient of external waste use		0.000
Waste transfer to other companies for the neutralization	tones	3.952
Coefficient of external waste neutralization		0.006
Waste disposal at facilities in operation	tones	0.000
Coefficient of internal waste disposal		0.000
Waste transfer to other companies for the storage	tones	0.000
Waste transfer to other companies for the burial	tones	268.543
Coefficient of external waste disposal		0.375

The analysis of Table 1 shows that the average annual availability of waste at the enterprise (227.194 tons) is significantly higher than the amount of waste at the beginning of the year at the company (10.351 tons), describing an active waste inflow during the year from internal and external sources. The indicator of specific waste generation characterizes the low waste content in manufacturing processes when 0.322 tons of waste are generated per 1 thousand of tons of output. Several calculated indicators have zero value: coefficients of internal waste neutralization, waste import, external use of waste and internal waste disposal. Two indicators are characterized by low values: the coefficient of internal waste use (3.49 10^{-5}) and the coefficient of external waste neutralization (0.006). This situation stems from the fact that the company has almost no waste use and disposal, including with the involvement of third-party companies. The comparison of the coefficients of external waste disposal (0.375) and internal waste disposal (0.000) shows that the company is actively uses the services of third parties, without developing its own infrastructure (sites and polygons).

The study conducted can be used to support the effective environmentally safe management decision-making to treat production and consumption waste and to develop the appropriate infrastructure in the enterprise [13, 14, 15].

Conclusions

The conducted study of environmental and economic evaluation of generation, flow and efficiency of use of production and consumption waste led to the following conclusions:

1. the analysis of different approaches to assessment and interpretation of environmental and economic indicators of the enterprise, including the study of waste flow, the efficiency of its use and the analysis of production waste volume is carried out;
2. the basic criteria and factors of formation of environmental and economic indicators, adapted to the conditions of a particular company are found;
3. the system of interconnected environmental and economic indicators that characterize the dynamics and the efficiency of production and consumption waste is developed;
4. the model calculations of the basic environmental and economic indicators of production and consumption wastes based on the data of industrial enterprises are done;
5. the proposals for improving the waste management system of the evaluated company on the basis of the assessment of macroeconomic indicators, taking into account the environmental characteristics obtained by the integrated monitoring system, are made.

References

[1] Shimova O S and Lopachuk O N 2005 Issues of cleaner production and consumption assessment Belarusian Economic Journal Issue 1 pp 113–120.
[2] Kuznetsov S N and Volynkina E P 2014 Integrated Model and Program for Strategic Planning of Solid Municipal Waste (SMW) Management at a Regional and Local Level. Ecology and Industry of Russia Issue 6 pp 43–47.
[3] Bezhenceva T V 2014 Express-assesment of "continuos improvement" as part of the environmental management system Herald of Omsk University Series "Economics" Issue 3 pp 4–10.
[4] Redina M M and Kalinin A R 2011 Economic criteria in the diagnostic of environmental sustainability of enterprises of oil and gas industry Mining informational and analytical bulletin Issue 9 pp 265–266.
[5] Dolzhenko E N, Monich A I, Kudryakov A G and Sazykin V G 2016 Model of expert evaluation the environmental management of the enterprise International Research Journal Vol 47 Issue 5 pp 75–78
[6] Yepifantseva Ye I 2015 Ecological economic mechanisms of accomplishing stable development of industrial companies Baikal Research Journal Vol 6 No 5 URL: http://brj-bguep.ru/reader/article.aspx?id=20374.
[7] Shuvalov Y V and Lobanov N Y 2000 Environmental and economic evaluation of efficiency of nature conservation activities of mining companies Mining informational and analytical bulletin Issue 2 pp 207–212.
[8] Rogachev A F, Shevchenko A A and Kuzmin V A 2013 Assessment of ecological and economic security of industrial enterprises by methods of fuzzy logic SPIIRAS Proceedings Issue 30 pp 77-87.

[9] Mikhailov V G, Koryakov A G and Mikhailov G S 2015 Ecological risk management in coal mining and processing Journal of Mining Science Vol 51 No 5 pp 930–936.
[10] Shinkevich A I, Lubnina A A, Koryakov A G, Mikhailov V G and Vodolazhskaya EL 2016 Economic aspects of risk management of stakeholders activities International Review of Management and Marketing Vol 6 Issue 2 pp 328–332.
[11] Golofastova N N, Mikhailov V G and Galanina T V 2016 Environmental safety management of a coal mining enterprise The 8th Russian-Chinese Symposium. Coal in the 21st Century: Mining, Processing and Safety pp 372–376.
[12] Kiseleva T V, Mikhailov V G and Karasev V A 2016 Management of local economic and ecological system of coal processing company IOP Conf. Series: Earth and Environmental Science 45 012013.
[13] Vyvarets A D, Distergeft L V and Naboichenko E S 2000 Economic incentives for processing anthropogenic formations Non-ferrous metals Issue 3 p 4.
[14] [Vyvarets A D, Karelov A S and Polyanskiy A M 2007 Conceptual bases of mechanism for evaluation of environmental and economic efficiency of waste-free production Russian Entrepreneurship Vol 86 Issue 2 pp 143-146 URL: http://bgscience.ru/lib/2022/.
[15] Pinaev V E 2013 Ecological-Economic Modeling of the Enterprises Bulletin of Moscow University Series 6 "Economics" Issue 3 pp 54-62.

Photolytic AND Catalytic Destruction of Organic Waste Water Pollutants

V F Torosyan[1, a], E S Torosyan[1, b], S O Kryuchkova[1], V E Gromov[2]

[1] Yurga Institute of Technology TPU Affiliate,
Leningradskaya 26, Yurga, Russian Federation, 652055
[2] Siberian State Industrial University,
Kirova st. 42, Novokuznetsk, Russian Federation, 654007

E-mail: [a]torosjaneno@mail.ru, [b]eltorosyan@mail.ru

Abstract. The system: water supply source – potable and industrial water – wastewater – sewage treatment – water supply source is necessary for water supply and efficient utilization of water resources. Up-to-date technologies of waste water biological treatment require for special microorganisms, which are technologically complex and expensive but unable to solve all the problems. Application of photolytic and catalytically-oxidizing destruction is quite promising. However, the most reagents are strong oxidizers in catalytic oxidation of organic substances and can initiate toxic substance generation. Methodic and scientific approaches to assess bread making industry influence on the environment have been developed in this paper in order to support forecasting and taking technological decisions concerning reduction of this influence. Destructive methods have been tested: ultra violet irradiation and catalytic oxidation for extraction of organic compounds from waste water by natural reagents.

Introduction

Food industry always consumes a considerable amount of raw materials and energy, its waste water even exceeds raw materials and is removed from production departments as polluted water. Discharge of some pollutants into the atmosphere is considered hazardous. However, territorial inspecting authorities of consumer rights protection and human wellbeing do not control discharge into the atmosphere and waste water produced by many of these enterprises in Russia.

The paper is aimed at development of methodic and scientific approaches to assess influence of bread making industry on the environment, which supports forecasting and taking technological decisions to reduce this influence; as well as at testing destructive methods: ultraviolet irradiation and catalytic oxidation for extraction of organic compounds from waste water.

The paper has revealed key factors of anthropogenic impact on the environment of a bread making enterprise in Kemerovo region. Those include the sum total of internal (equipment, efficiency, fuel type, technology, assortment) and external (raw material quality) characteristics of the enterprise. Plus, equipment and technologies used in this enterprise are interrelating to amount and quality of pollution affecting eco-systems.

For example, for the manufacture of parts Dough as a constructive material used alloy steel X18H10T. It is important to note that when selecting steels in contact with food environment, it is necessary to take into account their properties, operating conditions and design details of the nature of loads and stresses. Alloyed heat-treated steels have higher mechanical properties complex than carbon. They are better calcined possess higher strength and hardness. [1]

Pollutants emitted into the atmosphere by this enterprise have been detected in the following phases: delivery, storage and preparation of raw materials, dough making, baking, and transportation.[2] The analysis of data on emitted pollutants shows hazard "category" is made of delivery, storage, and raw material preparation (18.96%), dough making (0.02%), baking (60.65%), transportation (20.37%). Total value of specific discharge into the atmosphere over the complete technological cycle is given in Table 1.

Table 1 - Specific discharge into the atmosphere

Solids, tons per year		Gaseous substances, tons per year	
Ferric oxide	0.124990	Nitrogen dioxide	2.816618
Manganese	0.002289	Nitrogen oxide	0.457525
Black carbon (soot)	12.425382	Sulfurous anhydride	2.701164
Wood dust	0.160000	Carbon dioxide	33.225803
Flour dust	0.330800	Fluoric gaseous compounds	0.000102

Waste water is mainly produced in the following phases: delivering and washing raw materials and equipment, and cooling system utilization. [3] Therefore, waste water is polluted generally by raw material remains, half-products, ready-made products and cleaning agents. These are complex polydisperse systems consisting of suspended particles, organic and mineral substances. [4,5] As far as we know, concentration of suspended particles in waste water of many bread making enterprises in Russia exceeds the mean industry standard (170 mg/l) and is three times higher than the permissible concentration for waste water (212 mg/l). BOD rate exceeds by 38% the mean industry standard (326 mg/l) and is 3.5 times higher than the permissible concentration for waste water (35.3 mg/l) [Orenburg].

Organic substances include: xylene, butanol, ethanol, acetic acid, formaldehyde, ethylene glycol etc. Table 2 presents their components in waste water of bread making enterprise, their hazard class and removal efficiency in conditions of decontaminated waste water discharge into the objects of household and general water consumption.

Experiment

Two groups of methods are applied for waste water treatment: regenerative and destructive ones. Regenerative methods make it possible to decontaminate waste water and extract organic substances and then use them. Destructive methods are applied if extraction of impurities from waste water is impossible or commercially useless. Main destructive methods of dissolved organic substances extraction from waste water include thermal-oxidation and oxidation.

Photolytic destruction of organic substances via ultraviolet irradiation [7] and catalytic oxidation was used for research into waste water treatment of a bread making enterprise. [12,13] Sample solution was made of components presented in Table 2; their concentration in waste water exceeded significantly the permissible concentration.

IOP Conf. Series: Earth and Environmental Science **50** (2017) 012039 doi:10.1088/1755-1315/50/1/012039

Table 2 - Organic substances in waste water of a bread making enterprise

Pollutant	Concentration in biological treatment, mg/l	Removale fficiency, %	LHI	Permissible concentration	Hazard class	Discharge, t/year	Discharge, mg/l
Xylene (xylol)	1.0	50	org.	0.05	3	0.188600	10.46
Toluol	15	50	org.	0.5	4	0.080700	4.48
Ethanol	14	70	general	-	-	3.305800	183.6
Ethyl acetate	-	-	general	-	-	0.019200	-
Acetic acid	-	80	org.	1.0	3	1.420000	78.89
Nitrobenzene	-	70	Standard	0.2	3	0.125000	13.61
Formic acid	-	85	BOD	-	-	0.146120	15.922
Ethylene glycol	1000	65	Standard	1.0	3	0.084560	9.177
Formaldehyde	100	65	Standard	0.05	2	0.0345600	3.76
Petrol						0.040077	
Kerosene						0.001278	

Ultraviolet irradiation was carried out by 36 watt power lamp (wavelength 185.6 nm). 20 mg/l (concentrated) hydrogen peroxide and titanium chloride $TiCl_3$ (at a rate of 0.2 g titanium per 1 liter water) were gradually injected beforehand into 1.5 liter reference water (sample solution). The solution was aerated by a bubbler. Table 3 presents the results of experiments.

Table 3 - Ultra violet destruction of organic substances in reference solution

№	Volume of solution, mg	H_2O_2. concentration g/l	Ti3+ concentration mg/l	Chemical oxygen demand, mgO/dm3
1	200	-	-	886.8
2	200	10	0.1	547.3
3	200	20	0.2	408.3
4	500	30	0.3	463.6
5	400	40	0.3	450.2

Organic compounds were also destructed as the result of catalytic oxidation. We used the system clay-hydrogen peroxide in our experiment. The results are provided in Table 4.

Table 4 - Catalytic oxidation of organic substances

№	Volume of solution, ml	H_2O_2.concentration ml	Clay, g	Chemical oxygen demand, mgO/dm3		
				day	3 days	5 days
1	200	-	-	886.8	-	-
2	200	4.0	8.0	843.6	680.5	524.7
3	200	6.0	10.0	755.0	605.7	406.5
4	400	10.0	10.0	775.8	634.5	446.0
5	500	10.0	10.0	808.0	723.6	465.6

Discussion of results

As the result of ultraviolet irradiation oxygen dissolved in water transforms into ozone, which, in its turn, can oxidize organic substances if an active catalyst – titanium hydro peroxide is in solution. And the catalyst itself furthers decomposition of excessive hydrogen peroxide.

Presented in Table 4 results of investigation into destruction of organic substances under ultraviolet irradiation demonstrate the optimal correlation of H_2O_2.mg/l and Ti_3+mg/l concentrations, conditioning nearly double decrease in chemical oxygen demand, is 20 mg/l H_2O_2 and 0.2 mg/l Ti3+.

It is known that chlorine, sodium hypochlorite, chlorine dioxide and other reagents are used to obtain this correlation and can initiate generation of toxic substances. Fenton reagent (ferrum II sulphate – hydrogen peroxide) is a quite efficient oxidizer used for removal of organic substances from waste water [Fenton].

Clay is fine-dispersed according to its granulometric composition, and its chemical composition is presented in Table 5.

Table 5 - Chemical composition of fireproof clay

Clay	Oxide concentration, %									
	SiO_2	Al_2O_3	Fe_2O_3	TiO_2	CaO	MgO	K_2O	Na_2O	SO_3	III
Nizhe-Uvelsk deposit	54.27-63.09	22.63-29.04	2.60-3.28	1.35-1.53	0.65-1.05	0.62-0.95	0.65-1.00	0.24	-	7.62-10.03

Ferrum III, aluminum, titanium in the chemical composition of clay in the system clay – hydrogen peroxide facilitate processes of catalytic oxidation and coagulation [14]. Moreover, removal efficiency of organic substances from waste water calculated in experiment according to chemical oxygen demand (mgO/dm^3) depended on the process duration and order of reagent injection into the solution. If hydrogen peroxide was injected before clay, finally chemical oxygen demand was lower in comparison with another succession.

The data in Table 5 on catalytic oxidation of organic substances by natural reagent – clay show correlation 30mg/l H_2O_2.and 5% (mass) clay facilitates double reduction of chemical oxygen demand over 5 days.

Conclusion

1. Complex methodological approach introduced in this paper to assess bread making industry influence on the environment makes it possible to determine exposure degree of the enterprises on nature and develop first-priority measures of control over technological processes. The latter are to minimize anthropogenic impact at the most hazardous phases.
2. Intensification of in-receipt inspection of raw material quality can reduce pollutant formation.
3. Photolytic destruction of organic substances under ultraviolet irradiation and catalytic oxidation favours waste water treatment and decrease in chemical oxygen demand.
4. Clay – a natural reagent used for oxidative and destructive treatment of waste water is an efficient substance to shorten chemical oxygen demand in waste water without supplementary preparation.

References

[1] Chinakhov D A Gas Dynamic Control of Properties of Welded Joints from High Strength Alloyed Steels. China Welding. Vol. 23 No.3 (2014) pp. 27-31.
[2] Auerman Yu Ya Technology of bread making production: students' book Yu Ya Auerman; edit. by Yu I. Puchkova. – Saint-Petersburg: Profession, 2003. 416 p.
[3] Belov S. V. Environment protection S. V. Belov. M.: Higher school, 1983. 263 p.

[4] Volokhova L T Waste water of bread making enterprises as a source of pollutants L T Volokhova et al. Ecology and Industry. 1998. N 12. pp. 23-30.

[5] Kasperovich, V Yu Waste water of yeast and bread making enterprises and its utilization V Yu Kasperovich, G. B. Zinyukhin, A. V. Bykov Vestnik of Orenburg State University. 2001. No2. pp. 133-135.

[6] Bacherikova A K Influence of basicity of aluminum and ferrum sulphates on waste water treatment of fat-and-oil industry / Bacherikova A K, Tarasova T V et al. Proceedings of Higher Schools: Chemistry and Chemical Technology. 2000. V 43. Issue 1. pp. 85-87.

[7] Sosnina N A Photolytic destruction as measure to remove organic compounds from waste water N A Sosnina, A V Shtareva Vestnik of Tobolsk State University. Building and Architecture. 2011. № 3 (22) pp. 73-84.

[8] V F Torosyan, E S Torosyan. Development of mixture "clay – electro-smelting slag" for ceramic brick Applied Mechanics and Materials. 2014. Vol. 682. p. 480-484.

[9] Zaytseva S G Ultraviolet irradiation: environmental friendly method of disinfection S G Zaytseva Ecology of Production. 2007. 4.

[10] Pat. 2142915 of the Russian Federation, MPK 6C 02 F 1/32. Treatment of waste water containing organic impurities N K Zaytsev, D V Krasny, G M Zimina. № 99113308; Appl. 30.06.99. Publ. 20.12.99. Bull. № 35.

[11] Shcherbakova G S Destruction of organic impurities in water under ultraviolet irradiation: Thesis Shcherbakova G S; Russian Chemical and Technological University. – M., 2004.

[12] . Kashkovsky V.I. Catalytic Destruction of organic pollutants Kashkovsky V. I. Water treatment. 2012. 7. - pp. 13-16.

[13] Nikitin A A Introduction of local treatment facilities in bread making industry A A Nikitin Bread Baking in Russia. 2002. 2. pp. 38-39.

[14] Mal chik A. G. , Litovkin S. V. , Rodionov P. V. Investigations of physicochemical properties of bottom-ash materials for use them as secondary raw materials // IOP Conference Series: Materials Science and Engineering. 2015. Vol. 91, Article number 012081.

[15] Torosyan V. F. , Torosyan E. S. , Borovikov I. F. , Yakutova V. A. Oxidative activity of heated coal affected by antypirogens // IOP Conference Series: Materials Science and Engineering. - 2016. Vol. 127. 1, Article number 012022.

IOP Conf. Series: Earth and Environmental Science **50** (2017) 012040 doi:10.1088/1755-1315/50/1/012040

Integrated Approach To An Efficiency Assessment Of Self-Organizing Textile Materials Packages In The Subnormal Climate

M Rodicheva, A Abramov[a], P Kanatnikova, N Kanatnikov[b], G Kharlamov

Orel State University named after I.S.Turgenev,
Naugorskoe shosse 29, Orel, Russian Federation, 302020

E-mail: [a]Ant-lin88@mail.ru, [b]NVKanatnikov1989@gmail.com

Abstract. Design of heat-shielding clothes for subnormal climate still remains one of the unsolved problems of complex security. The solution is connected with the use of the self-organizing textile materials. Reasoning of textile package optimal selection requires the development of a comprehensive approach combining theoretical and experimental researches.

1.Introduction

One of aspects of complex security is creating of the demanded working conditions for workers including using the uniform. Its quality is estimated using the compliance of materials properties indicators and structure with regulatory requirements and is confirmed in the certification course for the majority of the exploited sets of overall. However, the statistic data shows that there is no essential decreasing of the level of professional-caused catarrhal diseases. In our opinion, it is connected with the shortcomings of standard requirements to the level of heat-shielding properties. For example, the size of the heat conductivity coefficient is established as a constant which is defined by the configuration of a set of materials; the uniform is effective only in the narrow range of environmental factors fluctuation. Compliance of uniform quality index with requirements under essential changes of values of external factors does not guarantee its efficiency. That is why regulatory requirements for the heat-shielding properties are established not for all climatic conditions. For example, they are not established for subnormal climate - in this case the temperature varies between -5 and +10 centigrade, fitful wind and possible precipitation are noted.

In these conditions uniform efficiency increase is connected with the development of new research techniques of a non-stationary heat mass exchange in textile materials packages. Special attention in scientific literature is paid to the solution of this problem. Methods of mathematical model construction of the heat mass exchange and its solving are developed within the theoretical researches. The essential contribution to this work was made by D. Ambesi, A. K. Haghi, M. Sobera et al [2, 6, 9].

On the basis of these methods a number of mathematical models of a heat mass exchange is offered [7,10,12]. R. Barauskas's model allows to optimize the structure of a textile materials package by parameters of water vapor and air permeability [7]. Model of C. Voelker et al allows to estimate the change of thermal comfort parameters of the person in clothes within non-stationary influences of environment [10]. Model of Y. Zhang et al allows to calculate the intensity of a complex heat mass exchange in "the person - clothes - environment" system [12].

During the course of experimental works the assessment of particular coefficients of quality of materials and an assessment of the influence of these indicators on a thermal condition of a person are carried out. Water vapour permeability as an important parameter is studied by W. Gibson, P.Yip et al on the example of modern membrane fabrics [4, 5]. M. Richards investigated the influence of humidity on the heat exchange between the person and environment at a temperature of +10°C [8].

For an assessment of efficiency of uniform under operating conditions a number of researchers develop accurate thermal models of the person - thermal manikin. I. Holmer, H. Nillson et al have made an important contribution to the development of methodology of such researches [7, 11].

Results of experimental and theoretical studies did not become widespread in Russia in a design of special uniform for conditions of subnormal climate. It is connected to complexity of coordination of theoretical and experimental results as well. This problem can be solved by using a single methodology of development of mathematical models of a heat mass exchange in the packages of materials and elements of an experimental complex on a single methodological foundation.

A similar approach is realized by authors on the example of research of complex processes of a heat mass exchange in the system "the person - clothes - environment" in the conditions of subnormal climate.

For this purpose, it is offered to use the experimental complex (fig. 1), allowing to investigate dynamics of thermal properties of textile fabrics at change of heat exchange conditions and intensity of a non-stationary heat mass exchange in packages of materials as well. At the same time on a thermal model of an element of a human body (position 1, figure 1) with an internal source of heat (position 2) the studied sample (position 3) is placed. Specialized devices (position 4-6) allow to model the influence of wind, precipitations and sweat on the intensity of processes of a heat mass exchange in the studied sample. Also the complex is equipped with devices for an assessment of a temperature field in a sample and environment for measuring the intensity of heat generation in a model of a human body element.

The methodical basis of an experimental complex allows to investigate dynamics of heat-shielding properties of packages of materials within the change of conditions of a heat mass exchange process. Therefore, with the help of this technique the efficiency of packages of the self-organizing textile materials can be investigated; this will allow to solve the problem of protection of the person from overcooling in subnormal climate.

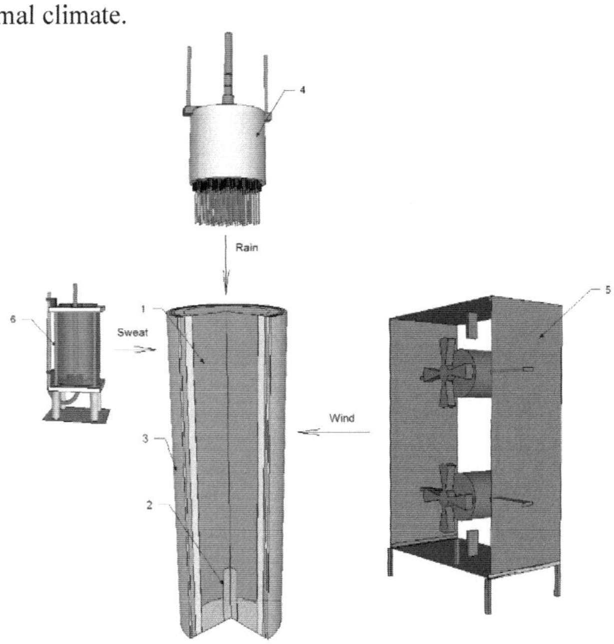

Figure 1 – Design of an experimental complex

2. Results of research of thermal properties of some samples of the self-organizing textile materials

Options of packages of the self-organizing textile materials in which optimum adaptive properties to change of conditions of a heat mass exchange of the person with the environment in the range of temperatures of 0°C ÷ +20°C are reached are presented in the Table 1. These temperatures correspond to the lower bound of a comfort zone and subnormal climate.

Table 1. Options of a selection of the structure of the materials package for conditions of subnormal temperatures

Recommended temperature range	Package configuration number	Package layers		
		Lower (underwear)	Medium	Upper
0 °C ÷ +10 °C	1	Knitted fabric of the Outlast brand	Knitted fabric of the Polartech Thermal Pro brand	Fabric of the Softshell brand
+5 °C ÷ +20 °C	2	Knitted fabric of the Outlast brand	-	Fabric of the Softshell brand

It is necessary to confirm the efficiency of packages by objective data for the use of these recommendations in the course of heat-shielding clothes production. The authors collected data on mechanical and thermal properties of the considered samples of the self-organizing textile materials (table 2).

Adaptive properties of the samples are formed owing to the original physical and chemical processes which are proceeding at the mesolevel of material structure. We have already analyzed the influence of these structural features on the character of a heat mass exchange in packages of materials earlier [1]. We offer to conduct the assessment of self-organization processes in the samples of materials within the change of heat exchange conditions by researching the dynamics of coefficients of heat conductivity and heat emission within the change of heat mass exchange conditions. Figure 2, a demonstrates the dynamics of heat conductivity coefficient of all considered samples of the self-organized materials; figure 2, b shows the results of research of a heat emission coefficient of Softshell fabric as a material of top.

Researches were conducted on the basis of the theory of similarity of thermal processes whereby the intensity of heat exchange on a border "a solid body - gas" is defined by a difference of temperatures between a surface and gas. In the considered range of temperatures (0 ÷ 20°C) the size of a temperature gradient is varied in the range of 10 ÷ 40°C. During the experiment these conditions were modelled using the heating of thermal model by this value concerning the environment. Therefore the temperature of thermal model is 32 - 72°C, temperature of samples 30 - 70°C.

Table 2. Samples of the self-organizing materials for clothes of subnormal climate conditions

№	Material trademark	Type of material	Fibrous structure	Weave	Area density, g/m^2	Thickness, мм	Specific heat, J/(kg·K)
1	Oultast	knitted fabric	100% PL*	Stockinette structure	190	0.5	1900
2	Polartech Thermal Pro	knitted fabric	100% PL	Lined	237	1.9	2000
3	Softshell	fabric	100% Ny*	Linen	128	0.3	1100

* - polyester fibers; ** - nylon fibers.

IOP Conf. Series: Earth and Environmental Science **50** (2017) 012040 doi:10.1088/1755-1315/50/1/012040

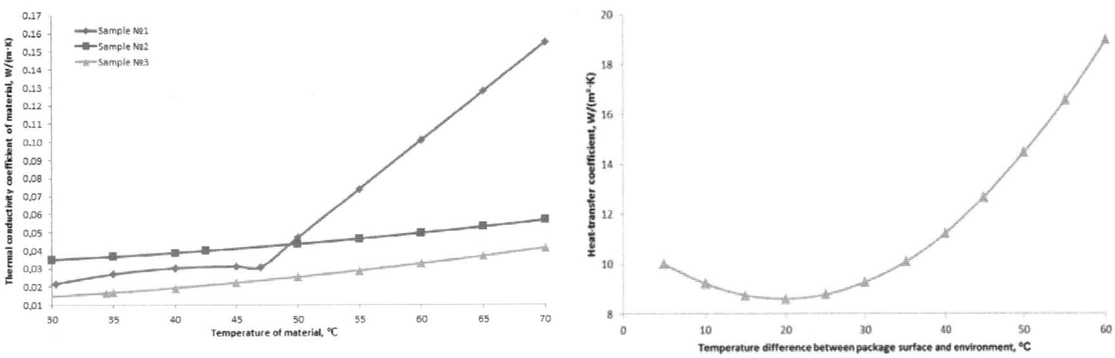

a) b)

Figure 2 – Results of research of thermal properties of the self-organizing textile materials

As follows from the obtained data, processes of self-organization are displayed in violation of linear change of thermal properties of a sample; that is especially remarkable on the example of heat conductivity coefficient of knitted fabric of the "Outlast" brand and heat emission coefficient of the "Softshell" brand fabric.

3. Finite element model of heat exchange in multilayer textile package

The package of materials is presented in the cover form (position 2, figure 3, a), which is evenly fitting the cylindrical element of a human body (position 1, figure 3, a) during the designing of mathematical model of a heat mass exchange. An impact of a wind, rainfall and sweating was not considered during the modelling. Therefore the heat transfer from a surface of a package is carried out by convection and thermal radiation.

Thermal energy is produced in a body element in an amount of Q_{met}, W and is transferred to a package of materials with a density of a heat flux q_{met}, W/m^2. Due to this, temperature t_{in}, °C is maintained on its internal surface. The vertical cut of a package of materials, which is perpendicular to a vector of a heat flux, is accepted as the design geometry of the model (position 3, figure 3, a; figure 3, b).

Textile materials are porous materials, that is why the heat in their structure is transferred by all mechanisms of transfer, but heat conductivity is dominating. Therefore, the equation (1) was chosen to describe the thermal processes in the structure of a package.

$$\rho C_p \frac{\partial T}{\partial t} + \nabla \cdot \left(-k\nabla T + \rho C_p T u\right) = Q \tag{1}$$

where T – is the temperature; ρ is the density; Cp – is the heat capacity at a constant pressure; k is the thermal conductivity; Q – is and heat source or heat sink; u is the velocity field.

The equation is written down in the form containing an environment speed vector that allows to connect the process of heat transfer in a package of materials with a heat emission. Value q (2) was expressed through the equation (1) for calculation of a field of heat flux density.

$$q = -k\nabla T + \rho C_p T u \tag{2}$$

In the solution of the problem of heat conductivity basic data are: value of temperature on an internal surface of a package of materials, values of coefficient of heat conductivity of layers and the law of heat exchange of a package with the environment in the form of heat transfer coefficient.

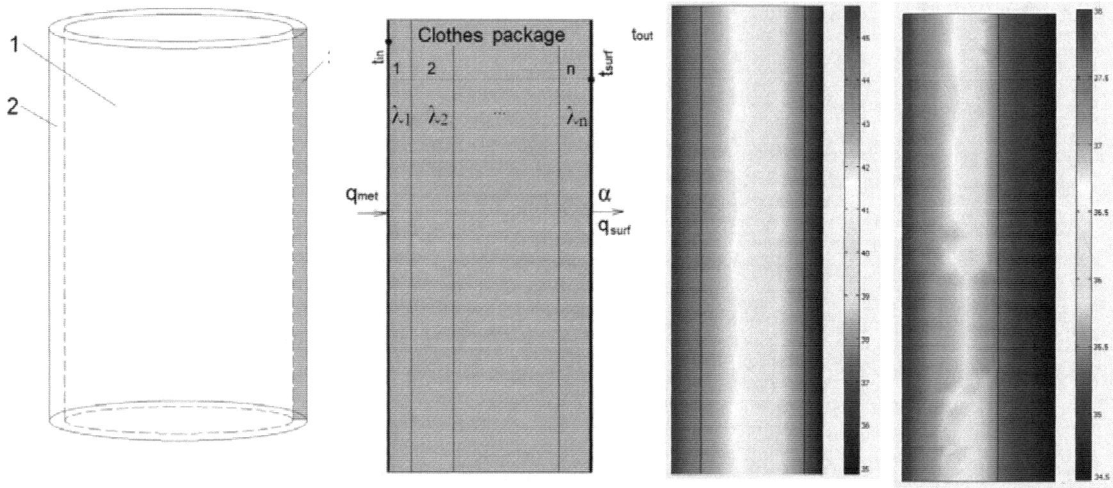

a) b) c) d)

Figure 3 - Stages of mathematical modeling of heat exchange in the packages of the self-organizing textile materials

Therefore the combined boundary conditions are used for its solving:

$$-n \cdot q = q_0 + h(T_{\text{inf}} - T) \tag{3}$$

where n – is the normal vector of the boundary; q_0 – is inward heat flux, normal to the boundary; h – is a heat transfer coefficient.

This form allows to connect an obvious heat flux with a heat transfer coefficient as functions of ambient temperature.

Author's method of experimental studies of a heat mass exchange in the system "the person - clothes - environment" allows to determine the value of heat transfer coefficient of the self-organizing materials. Therefore, in mathematical model the following form (3) of the equation was used for calculation of heat transfer intensity:

$$-n \cdot \left(-k\nabla T + \rho C_p uT\right) = q_0 + h(T_{\text{inf}} - T) + \varepsilon\sigma\left(T_{amb}^4 - T^4\right) \tag{4}$$

where ε –is the emissivity: T_{amb}– is the ambient temperature

The temperature of a surface of the package of materials is defined by solving the thermal conductivity equation. Therefore, boundary conditions for the heat transfer equation can be written in the following form (5):

$$q \cdot n = \left(\rho C_p uT\right) \cdot n \tag{5}$$

On the next step one solution (figure 3 c, d) was obtained for each configuration of sets of a materials package. For experimental verification of these solutions the package of materials was placed on the thermal model of an element of the human body. Primary converters of temperature were fixed on the each of layers. The thermal model of an element of a body was heated up to the working temperature determined by the size of the modelled gradient of temperatures. Thus, during the test of the first package the size of the temperature of a working surface was 45.8°C and 37.8°C for the second package.

For the next 90 minutes of experiment temperature of layers, thermodynamic parameters of air before and after heat exchange with a surface of a package of materials, and heat generation parameters in the thermal model of an element of a human body were measured. According the obtained results the comparison of experimentally determined distributions of a temperature field and field of heat flux density in a package with the results of the numerical decision was carried out (figure 4).

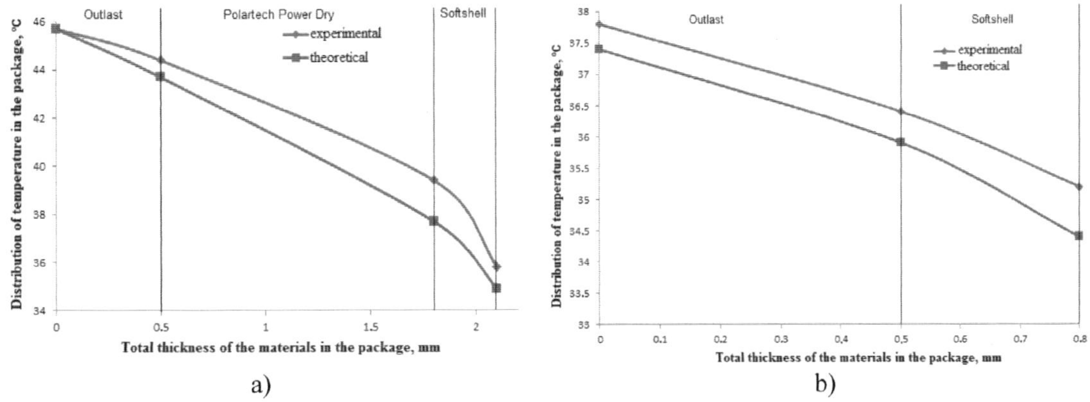

Figure 4 – Comparison of results of experimental and model researches

The maximum divergence between theoretical and experimental results made: on temperature no more than 4.3%; on the density of a heat flux is not higher than 3.5%. Therefore the mathematical model a heat mass exchange in the packages of the self-organizing materials can be used for the theoretical research of processes of a heat mass exchange in "person - clothes - environment" system.

4. An assessment of efficiency of packages of the self-organizing materials in the conditions of comfortable temperatures and subnormal climate

At the following stage of researches the complex of numerical decisions in the form of distribution of a field of temperature and density of a heat flux in a package of materials (figure 5) was found. Decisions are received under changing of ambient temperature with 5°C interval.

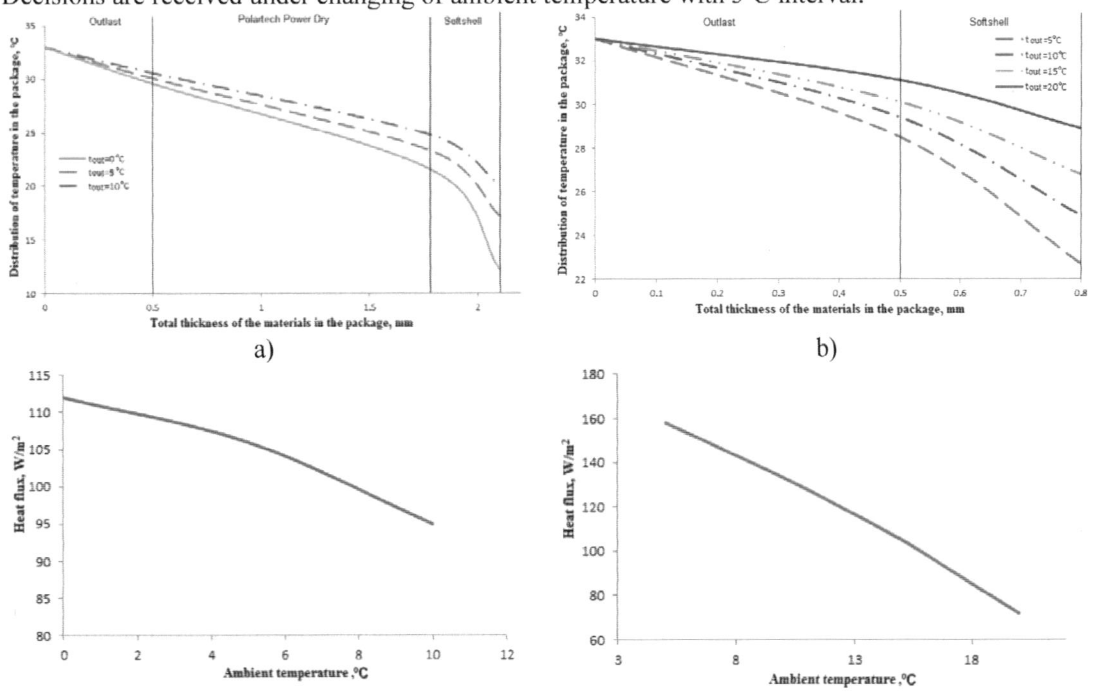

Figure 5 – Results of numerical solutions

IOP Conf. Series: Earth and Environmental Science **50** (2017) 012040 doi:10.1088/1755-1315/50/1/012040

According to the obtained data the ratio between the heat source from human metabolism during performing a physical activity of work difficulty categories IIa, IIb and III, and a heat emission in the environment through the packages of self-organizing materials for each configuration option from table 3 is calculated.

According to the SanPiN 2.2.2.540-96 requirements, continuous being in the conditions of the lowered temperatures should not exceed one and a half hours, after that time it is necessary to do the regulated break lasting 5-20 min in the warmed place. Therefore the estimated time made 60 and 90 min.

The received results need to be compared with requirements MR 2.2.7.2129-06; according to this document, maximum permissible change of heat content of an organism on cold is 5.2 kJ/kg during the work within 60 min and about 5 kJ/kg during the work within 90 minutes.

Table 3. The rate of gaining heat accumulated by an organism under exploited the clothes of the offered materials packages, kJ/kg

Work difficulty category	Ambient temperature, °C						
	0	5		10		15	20
	Package 1	Package 1	Package 2	Package 1	Package 2	Package 2	Package 2
After 60 minutes of continuous work							
IIa	0.9	1.0	0.1	1.2	0.6	1.0	1.6
IIb	1.4	1.5	0.6	1.6	0.5	1.5	2.0
III	2.4	2.5	1.6	2.6	2.5	2.5	3.0
After 90 minutes of continuous work							
IIa	1.4	1.5	0.4	1.8	1.7	3.1	4.9
IIb	2.0	2.2	1.7	2.5	3.0	4.4	6.1
III	3.5	3.7	4.7	4.0	6.0	7.4	9.1

5. Conclusion

The obtained results were assessed by the SanPiN 2.2.2.540-96 and MR 2.2.7.2129-06 requirements, therefore it is possible to make the following conclusions:

1. During operation of clothes on the basis of the recommended packages of the self-organizing materials, heat source from the metabolism of person prevails over a heat emission during working categories of IIa, IIb and III difficulty. Therefore the rate of gaining heat in an organism of the worker will accumulate.

2. The heat excess value meets the requirement to the maximum thermal condition of the person at the duration of a work of 60 minutes. The maximum thermal condition at the duration of a work of 90 minutes for configuration set №2 of clothes is upset in the case of:

- performance of the work of IIb difficulty category at ambient temperature +20°C;
- performance of the work of III difficulty category as the ambient temperature varies between 10°C and +20°C.

3. The recommended service conditions of packages of the self-organizing materials are crossed in the range of temperatures of +5 °C of ÷ +10 °C. During the work within 60 minutes the package No. 2 is preferable under all considered conditions since the rate of gaining heat accumulated in an organism in this case is less. During the work within 90 minutes efficiency of packages of materials may be different because of the effects of self-organization. The package No. 2 remains more preferable for carrying out the working category of IIa and IIb at the temperature of the air of -5 °C. The package No1 is more preferable in need of performance of work of category of weight of III, and also works of category of weight of IIb at the air temperature of +10 °C.

References

[1] Abramov A V, Rodicheva M V, Uvarov A V. Biotechnical bases of designing the clothes for protection against adverse temperatures. Biotehnicheskie osnovy proektirovaniya odezdy dlya zashchity ot neblagopriyatnykh temperatur [monograph], Orel, State University-ESPC., 2014, 137 p.

[2] Ambesi D, Flow, heat and mass transfer through CBRN protective clothing Proefschrift Politecnico di Torino, Italy 110 p.

[3] Barauskas R. etc. Computional Analysis and experimtnal investigation of the heat and Moistre transfer in multilayer textile package Materials Science (Medžiagotyra) Vol. 15, No.1. 2009 p. 80-85.

[4] Gibson P. Schreuder-Gibson H P. Yip B. Denker H. Bennadi Moistrue transport for reaction enhancement in fabrics / research article Journal of textiles Vol. 1 – 2013 – 8 p.

[5] Gibson P. Water-Repellent treatment on military uniform fabrics: physiological and comfort implications / report document 1.11.2006 – 6 p.

[6] Haghi A K. Heat & Mass transfer in textiles Published by WSEAS Press 2011 121 p.

[7] Nilsson H. Thermal comfort with virtual manikin methods / Building and Environment – Vol. 42, 2007 p 4000-4005.

[8] Richards M. Dry and wet heat transfer through clothing dependent on the clothing properties under cold conditions – International journal ofoccupational safety end ergonomics 2008 – Vol.14 №1. p 69-76

[9] Sobera M P. Flow, heat and mass transfer through protective textilles. Proefschrift Technische Univeriteit Czştochowa geboern te Iłza, Polen 174 p.

[10] Voelker C, Heat and Moistre transfer through clothing, IBPSA Building Simulation 2009, Univercity of Strathclyde Glasgow Scotland, p.1360-1366

[11] Wang F, Kuklane K., Holmer I. Development and validity of a universal ampirical equation to predict skin surface temperature on thermal manikins / Journal of thermal biology Vol. 35 2010 – p. 197-203

[12] Zhang Y. Noveto D Y. Ji Human envoironmental heat transfer simulation with CFD – the advances and challenges Building Simulation – 2009 Eleventh International IBPSA conference Glasgow Scotland – 2009 – p. 2162-2168

Analysis the Changes of the NGL Product Pipe Safety in the Process of Operation

V I Berg, V A Petryakov and A E Brand

Tyumen Industrial University
Volodarsky str. 38, Tyumen, Russian Federation, 625000

E-mail: Allbert-@mail.ru

Abstract. The paper presents to analyze the changes of the NGL product pipe safety in the process of operation. In the article the analysis and to mathematically calculated the margin of safety and durability according to change thickness of pipeline. The study of analysis the accident conditions risks during the pipeline operation have shown that such measure as the pipe wall thickness increase is appropriate. The proposed conditions on the application of the obtained results.

1. Introduction

The accidents in the process of hydrocarbon wide fractions transport are characterized by a high degree of danger in the parameters of their impact on the people and the environment. In the case of the hydrocarbon vapors ignition, the wave-front time and the hazardous exposure area may be of hundreds kilometers. Some specialists used to consider 1989 to be the starting date of the priority accident assessment risk, a year, when there was Ufa train disaster, the largest in the history of the Soviet Union and modern Russia. On the product pipeline "Western Siberia - Ural - Volga region" in the process of NGL transport, the 1.7 m slit has been formed, through which the hydrocarbons leak and the petroleum accumulation occurred. This product pipeline was located 900 meters to the Trans-Siberian Railway section. During the trains passing there was a spark, which caused the hydrocarbon vapors ignition. Then there was a large volume of hydrocarbons explosion with a capacity of about 250-300 tons of TNT. The fire covered the area of 250 hectares and the shock wave caused the substantial damage to the city of Ash, located 10 km from the explosion place. More than 1,000 people were injured as a result of this disaster.

The analysis of the causes and results of the disaster has changed the approach to the definition of the security and the accidents risk. The substantial changes were made to the SNIP 2.05.06-85* "Main pipelines", such as the increase of the minimal distance from the NGL transport subjects to the settlements and the 400 mm diameter NGL transport pipes ban. These changes provided the basis for the safety in the design and operation of the product pipelines. However, according to the technical and economic assessment, the given changes are extremely expensive and prevent the whole industry from the progress [1-3].

Currently, there is a necessity of petroleum industry intensive development due to the record level of hydrocarbon production and at the same time the active construction of additional product pipelines for petroleum derivatives transportation, including natural gas liquids. In view of the number of

IOP Conf. Series: Earth and Environmental Science **50** (2017) 012041 doi:10.1088/1755-1315/50/1/012041

economic difficulties and the necessity for the economic rationalization, it is important to analyze the accidents risk during the operation and to develop the optimal prescriptions and measures to increase the security level [4-6].

The aim of the article is to analyze the changes of the NGL product pipe safety in the process of operation.

2. Materials and methods

The analysis of the empirical and mathematical material: D = 700 mm with the pressure change from 4 MPa to 4.8 MPa, and cyclical voltage change Ng = 30 000 times should be produced to 11 mm. The technological linear section security level depends on the level and nature of the stress-strain state and the geometry of its components and the metal composition of the pipe. In the context of a large number of the short-term pressure changes in the pipeline, there is a tension increase in the pipe wall, which leads to the service life reduction.

During the transient regime, which is characterized by the pressure increase from P_1 to P_2, the hoop stresses occur, resulting in the strength and durability decrease. While reaching the significant strain indices in the pipe wall, the pipe bend appear which is accompanied by the pipeline and soil mass misalignment and the buckling failure. The result is the emergency situation of a breakthrough.

3. Results and Discussion

The differential equation of a curved surface should be applied:

$$\frac{\partial^4 w}{\partial x^4}+2\frac{\partial^4 w}{\partial x^2 \partial y^2}+\frac{\partial^4 w}{\partial y^4}=\frac{q}{D}, \quad (1)$$

where q – load distributed over the inner surface of the pipeline; w – pipe wall bend; D – bending rigidity.

Rework the differential equation of bending wall and write the final form of the common bend.

$$w=e^{\beta x}(C_1 \sin\beta x+C_2\cos\beta x)+e^{-\beta x}(C_3\sin\beta x+C_4\cos\beta x)+f(x), \quad (2)$$

In the context of solving the production problem for the pipe bend length L, the most important function is the rapidly damped function that characterizes a small area on the length L with a maximum bending which is able to bring about a breakdown [7-9].

$$w=e^{-\beta x}(C_3\sin\beta x+C_4\cos\beta x)+f(x), \quad (3)$$

where $f(x)=\frac{\Delta PR^2}{\delta E}$

Taking the second derivative we obtain the maximum bending moment:

$$M_{\max}=\frac{(P_2-P_1)R^2}{\delta E}2\beta^2 D, \quad (4)$$

The resulting bending stress:

$$\sigma(x)=\frac{6M(x)}{\delta^2}, \quad (5)$$

Considering the abovementioned formula, the maximum bending stress:

$$\sigma_{\max.eq}=\pm\frac{6(P_2-P_1)R^2}{\delta\cdot E\cdot\delta^2}\cdot 2\left(\sqrt[4]{\frac{3(1-\nu^2)}{(R\delta)^2}}\right)^2\cdot\frac{\delta^3 E}{12(1-\nu)^2}=\pm\frac{(P_2-P_1)R}{\delta}\cdot\sqrt{\frac{3}{1-\nu^2}}, \quad (6)$$

The maximum equivalent stress $\sigma_{eq\max}$ when v = 0.3:

$$\sigma_{eq\max}=\sigma\sqrt{\left[1+4.90\cdot\frac{P_2-P_1}{P_1}\cdot\left(1+1.73\cdot\frac{P_2-P_1}{P_1}\right)\right]}=\sigma\cdot K, \quad (7)$$

The safety factor by the ultimate stress and the flow characteristics:

$$n_b = \frac{\sigma_b}{\sigma_{max}} - \frac{\sigma_b \cdot \delta}{P_1 \cdot R \cdot K}, \tag{8}$$

$$n_t = \frac{\sigma_t}{\sigma_{max}} = \frac{\sigma_t \cdot \delta}{P_1 \cdot R \cdot K}, \tag{9}$$

To determine the durability of N $_{dur}$ when the pressure change is from P_1 to P_2, we use the deduced Manson experiments formula for the low cyclical failure, which connects the cycle total stress AM (elastic and plastic) with the number of cycles to fracture, and which is often used in practical calculations.

$$\varepsilon_a = \frac{1}{2}\left(\ln\frac{1}{1-\psi}\right)^{0.6} \cdot N_{dur}^{-0.6} + 1.75\frac{\sigma_b - \sigma_m}{E} \cdot N_{durr}^{-0.12}, \tag{10}$$

where ε_a - alternating stress; ψ - material necking; σ_b - material ultimate stress limit; σ_m - mean cycle stress; N_{dur} - number of cycles to fracture

The pipeline operation durability (in years) is as follows:

$$N_{dur} = \left[\frac{3.5\left[n_b - 0.5\left(1 + \frac{1}{K}\right)\right]}{1 - \frac{1}{K}}\right]^{8.33}, \tag{11}$$

The probability of an accident due to the resource exhaustion (conventional) with low cyclicality is inversely proportional to the operation durability:

$$P_a = \frac{1}{N_{dur}}, \tag{12}$$

According to the abovementioned formula the most important indicators with the initial conditions: P_1 = 4 MPa, D = 700 mm, σ_b = 520 MPa, Ng = 30 000, Steel – 17GIC, should be calculated.

Displayed in Table 1, P_2 more detailed pressure changes (and consequently $\frac{P_2 - P_1}{P_1}$ ratio) ranging from 0.3 to 0.4, which is equivalent to P_2 range from 5.2 MPa to 5.6 MPa at a constant wall thickness δ.

Table 1. Detailed pressure changes

$\frac{P_2 - P_1}{P_1}$	δ	R	K	n_b	N_{dur}	P_a
0.30			1.798	3.46	1.1E+11	2.7E-07
0.31			1.826	3.41	8.1E+10	3.7E-07
0.32			1.854	3.36	6.0E+10	5.0E-07
0.33			1.882	3.31	4.5E+10	6.7E-07
0.34			1.909	3.26	3.4E+10	8.9E-07
0.35	16	334	1.937	3.21	2.6E+10	1.2E-06
0.36			1.965	3.17	2.0E+10	1.5E-06
0.37			1.993	3.12	1.5E+10	2.0E-06
0.38			2.021	3.08	1.2E+10	2.6E-06
0.39			2.049	3.04	9.1E+09	3.3E-06
0.40			2.078	2.99	7.1E+09	4.2E-06

In Table 1, the ultimate indicator for counting is the failure probability calculation due to the resource end-of-life, and as a consequence, risk of mortality. Guided by the Federal Law № 123 amended on 10.07.2012, as well as "STR for the project documentation development for the object "The pipeline "AT GNS - THX ", point 2.1.1.1, where the established social risk is not more than 10 - 7 year - 1, the risks within these frames should be highlighted in red [10-14].

The safety factor percent growth with the wall thickness increase for different values pressure is displayed in Table 2. Step two and four mm are taken for more clarity.

Table 2. The safety factor percent growth with the wall thickness increase for different values pressure

$\frac{P_2 - P_1}{P_1}$	Increasing the safety factor		
	10-12 mm	12-14 mm	10-14 mm
0.1	17.180	14.800	29.440
0.2	17.160	14.790	29.410
0.3	17.140	14.800	29.410
0.4	17.150	14.740	29.360
0.5	17.130	14.810	29.410
The average value:	17.152	14.788	29.406

The data of the Table 2 are depicted graphically in the Figure 1.

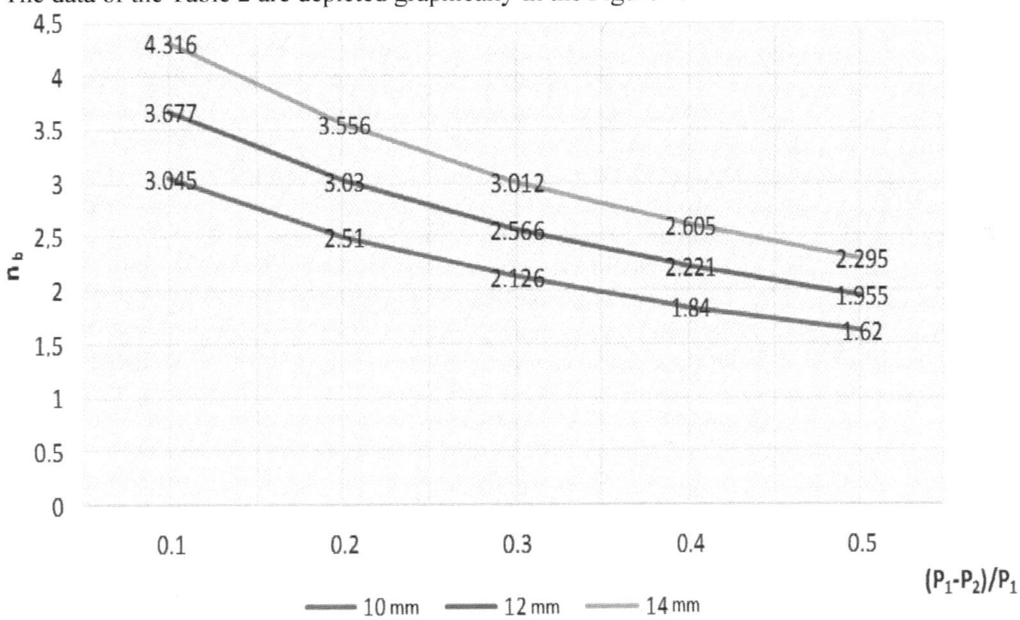

Figure 1 – Strength margin percent increase.

The graph illustrates that with the P_2 pressure boost at transient regime, the pipeline operation durability is reduced. Thus, the emergency possibility increases. In order to diminish this risk, the measures to enhance security should be used, especially the wall thickness increase within the established norms of social risk by STR and the Federal Law №123. The calculations imply that the thickness increase of the pipe wall diameter D = 700 mm with the pressure change from 4 MPa to 4.8

MPa, and cyclical voltage change Ng = 30 000 times should be produced to 11 mm.

4. Conclusion

The analysis of the accident conditions risks during the pipeline operation has shown that such measure as the pipe wall thickness increase is appropriate. It is calculated that the pipeline wall thickness increase by 1 mm improves the safety factor from 6 to 18%, by 2 mm - for 14-17%, and improves the durability from 3 to 10 times. The given data parameters decrease in the process of P_2 boosting and the initial wall thickness increase. The application of the abovementioned formulae and risk data can contribute to the technological objects security enhance during the dangerous substances transportation and preservation of human health and the environment.

References

[1] Yanbao G, Xiaoli M, Tao M, Deguo W, Shuhai L 2016 A novel method of risk assessment based on cloud inference for natural gas pipelines J.l of Natural Gas Science and Engineering 30 421–429

[2] Al-Owaisi S S, Becker A A, Sun W 2016 Analysis of shape and location effects of closely spaced metal loss defects in pressurised pipes J. Engineering Failure Analysis 68 172–186

[3] Boucharda P J, George D, Santistebanc J R, Brunoc G, Duttac M, Edwardsc L, Kingstonb E, Smithb D J 2005 Measurement of the residual stresses in a stainless steel pipe girth weld containing long and short repairs International Journal of Pressure Vessels and Piping 82 (4) 299–310

[4] Punit A,. Singha P K, Vivek B, Vazea K K, Ghosha A K, Pukazhendhib D M, Gandhib P, Raghavab G 2011 Predictions for fatigue crack growth life of cracked pipes and pipe welds using RMS SIF approach and experimental validation International Journal of Pressure Vessels and Piping 88 (10) 384–394

[5] Tchoquessi Diodjoa M R, Beleca L, Aragona E, Perrina F X, Bonnaudetb M, Lanardeb L, Meyerb M, Joliffa Y 2013 Numerical modelling of pipe internal stresses induced during the coating process – Influence of pipe geometric characteristics on stress state J. Materials & Design 52 429–440

[6] Denga D, Kiyoshimab S, Ogawac K, Yanagidad N, Saitoe K 2011 Predicting welding residual stresses in a dissimilar metal girth welded pipe using 3D finite element model with a simplified heat source J. Nuclear Engineering and Design 241(1) 46–54

[7] Sérgio B. Cunhaa, Theodoro A. Nettob 2012 Analytical solution for stress, strain and plastic instability of pressurized pipes with volumetric flaws International Journal of Pressure Vessels and Piping 89 187–202

[8] Bai Y, Bai Q 2014 Risk Analysis for Subsea Pipelines J. Subsea Pipeline Integrity and Risk Management 8 169–212

[9] Bai Y, Bai Q 2014 Pipeline Flow Risk Assessment J. Subsea Pipeline Integrity and Risk Management 8 307–330

[10] Shu-jiao Tong, Zong-zhi Wu, Ru-jun Wang, Hao Wu 2016 Fire Risk Study of Long-distance Oil and Gas Pipeline Based on QRA J. Procedia Engineering 135 369-375

[11] Pasquill F. Atmospheric Diffusion. 1974 J. Wiley 17 429

[12] Lisanov M V, Savina A V, Samuseva E A and Sumskoy S I 2010 Accidents on offshore oil and gas facilities Oil and Gas Journal Russia 5 (39) 48-53

[13] Sklavounos S and Rigas F 2006 Estimation of safety distances in the vicinity of fuel gas pipeline. J. of Loss Prevention in the process industries, 19 24–31

[14] Lisanov M V, Grazhdankin A I, Pchelnikov V A, Savin A V and Sumy S I 2006 Risk analysis of accidents on oil pipeline systems BTS and MN Druzhba J. Labour Safety in industry 1 34-40

The Use Of Foamed Concrete For Oil Spill Treatment And Carbon Capture

K A Akulov, V A Petryakov and N A Mostovaya

Tyumen Industrial University
Volodarsky str. 38, Tyumen, Russian Federation, 625000

E-mail: Allbert-@mail.ru

Abstract. The paper presents analyses the pollution and the spills distribution and presents the results of the experiment on the use of the oil spill elimination means with the foam concrete application. The studies have shown that in the medium term the pipeline transportation accidents, the number of spills and the environmental pollution increase is expected. In addition, studies have shown that the application of the foam concrete with a D 600 density is considered to be the most effective way of the oil spill elimination.

1. Introduction

One of the major environmental problem of our time is the oil-polluted land surface as a result of the low-quality operations implementation and the emergency situations in the hydrocarbon extraction and transportation. The oil pollutions influence the geological environment, causing a great damage to the soil surface layer, surface-and-groundwater, endangering people's health and jeopardizing the existence of entire ecosystems [1, 2].

According to the technical regulations, the soils and the subsoils are considered to be oil-contaminated if the oil and petroleum products concentration reaches the point when the negative ecological changes in the environment occur, such as: the violation of the ecological balance in the soil ecosystem, the soil biota loss, productivity reduction or loss of plants, change in the morphology and hydrophysical soils' properties and diminishing of soil fertility. The big danger is the pollution of groundwater and surface water as a result of oil leaching from the soil or ground and further dissolution in the water [3,4].

The article analyzes the pollution and the spills distribution and presents the results of the experiment on the use of the oil spill elimination means with the foam concrete application [5, 6].

2. Materials and methods

The analysis of the empirical material about methods oil spill treatment and carbon capture covering the period 1965 – 2015. Conducting experimentation with simulated oil spills have been covered by the foam concrete of different density: D 300, D 400, D 500, D 600, D 700, D 800.

3. Results and Discussion

In recent years, with the offshore petroleum production volume increase the oil pollution of the World ocean has grown. For the last 20 years, the number of the world's major offshore deposits discoveries exceeded the corresponding onshore index and the offshore petroleum production has reached 30% of the world total [7].

The main sources of the pollution include both the oil spills from the floating drilling units and the offshore fixed platforms in the process of continuous drilling, the oil production and transportation and the spills during the accidents. The overall spills volume is approximately 1% of the oil entering the World ocean total amount. Analysis of the World ocean pollution sources by 2014 is presented in the Table 1.

Table 1 – World ocean pollution sources

Source	mln. t/year
Marine freight	1.83
Accidental spill	0.30
River run-off including urban	1.90
Coastal area waste waters	0.80
Atmospheric fallout	0.60
Natural oil wells	0.60
Offshore petroleum production	0.08

Most of the oil and oil products losses and as a result the pollution are carried out in the process of water, railway, automobile and pipeline transportation [8,9]. The enormous length of pipelines, which is more than 200.000 kilometers in the Russian Federation, and the severe operating conditions provide for the high rates of the accidents and the pollution volume. Figure 1 is a graph showing the density (intensity) oil and oil products transportation through pipelines in the RSFSR and the Russian Federation.

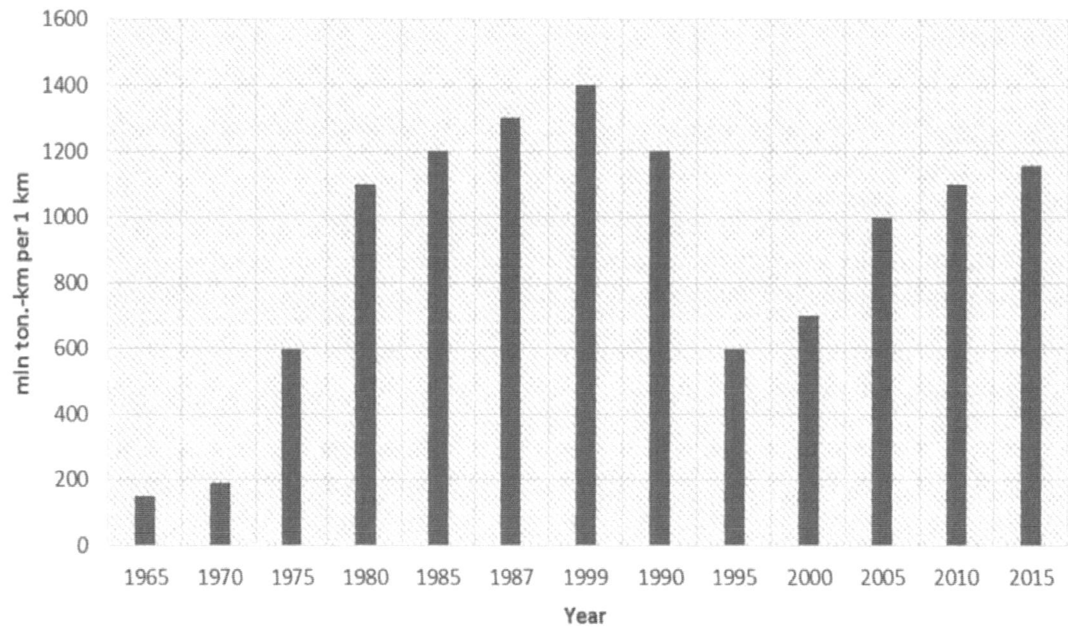

Figure 1 – The main pipelines and oil-product pipelines length and the pumping in the RSFSR and the Russian Federation.

As it can be observed, since 1995 the pump volume has had a clear increasing trend while the length of the main pipelines and oil-product pipelines almost unchanged, that means the operational piping load increase and, consequently, the risk of accidents and pollution increase.

The main causes of oil spills during the oil pipeline operation are: the underground corrosion, the construction works defects, the tube defect, mechanical damages, the operational imperfection, internal erosion and corrosion, natural disasters, equipment defect and others [10]. The main oil pipelines in the places of crossing the artificial and natural obstacles (roads, railways, rivers, lakes) are the most dangerous, for example the underwater lines. Lots of pipelines, located underneath the shipping lanes or in the channels, are more susceptible to the serious mechanical damages as a result of natural causes such as the erosion of the shallows, the bottom creep, and the anchor relocation in the process of dredging.

Table 2 – The pipeline system accident rates from 2010 to 2014

Parameter	2010	2011	2012	2013	2014	The average value 2010-2014
The number of accidents	13	17	21	12	8	14 ± 4
The long pipelines, 1000·km.	232.6	237.5	245.2	245.8	248.1	242 ± 6
The specific intensity of accidents (1/1000 km·year)	0.06	0.07	0.09	0.05	0.03	0.06 ± 0.02
The freight turnover, billion tons km.	2382	2422	2453	2513	2423	2439 ± 43
The specific frequency of accidents per year, 1/(trillion tons km)	5.5	7.0	8.6	4.8	3.3	5.8 ± 1.8
The total damage accidents,	146.0	161.5	154.8	318.9	96.5	174 ± 74
The average damage for one accident, mln. rub.	11.2	9.5	7.4	26.6	12.1	13 ± 7

According to the given table we can observe that the figures are not stable and do not have a clear trend, which is a negative factor in the prediction of further development of the petroleum industry. The vectors of the country development, mapped out by the Government of the Russian Federation in 2015 indicated the further oil production and transportation volume increase. However, the appropriations released for the purpose of petroleum industry development, show a downward trend, which may directly affect the equipment reliability rate and as a consequence, the number of accidents followed by the environmental pollution.

Thus, in the medium term the pipeline transportation accidents, the number of spills and the environmental pollution increase is expected.

In the process of the oil spill elimination the different methods and means are used, the choice of which is based on the performance and the contaminated area availability. The most common methods include the use of the foam concrete due to its low cost, the operating conditions and efficiency. In the process of the foam concrete behavior investigation and the effectiveness evaluation of the petroleum product removal, the sandy soil contaminations were made in the laboratory conditions. The simulated oil spills have been covered by the foam concrete of different density: D 300, D 400, D 500, D 600, D 700, D 800. The height of the foam concrete rise was measured visually after 1 and 4 days. On the 7th day it was possible to see how the foam absorbs the oil. The obtained data are represented in Table 3 and Figure 2 and 3. After the results analysis, it can be concluded that D 600 foam concrete has the maximum oil products removal efficiency.

Table 3 – Oil products removal efficiency

Foam concrete grade	The height of the oil products rise, cm				Oil products removal efficiency,%
	1st day	2nd day	3rd day	4th day	
D 300	0.5	0.86	1.0	1.0	70
D 400	0.7	1.00	1.5	1.9	80
D 500	0.8	1.30	1.5	2.6	90
D 600	0.9	1.20	2.7	3.6	99
D 700	0.6	1.00	1.8	2.6	99
D 800	1.0	1.30	1.4	1.6	98

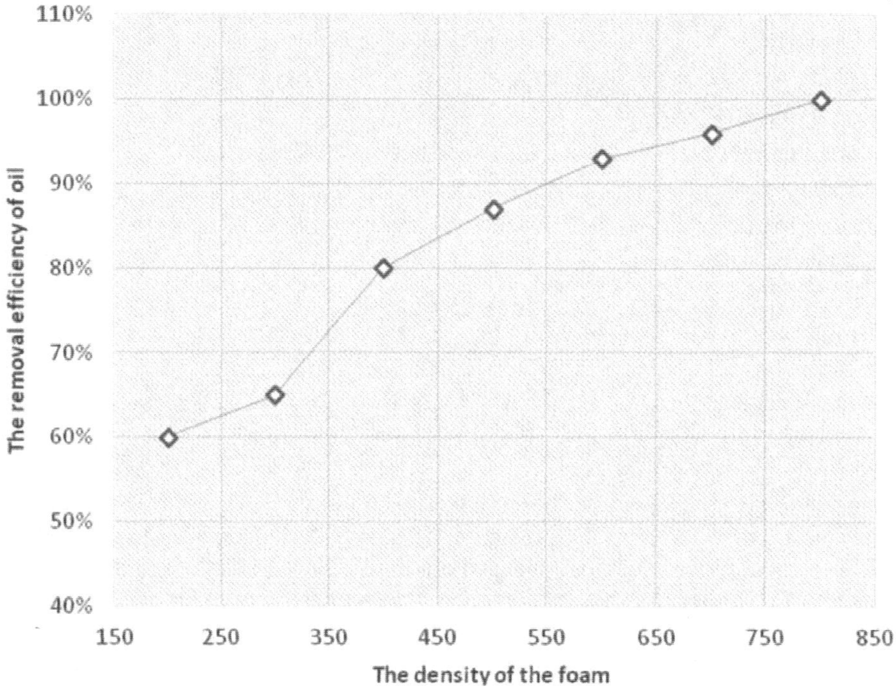

Figure 2 – The dependence of the efficiency of removal of oil from the density of foam concrete.

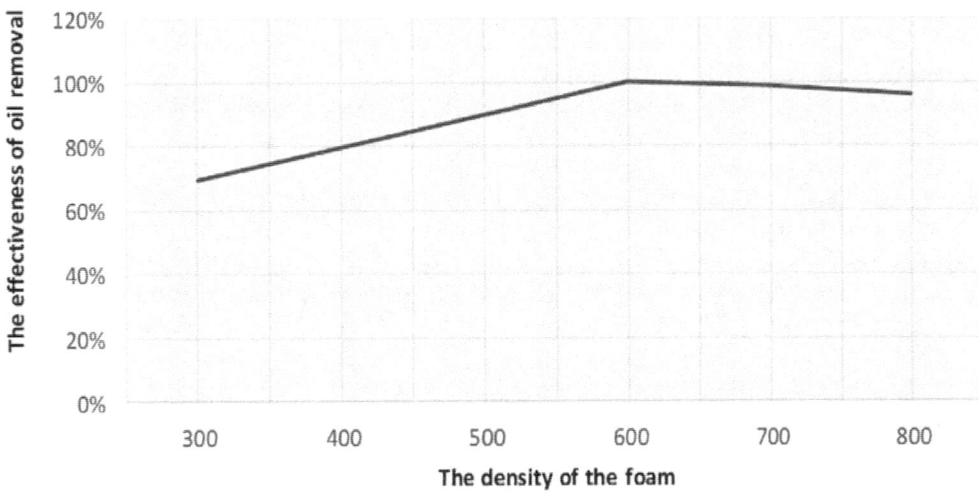

Figure 3 – Oil products removal efficiency dependence on the foam concrete density.

The experiments with the crude oil, exhaust and clean industrial oils, fuel oil have been conducted to find out what types of different fractions petroleum products can be removed from the surface the most effectively. In almost all cases, except the fuel oil, the spill has been removed successfully with the help of the foam concrete by 80-90%. The data of the experiments are represented in Figure 4.

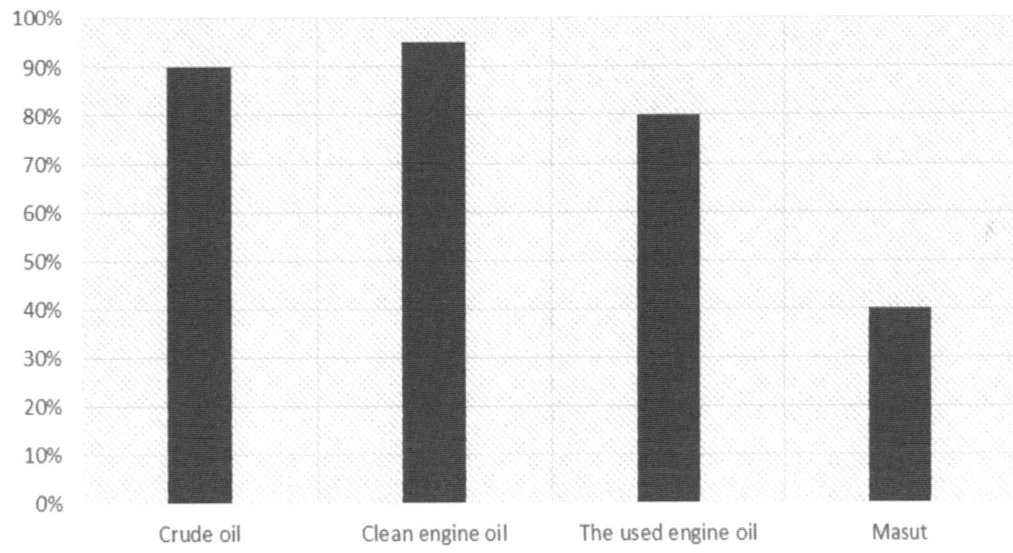

Figure 4 – Oil products removal efficiency dependence on the fractions.

4. Conclusion

Thus, conducted contaminants distribution analysis showed that the most emergency petroleum industry objects are the objects of pipeline transport infrastructure. Currently, the operating pipelines have a high extent of wear and a high risk of breakthrough. In the medium term the environmental pollution increase can be expected with the further increase in the production volumes and the investment decrease. In this case, the application of the foam concrete with a D 600 density is considered to be the most effective way of the oil spill elimination.

References

[1] Mohd Z J, Johnson A U, Mahmud H 2009 Shear strength of oil palm shell foamed concrete beams J. Materials & Design 30 (6) 2227–2236
[2] Mohammed Zuhear Al-mulali, Hanizam Awang, H.P.S. Abdul Khalil, Zaid Shaker Aljoumaily 2015 The incorporation of oil palm ash in concrete as a means of recycling: A review J. Cement and Concrete Composites 55 129–138
[3] Keun-Hyeok Yang, Kyung-Ho Lee, Jin-Kyu Song, Min-Ho Gong 2014 Properties and sustainability of alkali-activated slag foamed concrete Journal of Cleaner Production 68 226–233
[4] Lima S K, Tanb C S, Lima O Y, Leeb Y L 2013 Fresh and hardened properties of lightweight foamed concrete with palm oil fuel ash as filler J. Construction and Building Materials 46 39–47
[5] Miled K, Limam O 2016 Effective thermal conductivity of foam concretes: Homogenization schemes vs experimental data and FEM simulations J. Mechanics Research Communications 76 96–100
[6] Chomicz-Kowalskaa A, Gardziejczykb W, Iwańskia M M 2016 Moisture resistance and compactibility of asphalt concrete produced in half-warm mix asphalt technology with foamed bitumen J. Construction and Building Materials 126 108–118
[7] Vaitkus A, Čygas D, Laurinavičius A, Perveneckas Z 2009 Analysis and evaluation of possibilities for the use of warm mix asphalt in Lithuania Baltic J. Road Bridge Eng. IV (2) 80–86
[8] Jahromi S G 2009 Estimation of resistance to moisture destruction in asphalt mixtures Constr. Build. Mater. 23 2324–2331
[9] Remišová E 2004 Theory and measurements of bitumen binders adhesion to aggregate J. Komunikacie 6 (1) 58–63
[10] Van de Ven M F C, Jenkins K J, Voskuilen J L M, Ven den Beemt R 2007 Development of (half-) wam foamed bitumen mixes: state of the art Int. J. Pavement Eng. 8 (2) 163–175

Reducing The Risk Of Fires In Conveyor Transport

M S Cheremushkina, D A Poddubniy

Saint-Petersburg Mining University,
21 Line ave. 2, Saint-Petersburg, Russian Federation, 199106

E-mail: cheremushkina-ms@yandex.ru

Abstract. The paper deals with the actual problem of increasing the safety of operation of belt conveyors in mines. Was developed the control algorithm that meets the technical requirements of the mine belt conveyors, reduces the risk of fires of conveyors belt, and enables energy and resource savings taking into account random sort of traffic. The most effective method of decision such tasks is the construction of control systems with the use of variable speed drives for asynchronous motors. Was designed the mathematical model of the system "variable speed multiengine drive – conveyor – control system of conveyors", that takes into account the dynamic processes occurring in the elements of the transport system, provides an assessment of the energy efficiency of application the developed algorithms, which allows to reduce the dynamic overload in the belt to (15-20)%.

Introduction

In comparison with the fires on the earth surface, fires occurring in mines are more difficult to extinguish and represent a danger and threat to people's health. In addition, the analysis of the causes and circumstances of accidents that have occurred, shows that the vast majority of them occur in the mines during mining operations. Thus, of the 86 accidents reported in the coal industry of Kuzbass in the last 5 years, 82 have occurred in the mines, and 42% were caused by underground fires.

Statistics on accidents show that up to 70% of exogenous fires in underground coal mines occur in the excavations equipped with belt conveyors because of the fire of conveyor belts.

Investigations of accidents fixed a low efficiency of automatic fire suppression systems, currently used for fire protection of belt conveyors in coal mines. This is confirmed by numerous cases when the fire in conveyors belt had spread almost throughout the conveying line, equipped with a functioning automatic fire extinguishing installations[1].

Due to the lack of effectiveness of automatic fire suppression systems on conveyor belts resulting fire front from burning conveyors belt applies to the unworked coal and sometimes in the goaf. As a result, the liquidation of such fires and the consequences of them spent significant material and financial resources.

As the intensification of production and the steady increase in this regard, the load on a mechanized mining face, a sharp increase in intensity of production processes increases the likelihood of fires in mines equipped with belt conveyors.

The conveyor transport of mining companies consumes 1 km length of conveyor is about 1 million kW.hour of electricity per year with actually achievable consumption of about 0.3 million kW.hour. It is obvious that about 70% of the energy consumed on consumption of capital equipment.

IOP Conf. Series: Earth and Environmental Science **50** (2017) 012043 doi:10.1088/1755-1315/50/1/012043

Regulation of the conveyor not only provides a reduction in the cost of transportation of ore, but substantially increases the safety of operation. The most severe accident of the conveyors associated with the combustion of the conveyors belt by breaking its coupling with the belt pulley and the broken belt. The reasons in both cases are excessive force applied to the belt.

Disadvantages of existing drive systems of conveyor transport:

- need to create a high pre-tension to prevent slippage;
- if there are multiple drive pulleys occurs uneven distribution of forces on pulleys;
- in the processes of acceleration and deceleration there are higher torques when starting the engines.

This work is dedicated to the development of a control algorithm that meets the technical requirements of the mine belt conveyors, and enables energy and resource savings taking into account random sort of traffic. The most effective method of decision such tasks is the construction of control systems with the use of variable speed drives for asynchronous motors [2].

Proposed model of cross conveyors line

On the basis of the chosen scheme of conveyors location in the line with several cross conveyors (Fig.1) in the application Simulink MatLab has been developed the model of conveyor – multiengine electric drive in the form of separate blocks, the inputs of which represent the control and disturbance variables for a given element of the system, and outputs are state variables that are the subject of research or inputs to other blocks [3].

The object model was provided with the following components:

1. Subsystem «drive – conveyor» (konv1-konv4) consisting of a mechanical (belt pulley, belt, etc.) and electrical part of conveyor (multiengine variable speed electric drive).
2. Block which generate the load of conveyor (zagruzka1, zagruzka2).
3. Block of control system for cross conveyors line (Control).

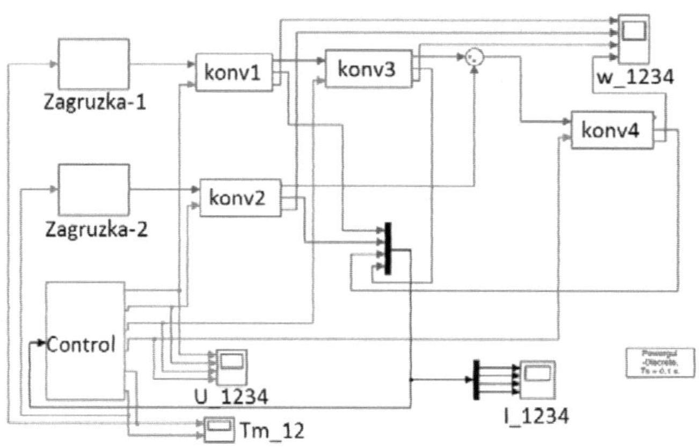

Figure 1 - Continuous transport system, consisting of four cross conveyors

Mathematical model of this continuous transport system is a model of four cross conveyors, three of conveyors are connected in series (konv 1, 3 and 4) and one parallel (konv 2). In addition to the subsystems, containing models of conveyors (konv 1, 2, 3, 4, Fig.2), also shown in the figure: blocks Zagruzka-1,2 (Fig.6), simulating the load and taking into account random type of traffic; control block of the transport system (CONTROL), the input of which receives signals (Is) from the conveyors, simulating their load, the output – control actions (U1-U4) to the conveyors, and the signals Tm1 and Tm2 coming on blocks Zagruzka1,2 and allows to take into account the current load on the continuous transport system.

Digital implementation of continuous transport system was created with the use of C language in the program Matlab. Using the model of continuous transport system it is possible to analyze operating modes of conveyors start-up in a predetermined order with rated torque up to rated speed, deceleration, conveyors adjustment of the predetermined load and to obtain functions (monitors w-1234, I_1234, Tm_12) determining the workload of individual conveyors in the system and showing the speed change of the conveyor depending on the load.

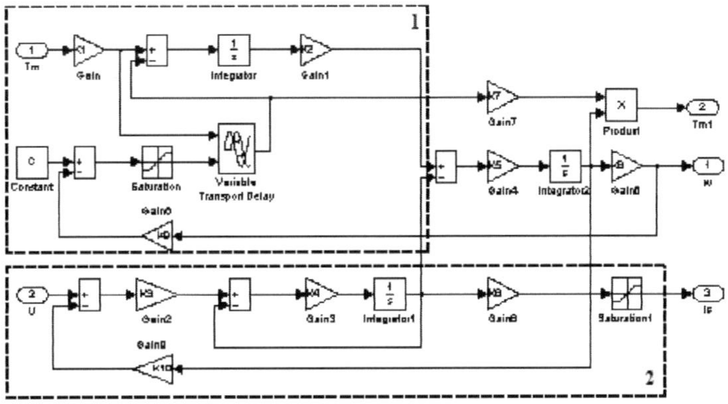

Figure 2 - Model of conveyor Simulink Matlab

Unit 1 simulates the variation of the load on the conveyor. The input signals "Tm" from the block "Zagruzka" (Fig.1), taking into account random type of traffic, and a constant value of load Constant. Unit 2 determines the response of the conveyor on the input signal U (control action from the control system of continuous transport system depending on the conveyors load). The output of the model Tm1 – load supplied to the next belt conveyor (Fig.1 blocks konv1-konv4), w is the speed of movement of the belt of conveyor, which is displayed on the monitor (Fig.1, monitor w1234), Is - stator current that determines the load level of the conveyor (Fig.1, monitor I-1234).

With the presented mathematical model the system of direct torque control (DTC) for multiengine asynchronous electric drive of the conveyor was adjusted.

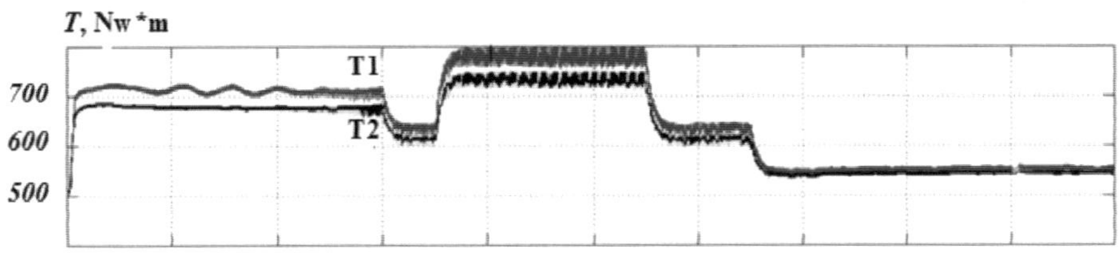

Figure 3 - Characteristics of torque at start-up, load-off and load-on (blue - existing system of drive control of conveyors; black - direct torque control (DTC)).

Figure 3 presents curves of moment change at start-up under rated load, the load-off and load-on, illustrating the decrease in dynamic moment (15-20)% compared to existing control systems for conveyor drives.

Designed control algorithm

The reference signals generation algorithm in the control system of the multiengine asynchronous electric drive allows to take into account probabilistic type of traffic, improves the uniformity of load transfer between the drive motors to 90% and eliminates slippage of the belt with the change of conveyor operating conditions [4].

As a local control algorithms for multiengine electric drive was used the algorithm of direct torque control (DTC) which provides maximum performance in the current (torque) loop and maximum value of the current (up to rated torque).

One of the main features of conveyor transport as a technological system is the random incoming of traffic from interconnected process equipment[5].

The model of load was created in Simulink Matlab (Fig.4), simulating random traffic arriving on the conveyor.

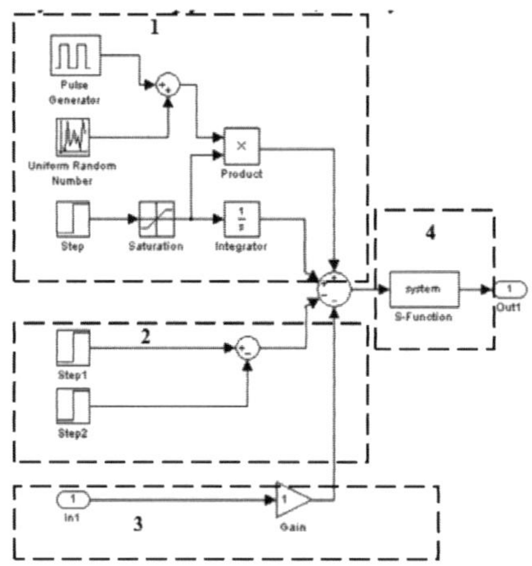

Figure 4 - The model of load traffic.

Block 1 simulates the traffic as a discrete sequence of pulses with random duration of incoming load and with random intervals of its absence. Within the duration of impulses the traffic is described as a continuous random process [6].

Block 2 simulates the load-off mode of conveyor.

Block 3 simulates the incoming of load from the other cross conveyors.

Block 4 is written in the Matlab language s-function that describes the distribution of traffic.

The reference signals generation algorithm in the control system of the multiengined asynchronous electric drive, taking into account the random type of the load is: the value of load mismatch (actual current of semiconductor converters) is used to adjust the speed of the individual drives [7].

The algorithm of reference signals adjustment is shown in Fig.5.

1. If $T_1 = T_2$, then $K \times (T_1 - T_2) = 0$, a corrective signal equal to 0.
2. If $T_1 > T_2$, then $K \times (T_1 - T_2) > 0$. The speed controller assignment (PC1) is reduced, and PC2 increases.
3. If $T_1 < T_2$, then $K \times (T_1 - T_2) < 0$. The speed controller assignment (PC1) is increased and PC2 decreases.

Coefficient K regulates the perfomance of torque adjustment [8].

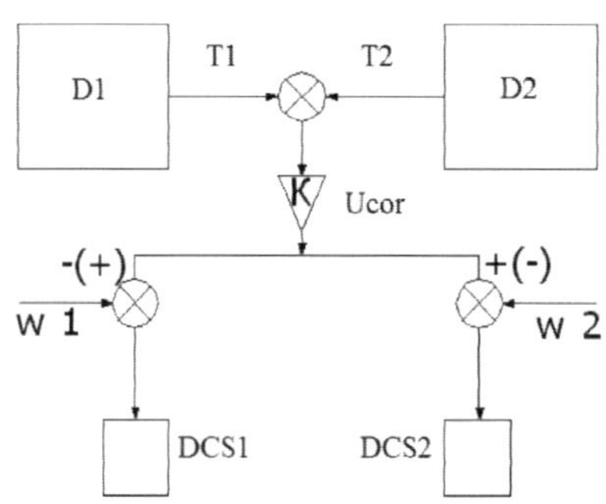

Figure 5 - Algorithm of reference signals adjustment (D1,2 – drives, Ucor – correction signal, w1,2 – speed assignment, DCS1,2 – drive control systems).

Fig.6 demonstrates curves of torque and speed of AC drive with the random load at start-up mode to full speed and subsequent braking to a complete stop and stepped load-on. From these curves it is seen that with the use of developed algorithm reduces the overshoot and torque oscillations, and therefore speed [9].

Researches of the drive operating modes showed that the implementation of speed adjustment algorithm of the individual drives have provided increased uniformity of load distribution between drive motors (mismatch is not more than 10%), limiting the overshoot of the engine torque (up to 15%), and therefore, the restriction of dynamic loads on the conveyor drive with the use of direct torque control algorithm and elimination of belt slippage when changing the conveyor operating conditions [10].

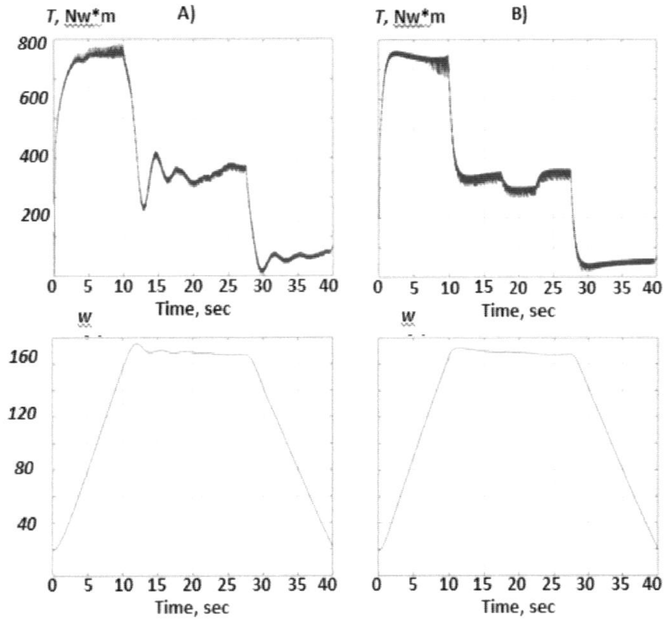

Figure 6 - Curves of drive torque and speed without adjusting the signals (A) and with the developed algorithm (B).

Conclusion

Improving safety operation, energy and resource savings of conveyor transport requires the installation of regulated asynchronous electric drive with semiconductor frequency converters and control of operating modes of cross conveyors line in consideration with the technological requirements and random type of traffic.

Mathematical model of electromechanical system of conveyor – multiengine variable speed drive with implementation in Simulink Matlab allows to perform researches of the conveyor operating modes taking into account the specifics of the mechanism operation at different control algorithms of electric drives.

Was demonstrated the necessity of applying the local digital torque control of induction motor with the developed algorithm of reference signals adjustment, providing increased uniformity of load distribution between drive motorsб elimination of belt dynamic loads (over-regulation until 15%) , and therefore, reduced risk of fires in conveyor transport.

References

[1] Baburin S V, Ustinov D A 2016 International Journal of Applied Engineering Research 11(1) 749-755.
[2] Baburin S V, Ustinov D A 2016 International Journal of Applied Engineering Research 11(9) 6402-6406.
[3] Branishtov S A, Tumchenok D A, Shirvanyan A M and Vershinin Y A 2016 International Conference on Mechanical Engineering, Automation and Control Systems (MEACS) 1 3-8.
[4] Abramovich B N, Sychev Yu A, Ustinov D A and Shkljarskiy A Ya 2014 Neftyanoe khozyaystvo - Oil Industry 8, 110-112.
[5] Liu Q, Peng L, Kang Y, Tang S, Wu D and Qi Y 2014 IEEE Transactions on Industrial Electronics 61(8) 4000-4010.
[6] Bardanov A I and Skamin A N 2015 International Conference on Mechanical Engineering, Automation and Control Systems (MEACS) 1 1-3.
[7] Araujo-Vargas I, Salas-Duarte S, Ramirez-Hernandez J, Del-Muro-Cuellar B and Rivera M 2015 IEEE International Symposium on Predictive Control of Electrical Drives and Power Electronics (PRECEDE) 1 79-84.
[8] Chen Z, Wang Z and Li M 2014 40th Annual Conference of the IEEE Industrial Electronics Society (IECON) 1 1041 – 1047.
[9] Jimenez F R, Salamanca J M and Cardenas P F 2014 IEEE 5th Colombian Workshop Circuits and Systems (CWCAS) 1 1-10.
[10] Kopteva A V 2015 Nondestructive measurement of the oil density by using radioisotope radiation Exposition Oil & Gas. Russia 3(42) 58 – 61.

IOP Conf. Series: Earth and Environmental Science **50** (2017) 012044 doi:10.1088/1755-1315/50/1/012044

Innovative Solutions Shockproof Protection In Occupations Associated With An Increased Risk Of Injury

O V Denisov [1, a], Y I Buligin[1, b], A E Ponomarev[1, c], I A. Ponomareva[2, d], V V Lebedeva [3, e]

[1] Don State Technical University, Rostov-on-Don, Russia
[2] Southern Federal University, Rostov-on-Don, Russia
[3] Far Eastern Federal University, Vladivostok, Russia

Email: [a]OVD63@mail.ru, [b]bulyur_rostov@mail.ru, [c]ae_ponomarev@mail.ru, [d]iaponomareva@sfedu.ru, [e]vikuuuuulya@yandex.ru

Abstract. An important direction in the development of the shockproof devices for occupations associated with an increased risk of injury is reducing their overall size with the preservation the ability of energy absorption. The fixture protection of large joints, with the brace in the coils of an elastic-plastic material with shape memory effect, can effectively protect people from injury and can be used in the domain of occupational safety to reduce injuries by shocks or jolts. In innovative anti-shock device as elastic-plastic material applied equiatomic Titanium-Nickel alloy which has acceptable temperature phase transitions that is necessary to restore shape. As an experienced model first approximation was adopted shockproof device, having in its composition a bandage in coils of elastic-plastic material with shape memory effect and with electric contacts at the ends. This solution allows the punches to plastically deform with the absorption of the impact energy, and then recover the original shape, including at the expense of electric heating.

Relevance and statement of a problem

In the Strategy of a homeland security of the Russian Federation till 2020 as one of the main strategic threats of a homeland security is specified the progressing lack of work. Concerns express that further decrease in the total number of the population of a share of able-bodied population is expected [1]. Currently, Russia is the implementation of the national state standard specification (GOST) R 54934-2012 "System of management of occupational safety and health protection", however, it is impossible to speak about decrease in professional incidence. In particular it is specified that one of basic reasons of growth of professional pathology are non-optimal working conditions of personnel, availability of workplaces with harmful and dangerous working conditions [2, 3]. Quite important place in professional pathology is taken by diseases and injuries of large joints. For example, about 15% among stationary patients of ortopedo-traumatologic departments patients with pathology of a knee joint and it constitute persons of working-age [4]. For the professions which are characterized by the increased risk of an injury rate there is a probability of obtaining on a workplace of an injury of a large joint. It is described that in case of jumps and falls at the time of a landing the organism is affected by forces exceeding its weight almost five times. The wrong landing, especially in total with twisting, as a rule, brings in a knee joint to micro and to macroinjuries of a joint that causes further development of pathology, including with decline in quality of life and working capacity [5].

There are a large amount of various protective devices, but the most convenient are kneecaps [6], which are applied to prevention of injuries. Currently using of such protection devices isn't widespread for wide use (except for athletes), besides, there are practically no devices for protection of large joints in professions which are characterized by the increased risk of an injury rate. Innovative technical solutions and materials of the increased degree of protection and efficiency are seldom applied [7].

In the analysis of the modern technical solutions which are the cornerstone of domestic shockproof devices [8, 9] it was established that there is quite a large amount of such devices, most of the described elastic elements, which are currently used elastic bandages, laminated materials, soft shapes. All of them are executed in the form of an elastic deformable element on the basis of soft polyurethane and/or gel. However, all existing samples for a more optimal use may be modified from the point of view of increase of efficiency of depreciation shocks or blows, increasing the power consumption, for example, by the operation of the damping element of the applied material is not only elastic, but also in the plastic regions of deformation, the inclusion of safety devices on rational energy-absorbing elements, along with the task of reducing the dimensions of the models.

Discussion and results.

The important direction of researches in development of shockproof devices for the professions which are characterized by the increased risk of an injury rate determined decrease in overall dimensions of these devices when preserving their characteristics and the power absorbing capability. The projectable result can be received by means of application of innovative technical solutions. The development of the technology of the new century dictates the importance of examining the use of occupational materials that are capable of deformation in elastic and plastic regions with the subsequent recovery of the original shape. Those are alloys with the form memory effect (FME). It should be noted that among such materials the undisputed leader acknowledged alloys on the basis of a Nickel-Titanium (TiNi). Their fundamental difference, for example, from alloys on the basis of copper, is that TiNi is a stable compound. For such alloys with a thermoelastic martensitic transformation characteristic feature is premartensitic instability of their structure to shift. A natural consequence of the development of the instability for Ti-Ni alloy (50:50) is an implementation of the martensitic transition below 60'C. It is established that the martensitic transition temperature decreases with increasing Nickel concentration. Therefore, by changing the percentage composition of the compound may change the transition temperature and the time of the manifestation of FME within a wide range. Obviously, this changes the nature of the martensitic transformation and memory properties. The analysis and study of alloys based on TiNi we need to understand that this class of materials is characterized by not only the general regularities of the structural properties and transformations, as the ability to control their changes in a wide range. It makes use of alloys based on TiNi for a wide variety of technical tasks, as well as their use for the development of new composites with new properties. It is therefore quite logical to use as elastic-plastic material equiatomic TiNi alloy, with the temperature of phase transitions for shape recovery, developed an innovative shockproof device.

In the research as a pilot model, it was considered the proposed fixture shockproof protection comprised of a bandage in the form of spirals made of elastic-plastic material with FME having electrical contacts at the ends. The proposed technical solution allows to be deformed plastically in case of pushes and blows, absorbing an impact energy then to recover the initial form, with availability of a possibility of recovery including due to impact of electric current (heating). The spiral of a prototype of the protection device is integrated on all length an elastic rod element from the soft polyurethane capable to deformation. It provides an opportunity not only implementation of relative fixing of a spiral, preserving required distance between spiral turns, but also that is important, allows to participate in distribution of an impact energy [10] (fig. 1).

To restore the original shape of the shock devices used heating made of TiNi alloy with FME brace wires to the temperature required for recovery of plastically deformed coils. The heated wire device is due to the proposed portable electric heater connected to the power source. It is important that the proposed fixture restoration of the shape of the element can also occur due to the use of other heating devices, including appliances (e.g., hair dryers, heaters) that can facilitate its application and increase efficiency of work performed.

For thermoelastic damping TiNi alloy with FME the range of temperatures at which the shape recovery material (martensitic transformation), is in the experiment 320...340 K.

After the shape recovery is a passive cooling device to the initial temperature. Shockproof protective device with reduced turns of the wire becomes fit to work, especially to the damping of jolts and bumps.

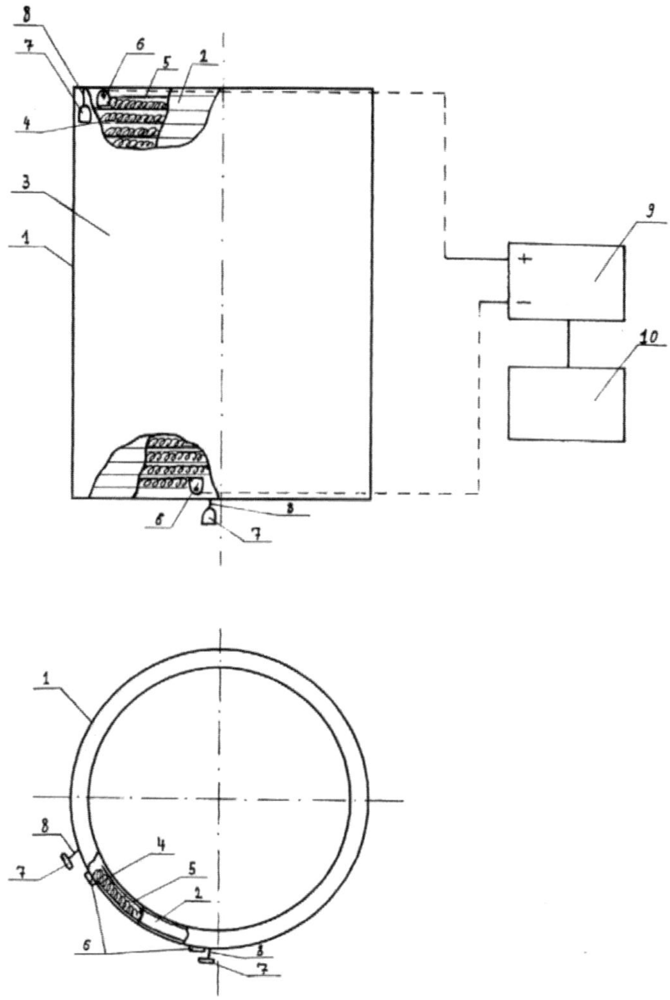

Figure 1 - Experimental model of shock devices:
1 – elastic bandage, 2 bandage of elastic plastic material, 3 – fabric,
4 – wire round with FME, 5 – elastic truss element, 6 – electrical connection, 7 – protective cap,
8 – threads, 9 – electric heater, 10 – power source

The pilot model of the device shockproof protection was carried out on the impact testing machine (GOST 10708-82), specially modified for this study under different temperature regimes accompanying the carrying out of works in different climatic and production conditions. In the course of the experiment at a given temperature was measured by a torque and the magnitude of shear deformation. The temperature of the spiral model was subjected to controlled changes by using the electric heater was measured using a digital thermometer (GOST R 8.625–2006). The signal transmission is carried out with the thermocouple junction by the conventional method were glued with glue BF-2 (GOST 12172-74) to the center of the spiral [12].

Features of methodology of the experiment: model shock devices, matched by size, was worn on the basis of imitating model of the human knee joint. Experimental device in this case is a tight fit protect "the joint" at the expense of the elastic properties of elastic bandage and brace.

As a model test device was made with the following dimensions: diameter d – 120.10-3 m (120 mm), length L – 210.10-3M (210 mm), the thickness of 5.10-3M (5 mm). The diameter DC of the spiral of TiNi alloy with FME amounted to 4.10-3M (4 mm) and wire diameter Dw – 1.10-3M (1 mm). The geometric characteristics of the model elements correspond to the normal linear dimensions (GOST 6639-69). The mass of experimental device was 0.5 kg.

The experiment plan was developed in accordance with the standard methodology of calculation of energy absorption for elastic-plastic elements [13, 14].

It is established that in case of blow to impact testing machine the working area U – 1.10-3M2 (1000 mm2) deformed by bending and twisting about 200 turns, which together absorbed approximately 150 J, and taking into account the deformation of the elastic shafts up to 200 J of impact energy.

For the purpose of determination of characteristics of power absorptivity at various temperatures one of investigation phases provided carrying out testing with measurement of dynamic characteristics in correlation with heating of deformable elements. Loading at each temperature level were carried out with the same speed, a certain angle of allotment of the impact testing machine.

Main results:

1. The power absorbing ability of these plastic elements with FME to a large extent correlates with a temperature (fig. 2).
2. The actual energy absorption is in the range of more than 100 J.
3. The nature of the curves is almost repeated at different shock loads up to critical (Fig. 3).

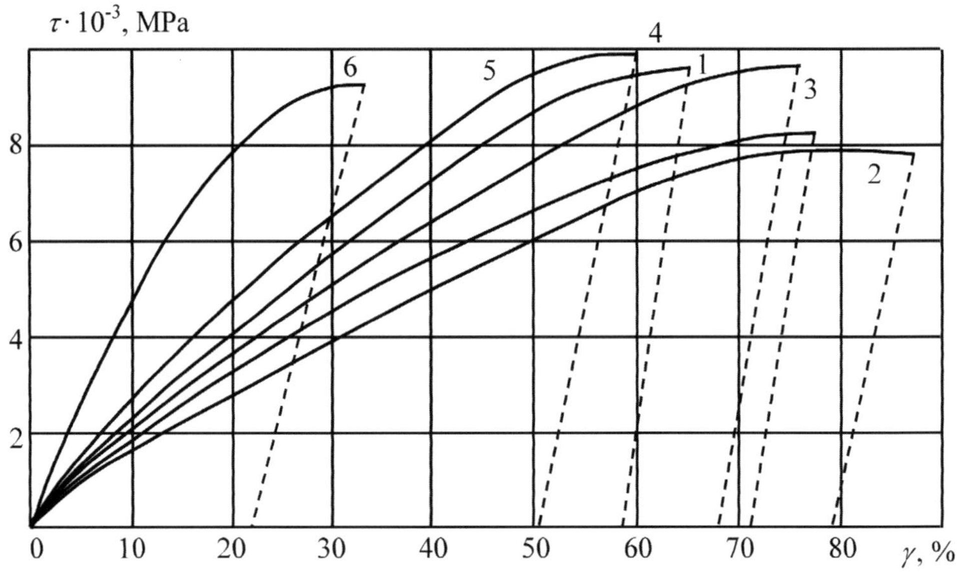

Figure 2 - Conditional charts of tension deformation
models based on the alloy TiNi-1 shear (torsion) and the increased temperatures:
1 – 298 K; 2 – 318 K; 3 – 323 K; 4 – 333 K; 5 – 353 K; 6 – 383 K (dotted line: elastic unloading)

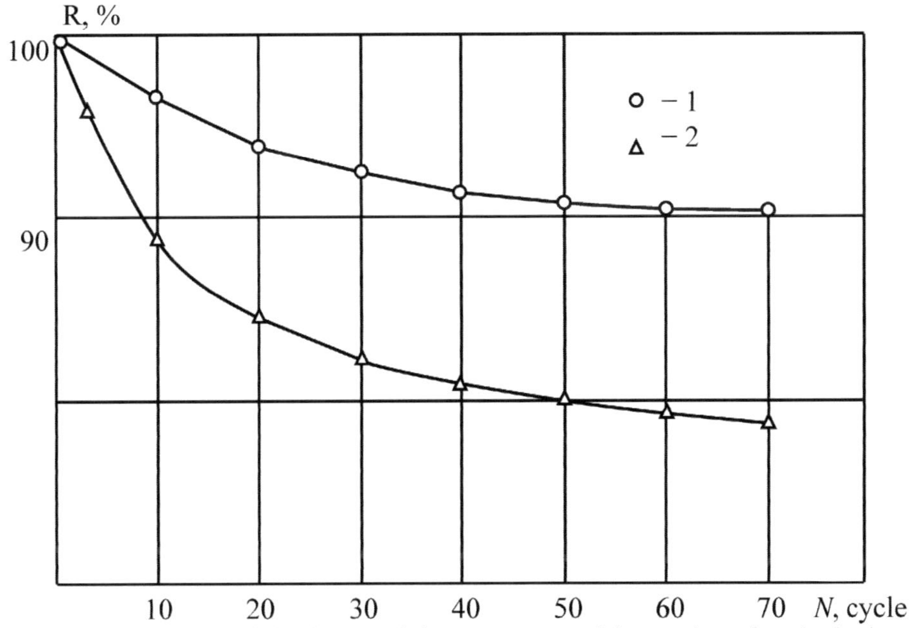

Figure 3 - The dependence of the degree of shape recovery of the number of cycles in the mode of FME (drums torsion at a temperature of 298 K and heated at 423 K) TiNi alloy:
1 – at constant strain of 10.3%; 2 – at constant strain of 15.4%.

In accordance with the regulations [15], the energy of over 100 J can be traumatic, consequently, for protective devices is sufficient absorption of energy of about 150-200 J.

According to the results of the experiment the features expected of the real device. In the event of a jolt or shock occur in the elastic deformation of the wire brace and elastic bandage, while the impact energy is absorbed and dissipated, thus preventing injury to the underlying tissues. When a significant impact that exceeds the level of capability of elastic deformation elements, the outer layer deforms an elastic bandage to limit state in the elastic zone, and rolled into a spiral wire brace initially deform in an elastic and then to plastic zones. It is established that the deformations of the spiral wire brace are made through twisting turns and plastic bending in the strike zone. During plastic torsion of a material is the highest energy absorption [16, 17].

When conducting experiments on the restoration of the shape of the experimental model showed a stable recovery rates spirals. It should be noted that the restoration of the deformed elements to a considerable extent depend on their degree of shear deformation. Thus, when shear strain to 10.3% observed degree of shape recovery in a first stroke reached 98-99 %, while at strains of 50 % (full collapse) the recovery was only about 80 %.

Figure 3 presents the simulated range of impacts in laboratory conditions, shows the evolution of the degree of shape recovery of a prototype. Shear strain in each cycle was carried out with a predetermined amplitude. The cycle consisted of the following stages:

– loading of the helical element at a temperature of 295...300 K;
– the removal of the load after reaching the desired deformation;
– subsequent heating to temperatures of shape recovery [18, 19].

Graphically, it is shown that with increasing number of cycles there is a gradual decrease of the degree of recovery of deformation until it reaches the stabilization value. With increasing amplitude of shear strain, the value of the degree of shape recovery decreases, and to achieve stabilization requires a greater number of cycles. In practice, most likely during the individual impact will be a sustainable recovery of the sample. Perhaps the explanation for this behavior TiNi elastic-plastic element redislocation mechanisms of deformation that correlates with the literature data and results of research on the effect of rote memory.

Conclusion

Thus, adaptation of shockproof protection of large joints is characterized not only reduction of dimensions in comparison with analogs (5 mm instead of 20-30 mm), but also allows to protect the employee from injuries very effectively. It can be recommended to application in the domain of occupational safety for decrease an injury in case of impact of blows or pushes.

References

[1] Babakov A N, Chernikova A G. Sostoyanie ohrany truda v Rossii [State of labor protection in Russia] // Bulletin of the IMSIT. 2013. No. 3-4. Pp. 21-23.
[2] Usikova O V, Sravnitel'nyj analiz organizacii sistem ohrany zdorov'ya i bezopasnosti truda (obzor zarubezhnogo opyta) [Comparative analysis of systems of health protection and labor safety (review of foreign experience)] // Sibbezopasnost' [Sibsecurity]. 2013. No. 1. P. 249-253.
[3] Gubaev F A, Nikitina N V. Analiz sostoyaniya bezopasnosti truda pri provedenii dorozhnyh rabot i predlozheniya po sovershenstvovaniyu trebovanij ohrany truda [Analysis of the state work safety during road works and proposals for improvement of requirements of labour protection] // Ohrana i ehkonomika truda [Protection and labor Economics]. 2011. No. 3 (4). P.20-21.
[4] Decaylo V P, Boloboshko K B. Struktura travm i zabolevanij kolennogo sustava [The Structure of injuries and diseases of the knee joint] // Novosti hirurgii [Surgery News]. 2007. V. 15. No. 1. P. 26-31.
[5] Denisov O V, Ponomareva I A, Zimenko V A. Zashchitnaya povyazka dlya krupnyh sustavov pri zanyatiyah sportom [Protective armband for large joints in sports] // Novye standarty modernizacii pedagogicheskogo obrazovaniya v formirovanii zdorovogo obraza zhizni i bezopasnosti zhiznedeyatel'nosti [The New standards of modernization of pedagogical education in the formation of healthy lifestyle and safety]: proceedings of the Third regional scientific-practical conference. Krasnodar: Kuban state University. 2015. Pp. 90-92.
[6] Orlova E V, Karateev D E, Kochetkov A V. Kompleksnaya reabilitaciya bol'nyh rannim revmatoidnym artritom [Complex rehabilitation of patients with early rheumatoid arthritis] // Nauchno-prakticheskaya revmatologiya [Scientific-practical rheumatology]. 2013. No. 4 (51). Pp. 398-406.
[7] Davydenko V N, Ponomareva, I A Inginov A Y. K voprosu o bezopasnosti trenirovok i profilaktike travmatizma pri zanyatiyah volejbolom [To the issue of safety training and injury prevention in the volleyball] // Fizicheskaya kul'tura, sport, zdorov'e i dolgoletie [Physical culture, sport, health and longevity]: materials of the third all-Russian with international participation scientific and practical conference. Rostov on/D: SFedU. 2014. Pp. 36-42.
[8] Patent: Nakolennik [Knee pad]: RF patent for the invention 2425658, Appl. 13.01.2010, publ. 10.08.2011.
[9] Patent: Nakolennik dlya zashchity kolena cheloveka [Knee pad for protecting the knee of a person]: RF patent for the invention 2308373, Appl. 29.11.2002, publ. 03.07.2003.
[10] Patent: Protivoudarnoe prisposoblenie [Shock-proof fixture]: RF patent for the invention RUS 2578997 20.08.2014.
[11] Effekt pamyati formy v splavah [The shape memory effect in alloys]: transl. from English. L M Bernstein. Moscow: Metallurgy. 1979. 472 p.
[12] Splavy dlya termopar [Alloys for thermocouples]: Reference book / I A Rogelberg, V M Balim. Moscow: Metallurgy. 1983. 360 p.
[13] Prikladnaya mekhanika [Applied mechanics]: Proc. for high schools / by the editorship of G B Iosilevich. – Moscow: Higher school. 1989. 350 p.

[14] Tikhomirov A G , Denisov O B , Denisov I V, Nazarov A Yu. Osobennosti uprugoplasticheskogo krucheniya stal'nyh obrazcov s razlichnoj iskhodnoj teksturoj [Peculiarities of elastic-plastic torsion of steel samples having different initial structure] // Izvestiya vysshih uchebnyh zavedenij [News of higher educational institutions]. The North Caucasus region. Series: Natural Sciences. 2004. No. 3. P. 43-44.

[15] GOST 12.4.128 – 83 Kaski zashchitnye. Obshchie tekhnicheskie trebovaniya i metody ispytanij [Protective Helmets. General technical requirements and test methods]. EN 1621-1 Motorcycle impact protection. Part 1. Test methods and requirements for protection.

[16] Patent: Ustrojstvo dempfirovaniya kolebanij ob"ekta [Device for damping vibrations of the object]: 2073142 patent of the Russian Federation, IPC 6F 16F1/14; Appl. 10.12.1993; publ. 10.02.1997.

[17] Patent: Upravlyaemaya sistema amortizacii avtomobilya [Controlled damping of a vehicle]: patent 2256831 of the Russian Federation, 2003.

[18] Topilin I V, Ponomareva I A. Ob ispytaniyah opytnoj modeli ehlementa protivoudarnoj ehkipirovki voditelej avtomototransporta [About the testing of the experimental model of element shockproof gear drivers of motor vehicles] (Part 1) // Inzhenernyj vestnik Dona [Engineering journal of Don]. 2015. Vol. 36. No. 2-2. P. 123.

[19] Eremin I I, Denisov O V. Ob ispytaniyah opytnoj modeli ehlementa protivoudarnoj ehkipirovki voditelej avtomototransporta [About testing of the experimental model of element shockproof gear drivers of motor vehicles] (Part 2) // Inzhenernyj vestnik Dona [Engineering journal of Don]. 2015. Vol. 36. No. 2-2. P. 124.

IOP Conf. Series: Earth and Environmental Science **50** (2017) 012045 doi:10.1088/1755-1315/50/1/012045

Monitoring Of Air Quality Parameters For Construction Of Fire Risk Detection Systems

I I Romancov[a], A G Dashkovky[b], V F Panin[c], D N Melkov[d]

National Research Tomsk Polytechnic University,
Lenina av. 30, Tomsk, Russian Federation, 634034

E-mail: [a] romaigor@yandex.ru; [b] agd53@mail.ru@tpu.ru; [c] vtkrjd@tpu.ru;
[d] dmitriimelkov@list.ru

Abstract. The analysis of fire developmental process is given, which showed that there are seven stages of fire development, a set of phenomena (factors, signs) of fire risk condition, characterized by a set of defined parameters, corresponds to each stage. Observed that the registration of high staging factors (high ambient temperature, content of CO_2, etc.) means the registration of actual low staging fire (thermal destruction of materials gases, fumes, etc.) - fire risk situation. It is shown that the decrease of registered factor staging leads to construction of fire preventive and diagnostic systems as the lower is registered stage, the more uncertain is connection between the fact of its detection and a fire. It is indicated that with development of electronic equipment the staging of fire situations factors used for detection is reducing in whole, and also it is noted that for each control object it is necessary to choose (identify) the optimal factor, in particular, in many ways the optimal factor for aircrafts are smokes and their TV image.

Introduction

Fire risk detection by monitoring of air quality parameters is widely used in equipment and depends on the character of the controlled object and its service conditions. Optical parameters of air quality allow estimating mediately air quality changes connected with signs indicative of development of the condition usually defined as signs of fire situation development.

In general the fire (F) development to the uncontrolled state passes up to seven stages [1]. Early low stages of F development can be determined as fire risk condition, a fire hazard situation, for various reasons such a state could not develop in a F.

Usually to the first stage is referred [1] the ingress of burning gases into the atmosphere of controlled room - due to the fault condition of gas main lines etc., fire stage (FS) registration is performed by means of gas sensors.

The second stage of F development is connected with temperature rise of substructure components' surfaces of equipment and apparatus, increasing of infrared radiation intensity, entering of combined-cycle thermal degradation products of non-metallic: structural, insulating, decorative and other materials into the atmosphere. Due to low concentrations of combined-cycle thermolysis products at this stage dominates the condensation mechanism of the formation of aerosols with particles of 0.01 - 0.1 micrometer, such aerosols mostly unobserved visually.

The third stage is associated with a further rise of warm-up temperatures of non-metallic materials and corresponding increase of flux densities of combined-cycle products of thermal (thermal-oxidative) decay materials. There is a coagulation of initial condensation particles along with steam products of thermolysis condensation; as a result the aerosol particle spectrum expands towards larger particle sizes. Thus, at the third stage the thermo destruction aerosols are optically active and as a rule can be observed visually.

Subsequent stages associated with the appearance of glow at the hotbed of fire, flame formation, production of considerable quantity of carbon dioxide, rise of environmental air temperature and so on, and so forth, also, as a rule, are characterized by stronger probability of process development to fire condition than mentioned initial stages of its development. All F detection methods are based on the registration of physical phenomena parameters (factors, fire risk signs) accompanying different developmental stages.

Thus, registration of early stages' factors (leakage of burning gases, rise of temperature and IR radiation of equipment surfaces, smoke generation, etc.) is referred, as noted above, to FS detection, and detection of factors of late stages, as a rule – to F detection.

In the construction of detection and alarm systems for fire risk condition of controlled object it is natural aspiration for detection of low F developmental stages - for FS detection.

About fire factor

About an optimal fire factor for construction of fire detection devices. Though the decrease in staging of recorded factor leads to increase in uncertainty of fire risk estimation. It is clear since the decrease in factor staging is an approximation to the normal state of controlled object, and the state arbitrary close to normal one is characterized by arbitrary small fire risk (FR) probability. The latter means that an attainment of threshold level by any low staging factor does not mean 100% probability of fire development after the event. In this case only some supposed FS is fixed which must be explored further to draw conclusions and implement the necessary protective measures.

In fact the decrease in staging of recorded factor leads to increase in probability of false actions of detection device, if any factor recording device is considered from the viewpoint of reliable F detection tasks. We understand the F detection device as traditional system of some sensors, where for F «determination» it is enough that the signal of at least one sensor to cross the threshold.

These devices more likely should be defined as F prevention or FS diagnostic devices.

The set forth describes only one aspect of constructing early F detection devices issue based on the registration of law staging factors - the probability of detecting a real fire risk situation decreases.

Another difficulty is that at the same level factor corresponding to certain fire developmental stage fire risk and fire safe situations are possible. This difficulty results from the fact that the factor level (intensity) is associated with the fact of real fire risk statistically, because object modes are extremely various in the general case and are implemented randomly in time.

Let us explain the last by example. Assuming that some object over a long period of time is controlled in temperature area of equipment and devices parts in different including the emergency states associated with a real FS. Then using collected statistics the minimum levels of fire-hazardous temperatures and the corresponding threshold settings for temperature sensors of FS detection system are determined. Seemingly now the system can prevent the F warning of FS occurrence. However in the majority of cases system action (actuation) doesn't mean FS occurrence, i.e. it is false. Moreover, in the object, particularly the complex one, such emergency situations unsuspected during the survey are possible, where FS occurs and quickly, irreversibly develops to F at temperatures less than the threshold ones, i.e. there is FS and F skip occurs.

With increasing threshold of the system decreases the possibility of false alarms and increases the probability of FS skip. Therefore for low staging factors in a certain range of their values an "overlap" of fireproof and fire risk modes (Fig. 1) takes place predetermining the ambiguity of such systems.

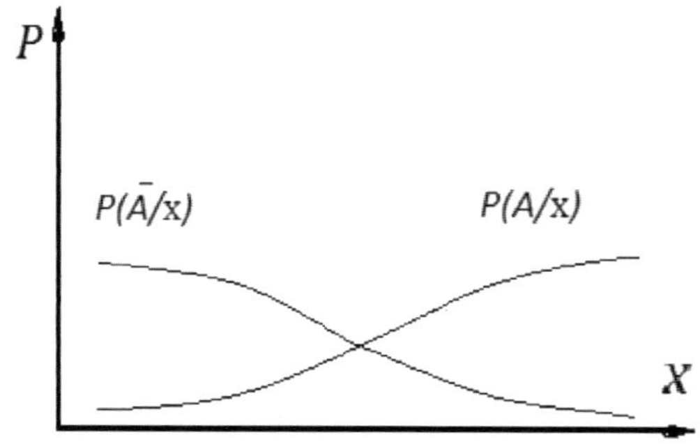

Figure 1 - "Overlap" of fire proof and fire risk modes: $P(\bar{A}/x)$ – probability of fireproof modes; $P(A/x)$ – probability of fire risk modes; x –factor of FS

False action and skip of FS equally reduce efficiency of the systems. The efficiency (E) of the system can be defined like this:

$$E = \frac{P(A) - F - P_P}{P(A)}, \tag{1}$$

where $P(A)$ – probability of FS; F – probability of system false action; P_P– probability of FS skip.

Obviously, in light of the above-mentioned, there is an optimal threshold level of low staging factor, where E is maximal. Dependence of E on threshold level has a shape shown at Fig. 2, i.e. E even at the maximum does not exceed 23 - 24%, and given the fact that it is rather critical to the threshold, the actual quantity of E is much lower which is confirmed by the statistics of right and wrong actions of such systems.

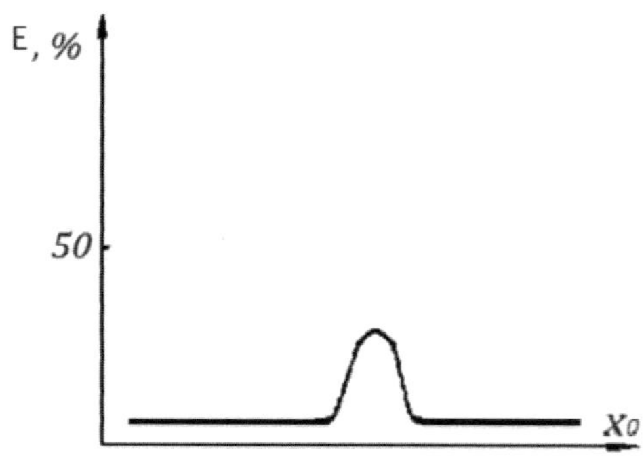

Figure 2 - Dependence of FS detection efficiency by traditional systems on threshold setting of registered factor x_o

Stated above applies to the registration of low staging factors (to FS registration) likewise relates to the registration of high staging factors (to actual F registration): the same impact of the threshold level on the probability of false action and F skip, the same dependence character of E on the threshold level etc.

For simplicity let's assume that the ways and means are found to eliminate inferiority of described systems and they work with 100% value of E. This means that the system based on the registration of factor associated with some stage, with a 100% probability detects exactly this stage of fire risk developmental process.

Hindsight at the development of fire detection means and alarm system in whole confirms the stated interpretation of their direction of development. Suffice it to say that the first fire sensors, fire alerts were based on the elements sensitive to factor of late stages - temperature rise of controlled environment (fusible alloys, bimetallic strips etc.). With the development of semiconductor optical detectors the sensors reacting to fire glow appeared. The development of gas- analytical instrument making put on the agenda the use of CO_2 gas analyzers. Currently intensively develops the "next" stage of FS development - smoke formation stage. There are also examples of the construction or attempts to construct FS detection systems based on registration of the second stage factors of F development: relatively moderate overheating, increasing of infrared radiation intensity, inflow of small concentrations of thermal destruction gases into the atmosphere.

Principles of design, construction and characteristics of the elements sensitive to different factors, F and FS devices and detection systems are described in numerous sources starting with the works of the 1960-s in the last century [2-5] and ending with such works as [6-8].

For specific transportation facilities, such as aircraft (AC) the F registration is unacceptable: too late would be the registration of a noticeable increase in ambient temperature, the presence of significant amounts of CO_2, flame, glowing fire source. Whereas reliable registration of fumes, submicron aerosol or thermal destruction gases, an abnormal temperature increase and heat radiation in certain areas of equipment and devices means, as noted above, the detection of FS when the fire hazard process is still controllable.

In the stated range of factors the smoke, being a FS factor, is the closest one to F, therefore, its definition is related to F. In light of the above mentioned the use of smoke factor assumes the smallest, compared with other factors, amount of additional preventive and diagnostic procedures and therefore it is more preferable of all low staging factors. In addition the smoke factor applied to the AC has a significant advantage: it can be detected by a relatively small number of sensors, as the smoke circulates throughout the space of the AC in an air stream (as aerosols and thermal destruction gases). While in order to control the radiation and surface and equipment temperatures even in the most critical areas an immeasurably larger number of radiation and temperature sensors are needed.

The stated material determines the predominant interest in the registration of fumes as the basis for the construction of reliable FS detection devices.

The results of researches and developments in the sphere of early FS detection in the world

The stated logic of the construction of FS detection systems predetermined, in fact, by the works similar to [1], is kept up to date. For example, in [9,10] there are descriptions of early FS detection devices, on the basis of the second stage of FS development (increase in temperature and intensity of the infrared radiation at the place of potential source of fire), in the [11-15] - on the basis of the third stage of FS development (smokes). However, in some cases, the term "early detection of fire" is used also for the systems based on registration of the later stages of the F development; in the [16-18], "early detection of F" is based on an open flame detection devices, for example, using the characteristics of flame flickering and moving, [17], [19] – based on the registration of CO content in a controlled environment. In the [20] the diamond sensor, developed in Japan for registration of the ultraviolet radiation from the potential fire source before the fire and smoke could reach dangerous levels, is discussed.

In the [21] it is reported on the F detection systems where after smoke content registration the intelligence systems for control of engineering equipment work automatically notify a fire fighting service, in which the flame detection is made using closed television (TV) system with subsequent transfer of an image to the control station of fire service. These measures may be regarded as mentioned preventive and diagnostic procedures allow clarifying the subsequent trajectory of FS development. In the [22] a multi-level approach to the development of automated fire alarm systems based on TV - images processing technology is offered. During high-speed scanning of controlled space, firstly the areas containing an image of burning objects with high probability are allocated.

Then the areas with the colorimetric characteristics corresponding to red color are tracked. Application of neural network provides a separation of burning and not burning areas.

At the same time more and more clearly appears the trend to use in the alarm F detection devices of complex of F features - signs of its early and late developmental stages. Using the signs of late stages intended to remove some uncertainty in the future F development after signs of early stages are found, for example, temperature rise or smoke.

So, already in 2004 in the [23] an overview of the entire spectrum of F alarm systems is provided, starting with systems based on the registration of one of the initial stages, for example, smoke stage, ending with flame detectors based on detection of IR - and UV - radiation, as well as systems using complex information and provided with intelligent functions. These intelligent systems for early F detection are presented in the [23-26]. Despite the complexity of different-type information merge technology and accounting of information-statistical characteristics of F developmental process, the use of complex information about F, in any case, should help to improve the reliability of F detection.

At the same time - in experiments with early FS detection based on the registration of its late-stage features or successive indication of early stages firstly, then the late ones, and subsequent merge of different-type information - almost always the question remains: what sign must be the "primer" starting the entire chain of events indicating the succession of signs of all the later stages of FS development - up to the fusion of different-type information and formation of an integrated solution about F? In other words, what feature of the stages to be understood as an optimum one: gases of initial thermal destruction of electrical and other materials, temperature, thermal (infrared) radiation, smoke, the presence of traces of carbon monoxide or, as in the [22], as such initial feature should then learn the detected (based on the technology of TV image) areas of controlled spaces containing moving images?

Analysis of more than thirty articles mentioned above and dedicated to the systems of early F detection based on the complex of signs shows that the most commonly the "primer" one the sign of smoke formation is called. And in combination with a sign of smoke formation in the complex systems are mentioned the results of TV images processing of smoke flows or moving flame.

Thus, at the present time as the most appropriate F sign to identify FS and F affirms a sign of smoke formation- including the systems built on the basis of indication of both early and late stages of F followed by the merging of different-type information and formation of integrated solution about FS of controlled object.

Conclusion

1. Fire-risk process – from the operating mode of the object to the F - develops in several stages.
2. A set of physical and chemical phenomena, or factors (signs) of fire risk condition, characterized by a set of certain parameters, corresponds to each stage.
3. Registration of high staging factors means, as a rule, the registration of F, low ones - FS.
4. Decrease in the staging of registered factor leads to the need of building, in fact, fire preventive and diagnostic systems: systems solutions for low staging factors should be analyzed additionally, as the lower is registered stage, the more uncertain is the connection between the fact of its detection and F. It is confirmed by current trends in the construction of intelligence F detection system systems.
5. Detection of the actual FS as situations described by a certain set of physical and chemical parameters, traditional systems implement ineffectively.
6. With the development of electronic equipment the staging of factors used to detect the FS is decreasing generally.
7. By many characteristics an optimal feature (factor) of F to identify FS for aircrafts in particular is smoke.
8. Analysis of the researches and developments of F and FS detection systems performed abroad in 2002- 2013 shows that the most appropriate initial ("primer") feature to detect the FS and F by a set of F signs is smoke.

References

[1] Leworthy L R. Automatic fire detection // WorkesEngjneering and factory service – Part 1, 1970, B,65, 766, –P.20-21; Part 2, 1970, B,65, 767, P.35-39; Part 3, 1970, B,65, 768, P.34-36.
[2] Iljinskaya LA. Elements of fire-prevention automatics. - M .: Energy, 1969. 72 p.
[3] Gerlovin Y N, Ivanov E N, Klimov G V. and others. Automatic fire detection and extinguishing devices. M .: Stroyizd, 1975. pp. 42-84.
[4] Bubyr N F. Fire automatic equipment. M .: Publishing department, 1977. P. 30-65.
[5] Sharovar F I. Devices and systems of fire alarm. M .: Stroyizd, 1979. - pp. 22 170.
[6] Sharovar F I .Principles of devices and systems of automatic fire alarms construction. M .: Stroyizd, 1983. - pp. 47 - 194.
[7] Chlenov A N. and others. New methods and technical means of fire detection:. Monograph. - M .: Russian Academy of FMS MES, 2007. - 175 p.
[8] Fedorov A V, Chlenov A N, Lukyanchenko A A, Butsynskaya T A, Denehin V F. Systems and technical means of early fire detection: Monograph. - M .: Russian Academy of FMS MES, 2009. - 160 p.
[9] Zhuiykov S. Novel sensor–actuator device for early detection of fire// Sens. and Actuators. A. – 2008–T. 141–№1. – C. 89–96.
[10] Technical article // IEN: Ind. Eng. News - Eur. – 2010, T. 36, N12. – C.16.
[11] Bomse D.S., HovdeD.Sh., Chen Shin-Juh, Silver J.A. // Proc. SPIE. – 2002, T. 4817. – C. 73-81.
[12] Wiseberg A. Keeping a close eye on fire // Fire – 2003. – T. 96. – №1182. – C. 34–35.
[13] Caught on camera // Fire Prev. and Fire Eng. J. – 2004, Aug. – C. 36–37.
[14] Johnson P.F. Fire Detection in Computer Facilities: 25 Years On // Fire Technol. – 2010. – T. 46. – №4. – C. 803–820.
[15] Jones M. Saving the tunnel customer // Tunnels and Tunnel. Int. – 2010, Oct. – C. 36–40.
[16] Engelhaupt D, Reardon P, Blackwell L, Warden L, Ramsey B. Autonomous long-range open area fire detection and reporting // Proc. SPIE. – 2005. – T. 5782. – C. 164–175.
[17] Sadok M, Zakrzewski R, Zeliff B. Video-based cargo fire verification system with fuzzy inference engine for commerical aircraft // Proc. SPIE. – 2005. – T. 5679. – C. 99–107.
[18] Cowlard A, Jahn W, Abecassis-Empis C, Rein G, Torero José L. Sensor Assisted Fire Fighting // Fire Technol. – 2010. – T. 46. – №3. – C. 719–741.
[19] Zhang R, Du Ji. Fuzzy clustering algorithm of early fire based on process characteristic // Key Eng. Mater. – 2010. – № 437. – C. 339–343.
[20] Diamond U V. Sensor used in fire detection system // Ind. Diamond Rev. – 2006. – №2. – C. 8.
[21] Jones C. Intelligent design // Fire Risk Manag. – 2010,Aug.-Sept. – C. 24–28.
[22] Kang M, Tung Truong X, Kim J.-M. Efficient video-equipped fire detection approach for automatic fire alarm systems // Opt. Eng. – 2013. – T. 52. – №1. – C. 17002.
[23] Scorfield S. Advanced views // Fire Prev. and Fire Eng. J. – 2004, Aug. – C. 28-31.
[24] Luo R C, Su Kuo L. Autonomous fire-detection system using adaptive sensory fusion for intelligent security robot // IEEE/ASME Trans. Mechatron. – 2007. – T. 12. – №3. – C. 274–281.
[25] Fireproof measure in electric systems // IEEE Trans. Power. Deliv. – 2008. – T. 23. – №2. – C. 625.
[26] Wang Sh-J, Jeng D-L, Tsai M-T Early fire detection method in video for vessels // J Syst. and Software. – 2009. – T. 82. – №4. – C. 656–667.
[27] Cheng C, Sun F, Zhou X. One fire detection method using neural networks // Tsinghua Sci. and Technol. – 2011. – T. 16. – №1. – C. 31–35.

IOP Conf. Series: Earth and Environmental Science **50** (2017) 012046 doi:10.1088/1755-1315/50/1/012046

Assessment of Potential Environmental Risks from Saline Soils Subsidence

L Ya Pobereznyi[1], L Ya Poberezhna[1], P O Maruschak[2], S V Panin[3,4]

[1]Ivano-Frankivsk National Technical University of Oil and Gas, Ivano-Frankivsk, 15 Karpatska str., Ukraine
[2]Ternopil Ivan Pul'uj National Technical University, 56 Ruska str., Ternopil, 46001, Ukraine
[3]Institute of Strength Physics and Materials Science RAS, 2/4, pr. Akademicheskii, Tomsk, 634021, Russia
[4]National Research Tomsk Polytechnic University, 30, Lenin Ave. Tomsk, 634050, Russia

E-mail: maruschak.tu.edu@gmail.com

Abstract. The nature and causes of soil subsidence in the areas of liquidated mining and chemical companies of the Carpathians are analyzed. Based on calculation results, was obtained dependences of salts concentration in the liquid, and the specific content of salt through the thickness of soil over time in cases of the dispersed and film salinity.

1. Introduction

Environmental safety of regions is one of the major problems in modern society. The deterioration of the ecological state of the environment and the worsening health of the population living in these areas determines the need to study environmental hazards and risks.

The problem of exploration and development of large land areas occupies a significant place in today's social conditions. High competition, high density of construction and insufficient land resources led to using structurally unstable soils for the construction. Saline soils often occupy large areas in technologically loaded regions.

The problem of preventing the destruction of civil and industrial buildings is one of the most important at the moment. When designing buildings in the area suffering from harmful effects of mining, we should take into account the current building regulations and international standards that are used in the construction of buildings [1].

Risk analysis should be an integral part of the procedure for adoption of design decisions and assessment of the impact of the projected activities on the environment at all stages of design. The main lines of further scientific research include elucidating the mechanism of modern morpho-dynamic processes, identifying the main natural and anthropogenic factors that contribute to their revitalization, city zoning by intensity, scope and nature of the process manifestation, a quantitative risk assessment for different types of objects, development of measures to reduce risks.

When choosing a construction site we must be sure that there are no deposits of potassium salts, there are deposits of non-economic reserves, the deposits are depleted, the process of deformation of the earth's surface was completed, the possibility of craters and cavities is excluded after stabilization of deformation, the excavation of pits is expected only after the completion of the amortization period of buildings [1].

The study of foreign experience enables the prediction of stability of the foundations of buildings erected in saline areas. In the current economic situation, residential housing develops at super-quick pace. Reducing the number of suitable building land leads to the construction on structurally unstable soils, including saline [2-3].

2. Parameter calculation methods

Subsidence of clay soils is characterized by a number of indicators: subsidence limit, soil skeleton density on the verge of subsidence, the relative linear and volumetric subsidence. Soil subsidence is often determined by calculation, using the method of M.J. Denisov, whereby the volume of soil subsidence V is calculated by the formula:

$$V = \left(1 - \frac{\rho_d^b}{\rho_d^c}\right) \cdot 100, \tag{1}$$

where ρ_d^b is the moist soil skeleton moisture; ρ_d^c is the dry soil skeleton moisture.
or the method of E.M. Sergeyev, where by the volume of completely saturated soil subsidence is described by the equation

$$V = V_c \cdot (1 + \beta_V \cdot W), \tag{2}$$

where: V is the volume of soil at humidity W; V_c is the volume of dry soil; β_V is the coefficient of volume subsidence equal to a relative increase in volume when humidity changes by 1:

$$\beta_V = \frac{V_1 - V_2}{V_2 W_1 - V_1 W_2}, \tag{3}$$

where V_1 and V_2 is the volume of soil at humidity W_1 and W_2, respectively.
Subsidence limit of the soil can be calculated from the ratio:

$$W_s = \frac{e_s \rho}{\rho_s}, \tag{4}$$

where: e_s is the soil porosity ratio on the verge of sinking; ρ is the density of wet soil; ρ_s is the density of solid soil.

In present-day study of soil mechanics there is no rigorous mathematical formulation of the stress-strain state of the array of saturated saline soils, and the variation of this state considering the processes of mass exchange and mass transfer between the components of soil. This greatly impedes further development of calculation methods for the base, which consist of saline soils. Salts of different quantitative composition, which are present in the soil, affect the deformation parameters of soils. When saline soils get wet, the salts contained in them are dissolved and removed, and soil porosity increases accordingly. An increase in the porosity of saline soils with moisture in turn leads to suffusion subsidence [1,4].
Subsidence of a homogeneous soil layer in the direction of axis X is calculated by formula (5):

$$S(t) = \int_{h_1}^{h_2} \alpha_0 (\sigma - \overline{\sigma_0})^{V_0} \frac{\xi_0 - \xi(t)}{\xi_0} dx \tag{5}$$

where: h_1, h_2 is the upper and lower limit of suffusion subsidence, respectively; $\sigma=\gamma_0 x$ is the change of the sealing load into the depth under natural stress state, MPa; γ_0 is the specific weight of soil in water-saturated state, N/cm^3; α_0, v_0 are the non-linear deformation parameters that are constant for a homogeneous layer of soil; $\xi(t)$ is the distribution function of the specific amount of salt in fractions of units; $\overline{\sigma_0}$ is the minimum pressure at which there is a suffusion of soil subsidence, MPa [4].

In general, the variation of the specific volume of soluble salts in the soil depending on depth and time is described by the system of equations (6):

$$\begin{cases} Dm_0 \dfrac{\partial^2 C}{\partial x^2} - v\dfrac{\partial C}{\partial x} - \delta \dfrac{\partial \xi}{\partial t} = m_0 \dfrac{\partial C}{\partial t}; \\ \dfrac{\partial \xi}{\partial t} = -\dfrac{\gamma}{\delta}(C_m - C)\xi^k \end{cases} \tag{6}$$

where: D is the coefficient of convective diffusion, cm^2/s; C is the concentration of salts in the liquid, g/cm^3; ξ is the specific volume of salts in the soil, in fractions of units. C_m is the concentration of water saturation with salts of a given composition, g/cm^3; γ is the generalized coefficient of salt output, s^{-1}; m_0 is the soil porosity; v is the groundwater filtration rate, cm/s; δ is the density of salt, g/cm^3; k is the salinization coefficient. In the system of equations (6), k accepts values 0; 0.5; 1.0. At $k = 0$, we obtain a case of the film salinization of soil; at $k = 0.5$ or $k = 1.0$, we obtain a case of the volume (disperse) salinization of soil [5].

Let us consider the mathematical model of calculating the concentration of salts in the filtered liquid, and the model of calculating changes in the specific amount of salt contained in saline soils, and the suffusion subsidence caused. To solve this problem, system (6) is used with the following initial and boundary conditions (7):

$$\begin{cases} C(\bar{x},0) = C_H \\ \xi(\bar{x},0) = \xi_0 \end{cases}, \quad \begin{cases} \dfrac{\partial C(0,t)}{\partial x} = \dfrac{v}{m_0 D}(C(0,t) - C_0) \\ \dfrac{\partial C(l,t)}{\partial x} = 0 \end{cases} \tag{7}$$

where: is the even distribution of salt concentration in the liquid at the initial time through the thickness of soil, g/cm^3; C_0 is the concentration of salts in the initial section of soil, g/cm^3; ξ_0 is the even distribution of specific volume of salts in the soil at the initial time, in fractions of units; l is the soil thickness, cm [6].

Let us solve system of equations (6) with the initial and boundary conditions (7) using the finite difference method. Let us consider system (6) at $k = 0$ and $v = \text{const}$. This is a case of the film salinization of soil. This problem is reduced to solving a system of differential equations:

$$\begin{cases} \dfrac{\partial C}{\partial \bar{t}} = \dfrac{1}{Pe}\dfrac{\partial^2 C}{\partial \bar{x}} - \dfrac{\partial C}{\partial \bar{x}} + N(C_m - C) \\ \dfrac{\partial \xi}{\partial \bar{t}} = -\dfrac{1}{\delta}N(C_m - C) \end{cases} \tag{8}$$

with the initial and boundary conditions (9),

$$\begin{cases} C(\bar{x},0) = C_H \\ \xi(\bar{x},0) = \xi_0 \end{cases}, \quad \begin{cases} \dfrac{\partial C(0,\bar{t})}{\partial \bar{x}} = Pe(C(0,\bar{t}) - C_0) \\ \dfrac{\partial C(l,\bar{t})}{\partial \bar{x}} = 0 \end{cases} \tag{9}$$

where: $\bar{x}=\dfrac{x}{l}$ is the dimensionless coordinate; $\bar{t}=\dfrac{v\cdot t}{m_0 l}$ is the dimentionless time; $Pe=\dfrac{v\cdot l}{m_0 D}$ is the Peclet number; $N=\dfrac{m_0\gamma\cdot l}{v}$ is the dissolution criterion.

Thus, problem (8)-(9) will depend only on two dimensionless criteria *Pe* and *N*.

3. Main results and their discussion

During calculations we used the following data: C_0 = 0.026 g/cm^3, C_H = 0.08 g/cm^3. Based on the calculation results we obtained dependences of salt concentration in the liquid (Fig. 1, a) and the specific content of salt through the thickness of soil over time (Fig. 1, b).

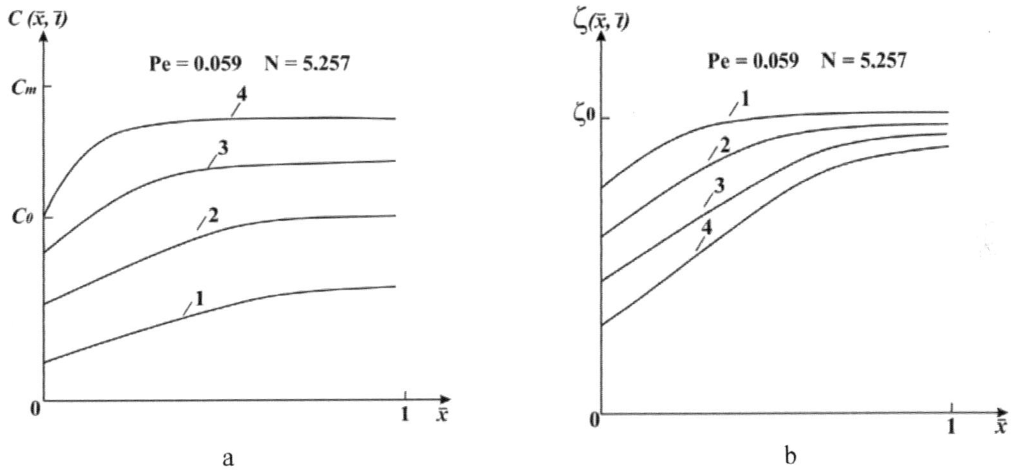

Fig. 1. Dependence of salt concentration in the liquid - a, specific content of salt in the soil – b: 1 - $\bar{t}$ = 0.09; 2 - $\bar{t}$ = 0.27; 3 - $\bar{t}$ = 0.45; 4 - $\bar{t}$ = 0.63

Using the obtained distributions of function $\xi(x,t)$, we can determine the suffusion subsidence of soil in time for the natural stress state in case of the film and disperse salinization by formula (5). Calculations were made for two specific cases of the gypsified saline soil: A = 10 % (slightly gypsified saline soil) and A = 40 % (strongly gypsified saline soil). As is seen from the above dependencies, slightly gypsified soils provide less deformation compared to strongly gypsified soils. Similarly, strongly gypsified soils are characterized by relatively rapid stabilization of deformation, at which suffusion subsidence stops. These studies were conducted for saline gypsified soils of the gypsum-clay cap of Stebnyk deposits, which protects saline clay rocks from infiltration of freshwater [3, 7]. However, in general, the obtained dependences can be used for other types of saline soils as well. Analyzing spatio-temporal dynamics of the major hazardous ecology forming process, we can conclude that within Kalush industrial and urban agglomeration there are irreversible violation processes of the geological environment stability. Moreover, the catastrophic ecological and technological situation turns to the crisis of the national, and eventually, cross-border level.

Subsidence of soil surface over mining deposits, which are flooded, is largely due to the deterioration of their geomechanical stability due to surface leaching of pillars and diffusion wetting of rocks in the zone of contact with the insufficiently saturated solutions. Experience in the construction of various buildings proved that they are all in any way susceptible to deformations that occur as a result of the movement of soil particles [8].

The main causes of deformation can be divided into two groups:
- general causes associated with the peculiarities of the physical and mechanical properties of soils, which are the basis of the building foundation;
- partial causes associated with the peculiarities of construction works and operating mode.

A key danger is the subsidence of soil under the foundations of residential and civil buildings. One of the main factors is the type of the foundation soil [9,10]. In calculating the possible subsidence we used the methods described in building codes.

Let us consider possible subsidence of buildings located on saline soils. We assume that we have a uniform diffusive salinity within each layer of soil. Let us consider a soil array divided into ***i*** layers under the isotropy of physical and mechanical properties of each layer [12]. Then subsidence under full leaching will be:

Case 1: salinization was caused by one chemical compound (kainite) – i.e. monosalinization. In this case, subsidence under full leaching (Fig. 2a), excluding soil compaction due to flooding, will be:

$$H = \sum_{i=1}^{n} \frac{D_{sal_i} \cdot \rho_{soil\,i} \cdot h_i}{\rho_{sal}}, \tag{10}$$

where: H is the subsidence of soil under full leaching of salts excluding flooding, (m); D_{sal} is the salt concentration in the ith layer of soil, (fraction from 1) ; ρ_{soili} is the density of the ith layer of soil, (kg/m^3); h_i is the height of the ith layer of soil, (m); ρ_{sal} is the density of salt, (kg/m^3)

Case 2: salinization was caused by j chemical compounds – i.e. polysalinization. In this case, subsidence under full leaching excluding soil compaction due to flooding (Fig. 2, b) for each soil layer will be:

$$H = \sum_{i=1}^{n} \left(\sum_{j=1}^{m} \frac{D_{sal_{ij}}}{\rho_{sal\,j}} \right) \cdot \rho_{soil_i} \cdot h_i, \tag{11}$$

where: H is the subsidence of soil under full leaching of salts excluding flooding, (m); $D_{sal_{ij}}$ is the concentration of the jth salt in the ith layer of soil, (fraction from 1); ρ_{soili} is the density of the ith layer of soil, (kg/m^3); h_i is the height of the ith layer of soil, (m); $\rho_{sal\,j}$ is the density of jth salt, (kg/m^3)

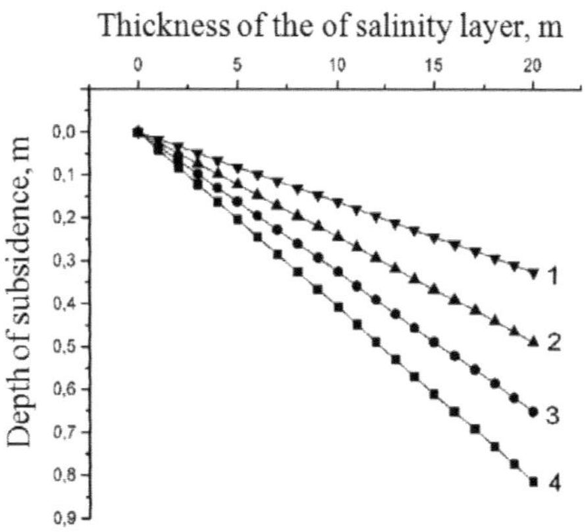

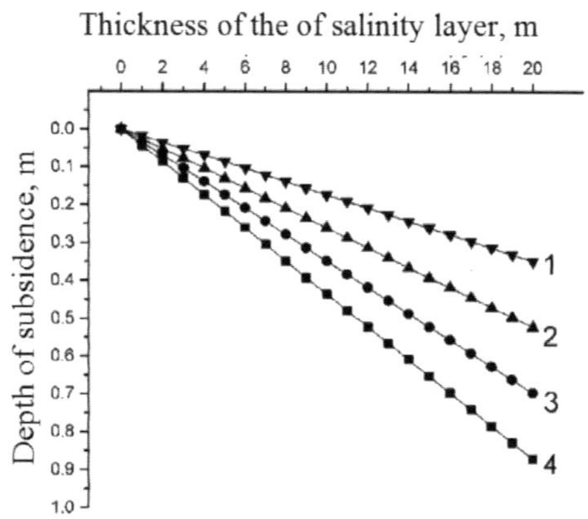

Fig. 2. Dependence of the depth of subsidence on the level and depth of salinity: a - in case of monosalinization; b - in case of polysalinization;
1 - D_{sal}= 0.02, 2 - D_{sal}= 0.03, 3 - D_{sal}= 0.04, 4 - D_{sal}= 0.05.

For polysalinization we considered soluble compounds most common in the territory in question. Low soluble compounds, and, therefore, low deformations in the leaching shall be neglected. Based on the analysis of the chemical composition of saline soils we identified key components of salinization: kainite, halite, carcallite, K_2SO_4, $MgSO_4$.

We divide salts into three types by their solubility in water:

I - $MgSO_4 + K_2SO_4$;
II - halite+ kainite;
III - carnallite.

We accept that the amount of the leached salts is directly proportionate to their solubility in water [11]. Then weighting coefficients will be: for I type salt 0.425; for II type salt 0.2; for III type salt 0.375, Fig. 3.

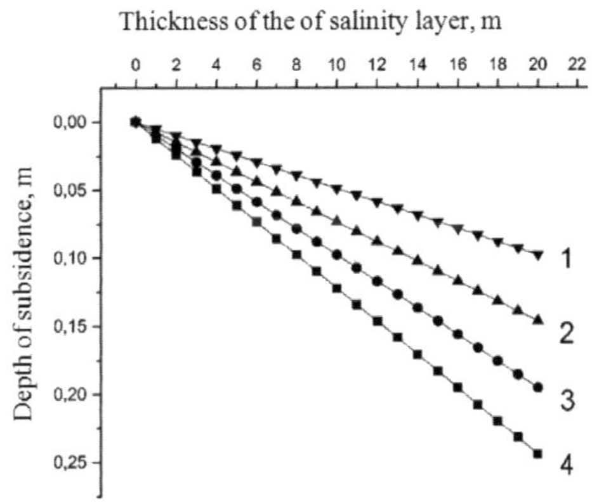

Fig. 3. Dependence of the depth of subsidence on the level and depth of salinity, taking into account weighting coefficients: 1 - D_{sal}= 0.02; 2 - D_{sal}= 0.03; 3 - D_{sal}= 0.04; 4 - D_{sal}= 0.05

Currently, there is no rigorous mathematical formulation of the problem of stress and strain condition in the area of water saturated saline soils. Therefore, it is necessary to improve the proposed model by taking into account soil subsidence as a result of flooding and compaction of soil particles.

It is also necessary to accumulate and organize theoretical and experimental data to create a database of weighting coefficients of the main salinization components to improve the quality of calculation of leaching processes with a view to a better prediction of suffusion subsidence of saline soils.

4. Conclusions

A simplified mathematical model of saline soils subsidence due to leaching in cases of mono- and polysalinity is proposed. The development of mathematical models with a view to making them universal is planned in the future. It is also planned to adapt calculations to the operating conditions of liquidated mining and chemical companies of the Carpathians.

Kinetic curves of suffusion subsidence are built for the disperse and film salinity, which will help predict and monitor subsidence of saline soils in technologically loaded areas. The relationship between the degree of salt dispersion and the rate of salt leaching is established, and the dissolution rate is shown to be proportional to the total area of the interface "salt-solvent".

References

[1] Mustafaev A.A. 1985 The deformations of saline soils in the foundations facilities, Moscow, Stroyizdat, 280 p.
[2] DSTU B V.2.1-5-96 Bases and foundations of building and structures. Soils. Statistical treatment of the test results, Kyiv: SC "Ukrarkhbudinform", 1997, 16 p.
[3] Hryshchenkov M.M. 2011 *Problems of Rock Pressure* **19** 6-26.
[4] Sadovenko I.A., Derevyagina N.I. 2013 *Transactions of Kremenchuk Mykhailo Ostrohradskyi National University* **81**(4) 129-134. (in Ukrainian)
[5] Poberezhna L., Poberezhnyi L. 2015 *Journal of the Ternopil National Technical University* **79**(3) 78-84.
[6] Cahill K. Physical Mathematics. Cambridge University Press, 1st Edition, 2013.
[7] Semchuk Ya. M. 2014 *Environmental Security and Sustainable Resource use* **9**(2) 164-165. (in Ukrainian)
[8] Harkushin P.K. 2001 *Ecology and Environmental Safety* 2 48-55. (in Ukrainian)
[9] Price D.G., 2009. Engineering geology principles and practice: De Freitas M.H (Ed.), Springer-Verlag Berlin Heidelberg, 450 p.
[10] DSTU B V.2.1-2-96 (GOST 25100-95). Bases and foundations of building and structures. Suils. Classification, Kyiv: SC "Ukrarkhbudinform", 1997, 43 p.
[11] Dean J.A. 1992 Lange's Handbook of Chemistry 14-th edn (New York: McGraw-Hill).

IOP Conf. Series: Earth and Environmental Science **50** (2017) 012047 doi:10.1088/1755-1315/50/1/012047

Improving Hygienic Characteristics of Coated Electrodes for Welding High-Alloy Steels

D P Il'yaschenko[1, a], D A Chinakhov[1, b], K V Ivanov[2], I D Sadikov[1]

[1] Yurga Institute of Technology, TPU
Leningradskaya 26, Yurga, Russian Federation, 652055

[2] Institute of Strength Physics and Materials Science SB RAS,
Academichesky pr. 2/4, Tomsk, Russian Federation, 634055

E-mail: [a]mita8@rambler.ru, [b]chinakhov@tpu.ru

Abstract: The article presents the results of experimental studies showing that the use of an inverter power supply instead of a diode rectifier provides:: fine-droplet electrode metal transfer which reduces generation time by 46% and transfer time by 28%; transfer of alloying elements from welding materials into the weld metal which reduces its loss from the welding line by 6% and the heat affected area by 3%; reducing the emission rate of welding fumes and their components by 23%; reducing specific emission of welding fumes and their components by 23%.

1. Introduction

In 2013 the American Conference of Governmental Industrial Hygienists (ACGIH) revised its recommendations on the quantitative composition of welding fumes and proposed to reduce the maximum permissible concentration (MPC). It is obvious that the above changes will become a regulatory standard that will make manufacturers of welding consumables to seek ways of adapting to the new conditions. In many countries and especially in the US, judicial proceedings between welders who suffer from occupational diseases and producers of welding materials take place. Thus, measure aimed at reducing the negative impact of welding fumes on human health are necessary. The above facts show that work at improvement of sanitary and hygienic characteristics of welding is going on at the international level [1].

An analysis of research works [2-4] showed that the quantitative and qualitative composition of welding fumes has a negative impact on human body, especially when welding high-alloy steels.

Arc welding of high-alloy chromium-nickel steels causes air pollution in the working area with welding fumes which contain harmful substances of various hazard categories (State Standard 12.1.005-88): I - hexavalent chromium, and nickel; II - manganese and soluble fluoride; III - trivalent chromium and insoluble fluorides.

Minimizing welding fumes and the most toxic components in their composition is an urgent problem that must be solved.

This paper deals with a hygienic assessment of coated electrodes of CL 11 type intended for welding high-alloy steels and finding ways to reduce the toxicity of welding fumes.

One of the ways to solve this problem is to use power supplies implementing various forms of energy conversion [5-7], which have an impact on the physical and chemical processes in welding.

IOP Conf. Series: Earth and Environmental Science **50** (2017) 012047 doi:10.1088/1755-1315/50/1/012047

Work [8] describes MMA model and proves that the amount of welding fumes in the air of the working area is reduced if the amount of alloying elements transiting into the weld and the slag crust is increased. Efficiency of transition of alloying elements from the electrode to the weld can be improved by reducing heat input into an electrode metal drop and minimizing size of droplets in droplet transfer.

2 Methods and techniques of experimental research:

Assessment of the stability of the welding process was performed using the statistical analysis of welding current and voltage oscillograms generated by power supplies having different dynamic characteristics. The main power supplies have been chosen: a power supply made as a common welding rectifier, and the power inverter. The study of dynamic characteristics was based on a comparison of current and voltage oscillograms obtained using the following devices: digital storage oscilloscope "AKIP-4122 / 1V"; Differential Probe "Pintek Electronics DP-50"; "current probe PR 1030" puller; OWON_Oscilloscope_2.0.8.26 software. The results of the study are presented in Tables 1, 2 and in Figure 1.

Table 1 - Welding modes with CL11 electrodes from various energy supplies

Power supply	Electrode type	Average values of a welding mode parameters	Number of short circuits
Diode rectifier	CL11	Current 86 A Voltage 24.5B Welding speed 0.27 m/min	12
Inverter			24

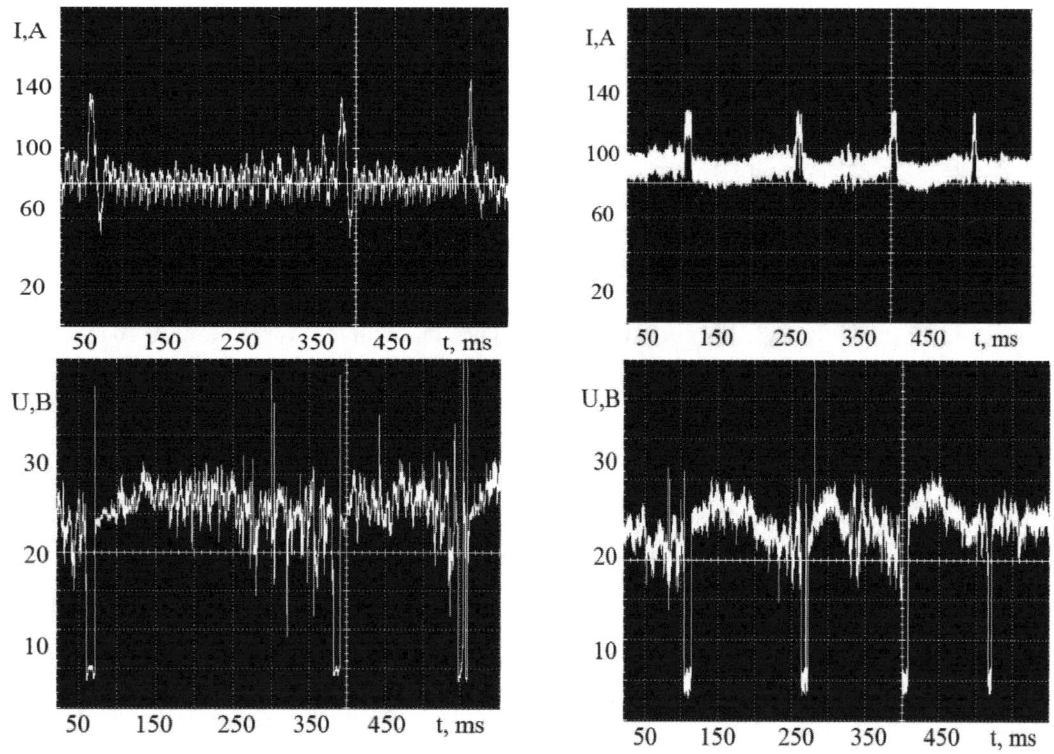

a b

Figure 1. Oscillograms of voltage and current of drop transfer cycle (Type CL 11 electrodes with a diameter of 3 mm): a) - diode rectifier; b) - inverter

IOP Conf. Series: Earth and Environmental Science **50** (2017) 012047 doi:10.1088/1755-1315/50/1/012047

3. Results

Analysis of the oscillograms shown in Figure 1 (Table 1.2) shows that when a power supply based on high frequency energy conversion (an inverter) is used, the droplet transition time decreases and the quantity of short circuits increases. In this case the droplets in electrode metal transfer are smaller in size which is proved to changes the chemical composition of the weld metal [9, 10]. To confirm this supposition microradiography of chemical composition at areas 1, 2, 3 shown in Fig. 3 and Table 3 was carried out (Fig. 2).

When MMA deoxidation - alloying of the weld metal occurs combined - through electrode coating.

Table 2 - Static analysis of dropt transfer parameters of CL11 coated electrodes metal

Parameters	Power supply	
	Diod rectifier	Inverter
The duration of short-circuit arc gap, m/s	12	8.7
The standard deviation of short-circuit time, m/s	3.76	3.52
The coefficient of time variation, %	31.3	28.9
Time of short-circuit cycle,., m/s	249	140
The standard deviation of short-circuit cycle time σ , m/s	76.3	45.8
Variation coefficient of short-circuit time, %	30.6	32.7

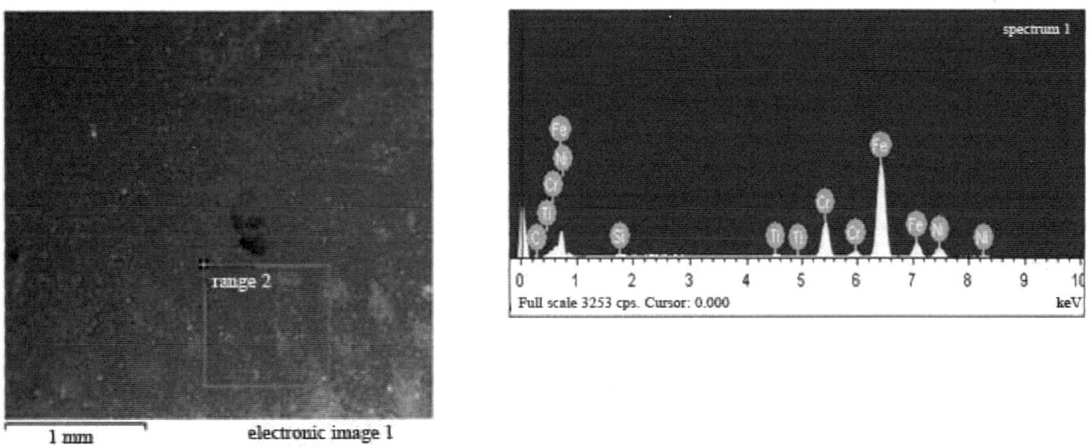

Figure 2. Electron microprobe analysis obtained by INCA consoles (Oxford Instruments)

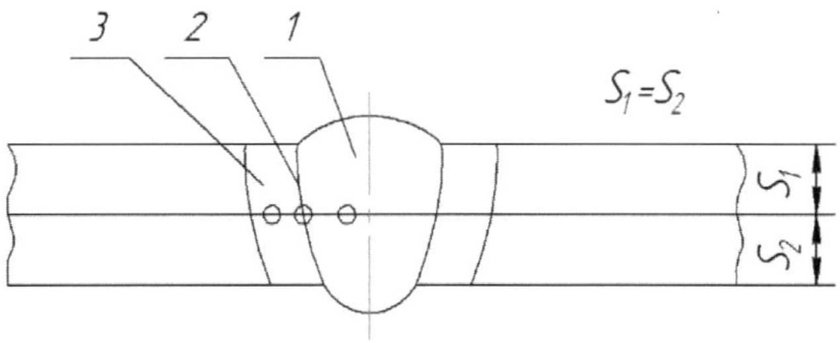

Figure 3. The scheme of electron microradiography: 1 – Weld; 2 – Weld line; 3 – HAZ

Table 3 - Chemical composition of weld metal made of steel H18N10T obtained electrodes brand CL 11

Elements content	Power Supply			
	Diode Rectifier		Invertor	
	Weight %	Atom%	Weight %	Atom%
Weld (1)				
Si K	0.36	0.71	0.38	0.75
Ti K	1.44	1.45	1.90	1.91
Cr K	18.54	19.66	19.21	20.37
Fe K	67.71	66.85	69.45	68.57
Ni K	10.19	9.57	10.81	10.15
Weld line (2)				
Si K	0.50	0.98	0.56	1.10
Ti K	0.60	0.69	0.65	0.75
Cr K	16.45	17.45	17.40	18.42
Fe K	70.52	69.53	71.44	70.77
Ni K	10.87	10.11	11.01	10.20
HAZ (3)				
Si K	0.50	0.97	0.53	1.05
Ti K	0.61	0.70	0.76	0.88
Cr K	17.19	18.21	17.41	18.45
Fe K	70.96	70.00	70.70	69.73
Ni K	10.63	9.98	10.71	10.05

An analysis of data in Table 3 shows an increase of the alloying elements in all the weld areas investigated when an inverter is used, which can be explained by fine-droplet electrode metal transfer (Table 2).

Samples of welding fumes for hygienic assessment of the electrodes were collected in the course of welding beads on plates of H18N10T steel with the electrode being tested. A diode rectifier and an inverter were used as the power supply. Bead welding was carried out with direct reverse polarity current at optimal conditions. Welding fumes were captured by means of a special cover insulating the weld zone.

At the same time, according to instructions [11, 12], welding fumes deposited on the filters of FPP-15-1.5 tissue when assessing the level of emission and on AFA-HA-18 filters when for subsequent chemical analysis of welding fumes samples. The weight of the emitted welding fumes and their components was determined by the gravimetric method. For this purpose, at least five welding fume samples were selected. The research results (Table 4, 5, 6) were subjected to statistical analysis [13].

Table 4 - The emission rate of welding fumes and their components (g/min)

Electrode type (diameter, mm)	Power supply type	CA	CrO_3	Cr_2O_3	Mn	Ni	F soluble	F insoluble
CL-11 (3)	Diode rectifier	0.257	0.012	0.006	0.007	0.006	0.012	0.013
	Inverter	0.2	0.09	0.004	0.005	0.0045	0.01	0.01

Table 5 - Specific emission (g/kg) of welding fumes and their components

Electrode type (diameter, mm)	Power supply type	CA	CrO_3	Cr_2O_3	Mn	Ni	F soluble	F insoluble
CL-11	Diode rectifier	10.48	0.48	0.26	0.31	0.26	0.50	0.51
	Inverter	8.07	0.37	0.20	0.24	0.20	0.42	0.42

Table 6 - Toxic level of TLVf welding fumes and NHL air exchange

Electrode type (diameter, mm)	Power supply type	TLV_f, mg/g^3	NHL, m^3/h
CL-11	Diode rectifier	0.15	1480
	Inverter	0.1	1000

According to the research results (Table 4, 5, 6) it has been found that the use of inverter rectifiers improves working conditions of welders in manual arc welding as compared with diode rectifiers. It reduces: a quantitative component of welding fumes from 0.257 to 0.2 (19%), manganese from 0.007 to 0.005 (23%), chromium oxide from 0.012 to 0.09 (24%).

Conclusion

The research has shown that the use of an inverter power supply instead of a diode rectifier provides:

1) fine-droplet electrode metal transfer which reduces generation time by 46% and transfer time by 28%;
2) transfer of alloying elements from welding materials into the weld metal which reduces its loss from the welding line by 6% and the heat affected area by 3%;
3) reducing the emission rate of welding fumes and their components by 23%;
4) reducing specific emission of welding fumes and their components by 23%.

Work carried out at the expense of the Russian Science Foundation grant, the project №16-19-10010.

References

[1] I.P.Gubenya, I.R. Yavdoschin Formation of solid components of welding fumes and the ways of their penetration into the human body/ Automatic Welding. No 1. 2016. 53-54.
[2] Cardiovascular effects in rats after intratracheal instillation of metal welding particle. Автор: Zheng, Wen; Antonini, James M.; Lin, Yen-Chang;. INHALATION TOXICOLOGY Volume: 27 Issue: P.: 45-53 Published: JAN 2015

[3] Home cage locomotor changes in non-human primates after prolonged welding-fume exposure Автор: Kim, Choong Yong; Sung, Jae Hyuck; Chung, Yong Hyun; INHALATION TOXICOLOGY Volume: 25 Issue: 14 P.: 794-801 Published: DEC 2013
[4] Chemical composition of welding aerosol in manual coated electrode arc welding. Il'yashchenko, D.P., Zernin, E.A., Shadskii, S.V. 2011 Welding International
[5] Improved Power-Quality-Based Welding Power Supply With Overcurrent Handling Capability. Narula, Swati; Singh, Bhim; Bhuvaneswari, G. IEEE TRANSACTIONS ON POWER ELECTRONICS Volume: Issue: 4 P.: 2850-2859 Published: APR 2016
[6] Power Quality Improvement Using Three Phase Modular Converter for Welding PowerSupply. Singh, Bhim; Narula, Swati; Bhuvaneswari, G. Group authors: IEEE. Conference: IEEE 5th Power India Conference Location: Murthal, INDIA публ.: DEC 19-22, 2012. Sponsors: IEEE. 2012 IEEE FIFTH POWER INDIA CONFERENCE Published: 2012
[7] Load-Resonant Converter with Changing Resonant Tanl Topology for WeldingApplications. Shklovski, Jevgeni; Janson, Kuno; Kallaste, Ants Group authors: IEEE Industrial Electronics Society. Конференция: 38th Annual Conference on IEEE-Industrial-Electronics-Society (IECON)Location: Univ Quebec, Ecole Technologie Superieure Montreal (ETS), Montreal, CANADAпубл.: OCT 25-28, 2012. 38TH ANNUAL CONFERENCE ON IEEE INDUSTRIAL ELECTRONICS SOCIETY (IECON 2012) Series: IEEE Industrial Electronics Society Pp. 518-524 Published: 2012
[8] MAW productivity development and reduction of its harmful effect on human organisms Ilyashchenko, D.P., Chinakhov, D.A., Gotovshchik, Y.M. Applied Mechanics and Materials 2014.
[9] Iljyashchenko D. P. , Chinakhov D. A. Investigating the Influence of the Power Supply Type upon the Weld Joints Properties and Health Characteristics of the Manual Arc welding. Materials Science Forum. 2012. Vol. 704. Pp. 608-611.
[10] Arc welding metallurgy: processes in the arc and electrode melting / Edited by I.K. Pokhodnya. 1990.
[11] Methodology guidelines 1927–78. Hygienic assessment of welding materials and methods of welding, weld overlay and metal cutting. Ministry of Health of the USSR, 1980.
[12] Methodology guidelines 4945–88. Determining hazardous substances in welding fumes. Ministry of Health of the USSR, 1990.
[13] Pollard, J. Reference book on statistical calculations. Russian translation. Moscow, 1982.

IOP Conf. Series: Earth and Environmental Science **50** (2017) 012048 doi:10.1088/1755-1315/50/1/012048

Acid-Base Properties Of Glass Substrate And $SiO_2 - Bi_2O_3$ Thin-Film Systems Obtained On It

A G Mal'chik[1, a], S V Litovkin[1], A V Filonov[1], O V Ulyanova[1, b], V E Gromov[2, c]

[1] Yurga Institute of Technology (branch) National Research Tomsk Polytechnic University, Leningradskaya st. 26, Yurga, Russian Federation, 652055
[1] Siberian State Industrial University
Kirova st. 42, Novokuznetsk, Russian Federation, 654007

E-mail: [a]ale-malchik@yandex.ru, [b]olgauljanova@mail.ru, [c]gromov@physics.sibsiu.ru

Abstract. The article describes an experimental research as a result of which $SiO_2 - Bi_2O_3$ films have been synthesized of film-forming solutions based on tetraethoxysilane and bismuth nitrate (III). Acid-base properties of a glass substrate and $SiO_2 - Bi_2O_3$ films obtained on it have been studied. The dependency of physical and chemical properties of $SiO_2 - Bi_2O_3$ composites on their percentage composition have been revealed.

Introduction

Efficient use of resources is the main task of the world community. First and foremost Resource conservation implies ensuring the most effective environmental management, choosing the best way of reproduction and protection of natural resources, taking into account environmental factors when solving economic problems. We should address such problems as advanced processing of natural raw materials and of industrial and consumer wastes, introduction of waste-free or low-waste resource and energy efficient technologies and environmentally friendly production, prolonging service life of equipment and its separate parts [11]. Many of these problems can be solved by obtaining and using new materials with desired properties, thin films and nanopowders being among them [8]. In this view it is important to study and control physicochemical properties on the surface of solids, to identify active centers responsible for its reactivity as well as for the depth and direction of solid phase processes [1,6]. One of important physicochemical characteristics of a solid surface is an acid-base parameter which depends on the nature of the substance, its preparation method, chemical composition, nature and amount of impurities on the surface [4]. The aim of this work is to study physical and chemical processes underlying properties and synthesis of thin films and to obtain thin films of $SiO_2 - Bi_2O_3$. The work investigates the acid-base properties of a glass substrate and studies physicochemical properties of the films obtained.

Experiments

To obtain thin-film systems with a predetermined composition film-forming solutions (FFS) made of 96% ethanol, tetraethoxysilane, bismuth nitrate. The films were prepared on silicon and glass substrates. Two methods were used for obtaining the films: centrifuging at the rate of 3000 - 5000 rev / min and drawing at the rate of 1 - 5 mm/s. The samples were incubated at a temperature of 333K for 30 minutes. The final film formation was performed at a temperature of 873 K for an hour. The refractive index and thickness of the films were measured at 5 points of each sample using a laser ellipsometers LEF-3M and "SE400advanced" . The phase composition of the films was studied with the help of DRON-3M diffractometer (using CuKa-radiation and at a shooting rate of 2 deg / min.). Acid-base properties of solid surfaces were determined by means of ion exchange adsorption using a pH- meter of pH - 673M type, as well as by spectrophotometric version of indicator method on the SF-20 spectrophotometer.

Discussion

SiO_2-Bi_2O_3 system is promising in terms of creating a gas-sensitive elements [9,10]. The presence of differently charged cations indicates a change in the adsorption and resistive properties. Analysis of the surface acidity found that in this system it is possible to obtain compound oxides with surface pH from 2.5 to 9.5 depending on the content of bismuth oxide in the samples and the firing temperature. This variation allows obtaining a substances with different types of surface-active centers. So the samples dried at 333K are strongly acidic on the surface. Data of samples total acidity are presented in Table 1.

Table 1 - Total acidity of samples $SiO_2 – Bi_2O_3$

Bi_2O_3 content, % mole	Sample acidity	
	Dried at T=333K	Fired at T=873K
0	–	5.40
10	2.49	5.40
20	2.11	5.50
30	1.95	5.80
40	1.75	6.40
50	1.70	7.05
60	1.68	7.46
70	1.66	7.49
80	1.5	7.49
90	1.48	–
Pure substrate	–	8.3

Figure 1 shows the dependence of surface acidity of SiO_2–Bi_2O_3 system samples on bismuth oxide (III) content. When the content of bismuth oxide in the system increases up to 30-35% surface acidity of the sample increases sharply (pH_{susp} decreases). When the content of bismuth oxide is more than 35% the acidity remains practically unchanged. The time of adsorption-desorption equilibrium stabilization changes symbatically with change in the acidity.

Annealing of the system at 873 K increases the pH_{susp} values (Figure 1, curve 2). The surface of the samples acquires weak acid properties when Bi_2O_3 content is not more than 30% and weak base properties when it is more than 60%. Values of surface acidity of samples consisting of 100% of SiO_2 and Bi_2O_3 indicate that if the mass fraction of Bi_2O_3 in the system is below 30% , there are no significant quantities of Bi_2O_3, to affect the density of the samples. When the content of Bi_2O_3 in the system is more than 30% of a composite structure SiO_2–Bi_2O_3 is being rapidly developed. Thus compounds of varying composition $mSiO_2$:n Bi_2O_3 are generated.

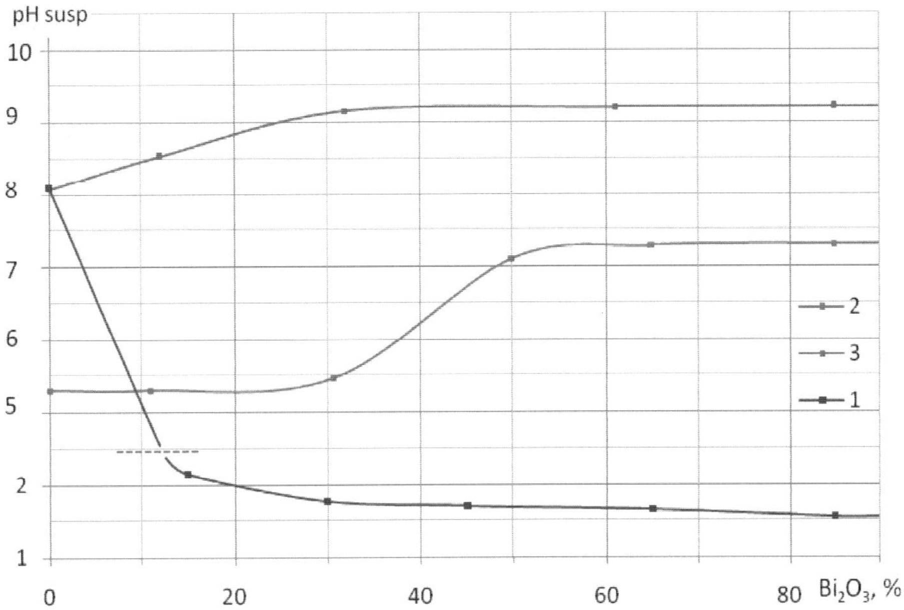

Figure 1 – Dependence of surface acidity of SiO_2–Bi_2O_3 samples on bismuth oxide (III) content. (1 – dried at 333K; 2 - annealed at 873K; 3 – three layer surface annealed at 873K)

Equal weight percentage interval of $mSiO_2$:nBi_2O_3 system (less than 30%, from 30% to 60%, and more than 60%) in which the change in samples acidity and eutectic state occur indicate correlation of the two parameters: the higher Bi_2O_3 concentration in a sample, the less acidic it is. Thus an intensive generation of a composite material with a high content of Bi_2O_3 is taking place. Moreover, increased concentrations of $Bi(NO_3)_3 \cdot 5H_2O$ in the system of film-forming solutions (FFS) based on tetraethoxysilane (TEOS) and $Bi(NO_3)_3 \cdot 5H_2O$ lead to increase in the temperature at which the materials are generated.

Decrease of the eutectic temperature in the film-forming solutions system Si-Bi-O is associated with generation of film-forming solutions of intermediate organic associates. So, matching of the curve break points showing changes in the acidity of the samples depending on their composition with eutectic points proves a decisive influence of Bi_2O_3 phase on the surface acidity of the synthesized film structures.

As mentioned above, decrease of the object acidity from pH = 5.5 to pH =7.5 is consistent with amphoteric properties of Bi_2O_3 in the presence of SiO_2.

According to acidity changes, drying does not lead to Bi_2O_3 phase in the in the entire investigated content interval. The surface of the objects is strongly acidic in nature due to Bi nitrite occurrence in polycondensed silicone compounds.

Anealing at 873 K causes decomposition of bismuth nitrate and generation of silicate Bi, as evidenced by the decrease in acidity from pH = 2 to pH = 5.5. Increase in the mass content of bismuth oxide (III) up to 30% does not lead to changes in surface acidity. This also corresponds to the state chart data. A further increase in concentration induces a sharp decrease of acidity and is due to $2Bi_2O_3 \cdot 3SiO_2$ phase developed in all the samples. When the mass fraction of Bi_2O_3 is 60% pH is being stabilized and is close to bismuth oxide surface acidity of pH = 7.5.

Another dependence of sample surface acidity on the SiO_2: Bi_2O_3 ratio can be observed when the samples are applied onto a glass substrate. In this case, the surface basic capacity increases drastically in samples containing up to 30% of bismuth and does not change when Bi_2O_3 content is increased (Figure 1, curve 3).

IOP Conf. Series: Earth and Environmental Science **50** (2017) 012048 doi:10.1088/1755-1315/50/1/012048

The study of a set of acid-base surface centers by means of an indicator method showed that there is of a large number of centers with an acid-base strength of pH = 8. Kinetic curve of acidity changes for the sample in which Bi_2O_3 - 70% is shown as an example of an established equilibrium (Figure 2).

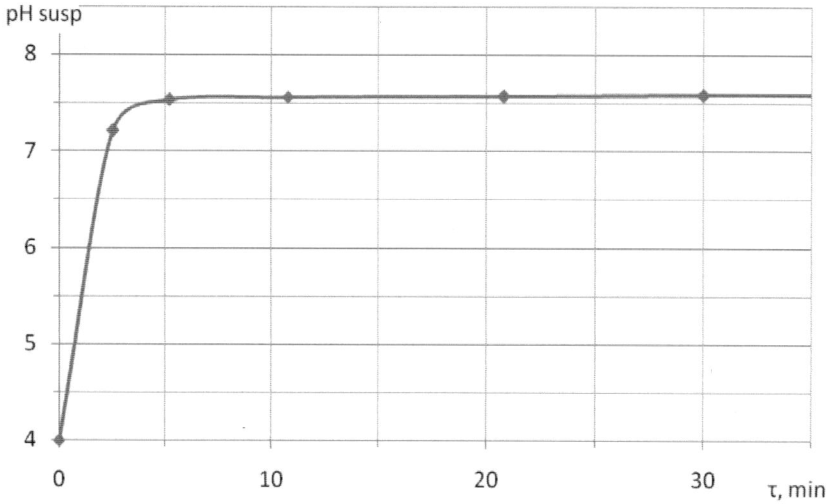

Figure 2 – Kinetic curve of acidity changes in 70% Bi_2O_3 – 30% SiO_2 sample

In the sample composed of 70% Bi_2O_3- 30% SiO_2 the presence of basic Bronsted centers is proved by differentiation of surface active centers with respect to strength and concentration by means of Hammett method of acid-base indicator adsorption. A dependence of distribution of adsorption centers on the sample surface shows that basic Bronsted centers characterized by pKa = 7 predominate on the 70% Bi_2O_3 – 30% SiO_2 sample surface but there are also acid centers with pKa = 2.8 (Figure 3).

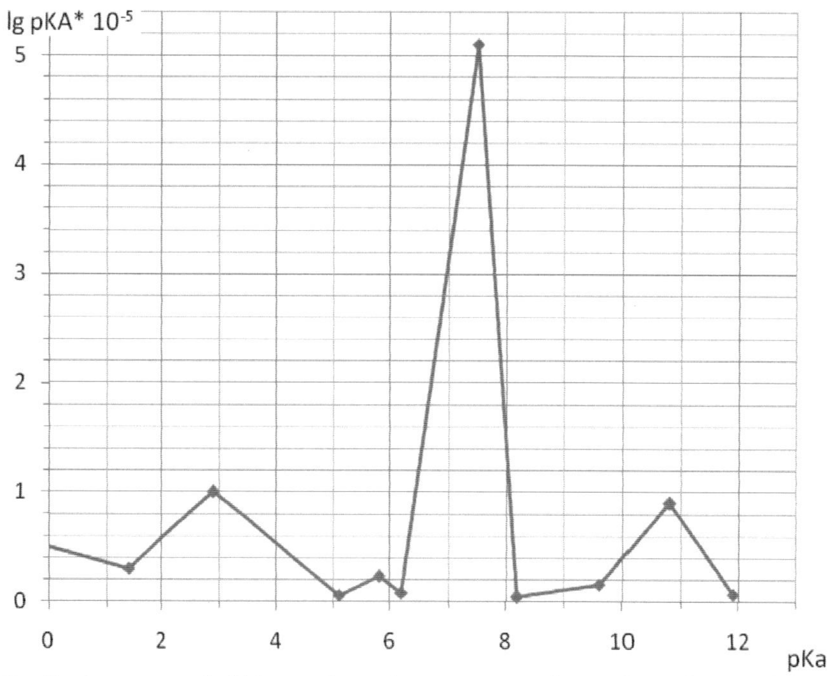

Figure 3 – Distribution curve of aid-base adsorption centers on the surface of 70% Bi_2O_3 – 30% SiO_2 sample

However, the basic centers are ten times more numerous, their number increasing alongside with an increase in Bi_2O_3 content in the system. The nonlinear dependence is stipulated by chemical compounds found in the system. The predominance of the basic centers may be due to the fact that the water proton is attracted to oxygen greater than the hydroxide ion is attracted to Bi^{3+} cation, and the basic centers predominate on the surface [3, 5, 7].

Thin-film structures produced from film-forming solutions containing bismuth and silicon compounds as well as powder materials made from film-forming solutions can be used to create gas-sensing thin-film and ceramic materials for gas sensors.

For films of SiO_2 - Bi_2O_3 system, a state diagram showing composition-property relations was constructed in which the property component is presented by a structure-sensitive refractive index (Figure 4).

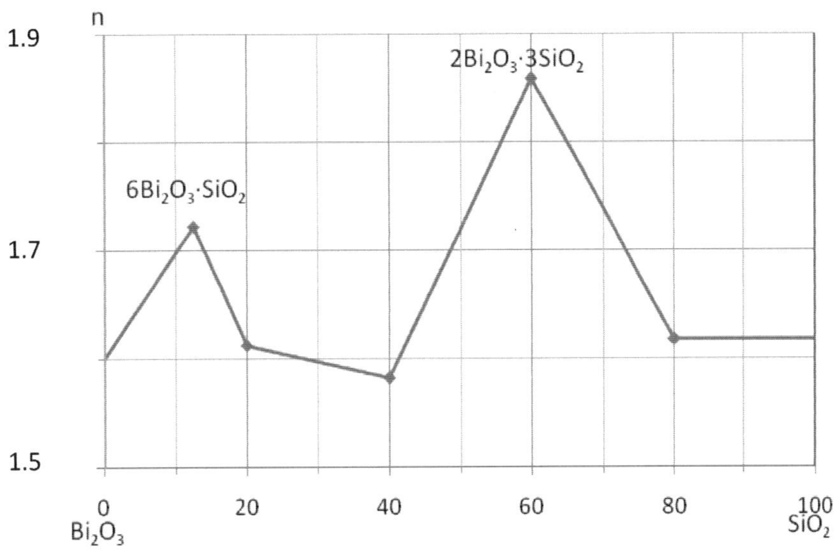

Figure 4 – Composition-property diagram for SiO_2 – Bi_2O_3 system

Two data points showing generation of the corresponding silicates ($xSiO_2$ – yBi_2O_3) are located on the curve, the existence of silicates is confirmed by X-ray analysis (Table 2).

Table 2 - X-ray phase analysis of SiO_2- Bi_2O_3samples

Compound	d, A° (theory)	J (%)	2Θ	d, A° (experiment)
SiO_2	4.1103	100	30.615	4.1123
	4.0586	60.3	27.16	3.7901
	3.8245	100	50.72	3.8244
	4.2734	100	35.48	4.1975
	1.8211	75.4	60.81	1.5673
	1.0119	20.1	74.44	1.0118
Bi_2O_3	2.6901	100	33.47	2.7508
	2.7631	71.4	32.40	2.6183
	1.6656	33.6	23.5	1.6341
	1.46127	27.0	63.69	1.45607
	1.23483	10.2	77.25	1.23483
	1.21607	23.5	78.69	1.21607
	1.0497	10.7	66.44	1.0372

Compound	d, A° (theory)	J (%)	2Θ	d, A° (experiment)
$6SiO_2 \bullet Bi_2O_3$	1.2103	64.8	42.875	1.2103
	1.7134	70.4	44.75	1.5837
	1.77058	25.5	51.63	1.75046
	1.63347	46.9	56.31	1.62345
	2.1090	64.8	42.89	2.1090
	2.5952	24.5	34.56	2.5952
	1.19606	19.9	80.25	1.2016
$2SiO_2 \bullet 3Bi_2O_3$	2.7549	100	32.50	2.6958
	3.2008	88	27.19	3.1890
	4.2422	59.7	20.94	4.2003
	4.1329	36.7	21.44	4.0975
	1.5229	24.5	60.81	1.5231
	2.51222	12.8	42.875	2.51171
	1.21607	78.69	23.5	1.3887

Electrical properties listed in Table 3 show that the compounds of $SiO_2 \bullet Bi_2O_3$ system are dielectrics with highly ionic bonds [2].

Table 3 - Physical and chemical properties of $SiO_2 - Bi_2O_3$ system films

Property	Content of Bi_2O_3 in the film, %								
	10	20	30	40	50	60	70	80	90
Refractive index, n	1.73	1.52	1.50	1.48	1.74	1.85	1.73	1.53	1.53
Dielectric capacity, ε	2.43	3.98	4.21	4.62	5.61	4.02	5.83	6.51	7.89
Adhesion strength, F, $кг/мм^2$	0.99	0.99	0.99	0.98	0.99	0.99	0.98	0.99	0.99

A coating method is of value only if it allows obtaining layers of different materials with a wide range of physical properties to meet technical requirements.

As to the physical characteristics of such films, of the greatest interest are their optical properties, particularly a refractive index and an absorption coefficient. Electrical and chemical protective properties of the films are also important.

High chemical stability and thermal resistance of $SiO_2 - Bi_2O_3$ oxide films as well as unique technological advantages provided by film-forming solution synthesis allowed the use of $SiO_2 - Bi_2O_3$ film as protective coatings for miniature incandescent lamps for medicine which have to be resistant to corrosive environments. It is proved that in acidic and alkaline medium, coatings provide light transmittance up to 93% in the visible range and have good dielectric properties [12, 13]. Chemical resistance of the films to water and to corrosive environments is presented in Table 4.

Table 4 - Chemical resistance of $SiO_2 - Bi_2O_3$ films to water and to corrosive environments

Test conditions		Test results
Environment	Time	
Distilled water	2 years	No changes
Piped water	2 years	Initiation of destruction in 14 months
Acid solution	6 months	Initiation of destruction in 3 months
Alkaline solution	6 months	Initiation of destruction in 4 months

Conclusions

The results of the research let us draw the following conclusions: acid-base properties of a glass substrate and the thin films formed on these structures have been studied, as well as properties of powder materials made from film-forming solutions with addition of $Bi(NO_3)_3{\bullet}5H_2O$. It is proved that major Bronsted centers prevail. By means of X-ray diffraction, optical analysis and other methods, relationship between physical and chemical properties of SiO_2–Bi_2O_3 composites and their composition have been revealed. Silicate compounds Bi in a thin film form are found to generate. The refractive index of SiO_2–Bi_2O_3 films changes in the range from 1.48 to 1.85; film thickness does not depend on the solution ripening time. The films with this composition can be recommended for use as a dielectric coating with high thermochemical stability, which significantly extends the equipment life and meets requirements for resource-saving.

References

[1] Borilo L P. Thin-film inorganic nanosystems / ed. Dr. tehn. Sciences prof. Kozik V V - Tomsk: Publishing house of Tomsk State University, 2003. – 134.
[2] Synthesis and properties of films in the SiO_2-Bi_2O_3 system. Borilo L P, Kuznetsova S A, Kozik V V, Zabolotskaya A V, Mal'chik A G. Russian Journal of Inorganic Chemistry. 2014. T. 59. 10. P. 1065-1068.
[3] Physicochemical investigation of formation of SIO_2 powders and films from film-forming solutions. Mal'chik A G, Borilo L P, Kozik V V. Russian Journal of Applied Chemistry. 1996. T. 69. 2. P. 198-201.
[4] Ikonnikova L F, Minakova T S, Korochenko N M, Mal'chik A G, Egorova L A. Acid-base characteristic of a solid surface as a criterion for evaluating the suitability of materials for product. Abstracts VIII international conference. Theory and practice of adsorption processes. - Moscow. - 1996.
[5] Bivol V G, Burchakova V I. Bismuth film structure and its analogs. - Chisinau: Shtintsa, 1978.
[6] Sergeev A N, Serebrennikov V. Material studies in refractory oxide films. - Tomsk: Publishing house of Tomsk University Press, 1981. 145 p.
[7] Pout G, Tu K, Meyer G. Thin films. Mutual diffusion and reaction. / Ed. Kiselyov V F - M.: Mir. - 1982. - 382s.
[8] Shabanova N A, Popov V V, Sarkisov P D. Chemistry and technology of nano-dispersed oxides. - M.: ECC "Academy", 2006. 309 p.
[9] Chinakhov D A, Chinakhova E D, Gotovschik Y M, Grichin S V. Influence of welding with two-jet gas shielding on the shaping of a welding joint. IOP Conf. Series: Materials Science and Engineering 125 (2016) 012013. doi:10.1088/1757-899X/125/1/012013.
[10] Chinakhov D A. Gas Dynamic Control of Properties of Welded Joints from High Strength Alloyed Steels. China Welding. Vol. 23. 3 (2014) pp. 27-31.
[11] Mal'chik A G, Litovkin S V, Rodionov P V. Investigations of physicochemical properties of bottom-ash materials for use them as secondary raw materials. IOP Conference Series: Materials Science and Engineering. 2015. Vol. 91, Article number 012081.
[12] Torosyan V F, Torosyan E S, Borovikov I F, Yakutova V A. Oxidative activity of heated coal affected by antypirogens. IOP Conference Series: Materials Science and Engineering. 2016. Vol. 127 . 1, Article number 012022.
[13] Torosyan V F, Torosyan E S. Development of mixture "clay – electro-smelting slag" for ceramic brick. Applied Mechanics and Materials. 2014. Vol. 682. p. 480-484.

Object-Oriented Programming When Developing Software in Geology and Geophysics

R K Ahmadulin and L N Bakanovskaya

Industrial University of Tyumen, 38 Volodarskogo st., Tyumen, 625000, Russia

E-mail: ark@tsogu.ru

Abstract. The paper reviews the role of object-oriented programming when developing software in geology and geophysics. Main stages have been identified at which it is worthwhile to apply principles of object-oriented programming when developing software in geology and geophysics. The research was based on a number of problems solved in Geology and Petroleum Production Institute. Distinctive features of these problems are given and areas of application of the object-oriented approach are identified. Developing applications in the sphere of geology and geophysics has shown that the process of creating such products is simplified due to the use of object-oriented programming, firstly when designing structures for data storage and graphical user interfaces.

1. Introduction

Modern information technologies have become part of all spheres of human activity, from organizing the educational process to optimizing the duration of production cycles[1, 2]. Nowadays, it is hard to imagine the area of activity in which a person would not have come to the aid of modern information technology in general and software in particular.One such area is geology and geophysics.

Design considerations of geophysical research, processing of the results, geological modeling, forming geological engineering reports, etc., often require significant amounts of data processing. Thedevelopmentsthemselvescanbeperformedbothcompletelyindependently[3, 4], andon the basis of the existing earlier developments realized using the methods of structural and modular programming [5, 6].However, in order to effectively use the earlier results (models, algorithms and software) it is advisable to use a more modern object-oriented programming paradigm (OOP).

Thegivenpaperbasedon [7]reviewstheroleoftheprinciplesofobject-oriented programming when developing software in geology and geophysicsand stages, at which it is worthwhile to apply the sated principles in the first place.

2. Research

In the period from 2002 to 2015 the authors took part in a number of projects based on the OOP technology.In the course of these projects the following was developed:

- information system of storage, processing and analysis of geotechnical data[3];
- input module for the geotechnical reports from computers that are not connected to the corporate network [4], with the possibility of uploading them to the database;

- software package designed for use in the design, quality assessment and optimization of field geophysical studies for oil and gas[5];
- program for solving the inverse problem of the near-field transient electromagnetic sounding method, based on the existing functional [6], etc.

Common properties possessed by the abovementioned developments are the following.

1. In most cases, as a basis has been accepted an existing mathematical apparatus implemented in an earlier programming language not compatible with the OOP paradigm (particularly - FORTRAN and C, for example [5, 6]).

2. Processed data generally have a hierarchical structure (for example [3, 4, 6]).

3. It is required to represent data in different forms, depending on the user's needs at a particular time(forexample[4, 6]).

4. There are cases where, depending on the specific situation, the data set should be significantly or slightly different(forexample[3, 4, 6]).

5. For the user the most convenient is the graphical user interface including the "tree" that displays a set of hierarchical data(forexample[4, 6], figure 1).

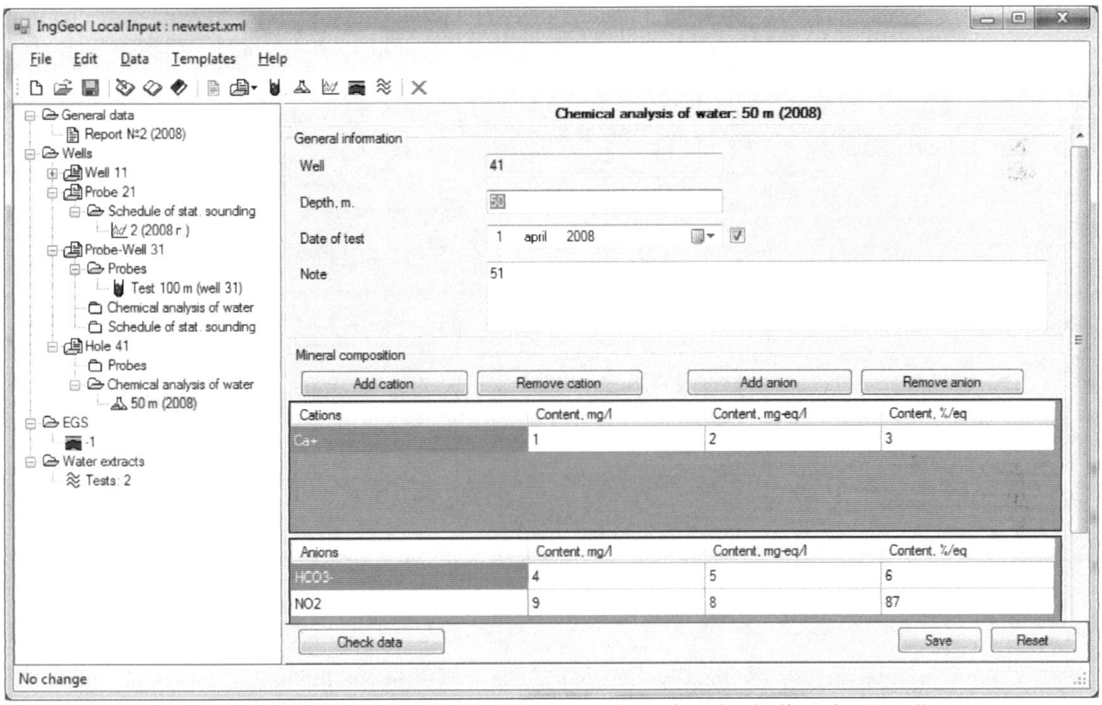

Figure 1. An example of the program interface including the "tree".

Based on the work [7], we will consider the experience of using object-oriented programming principles to:

a) process data;

b) store and transmit data;

c) display data on the screen.

To process data in geological and geophysical problems experts frequently use complex algorithms often developed on the basis of the author's techniques and already implemented in the form of programs. For example, the result of [8] became programs implemented in FORTRAN using the structural and modular approaches. As noted in [5, 6], a minus of such developments is the lack of a user-friendly graphical interface, which urges their re-implementation (development) with the use of modern tools.

At the same time, the algorithms themselves are usually complex and require considerable time for their transcoding to modern programming languages, and further debugging.In the case of the problem in question were limited to time resources, so it was decided to leave their implementation "as is" by connecting them to developing products in the form of a library.Thisapproachwastested, forexample, in [5, 6].

Recoding the existing algorithms in modern programming languages is expedient in cases where the need to accelerate the work of programs, for example, by the methods of parallel programming, is more important than the time spent on development.

An important component of the geological and geophysical program is data storage and transmission. As noted above, such data may have a complex hierarchical structure (Figure 2).An example of such a structure is presented in [7].

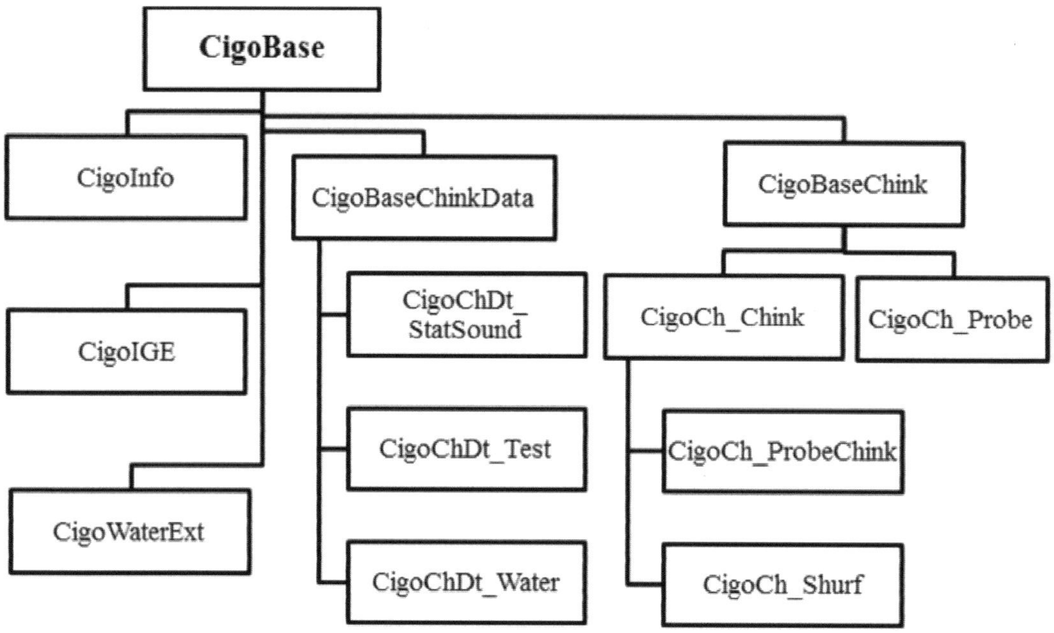

Figure 2. An example of a class hierarchy implemented for data storage and transmission.

It is for the implementation of the hierarchical structures that the OOP paradigm was applied in the first place. Within the framework of these structures the works have used such principles as:

- inheritance, to create structures obtained on the basis of others;
- encapsulation, to combine geological and geophysical data with the methods necessary for their support and ensuring their correctness, as well as to conceal the internal service fields and methods;
- polymorphism, due to which data "variability" is provided depending on a particular task.

The program's flexibility when working with heterogeneous data was also guaranteed through the use of virtual and abstract methods.

Also, the use of classes for storage allowed a convenient data transmission to different modules for display in different formats, which was also important in the case of geological and geophysical programs.

We also used the methods of object-oriented programming to develop graphical interfaces required to display data on the screen.

It should be noted that the use of OOP to create a graphical interface is widely used in various domains, such as [7, 9]. The considered problems have some distinctive features.Geological and geophysical applications typically have a variety of graphical interfaces presenting data in different formats, often slightly differing. Another situation is the need to present different data in the same format (Figures 3, 4).

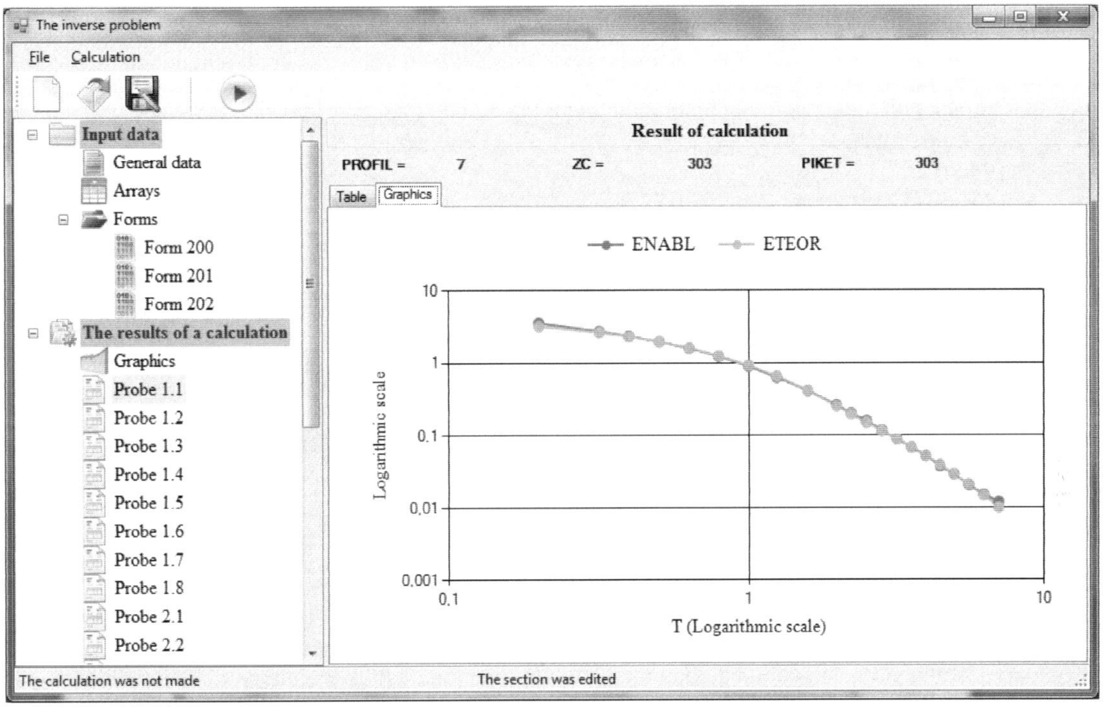

Figure 3. An example of a calculation program interface for solving the inverse problem of the near-field transient electromagnetic sounding method.

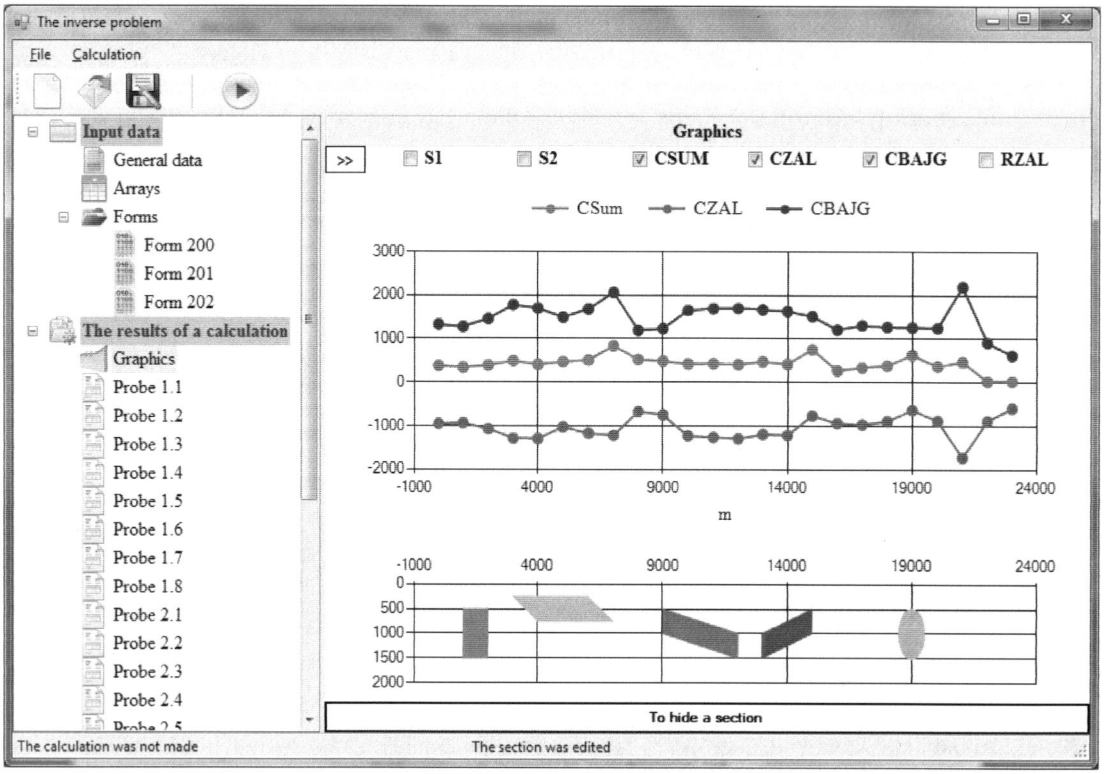

Figure 4. An example of a calculation program interface for solving the inverse problem of the near-field transient electromagnetic sounding method.

OOP capabilities are very useful in the development of a set of graphical interfaces, namely:

- inheritance allows you to create a hierarchy of graphical interfaces with similar features in a set of controls and / or software implementation;
- polymorphism and abstract methods provide the ability to use a single graphical interface to represent different data, different graphical interfaces to represent the same data and also to create a hierarchical tree, each node of which allows you to open the data using a particular graphical user interface;
- virtual methods allow you to specify the desired functionality for both the methods defined in parent classes, and for the standard forms methods. And the ability to override them provides the developer with a powerful tool for rapid creation of interfaces on the basis of ready-made components, but with their own functions.

Developments carried out on the basis of the OOP principles and addressed in this paper help solve a variety of problems in geology and geophysics.

The information system of storage, processing and analysis of geotechnical data [3, 4] enabled automatic provision of the urban development activity with geotechnical data. A unified electronic bank of geotechnical data, software and methodological support was created for the evaluation, monitoring and forecast of geotechnical conditions. As a result, access to the right information on the desired object can be accessed from any computer that has access to the database and installed developed software. The system has practical application.

The software package of design and quality evaluation and optimization of field geophysical studies for oil and gas [5] allows solving a wide range of problems related to calculation and analysis of the interference systems and observation networks during seismic survey. The package was used to solve many practical problems. These include construction of a model (structural map) of a reflecting horizon, study of the constructed model stability and the observation network redundancy, construction of the generalized indicator of the quality of object selection, etc. Another field of software system application is its use to train geophysics students.

The program for solving the inverse problem of the near-field transient electromagnetic sounding method [6] was also used to solve industrial problems and demonstrated high efficiency on a large geological material. The algorithm was applied to interpret field data of the near-field transient electromagnetic sounding method, in the search for electrically polarized objects, etc.

3. Conclusion

Thus, the experience gained in the development of applications in the field of geology and geophysics, showed that object-oriented programming simplifies the process of creating similar programs. The main stages at which it is advisable to use these methods in this domain in the first place is the development of structures for data storage and graphical user interfaces.

Using the experience described in the work can be extended to develop programs to address environmental challenges.

References

[1] Mokeeva N S and Bakanovskaya L N 2009 Creation automatic worker of the place of the technologist fordesigning the technological process *Programmnye produkty i sistemy* **4** 52

[2] Bakanovskaya L N 2015 On the question of the development of neural network expert systems *Materials of the XI International scientific and practical conference "Conduct of modern science" Economic science* (Sheffield; Science and education LTD) **3** 11-4

[3] Turenko S K, Aksarin M Yu, Morozov M V and Parhimovich R V 2008 Information system of storage, processing and analysis of geotechnical data *Proc. of Institutions of Higher Education. Oil and Gas* **5** 12-20

[4] Ahmadulin R K, Turenko S K and Morozov M V 2010 The program for the local data input for AS INZHGI *Crystals of creativity: Proceedings of the Student Academy of Sciences* (Tyumen; TyumGNGU) 300-3

[5] Turenko S K and Ahmadulin R K 2015 Development of software to optimize the field geophysical studies for oil and gas *Geology, geophysics and development of oil and gas fields* **8** 30-3

[6] Ahmadulin R K and Dmitriev A N 2015 Implementation of the program to solve the inverse problem by the ZSB method with a GUI based on the existing functionality *Oil and gas of West Siberia: Materials of the International scientific and technical conference dedicated to the 90th anniversary of Kosukhin Anatoly Nikolaevich* (Tyumen; TyumGNGU) **1** 73-6

[7] Ahmadulin R K 2016 Experience in applying the principles of object-oriented programming in the development of applications in geology and geophysics *Automation, telemechanics and connection in the oil industry* **1** 38-42

[8] Turenko S K 1995 *Computer technology for optimization of field geophysical studies for oil and gas (on the example of seismic survey)* Author's abstract of Cand. Sc. Eng. Disseration (Moscow) 36

[9] Mokeeva N S and Bakanovskaya L N 2009 Development of an automated system software module for design process *Molodoy uchenyi* **3** 38-41

Modeling hydrodynamic flows in plasma fluxes when depositing metal layer on the surface of catalyst converters

D A Chinakhov[1, a], V D Sarychev[2], A Yu Granovsky[2], S A Solodsky[1], S A Nevsky[2], S V Konovalov[2]

[1] Yurga Institute of Technology of Tomsk polytechnic university, Russia, Yurga, Leningradskaya str., 26
[2] Siberian State Industrial University. Russia, Novokuznetsk, Kirova str.,42

E-mail: [a]chinakhov@tpu.ru

Abstract. Air pollution with harmful substances resulting from combustion of liquid hydrocarbons and emitted into atmosphere became one of the global environmental problems in the late 20th century. The systems of neutralization capable to reduce toxicity of exhaust gases several times are very important for making environmentally safer combustion products discharged into the atmosphere. As revealed in the literature review, one of the most promising purification procedures is neutralization of burnt gases by catalyst converter systems. The principal working element in the converter is a catalytic layer of metals deposited on ceramics, with thickness 20-60 micron and a well-developed micro-relief. The paper presents a thoroughly substantiated new procedure of deposing a nano-scale surface layer of metal-catalyst particles, furthering the utilization of catalysts on a new level. The paper provides description of mathematical models and computational researches into plasma fluxes under high-frequency impulse input delivered to electrode material, explorations of developing Kelvin-Helmholtz, Marangoni and magnetic hydrodynamic instabilities on the surface of liquid electrode metal droplet in the nano-scale range of wavelengths to obtain a flow of nano-meter particles of cathode material. The authors have outlined a physical and mathematical model of magnetic and hydrodynamic instability for the case of melt flowing on the boundary with the molten metal with the purpose to predict the interphase shape and mutual effect of formed plasma jet and liquid metal droplet on the electrode in the nano-scale range of wavelengths at high-frequency impact on the boundary "electrode-liquid layer".

1. Introduction

Emission cut of harmful combustion products is a vital problem relevant to the ecological state of our planet, population health, gene pool and cultural treasure preservation, environment safety of foodstuffs. The ecological problem is as important as preservation of peace, and providing world population with foodstuffs. Therefore, each particular measure aimed at cutting harmful emissions of internal combustion engines into the atmosphere is urgent and significant. When manufacturing catalyst converters a number of difficulties are to be overcome; among them there is coating the working surface with a thin layer of metal-catalysts. As a consequence, attempts are made all over the world to design new efficient procedures of depositing such layers.

The regularities describing generation and evolvement of structure and phase states in materials exposed to concentrated energy flows have been in the focus of researches for a long time [1]. The surface layers of materials modified with the help of these flows are distinguished by high hardness and wear resistance [2, 3]. Heterogeneous plasma fluxes are one of the most efficient modification procedures [4]. This is a quite complex process including formation of droplets on the electrode tip; movement of droplets in the jet; interaction between plasma and droplet jet with the product; processes occurring in the molten layer of the product enriched with droplet component of the jet [5, 6].

The technology of plasma arc finds wide application in metallurgy, powder spraying, surfacing etc. Thermal conducting plasma is formed when consumable electrode surfacing. Its properties and operating conditions are of great importance for processes of weld pool and weld formation, influencing this way on properties of the modified surface. Therefore, computational modeling of hydrodynamic processes and thermal fields are relevant to the development of surface modification procedures [7]. Some researchers [8] explored the effect of cathode casing on the thermal field. It's revealed that different potential boundary conditions on the back surface of the anodic plate can hardly effect on the heat flow in cathode or within the arc, affecting, however, current density and temperature field near the anodic plate. For the case the same computational domain includes cathode and anode arcs, additional conditions on the boundary "plasma - electrode" are required to take into consideration the energy flowing through it. The calculation of isotherms for electric current 200 detects a satisfactory fit with the experiment. The authors [9] investigated formation of self-organizing structures in electrodes. As a result, a three-dimensional double-temperature imitation model was developed and revealed that a dotted structure arises gradually due to the increasing level of anode cool-down: from one fuzzy spot at low cooling levels to small spots covering anode zone at intense cool-down. The authors [10] compared models of turbulent flows for free burning of high-intense argon arc. They modified Navier-Stocks equations via considering the transfer of emission, power consumption and electromagnetic forces. To model the turbulence, which is relevant to the arc boundary because of sharp temperature gradient between the arc column and gas around it they referred to zero and two equation of viscosity. It was found out that the proposed model can provide better description of temperature profiles in the arc. The model of plasma flux in anode zone of plasma gun at constant current is presented in [11]. Maxwell and heat transfer equations are solved within this model both for liquid and solid parts of the electrode. The mean temperature of tungsten and copper parts of anode can be assessed with the help of this model. The second important feature of the model is that it takes into account injection of gas into the plasmatron. The researchers [12] fulfilled computational modeling of high pressure in arc discharge relying on self-congruent simulation of the most components, e.g. electrodes and their interaction. In particular, the arc column and cathode part of the discharge are modeled on the base of the double-temperature hydrodynamic model and the model of nonlinear heating surface, respectively. The results of modeling are presented for free burning arc in the atmosphere pressure of argon, covering the arc currents from 10 to 200 A. It was revealed that electric power delivered to the near cathode layer is given to cathode and arc column, as well. This effect can't be described in terms of local thermodynamic equilibrium (LTE). The enthalpy of electron transfer can significantly increase the thermal input caused by the conductivity of electrons and heavy particles, dominating, therefore, the process of power transfer from the near cathode layer into the arc column. Hence, temperatures calculated in terms of LTE-model are much higher than those taking into account the input of electrons and heavy particles [12].

The processes occurring in the molten pool under the action of plasma are investigated in works [13-17]. The authors [13] addressed to the issue of heat exchange and liquid flows in conditions of pulse arc welding the stainless steel and solved this problem relying on the finite elements method. The proposed model made it possible to explore the time evolvement of the weld pool at constant and pulsed current. Having compared the results, it's concluded that pulsed current furthers formation of a better joint weld. Computational simulation was supplemented with experiments in [14], including high-speed recording the infra-red range of the weld pool surface. While processing and analyzing the images a satisfactory fit with the computational model was revealed. The effect of various parameters on the weld pool geometry was analyzed in [15]. According to the results, the gradient of surface tension is the most relevant factor influencing the weld pool geometry, whereas magnetic conductivity, coefficient of thermal expansion of material are less important. The paper [16] is dedicated to modeling processes in electric arc and weld pool with the dividing electrode. This electrode influences significantly on temperature and flow of arc plasma but is irrelevant at peak temperature and arc voltage. Having increased the dimensions of this electrode, a conclusion is made that one-peak profile of current density distribution, heat flow and temperature on anode become double-peak. The overall heat release in anode changes a little in spite of some oxygen. A standardized model of the system "welding arc – weld pool" is presented in [17]. It's ascertained that temperatures and velocities of distributions in the arc are not rotationally-symmetric.

Gas (argon) streamline and heating in the diaphragmatic channel are calculated according to the position and dimensions of the diaphragm, consumption and intensity of gas rotation in terms of equilibrium magnetic and gas-dynamical model [18]. The specifics of arc, changing in channels with alternating cross-section was determined for rotational and non-rotational flows. It is indicated that diaphragming of the channel together with intense gas rotation provides localization of plasma flux before the diaphragm in the narrow paraxial zone with the distinct thermal interface and the outer gas flow (effect of vortex thermal insulation); the length of stabilization zone depends on the size of a diaphragm opening and its position in the channel. The authors [19] dealt with the computational investigation of a micro-wave discharge in argon jet, injected axially into the coaxial with the shortened inner electrode. It's ascertained that the inner electrode exit section is of great importance for formation of a micro-wave discharge, the flow of cold gas, moving in its channel, cools its walls down, stabilizing, therefore, the burn of a micro-wave discharge and furthering formation of the sharp front heat edge.

The present day research into modeling the plasma fluxes is focused on the development of a single mathematical model describing processes both in the zone adjacent to cathode and in the weld pool. Therefore, this work aims at modeling hydrodynamic flows in the diaphragmatic electric arc.

2. Problem formulation

It's quite reasonable to begin exploration of magnetic and hydro-dynamic flows in plasma with calculation of electromagnetic force. We write Maxwell equation for its calculation using scalar potential V and vector potential $\vec{A}$:

$$\nabla\cdot\left(\sigma\nabla V+\sigma\frac{\partial\vec{A}}{\partial t}\right)=0,\ \sigma\frac{\partial\vec{A}}{\partial t}+\frac{1}{\mu_0}\nabla\times\left(\nabla\times\vec{A}\right)+\sigma\nabla V$$
$$\vec{E}=-\nabla V-\frac{\partial\vec{A}}{\partial t},\quad \vec{j}=\sigma\vec{E},\ \vec{B}=\nabla\times\vec{A} \tag{1}$$

The model based on Navier-Stocks equation and that of thermal conductivity is used to model the plasma flow and temperature distribution in it:

$$\rho\left(\frac{\partial\vec{u}}{\partial t}+\vec{u}\cdot\nabla\vec{u}\right)=-\nabla p+\mu\Delta\vec{u}+\vec{F_V},$$
$$\nabla\cdot\vec{u}=0, \tag{2}$$
$$\rho C_p\left(\frac{\partial T}{\partial t}+\vec{u}\cdot\nabla T\right)=\nabla\cdot\left(k\nabla T\right)+S_V$$

where $\vec{u}$ - velocity vector, p - pressure, ρ - density, μ – dynamic viscosity, $\vec{F}_V$ - volumetric forces, resulting from the sum of Lorenz and gravitation forces, T- temperature, C_p – specific heat, k – thermal conductivity coefficient, S_V – volumetric heat sources.

The volumetric heat source for anode and cathode is Joule effect $S_V = \vec{j} \cdot \vec{E}$, whereas for plasma there is additional heating from electron enthalpy and losses caused by radioactive emission.

Boundary conditions for formula (1) and (2) are formulated. We consider a cylindrical electrode, one of its tips is semi-spherical and the other one is flat (Figure. 1).

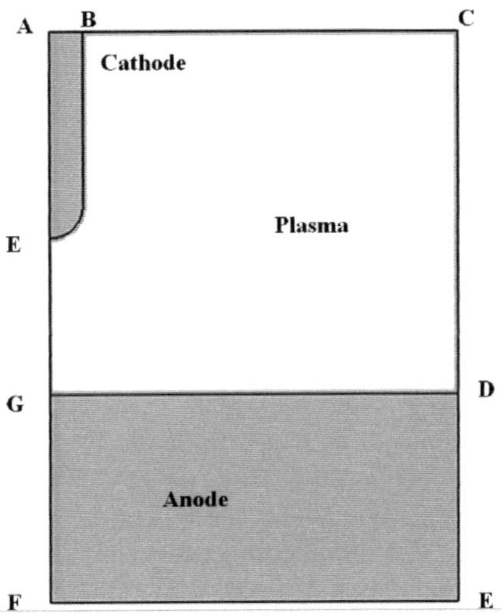

Figure 1. The geometry used in the calculation of the plasma arc

The heat flow on line GD (plasma/anode) is set as a total of plasma heating, and heat arising from electron condensation $|\vec{j} \cdot \vec{n}|\varphi_a$ (where φ_a anode work function of electrons) and radiation losses $\varepsilon\sigma_B T^4$ (where ε – emissivity of anode, σ_B – Stefan-Boltzmann constant). On line EB (plasma/cathode) the heat flow depends on plasma, ion heating $j_i V_i$, radiation losses $\varepsilon\sigma_B T^4$ and thermo-ion cooling $j_e \varphi_c$ (power required for electron emission from cathode, which is a product of electron flow density and cathode work function of material). Boundary conditions for Maxwell and magnetic hydrodynamic equations are provided in Table 1. The properties of materials necessary for calculations are given in Table 2. The kinetic parameters of plasma (heat capacity, thermal conductivity, viscosity, electric conductivity) are dependent on temperature and formulated in the table.

Table 1. Boundary conditions

	AB	BC, CD	DE	EF	AE, FG	EG
T, q	$-\vec{n} \cdot \vec{q} = 0$	T_0	T_0	T_0	$\vec{n} \cdot \vec{q} = 0$	$\vec{n} \cdot \vec{q} = 0$
(u, v, w)	-	$p = p_0$	-	-	-	$\vec{n} \cdot \vec{v} = 0$
$(V, \vec{A})$	$\vec{j} \cdot \vec{n} = \frac{I_c}{\pi R_c^2}$	$\vec{j} \cdot \vec{n} = 0$ $\vec{A} \times \vec{n} = 0$	$\vec{j} \cdot \vec{n} = 0$ $\vec{A} \times \vec{n} = 0$	$V = 0$ $\vec{A} \times \vec{n} = 0$	$\vec{j} \cdot \vec{n} = 0$ $\vec{B} = 0$	$\vec{j} \cdot \vec{n} = 0$ $\vec{B} = 0$

Table 2. Material properties of cathode, anode and gas

Symbol	Nomenclature	Value
ρ_m	Density anode and cathode metal	7500 kg m^{-3}
C_m	Specific heat of metal	602 J kg^{-1} K^{-1}
k_m	Thermal conductivity of metal	26 W m^{-1} K^{-1}
σ_m	Electrical conductivity of metal	7.7×10^5 Ω^{-1} m^{-1}
φ_c	Surface work function of cathode	4.52 V
V_i	Plasma's ionization potential	15.68 V
φ_a	Surface work function of anode	4.65 V
ε	Surface emissivity	0.9

3. Results and discussion

System (1), (2) with boundary conditions was solved with the help of the finite elements method; the diaphragm was ignored and taken into consideration. Cathode had a form of a cylinder, limited by a semi-sphere of the same radius; i.d. 0.8 mm, in the first case anode had a form of a plane, and in the second – a plane with the opening, the radius of which was 2 mm. Cathode-anode distance (EG) was 3.8 mm. The density of current was set on cathode to provide the overall current 100 A. The potential of anode was assumed zero. The initial conditions are in line with gas at rest and material of the electrodes. The temperature at the start time is constant (300K) for the whole zone. The initial density was set constant for each zone: for cathode and anode – the density of steel, for the inter-electrode space – the density of air.

The space-wise step was 0.1 mm, and the time step was set automatically according to Courant condition.

Streamline analyzed for the case without opening in anode.

Streamline is non-stationary in plasma jet at points of time below 1 μs, the heat wave propagates from cathode to anode, and three vortex zones are formed in the area of streamline. One zone caused by the opposite jets is close to cathode and symmetry axis. These jets arise from the action of radial component of electromagnetic force directed towards the axis. The second vortex zone is located near the symmetry axis, approximately in the middle between anode and cathode. The interaction of these zones results further in formation of one stable vortex zone, located at the distance equal to two radii from the axis. The third vortex zone is formed near anode at point of time 0.1 μs and exists for the whole period of calculation.

Therefore, the essentially non-stationary streamline in plasma transforms into the quasi-stationary state, induced by viscous and electro-dynamic forces. The streamline has a complex structure comprising some zones: one close to the electrode, the paraxial flow with the impact transition, and two vortex zones. The distribution of velocity on the axis is presented in Fig. 2, b, which shows that arc plasma velocity increases from zero (E – cathode surface), reaching its peak ~500 m/s at the distance 1 mm from cathode surface, reducing afterwards to zero and reversing the sign. Analytical models or numerical computations with a significant spatial step can hardly facilitate identifying the above specifics.

From practical perspective, the analysis of hydro-dynamic situation furthers making two important conclusions. First, the vortex near cathode and its existence for 1 μs facilitates formation of tiny particles approximately 100 nm of the molten material. Second, the vortex zone close to anode causes a counterflow of plasma, preventing small particles from anode surface and complicating deposition of the layer. Therefore, we suggest using anode with an opening. The calculations for the case with the opening are given in Fig. 3, demonstrating that vortex zones do not arise; this fact is relevant for the development of surfacing procedure.

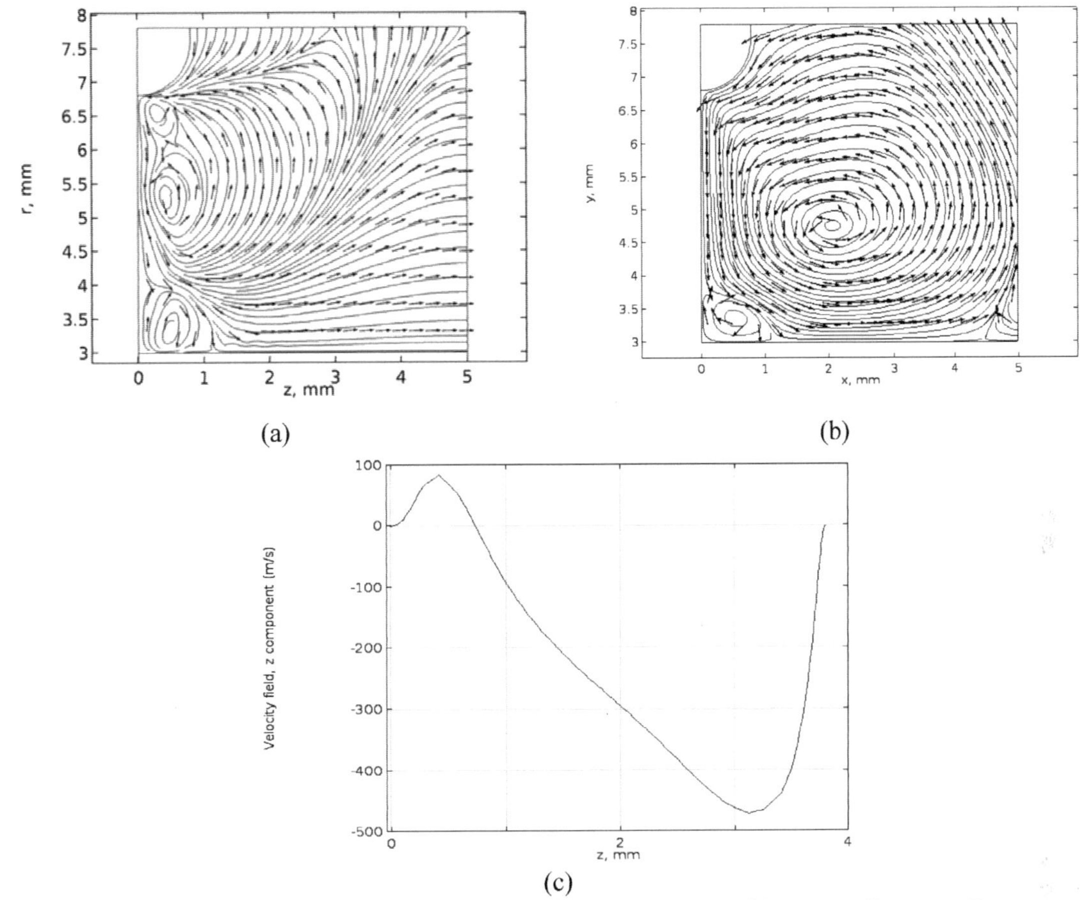

(c)

Figure 2. Velocity field of arc: (a) - streamlines at t=1.1 μs; (b) - streamlines at t=8 ms; (c) - v_z on the axis of symmetry GE (t=8 ms)

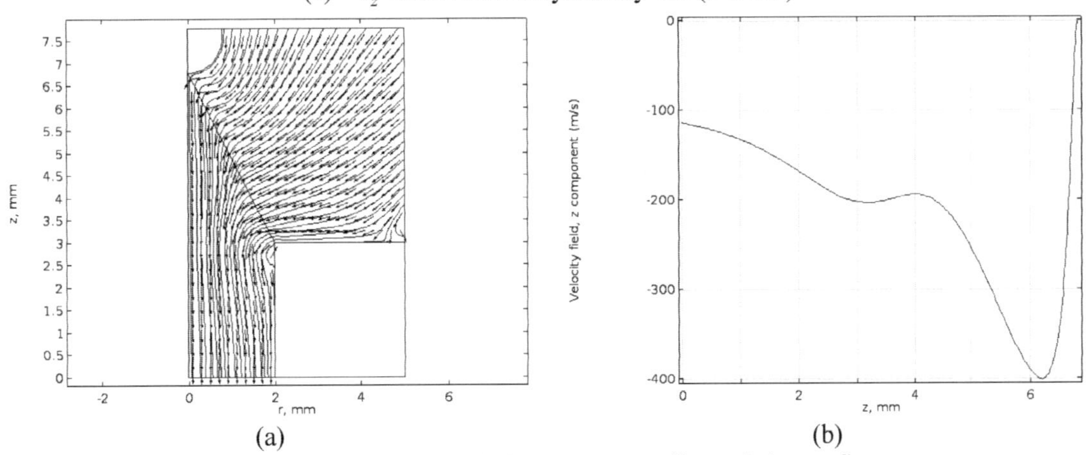

(a) (b)

Figure 3. Plasma velocity fields: a) – streamlines of plasma fluxes; b) – velocity of arc plasma on the axis.

As stated before [20], three vortex zones of hydro-dynamic flows in the jet of electric arc plasma arise in the arc plasma and other injection sources of charged particles generated at the consumable electrode tip (between a rod electrode and a flat conductive target located perpendicularly to it). The interaction of these zones results in formation of two stable vortex zones; one of which is at the distance of two electrode radii from the axis and in the middle between the electrodes, the other vortex zone is immediately close to the solid wire electrode. A counterflow of plasma is generated in this zone and emits small particles. The vortex zones do not occur when using a flat electrode with the opening. Therefore, small particles can pass through it and get consolidated. The hydro-dynamic situation analyzed in previous investigations revealed formation of the first vortex close to cathode, which existence for 1 μs furthers formation of small particles (approximately 100 nm) of molten material.

As reported in [21], small particles of molten metal leave the surface due to high-frequency impact of Rayleigh waves on the boundary "solid-liquid" in conditions of plasma fluxes. The ultrasound impact on the electrode with diameter 0.8 mm and molten tip (material - Fe, Ni, Co) was tested in laboratory conditions and its relative extension was identified $\Delta l/l \approx 10^{-6} \ldots 10^{-2}$ m at oscillations 2...50 MHz.

In consideration of elastic anisotropic medium, e.g. metal, it's noteworthy that two types of bulk waves arise in this medium under point energy deposition: longitudinal waves, where particles spread in the same direction as the wave propagates, and transverse waves with particles spreading perpendicularly to the direction of wave propagation. Longitudinal and transverse waves might be accepted bulk free oscillations.

As a magnetic impulse is delivered to the welding wire, consideration of transverse waves can be omitted because a point impulse is applied throughout the diameter; therefore, arising transverse waves are mutually absorbed in each point. The velocity of longitudinal waves is designated (c_n). So the velocity of longitudinal wave propagation through elastic materials is formulated as follows [21]:

$$c_n = \sqrt{\frac{E(1-\nu)}{\rho(1+\nu)(1-2\nu)}}, \tag{3}$$

where ρ – density of material; ν – Poisson ratio; E – Young's modulus.

Thus, surficial drift propagates along axis x in form of an elastic wave and is exponentially damped deep into the wire. This surficial wave is called Rayleigh wave R(c), its velocity is written:

$$R(c) = (2-m^2)^2 - 4\sqrt{1-m^2} = 0, \tag{4}$$

where m=c/c$_n$, where c_n – sound speed in the material.

As a consequence, a required frequency of the welding wire can be determined via calculating Rayleigh wave velocity, since we know that the wave becomes concave at velocities approximating but not exceeding R(c), i.d. $c_n<R(c)$, and convex at $c_n>R(c)$.

Provided that a welding wire of ferromagnetic materials is used, a surficial wave is generated by the applied electromagnetic impulses and spreads in both directions from the impact center, i.d. electromagnetic impulse. When mechanized shielded welding such electromagnetic impulses are applied to the wire moving in the welding arc zone. As the velocity of surficial Rayleigh waves exceeds significantly the velocity of wire displacement, the latter can be ignored in calculations.

Therefore, Rayleigh waves approach the liquid layer of anode tip, propagating along the welding wire surface, and on the solid-liquid boundary there are waves, regarded as Love's waves with vertical polarization and quite different structure and velocity as compared with Rayleigh waves. They include a weak inhomogeneous wave in the liquid, the amplitude of which decreases slowly moving from the boundary of two essentially inhomogeneous (longitudinal and transverse) waves in the solid. That is why wave energy and motion of particles are localized mostly in the liquid but not in the solid [21].

Rayleigh wave is generated on anode material due to the applied high-frequency impulses, the amplitude of which decreases slowly moving from the boundary of two essentially inhomogeneous (longitudinal and transverse) waves in the solid. So wave energy and motion of particles are localized mostly in the liquid but not in the solid. When propagating, these waves constantly emit energy into the liquid, generating there an inhomogeneous wave, spreading in both directions from the phase boundary. The phase velocity of this surficial acoustic wave is determined, attenuation constant on the wavelength $\lambda \sim 0.1$. As a consequence, energy of a high-frequency electromagnetic field is delivered through the electrode into the liquid droplet due to Rayleigh and Love's waves. The liquid layer of the electrode surface gets more active because of the obtained energy, as the result, the increase is registered in forces of all-around compression of nano-scale heterogeneity bridges; they get destructed faster, initiating the detachment of droplets from the electrode surface. As identified in previous research [22, 23], heat and mass transfer in the welding zone, frequency and dimensions of formed droplets can be controlled via changing the shielding conditions of molten electrode metal droplet and arc plasma column.

Therefore, conditions required for formation of nano-scale waves on the liquid anode layer are to be provided with specially designed equipment, and detachment of particles is possible due to high-frequency impact on the electrode, making the particles leave the phase interface "electrode – liquid layer". High-frequency (ultrasound) electromagnetic impact on the metallic electrode generating elastic longitudinal oscillations within range 2.5...25 MHz is accepted for the detachment of nano-scale waves from the electrode surface. As demonstrated in experiments, plasma flux treatment results in a structure with cellular crystallization, cells dimensions range 100 to 300 nm (Fig. 4).

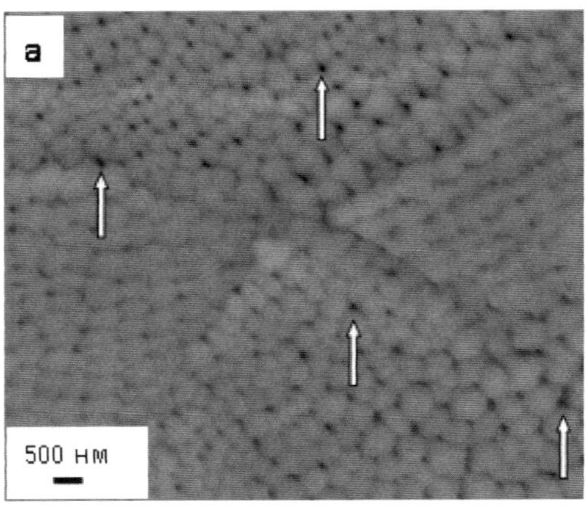

Figure 4. – Microstructure Ni, deposited via high-frequency plasma spraying

Conclusions

1. The conditions are ascertained which are necessary for origination and evolvement of thin layers with surficial-periodic micro-and nano-scale relief on the surface of liquid metal when supplying metallic wires into the zone of heterogeneous plasma of the electric arc in conditions of Kelvin-Helmholtz and thermo-capillary (Marangoni) instabilities. The supplementary magnetic and hydro-dynamic impact of oriented plasma fluxes on thin molten metal layer occurs when using a working element to form required directions of plasma fluxes together with high-frequency (ultrasound) impact, as the result, the flow of micro-and nano-scale droplets is deposited on the substrate (target).

2. Generating elastic Rayleigh waves deliver the longitudinal oscillations to the phase boundary "electrode – liquid layer", creating conditions this way for detachment of particles formed by the vortex flux of the plasma jet. The amplitude of Rayleigh wave decreases slowly (exponentially) moving from the boundary between two significantly inhomogeneous (longitudinal and traverse) waves in the solid. It is the reason for the localization of wave energy and its constant emission into

the liquid, but not into the solid, an inhomogeneous wave is generated in the liquid and propagates from the boundary, increasing the detachment effect of the micro-waves.

3. The proposed deposition procedure of a catalytic layer using the ultra-disperse metal furthers generation of the electricity-conductive foil of good quality with higher catalytic activity of metal due to the nano-scale particles. The above procedure makes it possible to form coatings distinguished by good adhesion on various surfaces, as well as with good continuity, homogeneity and reproducibility of the layer composition.

References

[1] Hong Ma, Guoliang Qin, PeihaoGeng, Fei Li, Banglong Fu, XiangmengMeng. 2015 Materials & Design.**86** 587.
[2] Ivanov Y F, Kolubaeva Y A, Bessonov D A, Vorobev S V, Gromov V E, Tsellermaer V Y, Ivanov K V. 2013 J. Surf. Investig. X-ray, Synchrotron Neutron Tech. **7** 90.
[3] Grishunin V A, Gromov V E, Konovalov S V, Ivanov Y F, Teresov A D. 2013 J. Surf. Investig. X-ray, Synchrotron Neutron Tech.**7** 990.
[4] Kapralov E V, Raykov S V, Budovskikh E A, Gromov V E, Vashchuk E S, Ivanov Yu F. 2014 Bulletin of the Russian Academy of Sciences: Physics **78** 1015.
[5] Boronenkov V, Zinigrad M, Leontiev L, Pastukhov E, Shalimov M, Shanchurov S. 2011 Phase Interaction in the Metal - Oxide Melts - Gas –System (Berlin: Springer).
[6] Słoma J, Szczygieł I, Sachajdak A. 2011 Archives of Civil and Mechanical Engineering **11** 437.
[7] Shih T -M, Zheng Y, Arie M, Zheng J -C. 2013 Numerical Heat Transfer; Part A: Applications **64** 435.
[8] Xi Chen, He-Ping Li. 2001 International Journal of Heat Mass Transfer **44** 2541.
[9] Trelles J P. 2014 Plasma Sources Science and Technology**23** 054002.
[10] Kim Y-J, Lee J-C. 2013 Journal of the Korean Physical Society**62** 1252.
[11] Bolot R, Coddet C, Allimant A, Billieres D. 2011 Journal of Thermal Spray Technology **20** 21.
[12] Li H-P, Benilov M S. 2007 Journal of Physics D: Applied Physics40 2010.
[13] Traidia A, Roger F, Chidley A. 2011 World Academy of Science, Engineering and Technology**73.**
[14] Traidia A, Roger F. 2011 International Journal of Heat Mass Transfer**54** 2163.
[15] Tong L G, Gu J C, Wang L, Yin S W. 2015 International Journal of Heat Mass Transfer**90** 968.
[16] Xinxin Wang, Ding Fan, Jiankang Huang, Yong Huang. 2015 International Journal of Heat Mass Transfer**85** 924.
[17] Zhang Tong, ZhengZhentai, Zhao Rui. 2013 Journal of Materials Processing Technology**213** 2329.
[18] Lelevkin V M, Semenov V F. 2002 Technical Physics Letters **28** 722.
[19] Gritsinin S I, Davydov A M, Kossyi I A, Kulumbaev E B, Lelevkin V M. Plasma Physics Reports **39** 579.
[20] Granovskii A Y, Sarychev V D, Gromov V E. Model of formation of inner nanolayers in shear flows of material // Technical Physics. –2013. –V.58 (10), –P. 1544-1547.
[21] Solodsky S A, Sarychev V D, Borisov1 I S. Technology of ultrasonic control of gas-shielded welding process. IOP Conf. Series: Materials Science and Engineering 125 (2016) 012039.
[22] Chinakhov D.A. Gas Dynamic Control of Properties of Welded Joints from High Strength Alloyed Steels. China Welding. Vol. 23 No.3 (2014) pp. 27-31.
[23] Chinakhov D.A., Grigorieva E.G., Mayorova E.I., Kartsev D.S. The influence of shielding gas flow rate on the transfer frequency of electrode metals drops. IOP Conf. Series: Materials Science and Engineering 142 (2016) 012005 doi:10.1088/1757-899X/142/1/012005.

IOP Conf. Series: Earth and Environmental Science **50** (2017) 012051 doi:10.1088/1755-1315/50/1/012051

Structure and Properties of Detonation Coatings of Mechanocomposite Based on Bazalt-NiAl

V I Yakovlev[a], A A Sitnikov[b], A G Ovcharebko[c], T A Golovina[d], T A Sitnikova[e]

Polzunov Altai State Technical University, 656038, 46, Lenin prospect, Barnaul, Russia

[a] anicpt@rambler.ru, [b] sitalan@mail.ru (corresponding author), [c] tmk@bti.secna.ru
[d] gol-an-ta@mail.ru, [e] sitalan@mail.ru

Abstract. The chemical composition, structure and microhardness of detonation coatings based on mechanically activated mixture of basalt powder and NiAl have been studied. It has been shown that the application of mechanomixture by detonation method results in the formation of composite coating similar to the composition of starting mixture. It has high microhardness and temperature resistance that will increase the wear resistance of parts coated with basalt at elevated temperatures.

Introduction.

Basalt is a natural material of volcanic origin, is increasingly used in construction, composite materials manufacture, as protective coatings, thin fibers, special fabrics, etc. The following fact is important: rocks-basalts make up about 30% of the Earth's crust that makes it virtually unlimited source of raw materials for the manufacture of products with the use of mineral resources. Only in Russia there are more than 150 explored and nearly 70 mastered fields. Because of its availability and therefore cheapness, it is used as crushed stone in gravel roads construction and dumping in the form of crushed stone of various fractions. Material, which are cheaper than basalt don't exist in principle. If we compare it with the other plate-fillers used for the production of protective coatings and having comparable properties, basalt is 10 times as cheap as they.

The simplicity of manufacture, availability and raw materials cheapness, as well as valuable functional characteristics of basalt fiber - all of this cause an increased interest to coatings, basalt fiber, wool, scales and other materials on its basis.

Today we can distinguish some of the valuable characteristics of composites and coatings made of basalt, such as for example:

- the strength of basalt fiber is 15 times as high as the strength of steel one of the same section and material density is 4 times less [1,2].
- Amazing corrosion resistance allows manufacturers to provide a guarantee on the basalt coated pipe, reinforcement, and other products - up to 50 years or more, they do not rust, as basalt products are totally inert to acidic and alkaline conditions [5].
- In basalt coated pipes and boilers scum and precipitation are hardly deposited [6].
- basalt coated products can "breathe" and maintain thermal cycles at temperatures from -200 °C to + 200 °C [3].

The chemical composition of gabbro-basalt group rocks is shown in Table 1.

Materials and methods.

The methodology of the study the process of basalt cover applying by detonation-gas method. At Problematic Research Laboratory (PRL) SHS Materials ASTU for spraying coatings detonation spraying installation "Katun-M" has been used.

Basalt grit with particles of 3-4 mm. size was subjected to treatment in a planetary ball mill AGO-2C for 10 minutes and sieved into 3 fractions (<63 microns 63 microns 100...200 mm). Then, obtained powder was dried for one hour and a half at 200 °C temperature. Mehanokompozit composition of basalt powder with additions of 30% intermetalida NiAl was used as sprayed powder.

Table 1 - Chemical composition of basalt

Mass fraction of elements, %									
SiO_2	Al_2O_2	FeO + Fe_2O_2	CaO	MgO	K_2O + Na_2O	TiO_2	MnO	P_2O_2	unavoidable impurities
from 39 to 51	from 10 to 19	from 10 to 18	from 8 to 13	from 4 to 12	from 2 to 5	from 0.2 to 3	not more than 0.4	not more than 2.5	not more than 5

Detonation spraying is made on the surface of the part - a steel plate of 50×40 mm, which was subjected to sandblasting.

Studies of the chemical composition of basalt powder with NiAl addition was conducted with the help of Carl Zeiss EVO 50 scanning electron microscope with console for energy-dispersive chemical analysis EDS «INCA X-ACT» made by «Oxford Instruments».

The chemical analysis held during the study of powder particle (Figure 1) showed the content of elements that is presented in Table 2.

The studies of the chemical composition of the coating on the basis of basalt powder (Figure 2) of the region "Spectrum 1" showed the presence of elements of the proportions indicated in Table 3.

The coating chemical composition showed that weight and atomic ratio changes unlike the powder chemical composition practically didn't happen.

Table 2 – The chemical analysis of elements contained in the basalt powder particle

An element	Weight, %	Atomic, %
O	55.53	71.34
Mg	3.88	3.66
Al	7.06	5.38
Si	16.63	12.17
Ca	2.74	1.43
Fe	8.22	3.03
Ni	6.01	4.11

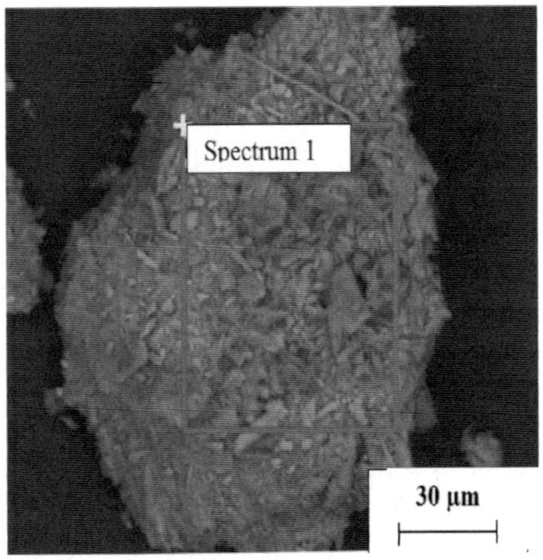

Figure 1 - Basalt powder particle. Spectrum 1 - the study area of chemical composition

Table 3 - Chemical composition of the coating

An element	Weight, %	Atomic, %
O	52.26	68,28
Mg	6.08	5.23
Al	7.27	5.63
Si	17.27	12.86
Ca	5.68	2.96
Fe	9.04	3.38
Ni	3.4	2.7

To investigate the general coating structure microsections were prepared. The total structure was observed after etching in 3% alcohol solution of nitric acid, and in a study of fine structure of coating chemical etching with an aqueous solution of 20% red blood salt K3 [Fe (CN) 6], and 20% KOH was used for 5 minutes.

Figure 2 - Basalt coating with the SHS composite NiAl addition. Spectrum 1 - the study area of chemical composition.

The study of overall (Figure 3) and fine (Figure 4) structures were carried out on optical microscope Carl Ziess AxioObserver Z1. The coating is dense with no delaminations. It means a good coating adhesion to the substrate (adhesion). As it is seen from the figure the coating is multilayer without visible pores and cracks.

Micro Durometer 402 MVD was used to measure microhardness and for Vickers testing with the possibility of operating in automatic and manual modes in the load range of 0.01 - 2 kg.

Hardness measurement was carried out on the coating layer, the indenter load was 100 g. 8 injections were made along the entire length of the coating (Table 4).

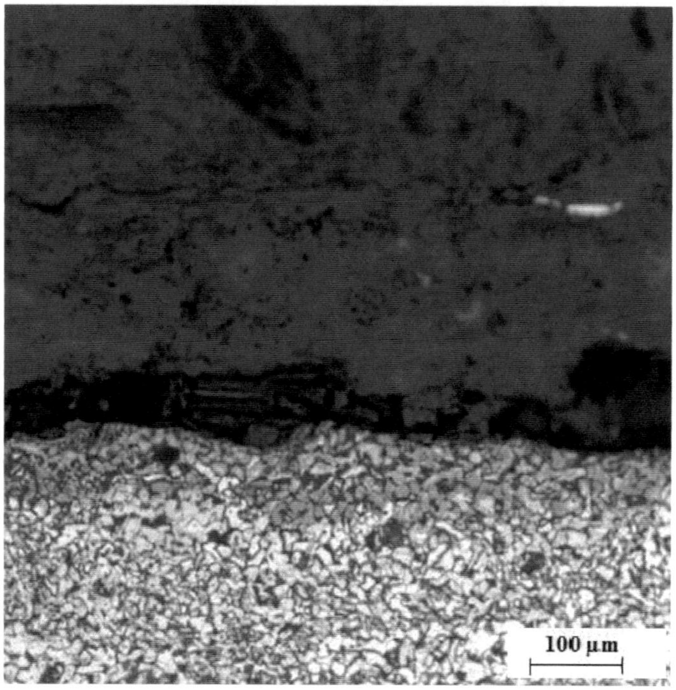

Figure 3 - The coating general structure

IOP Conf. Series: Earth and Environmental Science **50** (2017) 012051 doi:10.1088/1755-1315/50/1/012051

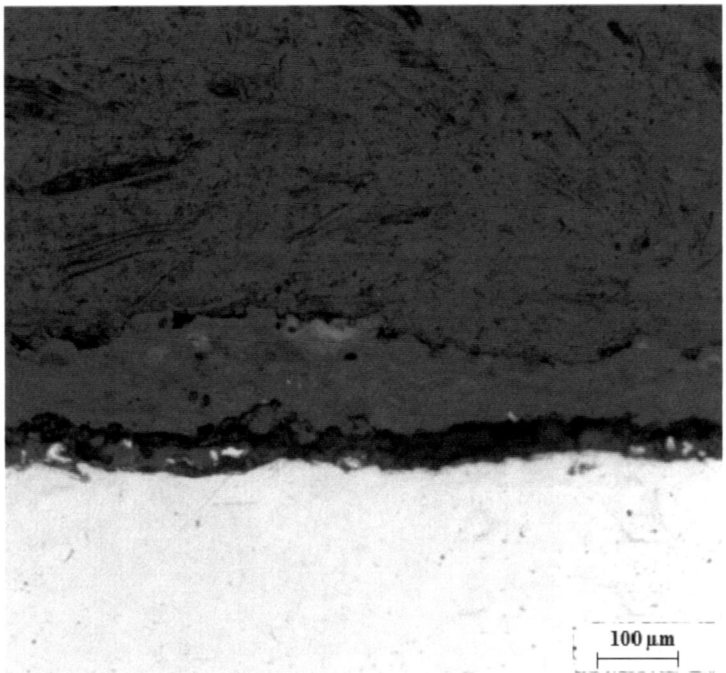

Figure 4 - The fine structure of coating

Average coating microhardness value was 473 HV, it is 3.5 times as high as base metal microhardness (134 HV), and in some areas of coating microhardness is up to 600 - 670 HV (Figure 5), it is 4.5 - 5 times as high as base metal microhardness, it will increase the wear resistance of the coated parts.

Table 4 - Tests for microhardness

Injection №	Vickers indications
1	393 HV
2	497 HV
3	512 HV
4	428 HV
5	467 HV
6	482 HV
7	404 HV
8	602 HV
Average value	473 HV

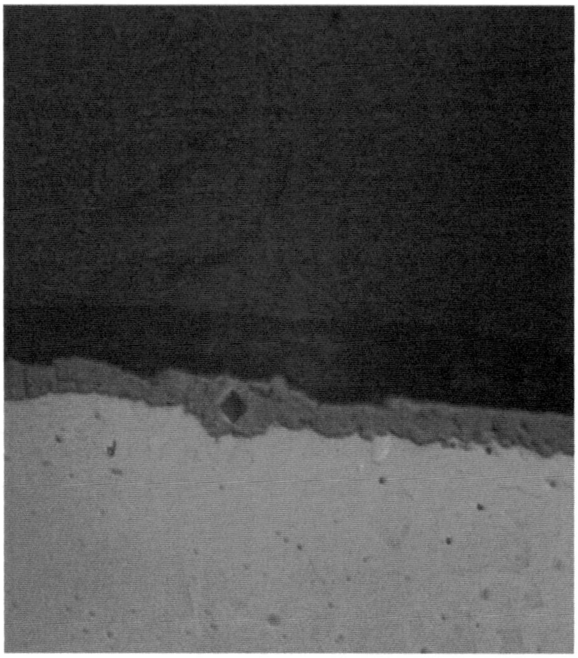

Figure 5 - Vickers microhardness tests

Conclusions.

Basalt coating of mechanically activated powder was obtained. It was deposited on the steel plate by detonation-gas method. The structure research showed that the coating is dense and multi-layered. The chemical composition of the coating showed that the weight and an atomic ratio of the elements was practically unchanged with respect to the chemical composition of the powder . Formed coating has high microhardness. This fact allows to assume an increase in wear resistance of basalt coated parts.

References

[1] Tatarintseva O.S. Basalt technology today: state and prospects. Proceedings of the 3rd All-Russian scientific-practical conference of students, graduate students and young scientists with international participation. Biysk: Publishing House of the ASTU 2010. In 2 parts: P.1. p. 7-12 (in Rusian).

[2] Dzhigiris, D.D. Fundamentals of production of basalt fibers and products / D.D. Dzhigiris, M.F. Mahova. - M .: Teploenergetic [Heat and Power Engineer], 2002. 416 p. (in Rusian).

[3] Zemtsov A.N. Basalt wool: Past and Present/ A.N. Zemtsov, S.I. Ogaryshev Perm 2003. 124 p. (in Rusian).

[4] Boguslayev, V.O. Detonation applying of coatings on parts of aircraft engines and technological equipment, followed by a magneto-abrasive treatment (monograph)/ V.O. Boguslayev, A.I. Dolmatov, P.D. Zhemanyuk, A.I. Kulagin, V.G. Mihaylutsa, V.A. Simonenko. Zaporozhye: Deca, 1996. 366 p. (in Rusian).

[5] National Research Council (U.S.) Advanced fibers for high-temperature ceramic composites: advanced materials for the twenty-first century // Washington D.C.: National Academy Press. 1998. 94p.

[6] Evaluation of wear resistance of products on the basis of mechanically activated materials / A.V. Sobachkin, A.A. Sitnikov, A.P. Sviridov // Applied Mechanics and Materials. 2015. V. 698. p. 374–377 (in Rusian).

Technological Support of Critical Parts for Railway Transport Working Properties

A V Gabets[1, a], D A Gabets[2, b], A M Markov[2, c], M V Radchenko[2, d], S L Leonov[2, e]

[1] Limited liability company "Altai steel plant",656038, 116/52, Kalinina prospect, Barnaul, Russia
[2] Polzunov Altai State Technical University, 656038, 46, Lenina prospect, Barnaul, Russia

[a] gabeca@mail.ru, [b] gabets22@mail.ru, [c] andmarkov@inbox.ru (corresponding author), [d] mirad_x@mail.ru, [e] sergey_and_nady@mail.ru

Abstract. The materials of complex research of operational properties of a new brand cast iron CHMN-35M. Optimal chemical composition was determined. The obtained results allow to conclude about possibility of its use for the manufacture of critical parts of rolling stock of railway transport, in particular of a side bearing cap

Introduction

The solution on improving the speed and volume of cargo transportation by railway transport is connected with increased requirements to critical parts of the rolling stock. One of such details is a side bearing cap. The cap is installed on the support in the truck bolster and designed to hold a car body of the lateral forces. It is a box construction usually made of steel. Operating conditions of the slide bearing cap are that at a certain speeds on curve space interval, arising lateral forces can lead to fragile and fatigue break downs of the cap, that in turn cal lead to the rolling stock descent from the ways.

In most cases, the main material for the slide bearing cap manufacturing is steel (steel 20GL, GOST 977-88). At the same time, the long-term experience of the rolling stock use in terms of increased speed and freight volume indicates that this material cannot fully provide truck van specified run. According to the regulatory documents the run is no less than 160 thousand km or 3 years of operation.

The structure of refusals (defects) of the freight-car truck is received on the basis of data collected during the first depot repair at the JSC "Carriage Repair Company – 2" [3, 4].

The reasons of the freight-car truck refusals can be divided into two main groups: fatigue break-down defects and ware surface damage. Besides, up to 43% of the freight-car truck cuts to the current repair are connected with the slide bearing cap: 9% of them is slide bearing cap defect and 34% is exceeding of the permitted clearance points between the effective areas.

In use of the slide bearing cap, the bearing area is worn down (Figure 1). The amount of wear is defined as the distance between its effective area and the center of the bottom opening. The excessive wear is considered to be more than 3 millimeter.

Figure 1 – Wear of the slide bearing cap's effective area

Materials and methods

The previous researches [3, 7] demonstrate that the effective material, shown satisfactory operational characteristics in case of high load is gray cast iron, in particular SCH35 iron. In comparison with the steel, the iron hardness is much higher (steel hardness 20GL – 140…160 HB; iron – SCH35 – 210…275 HB), gray cast iron contains graphite which in the frictional couples can serve as cutting oil. However, these properties are not always sufficient for the buffing loading conditions, subjected to the slid bearing cap. In this regard the task on the development of particular iron brand, providing the following operational indicators becomes urgent:

– hardness in the range 250…300 HB, providing increased wear resistance, in comparison with existing, which will extend slide bearing cap wear-life;

– high margin of strength for reducing construction material intensity (temporary resistance in case of stretching (σ_B) at least 380 MPa);

– high level of lamellar perlite in the structure, improving tribological behavior in the formed frictional couple.

The research results, carried out by Russian and foreign scientists N.G. Girshovich, N.N. Aleksandrov, B.S. Milman, KI. Vatchenko, V.S. Shumikhin, Thorbjorn, J. Bolton, K.D. Millis and others indicate that the stated requirements can be achieved on the basis of alloying and modification technologies use of existing iron brands [1, 2, 6, 8]. Herewith, the use of ligatures based on manganese, molybdenum, chromium, nickel and vanadium, and also complex modifiers comprising strontium, zirconium, barium, selenium, manganese, and rare-earth elements alloys is of interest.

The basis for a new brand ware-resistant cost iron (base material) development has become gray cast iron SCH35 (GOST 1412-85). This iron is currently used for manufacture such details of freight-car truck as "Friction Wedge" (TU 3183-234-01124323-2007), and can also be used for manufacturing slide bearing cap. The feedstock has the following chemical makeup, presenting in the table 1.

Table 1 – Chemical makeup of the gray cast iron SCH35 GOST 1412-85

Brand	Weight content, %				
	C	Si	Mn	P	S
				No less	
SCH35	2.9…3.0	1.2…1.5	0.7…1.1	0.2	0.12
					Rest - Fe

For determining the optimal chemical makeup of the new ware-resistant cost there was conducted varying percentage of chemical elements in the feedstock. Within the LLC "Altai Steel Plant" (Barnaul, Russia) conditions experiment samples with various chemical makeup were received (Table 2):

– experiment samples 1 – iron SCH35 (base material);

- experimental sample 2 – iron alloyed with nickel and vanadium;
- experimental sample 3 – iron alloyed with molybdenum;
- experimental sample 4 – phosphor-doped iron.

Results and discussions

The research of the received experimental samples structure has shown that existence of alloying elements of enhanced concentration has led to austenite stabilization and receiving ledeburite structure. Conceivable, this is due to the rapid cooling of test bars, as indicated by the material structure; sample 2 and 4 (Figure 2 and Figure 3).

Figure 2 – Honeycomb ledeburite and perlite (experimental sample 2)

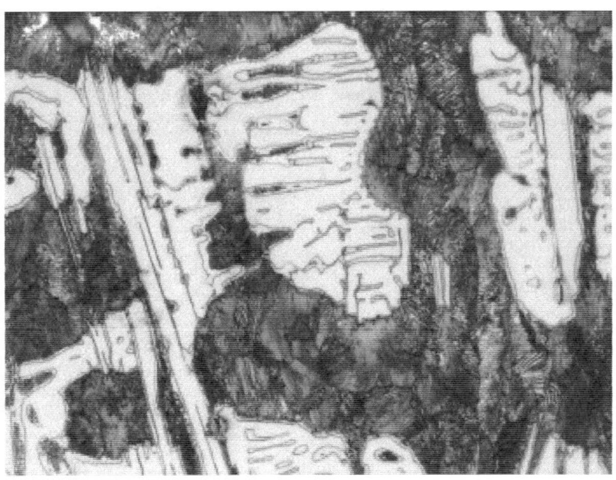

Figure 3 – Lamellar ledeburite and perlite (experimental sample 2)

To evaluate the working properties of experimental samples the mechanical tests were carried out. The experiments show (table 3) that the higher values of mechanical characteristics were received in the alloy cast iron, containing alloying elements as ferromolybdenum FeMo 60, ferronickel FeNi 75, and ferrovanadium FeV 50.

Table 2 – Chemical makeup of alloyed iron experimental samples

Experimental sample	Weight content, %									
	C	Si	Mn	P	S	Cr	Ni	Cu	Mo	V
1	3.24	1.59	0.69	0.040	0.014	0.11	0.04	0.08	0.03	0.002
2	3.36	1.54	0.82	0.036	0.012	0.15	1.56	0.34	0.005	0.60
3	3.12	1.56	0.83	0.045	0.012	0.17	0.39	0.19	1.18	0.006
4	2.57	1.44	0.74	0.754	0.009	0.12	0.39	0.14	0.03	0.010

Table 3 – Mechanical characteristics of alloy cast iron experimental samples

Experimental sample	Hardness, HB	Temporary resistance, MPa
1	324…358	140…158
2	430…378	165…179
3	407…450	220…235
4	372…403	145…162

Influence of modifiers chemical makeup on the iron mechanical characteristics is estimated during the series of experiments with the use of various complex modifiers. As it was distinguished above [3, 4] for the required iron tribological behavior as a modifying agents, containing zirconium and silicium it is required to use zirconium dioxide (ZrO_2) and silicon dioxide (SiO_2). Zirconium interaction with free nitrogen creates zirconium nitride inclusions that lead to decrease the formation of core blows. Therefore, it was chosen the modifier of Z-GRAPH®T group, containing the following chemical elements: silicium (60…70%), manganese (6.0…7.0%), zirconium (6.0…7.0%), barium (2.0…4.0%), aluminium (1.0…2.0;), calcium (1.0..2.0%), ferrum (76.0…92.0%).

Consequently, to decrease the chill effect, increase strength, and wear resistance of the iron, the technology of its receipt should include operations of alloying cast iron SCH35 with molybdenum and nickel, and the subsequent modification by the GRAPH®T modifier. The two types of this technology are worked out:

– the first variant (samples 1, 2): the whole weight of the liquid synthetic iron, alloyed with molybdenum and nickel, modified by complex modifier GRAPH®T;

– the second variant (samples 3, 4): the whole weight of the liquid synthetic iron, alloyed with molybdenum and nickel, modified by complex modifier GRAPH®T.

The chemical makeup of experimental samples, received on the stated variants is shown in the table 4.

Table 4 – Chemical makeup of the alloy cast and modified iron

Experimental sample	Elements weight content, % Fe – the rest								
	C	Si	Mn	P	S	Cr	Ni	Cu	Mo
1	3.00	1.82	0.86	0.062	0.032	0.06	0.06	0.16	0.55
2	2.91	1.41	0.85	0.066	0.029	0.06	0.06	0.17	0.59
3	2.95	1.50	0.87	0.059	0.034	0.06	0.47	0.17	0.57
4	2.99	1.49	0.88	0.066	0.036	0.06	0.47	0.17	0.57

IOP Conf. Series: Earth and Environmental Science **50** (2017) 012052 doi:10.1088/1755-1315/50/1/012052

Investigation of mechanical properties of experimental samples has shown that the second variant of iron receiving provides much higher operation factors: samples 3 and 4 have much higher hardness and temporary tear resistance in comparison with samples 1 and 2 (table 5). At the same time, the value of temporary resistance in all four samples significantly exceeds the value of temporary resistance of the iron SCH35 GOST 1412-85 (feedstock).

Table 5 – Mechanical characteristics of the alloy cast and modified iron samples

Experimental sample	Hardness, HB	Temporary resistance, MPa
1	276…302	540…558
2	253…286	400…424
3	291…312	490…411
4	308…330	549...570

Achieved results and the patent search allowed declaring the resulting iron as an invention (patent number 2562554). The material assigned an index CHMN-35M. Its chemical makeup and main mechanical characteristics are described in the tables 6 and 7.

Table 6 – Chemical makeup of the modified iron CHMN35

Brand	Main elements of chemical makeup, %												
	C	Si	Mn	Mo	Ni	Zr	Al	Ca	Ba	Cr	Cu	S	P
CHMN-35M	2.5 … 2.8	1.3… 1.5	0.7… 1.0	0.6 … 0.9	0.5 …0.8	0.0005… 0.1	0.0005… 0.1	0.0005 … 0.1	0.0005… 0.1	0.05… 0.3	0.05 …0.3	0.05 …0.1	0.05 … 0.1
The rest - Fe													

Table 7 – Mechanical characteristics of the modified iron CHMN35

Brand	Temporary tensile resistance, MPa, at least	Brinnel hardness, HB	
		At least	At most
CHMN-35M	350	250	300
The minimum value of the temporary tensile resistance can exceed not more than 100 MPa			

The quality of the new iron and its characteristics are assessed within laboratory and stand tests. Polished sections analysis of non-etching type allowed identifying graphitic phase. Study of metal matrix has shown that metal backing of the alloy cast iron has perlite alongside with acicular perlite and iron carbide.

Besides, this iron CHMN-35M metal matrix has perlite-ferritic metal backing, 85% of which are occupied by perlitic ranges, and 15% occupied by ferrite, saturated by the net of carbide inclusions (Figure 4-6).

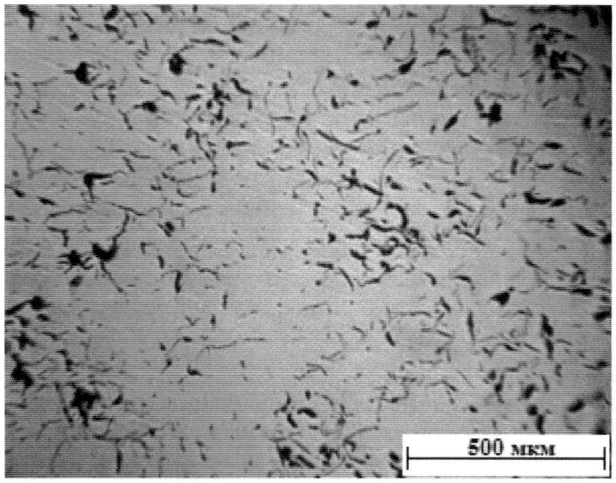

Figure 4 – Lamellar swirling form of graphitic inclusions

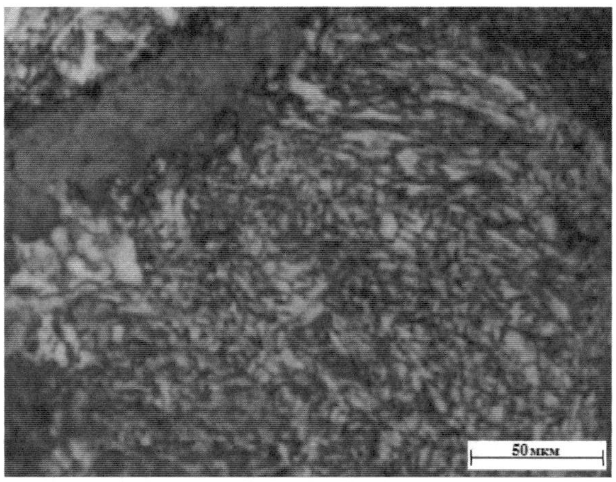

Figure 5 – Perlite and acicular ferrite

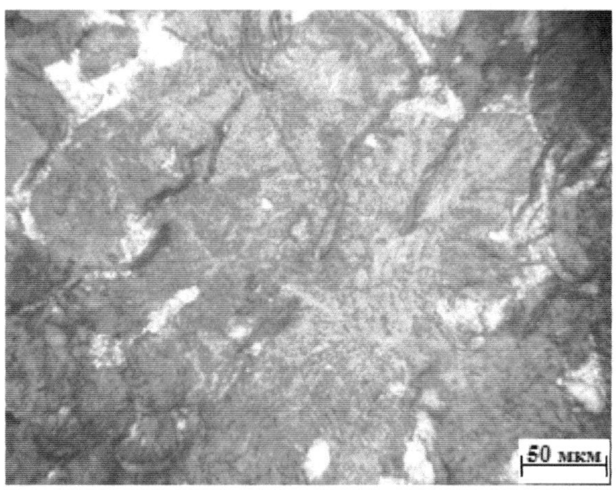

Figure 6 – Carbide inclusions

To evaluate working properties of the iron CHmn-35M it was conducted its tribological tests with serial gray cast iron SCH35 by simulating work of "Car body – slide bearing cap".

As research equipment it was chosen a friction machine of 2168 UMT type. Experimental samples were made in the form of frictional couple "hob – shaft". Shafts material is iron of CH35 and CHMN-35M types, hobs material is steel 20GL, 30KHGSA, and 09G2S. During the experiment, we controlled the change of weight wear during the various combining of hubs and shifts materials. The achieved results are presented in the table 8.

Table 8 – Results of ware resistance study

№	Sample	Material	Weight wear mass, gr	Accumulative wear of friction couples, gr.
1	Hob	30KHGSA	0.12	1.13
	Shaft	SCH35	1.01	
2	Hob	30KHGSA	0.05	0.56
	Shaft	CHMN-35M	0.51	
3	Hob	20GL	0.04	1.02
	Shaft	SCH35	0.98	
4	Hob	20GL	0.12	0.78
	Shaft	CHMN-35M	0.66	
5	Hob	09G2S	0.45	0.80
	Shaft	20GL	0.35	
6	Hob	09G2S	0.13	0.69
	Shaft	CHMN-35M	0.56	

As reflected by the results, in all combinations iron of CHMN-35M brand has shown the best (approximately in 1.5 times) wear resistance in comparison with SCH35 iron. Therefore, the total value of accumulative wear (hob and shaft wear) in couples with CHMN-35M iron is also lower.

Conclusion

Summing up what has been said, the developed new composition of the alloyed modified cast iron CHMN-35M, different from analogs by the distribution of graphite inclusions (combination of equilibrium and nonequilibrium distribution), and the presence in its structure free ferrite, has increased wear resistance. It can be recommended for manufacturing details of heavy-duty freight-car truck.

References

[1] Aleksandrov N.N. Ways of improving quality of modifying gray cast iron / N.N. Aleksandrov, B.S. Milman, L.S. Kapustina // Manufacturing of new lagatures and their use in the foundry engineering in order to save metal. Chelyabinsk. 1979. p. 88-89.

[2] Vatchenko K.I., Shumikhin V.S. Melting and secondary metallurgy of iron for castings: Textbook. Kiev: High school, 1992. p 246.

[3] Gabets A.V. Study of properties and development of an efficient car body design / A.V. Gabets, A.M. Markov, D.A. Gabets. Barnaul, 2016.

[4] Gabets A.V. Modelling of details working properties manufactured of particular iron CHMN-35M/ A.V. Gabets, A.M. Markov, D.A. Gabets // Polzunovskiy vestnik. 2016. 2. p. 36-44.

[5] Gabets A.V. Friction wedge ware evaluation of freight-car truck oscillation suppression / A.V. Gabets, I.V. Levkin, M.V. Sapetov, A.V. Semenov // Polzunovskiy vestnik. 2015. 41. p. 16-18.

[6] Girshovich N.G. Irin casting. L.-M. State scientific and technical publishing house for ferrous and non-ferrous metallurgy. 1978. P 708.
[7] Sukhov A.V. Increased wear resistance of friction couple friction wedge – friction plate of freight-car tuck wedge-type shock absorber / A.V. Sukhov, B.V. Bortch, A.V. Gabets // Vestnik of research institute for rail way transportation. 2015. 5. p. 25-29.
[8] Torbjørn Skaland Study of duration of modifiers action in the iron with graphite of different form // Foundry engineering. 1999. 6.

Forecasting of Machined Surface Waviness on the Basis of Self-oscillations Analysis

E B Belov [a], S L Leonov [b], A M Markov [c], A A Sitnikov [d], V A Khomenko [e]

Polzunov Altai State Technical University, 656038, 46, Lenina prospect, Barnaul, Russia

[a] jeka-beliy@yandex.ru, [b] sergey_and_nady@mail.ru (corresponding author), [c]andmarkov@inbox.ru, [d] sitalan@mail.ru, [e] agtu-tap@mail.ru

Abstract. The paper states a problem of providing quality of geometrical characteristics of machined surfaces, which makes it necessary to forecast the occurrence and amount of oscillations appearing in the course of mechanical treatment. Objectives and tasks of the research are formulated. Sources of oscillation onset are defined: these are coordinate connections and nonlinear dependence of cutting force on the cutting velocity. A mathematical model of forecasting steady-state self-oscillations is investigated. The equation of the cutter tip motion is a system of two second-order nonlinear differential equations. The paper shows an algorithm describing a harmonic linearization method which allows for a significant reduction of the calculation time. In order to do that it is necessary to determine the amplitude of oscillations, frequency and a steady component of the first harmonic. Software which allows obtaining data on surface waviness parameters is described. The paper studies an example of the use of the developed model in semi-finished lathe machining of the shaft made from steel 40H which is a part of the BelAZ wheel electric actuator unit. Recommendations on eliminating self-oscillations in the process of shaft cutting and defect correction of the surface waviness are given.

Introduction

Increasing requirements on the operating capabilities of mechanical engineering industry products lead to the corresponding tightening of requirements on geometrical parameters of the part surface quality - accuracy of sizes, shape deviations, relative positions of surfaces, and their waviness and roughness [1].

The main influence on these parameters is caused by oscillations in technological systems which occur during mechanical treatment [2].

There are very detailed studies of fluctuation causes such as a regenerative effect, a coordinate link, and the system nonlinearity [3]. There are also special models to study these causes. [4] However, there are no integrated and comprehensive records of all these causes as well as their relation with forced oscillations in contemporary studies. This is mostly associated with the fact that both forced and self-oscillations are described by a system of differential equations. Their numerical solution is time-consuming because of the need to obtain steady-state oscillations [5, 6].

On the other hand, there are quite effective approximate methods for forecasting steady-state oscillations, one of which is the method of harmonic linearization [7]. The use of this method significantly reduces the time of predicting self-oscillations and the time of developing engineering software for the design of machining operations with certain requirements on the waviness of machined surfaces [8].

Thus, it is relevant to do research aimed at solving the problem of predicting waviness of the machined surface during metal turning based on the analysis of self-oscillatory phenomena.

The research objective is to predict the waviness of machined surfaces during metal turning by analyzing self-oscillations using the harmonic linearization method.

Research tasks include:

1. To identify the key features of the waviness formation.
2. To develop a simulation dynamic model to predict self-oscillations in turning taking into account the parameters of the technological system.
3. To create an algorithm for calculating steady self-oscillations by applying the method of harmonic linearization.

Theory

A simulation dynamic model for calculating waviness parameters in turning is based on the analysis of the self-oscillation process.

The main sources of self-oscillations in turning are a coordinate link bond and a nonlinear dependence of the force on the cutting speed [9]. Figure 1 shows a block diagram of a nonlinear system with due regard for the coordinate link.

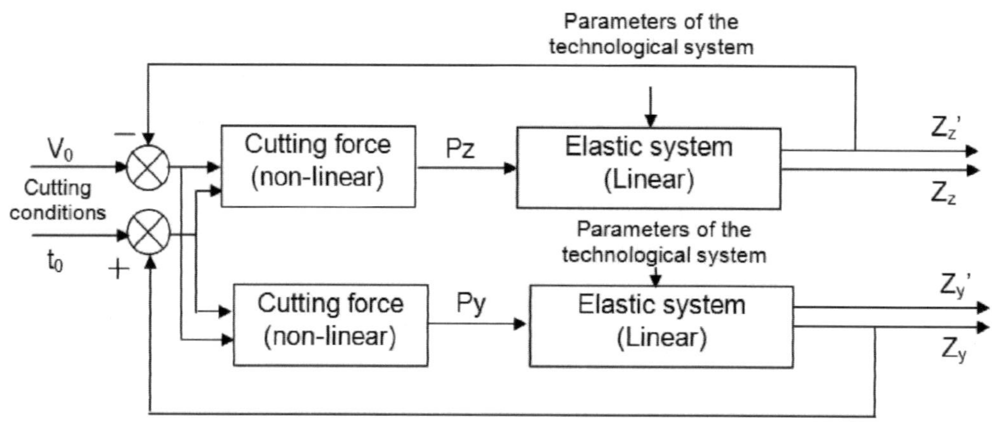

Figure 1. A block diagram of a nonlinear system with regard to the coordinate link

In the falling section of the cutting force-speed dependence (Figure 2) the elastic system described by the second order differential equation (1) may become unstable, which corresponds to the appearance of self-oscillations in a nonlinear system.

In this case, the source of self-oscillations is the cutting force - cutting speed dependence. Let us consider the following example of a single-mass system:

$$m\ddot{z} + G\dot{z} + Cz = P_z(V_F) \tag{1}$$

Dependence P_z (VF) is as Fig.2.

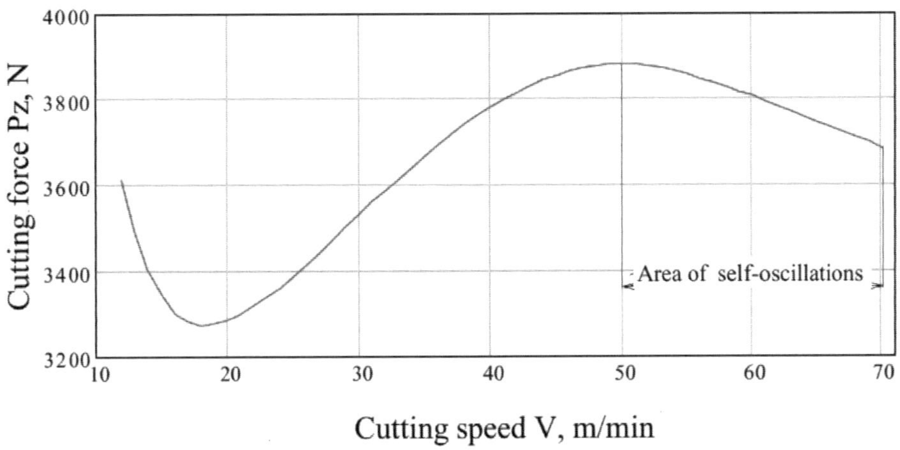

Figure 2. Dependence of the force Pz on the cutting speed

Where $V_F = V_0 - \dot{z}$. Therefore, in the area of possible occurrence of self-oscillations (for example, when Vo = 60 m / min) can be taken approximately $P_Z = P_0 - kV_0 + k\dot{z}$. Moreover, the more abrupt descending section is, the greater is the value of k. Then the differential equation looks as follows:

$$m\ddot{z} + (G - k)\dot{z} + Cz = P_0 - kV_0. \quad (2)$$

when G-k<0 the system becomes unstable. A nonlinear dependence (Figure 2) near the speed of V = 50 m/min and at higher speeds restricts the amplitude of oscillations making them stable.

The coordinate link manifests itself in the fact that in general under the influence of an applied force the cutter vertex tends to shift not only in the direction of the force vector, but also in the orthogonal direction. This is due to the fact that the rigidity of the tool unit varies in different directions, i.e. there exist principal axes of rigidity (axes V_1 and V_2) which usually do not coincide with the coordinate axes of the machine. Figure 3 shows the calculation scheme for the mathematical description of the oscillating system.

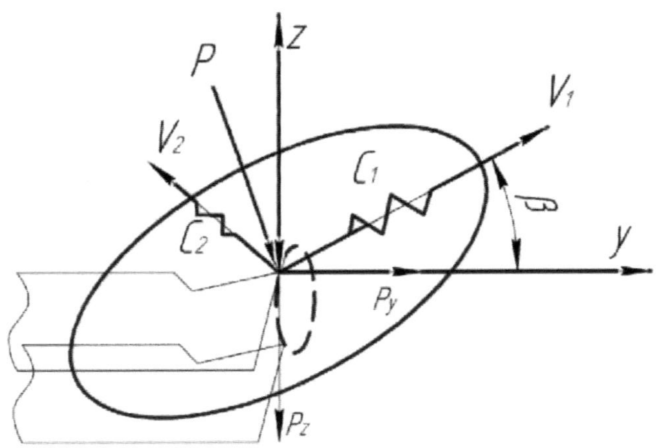

Figure 3. The calculation scheme for the mathematical description of coordinate links.

Let us consider a single-mass system where oscillations occur only of one element - the turning tool - along the principal axes V1 and V2. Then the equation of motion is a system of two nonlinear second order differential equations:

$$\begin{cases} m_1\ddot{V}_1 + G_1\dot{V}_1 + C_1V_1 = P_y(V_F;t_F)\cdot\cos\beta - P_Z(V_F;t_F)\cdot\sin\beta = P_1(V_F;t_F) \\ m_2\ddot{V}_2 + G_2\dot{V}_2 + C_2V_2 = -P_y(V_F;t_F)\cdot\sin\beta - P_Z(V_F;t_F)\cdot\cos\beta = P_2(V_F;t_F) \end{cases} \quad (3)$$

where: m_1, m_2, G_1, G_2, C_1, C_2 are reduced masses of the cutter, damping coefficients and the rigidity along the axes V_1; V_2, respectively; V_F is an actual cutting speed; t_F is an actual depth of cut.

Earlier studies [9, 10] obtained the formula for calculating the tangential and normal components of the cutting force (4).

$$P_Z = K_t \cdot K_0 \cdot \left(\frac{a\cdot V\cdot k_1 + b\cdot k_2}{V^2 + f\cdot V\cdot k_5 + g\cdot k_6} + \frac{c\cdot k_3}{V} + d\cdot k_4 \right) \quad (4)$$

Analytical expressions for the calculation of the components Pz and Py, were approximated by the experimental data with the help of the least squares method for steel 40H.

To calculate self-oscillations it is necessary to use numerical methods to solve the differential equation (3) with the power obtained by the formula (4). Initial conditions for the equation (3) can be chosen randomly, but for the steady self-oscillations it is necessary to set longer periods of time, which greatly increases the length of calculations. This is especially important with oscillations on the stability boundary when oscillations are adjusted slowly.

Second order differential equations possess filtering properties [11] and even non-linearity of the cutting force oscillations are described by the law close to the mono harmonic law for each of the axes. To test this hypothesis, we used a numerical method to solve the system (3) and the solution was decomposed into a Fourier series. The experience of such numerical calculations shows that the first harmonic of oscillations has the highest amplitude.

Taking this into account, an algorithm for obtaining the first harmonic steady-state oscillations was developed that allows excluding the transition process from calculations, and significantly reduces the solution time. The method of harmonic linearization (MHL) is the basis for this algorithm [12].

A linear part of the system has filtering properties and actually cuts out the power of spectrum cuttings of the first harmonics. (Shown on the example of a single-mass system). To solve the equation the parameters of this harmonic must be found, namely, the amplitude, frequency and a constant component.

We assume the oscillations along the corresponding axes to be monoharmonic with different frequencies:

$$\begin{aligned} V_1 &= A_1\sin(\omega\tau) + V_{10} & V_2 &= A_2\sin(\omega\tau) + B_2\cos(\omega\tau) + V_{20} \\ \dot{V}_1 &= A_1\omega\cos(\omega\tau) & \dot{V}_2 &= A_2\omega\cos(\omega\tau) - B_2\omega\sin(\omega\tau) \\ \ddot{V}_1 &= -A_1\omega^2\sin(\omega\tau) & \ddot{V}_2 &= -A_2\omega^2\sin(\omega\tau) - B_2\omega^2\cos(\omega\tau) \end{aligned} \quad (5)$$

where V_{10}, V_{20} are average values of the elastic release (a constant component); A_1, A_2, B_2, ω the amplitude and frequency of oscillations along the axes V_1 and V_2 respectively.

To do this, the cutting force will be reduced by the linear part (Fig. 1)

$$\begin{aligned} P_z(V_F;t_F) &= a_{10} + a_1\sin\omega\tau + b_1\cos\omega\tau \\ P_y(V_F;t_F) &= a_{20} + a_2\sin\omega\tau + b_2\cos\omega\tau \end{aligned} \quad (6)$$

The expansion coefficients generally depend on the values of A and ω. To arrive at a solution some parameters of this harmonic must be found, namely, amplitudes (A_1, A_2, B_2), frequency (ω) and constant components (V_{10}, V_{20}). The method of calculating the coefficients requires the use of numerical methods for the computation of integrals and the solution of the equation, but it is much easier and faster than to solve a differential equation.

The use of the harmonic linearization (MHL) method allows calculating the parameters of the first self-oscillation harmonic, and significantly reduces the calculation time. When it is necessary to obtain higher harmonics or to calculate oscillations of multimass systems, the algorithm remains practically unchanged, only the number of required parameters increases. The waviness of the machined surface is determined by the oscillation parameters of the technological system.

High altitude parameters of waviness are determined by oscillations in the normal direction to the treated surface. Wherein the waviness parameter Wz is equal to the double amplitude of the oscillation (Az) along the axis z:

$$W_Z = 2\ A_Z \tag{7}$$

Stepwise waviness parameters depend on the oscillation frequency ω and the cutting speed V_F. Fluctuations in the tangential direction have some influence on the step of waviness making it uneven. Figure 4 shows a fragment of the calculated polar diagram of the surface of the part in the presence of self-oscillations. Due to the oscillations along two coordinates, waviness is of a polyharmonic character.

At the Barnaul Mechanical Presses Plant during the semifmish turning of the shaft (steel 40H), which is a part of the BelAZ electric motor drive, oscillations occurred on the CNC machine from the Japanese company MAZAK, model Integrex-e500 II 3000U. As a consequence, tool wear increased and the waviness of a machined surface appeared.

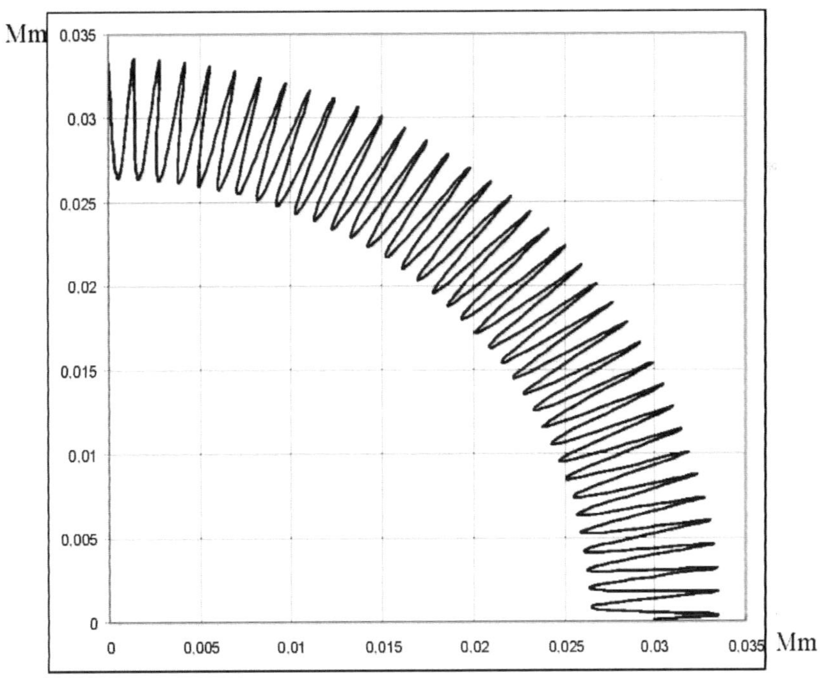

Figure 4. The polar diagram fragment of the surface where ω = 12000 Hz, A_1 = 0.003 mm, A_2 = 0.0009 mm, B_2 = -0.07 mm

In the course of solving this problem with the use of the developed software, it was proposed that the mode of cutting should be changed, namely, the previously used cutting speed from V = 200 m/min to V=150 m/min by increasing the feed S = 0.4 mm/rev up to S = 0.55 mm/rev should also be reduced. Through adjustments in the cutting mode the vibrations (oscillations) were reduced and waviness defects of the machined surface were eliminated. Thus the tool life was increased by 1.9 times, with all the design parameters of the drawing being unchanged.

Conclusions

1. The developed simulation model for predicting self-oscillations in turning takes into account non-linearity and coordinate links in the technological system and allows predicting oscillations and related waviness of the machined surface with a sufficient accuracy.

2. The use of the harmonic linearization method allows reducing the calculation time of basic harmonic oscillations, which enables using this approach in the turning operation design.

3. The mathematical model of self-oscillations supplemented by the harmonic linearization method allows controlling oscillations and makes it possible to predict waviness of the machined surface during turning already at the phase of technological process design.

References

1. F. Mason, Modern trends in the measurement of the geometric parameters of parts, Tooling & Production. 4 (2003) 24-26.

2. A.I. Kashirin, Research of vibrations in metal cutting, AN SSSR Publ., Moscow, 1964.

3. V.M. Svinin, Research of conditions of excitation and damping regenerative of os dilations in the process of cutting, Obrabotka metallov (tekhnologiya, oborudovanie, instrumenty). 1 (2005) 2931.

4. G.S. Lazarev, Self-oscillations in metal cutting, Graduate school, Moscow, 1981.

5. V.A. Kudinov, Self-oscillations at low and high frequencies at cutting, STIN Publ. 2 (1997) 16-21.

6. S. L. Leonov, A. M. Markov, E. V. Smirnov, Calculation of self-oscillations in the metal cutting, Certificate for official registration program. № 2004610835 (2004).

7. B. N. Bublik, N. F. Kirichenko, Fundamentals of control theory, Graduate school, Kiev, 1985.

8. V. A. Prilutskii, Technological methods of reducing the wavy surface, Machine-building, Moscow, 1982.

9. S. L. Leonov, A. T. ZinoVev, The basics of creating imitating precision forming, AltGTU Publ., Barnaul, 2006.

10. E.B. Belov, S.L. Leonov, A.B. Belov, V.D. Goncharov, Forecasting of self-oscillations in turning, NTU Publ., Kharkiv, 2011.

11. S. I. Hiukhin, Mathematical foundations of the modeling process forming surfaces of the cutting tool, STIN Publ. 12 (2006) 23-25.

12. V. A. Besekerskii, E. V. Popov, The theory systems of automatic regulation, The Science, Moscow, 1982.